CELLULAR AND MOLECULAR NEUROPHYSIOLOGY

THIRD EDITION

Legend for the cover image: Culture of hippocampal neurons (14 DIV). The green cells were marked for phosphoCREB (alexa 488), the blue cells for MAP2 (cyanin 5) and the red (and blue) cell was filled with rhodamine during whole cell patch clamp recording. (Courtesy of Christophe Porchet INMED France.)

CELLULAR AND MOLECULAR NEUROPHYSIOLOGY

THIRD EDITION

Constance Hammond

AMSTERDAM • BOSTON • HEIDELBERG • LONDON • NEW YORK • OXFORD
PARIS • SAN DIEGO • SAN FRANCISCO • SINGAPORE • SYDNEY • TOKYO
Academic Press is an imprint of Elsevier

Academic Press is an imprint of Elsevier
84 Theobald's Road, London WC1X 8RR, UK
Radarweg 29, PO Box 211, 1000 AE Amsterdam, The Netherlands
30 Corporate Drive, Suite 400, Burlington, MA 01803, USA
525 B Street, Suite 1900, San Diego, CA 92101-4495, USA

First edition published as Cellular and Molecular Neurobiology 1996
Second edition 2001
Reprinted 2003, 2005
Third edition 2008
Reprinted 2008

Notice
No responsibility is assumed by the publisher for any injury and/or damage to persons
or property as a matter of products liability, negligence or otherwise, or from any use
or operation of any methods, products, instructions or ideas contained in the material
herein. Because of rapid advances in the medical sciences, in particular, independent
verification of diagnoses and drug dosages should be made

British Library Cataloguing in Publication Data
A catalogue record for this book is available from the British Library

Library of Congress Cataloging-in-Publication Data
A catalog record for this book is available from the Library of Congress

ISBN: 978-0-12-374127-1

For information on all Academic Press publications
visit our website at books.elsevier.com

Printed and bound in *China*

08 09 10 10 9 8 7 6 5 4 3 2

Working together to grow
libraries in developing countries
www.elsevier.com | www.bookaid.org | www.sabre.org

ELSEVIER   BOOK AID International   Sabre Foundation

Contents

Contributors

Ben Ari (Chapter 20)
Directeur de Recherche Inserm
Institut de Neurobiologie de la Méditerranée
163 route de Luminy, BP 13
13273 Marseille 9
France

Monique Esclapez, PhD (Appendix 6.2)
INSERM U751
Faculté de Medecine Timone
27, boulevard Jean Moulin
13385 Marseille Cedex 05
France

Laurent Fagni, PhD (Chapter 12)
Institut de Génomique Fonctionnelle
Dépt. de Pharmacologie Moléculaire
UMR 5203 CNRS – U 661 INSERM –
   Université Montpellier I & II
141, rue de la cardonille
34094 Montpellier Cedex 05
France

Constance Hammond (Chapters 1–10, 13–19)
Directeur de Recherche Inserm
Institut de Neurobiologie de la Méditerranée
163 route de Luminy, BP 13
13273 Marseille 9
France

Roustem Khazipov (Appendix 9.2)
Directeur de Recherche Inserm
Institut de Neurobiologie de la Méditerranée
163 route de Luminy, BP 13
13273 Marseille 9
France

François Michel (Appendices 5.1 and 18.1)
Ingénieur
Université Aix-Marseille II
Directeur de Recherche Inserm
Institut de Neurobiologie de la Méditerranée
163 route de Luminy, BP 13
13273 Marseille 9
France

David D. Mott, PhD (Chapter 11)
Assistant Professor
Department of Pharmacology, Physiology and
   Neuroscience
School of Medicine
University of South Carolina
Columbia, SC 29208

Jean-Philippe Pin, PhD (Chapter 12)
Directeur de Recherche – CNRS
Institut de Génomique Fonctionnelle
Dépt. de Pharmacologie Moléculaire
UMR 5203 CNRS – U 661 INSERM –
   Université Montpellier I & II
141, rue de la cardonille
34094 Montpellier Cedex 05
France

Foreword

This excellent and highly acclaimed textbook by Hammond and co-authors, now in its third edition, remains faithful to its central philosophy that teaching science cannot simply rely on the presentation of facts, but must also include the intellectual journey that gives birth to key discoveries and results in the solution of long-standing puzzles. Science is in constant motion, and neuroscience, in particular, cannot be understood without appreciating how ground-breaking experiments were designed to test novel hypotheses using a combination of intuition, knowledge and experience. Readers will greatly appreciate this book for several reasons. Chief among them is that the text introduces them to the core scientific process, beyond simply presenting solutions to problems. A unique aspect of this textbook is that most figures are reproduced from the original papers that first demonstrated a given finding. This is an excellent didactic approach, because these original figures not only convey information concerning particular experimental arrangements and results, but they also introduce students to the history of neuroscience first hand. At the bottom of each figure legend, for example, readers will find the full reference including the authors, title, and journal for the paper that reported the particular discovery. There is no better way to teach students integrity and self-confidence than to introduce them to the original papers.

Because the current edition contains many updated chapters and appendices, students will learn about experiments performed by pioneering authors whom they can actually meet in person at scientific conferences. This is where the true power of the book lies, as this approach enables students to familiarize themselves with the ever-changing, flesh-and-blood frontlines of cutting-edge research, while they are still in the process of studying the key concepts of molecular and cellular neuroscience. The topics range from elementary properties of excitable cells to detailed discussions of ion channels, receptors, and synaptic transmission, all the way to dendritic integration and various forms of neuronal plasticity. This book is a concise, yet in-depth, highly informative text that will continue to inspire present and future practitioners of neuroscience.

Ivan Soltesz, PhD
University of California, Irvine

Acknowledgements

The authors would like to thank all those who have contributed to this edition and the following individuals who have contributed to previous editions of the book:

Andrea Nistri, International School for Advanced Studies (SISSA), Trieste, Italy and Aron Gutman (deceased) formerly Kaunas Medical Academy, Kaunas, Lithuania (previous Chapter 4, now part of Chapter 3); Gautam Bhave and Robert Gereau, Baylor College of Medicine, Houston, TX, USA (Chapter 12); Yusuf Tan, Bogazici University, Istanbul, Turkey (Appendix 5.1); Charles Bourque.

1

Neurons

By using the silver impregnation method developed by Golgi (1873), Ramon y Cajal studied neurons, and their connections, in the nervous system of numerous species. Based on his own work (1888) and that of others (e.g. Forel, His, Kölliker and Lenhossék) he proposed the concept that neurons are isolated units connected to each other by contacts formed by their processes: 'The terminal arborizations of neurons are free and are not joined to other terminal arborizations. They make contacts with the cell bodies and protoplasmic processes of other cellular elements.'

As proposed by Cajal, neurons are independent cells making specific contacts called *synapses*, with hundreds or thousands of other neurons sometimes greatly distant from their cell bodies. The neurons connected together form circuits, and so the nervous system is composed of neuronal networks which transmit and process information. In the nervous system there is another class of cells, the glial cells, which surround the various parts of neurons and cooperate with them. Glial cells are discussed in Chapter 2.

Neurons are *excitable* cells. Depending on the information they receive, neurons generate electrical signals and propagate them along their processes. This capacity is due to the presence of particular proteins in their plasma membrane which allow the selective passage of ions: the ion channels.

Neurons are also *secretory* cells. Their secretory product is called a *neurotransmitter*. The release of a neurotransmitter occurs only in restricted regions, the synapses. The neurotransmitter is released in the extracellular space. The synaptic secretion is highly focalized and directed specifically on cell regions to which the neuron is connected. The synaptic secretion is then different (with only a few exceptions) from other secretory cells, such as from hormonal and exocrine cells

which respectively release their secretory products into the general circulation (endocrine secretion) or the external environment (exocrine secretion). Synapses are discussed in Chapter 6.

Neurons are *quiescent* cells. When lesioned, most neurons cannot be replaced, since they are postmitotic cells. Thus, they renew their constituents during their entire life, involving the precise targeting of mRNAs and proteins to particular cytoplasmic domains or membrane areas.

1.1 NEURONS HAVE A CELL BODY FROM WHICH EMERGE TWO TYPES OF PROCESSES: THE DENDRITES AND THE AXON

Although neurons present varied morphologies, they all share features that identify them as neurons. The cell body or *soma* gives rise to processes which give the neuron the regionalization of its functions, its polarity and its capacity to connect to other neurons, to sensory cells or to effector cells.

1.1.1 The somatodendritic tree is the neuron's receptive pole

The soma of the neuron contains the nucleus and its surrounding cytoplasm (or *perikaryon*). Its shape is variable: pyramidal soma for pyramidal cells in the cerebral cortex and hippocampus; ovoid soma for Purkinje cells in the cerebellar cortex; granular soma for small multipolar cells in the cerebral cortex, cerebellar cortex and hippocampus; fusiform soma for neurons in the pallidal complex; and stellar or multipolar soma for motoneurons in the spinal cord (**Figure 1.1**).

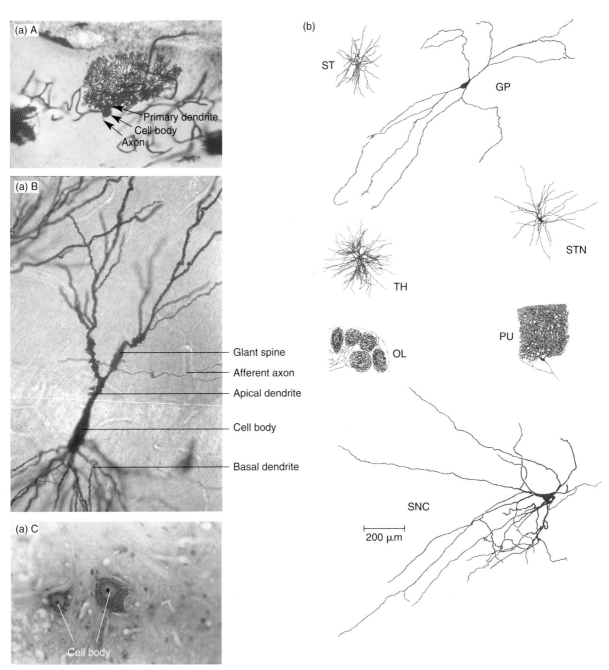

FIGURE 1.1   The neurons of the central nervous system present different dendritic arborizations.
(a) Photomicrographs of neurons in the central nervous system as observed under the light microscope. A – Purkinje cell of the cerebellar cortex; B – pyramidal cell of the hippocampus; C – soma of a motoneuron of the spinal cord. Golgi (A and B) and Nissl (C) staining. The Golgi technique is a silver staining which allows observation of dendrites, somas and axon emergence. The Nissl staining is a basophile staining which displays neuronal regions (soma and primary dendrites) containing Nissl bodies (parts of the rough endoplasmic reticulum). **(b)** Camera lucida drawings of neurons in the central nervous system of primates, revealed by the Golgi silver impregnation technique and reconstructed from serial sections: ST, medium spiny neuron of the striatum; GP, neuron of the globus pallidus; TH, thalamocortical neuron; STN, neuron of the subthalamic nucleus; OL, neurons of the inferior olivary complex; PU, Purkinje cell of the cerebellar cortex; SNC, dopaminergic neuron of the substantia nigra pars compacta. All these neurons are illustrated at the same magnification. Photomicrographs by Olivier Robain (aA and aB) and Paul Derer (aC). Drawings by Jérôme Yelnik, except OL and PU by Ramon Y Cajal (1911).

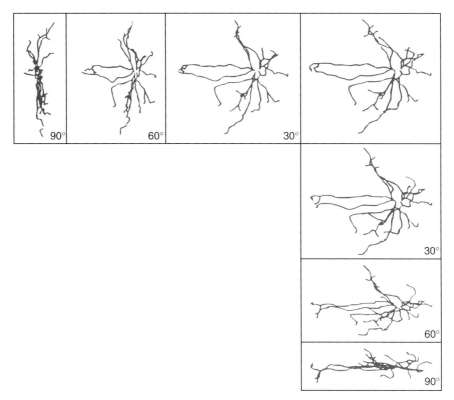

FIGURE 1.2   **Tridimensional illustration of a dendritic arborization.**
Computer drawing of a neuron of the subthalamic nucleus injected intracellularly with horseradish peroxidase (HRP) and reconstructed in three dimensions from serial sections. At 0°, the dendritic arborization of this neuron is represented in its principal plane; i.e. in the plane where it has its largest surface. In this plane, the dendritic field is almost circular (859 μm long and 804 μm wide). 30°, 60° and 90° rotations from the principal plane around the horizontal (horizontal column) and vertical (vertical column) axis show that the dendritic field has a flattened ovoidal form (230 μm thick). From Hammond C and Yelnik J (1983) Intracellular labelling of rat subthalamic nucleus with horseradish peroxidase: computer analysis of dendrites and characterization of axon arborization. *Neuroscience* **8**, 781–790, with permission.

One function of the soma is to ensure the synthesis of many of the components required for the structure and function of a neuron. Indeed, the soma contains all the organelles responsible for the synthesis of macromolecules. Most neurons in the central nervous system cannot further divide or regenerate after birth, and the cell body must maintain the structural integrity of the neuron throughout the individual's entire life. Moreover, the soma receives numerous synaptic contacts from other neurons and constitutes, with the dendrites, the main receptive area of neurons (see **Figure 1.5** and Section 6.2).The neurons have one or several processes emerging from the cell body and arborizing more or less profusely. The two types of neuronal processes are the dendrites and the axon (**Figures 1.1** and **1.3**). This division is based on morphological, ultrastructural, biochemical and functional criteria.

The dendrites, when they emerge from the soma, are simple perikaryal extensions, the primary dendrites. On average, between one and nine primary dendrites emerge from the soma and then divide successively to give a dendritic tree with specific characteristics (number of branches, volume, etc.) for each neuronal population (**Figures 1.1** and **1.2**). The dendrites are morphologically distinguishable from axons by their irregular outline, by their diameter which decreases along their branchings, by the acute angles between the branches, and by their ultrastructural characteristics (**Figures 1.1**, **1.3** and **1.7**). The irregular outline of dendrites is related to the presence of numerous appendices of various shapes and dimensions at their surface. The most frequently observed are the dendritic spines which are lateral expansions with ovoid heads binding to the dendritic branches by a peduncle that is variable in length (**Figure 1.3**). Some neurons are termed 'spiny' because there are between 40,000 and 100,000 spines on the surface of their dendrites (e.g. pyramidal neurons of the cerebral cortex and hippocampus, the medium-sized neurons of the striatum, and the Purkinje cells of the cerebellar cortex). However, other neurons with only

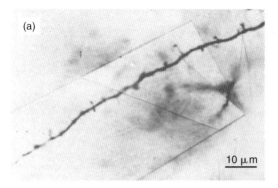

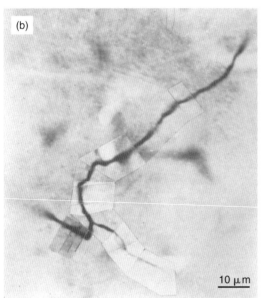

FIGURE 1.3 **Dendrite and axon of a rat subthalamic nucleus neuron.**
(a) A distal dendrite: dendritic spines of various shapes are present on its surface. **(b)** The axon: it has a smooth surface and gives off an axonal collateral. The processes of this neuron are stained by an intracellular injection of horseradish peroxidase. To follow the dendrites and axon along their trajectories, each figure is a photomontage of numerous photomicrographs of serial sections. From Hammond C and Yelnik J (see Figure 1.2), with permission.

a few spines on their dendritic surface are termed 'smooth' (e.g. neurons of the pallidal complex) (**Figure 1.1**). The transition from the cell body to proximal dendrites is gradual, and the cytoplasmic architectures of proximal dendrites and the cell body are similar. In particular, the endoplasmic reticulum and ribosomes are almost as abundant in the proximal dendrites as in the cell body. Moreover, even distal dendrites contain ribosomes and endoplasmic reticula.

Dendrites and soma receive numerous synaptic contacts from other neurons and constitute the main receptive area of neurons (see **Figure 1.5** and Section 6.2). In response to afferent information, they generate electrical signals such as postsynaptic potentials (EPSPs or IPSPs; see **Figure 1.5** left 1, 2) or calcium action potentials, and integrate the afferent information. Chapters 8–10 look at the mechanisms underlying the excitatory (EPSP) and inhibitory (IPSP) postsynaptic potentials generated in the postsynaptic membrane in response to transmitter release. Chapters 13–16 discuss how these postsynaptic responses are integrated along the somato-dendritic tree. Although dendrites are generally a receptive zone, there are certain exceptions: some dendrites are connected with other dendrites and act as a transmitter area by releasing neurotransmitters (see **Figure 6.2d**).

1.1.2 The axon and its collaterals are the neuron's transmitter pole

The axon is morphologically distinct from dendrites in having a smooth appearance and a uniform diameter along its entire extent, and by its ultrastructural characteristics (**Figures 1.3**, **1.6** and **1.7**). Axons are narrow from their origin, and do not usually contain ribosomes or endoplasmic reticula. The transition from the cell body to axon is distinct; the region of the cell body from which an axon originates is called the axon hillock and it tapers off to the axonal initial segment, where action potentials begin. Although most parts of the cell body are rich in endoplasmic reticula, the axon hillock is not. At the axon initial segment, the plasma membrane has thick underlying structures, and there is a specialized bundle of microtubules. In some neurons the axon emerges at the level of a primary dendrite.

The axon is not a single process; it is divided into one or several collaterals which form right-angles with the main axon. Some collaterals return toward the cell body area; these are recurrent axon collaterals. The axon and its collaterals may be surrounded by a sheath, the myelin sheath. Myelin is formed by glial cells (see Sections 2.2 and 2.3). The length of an axon varies. Certain neurons in the central nervous system have axons that project to one or several structures of the central nervous system that are more or less distant from their cell bodies (**Figure 1.4**), whereas other neurons have short axons (a few microns in length) that are confined to the structure where their cell bodies are located; these are interneurons or local circuit neurons (see **Figure 1.13**).

Thus projection (Golgi type I) neurons and local-circuit (Golgi type II) neurons can be differentiated. In Golgi type I neurons, the length of the axon is variable: certain projection neurons are directed to one structure only (e.g. corticothalamic neurons; see **Figure 1.14**) whereas other projection neurons have numerous axon collaterals which project to several cerebral structures (**Figure 1.4**).

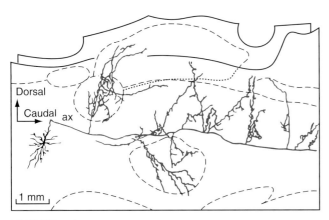

FIGURE 1.4 Neuron of the cat reticular formation (brainstem) showing a complex axonal arborization.
This reticulospinal neuron has been stained by intracellular injection of peroxidase and drawn in a parasagittal plane obtained from serial sections. The axon (ax, black) gives off numerous collaterals along its rostrocaudal trajectory, making contacts with different neuronal populations (delimited by broken lines). Scale: 7 mm = 1 μm. From Grantyn A (1987) Reticulo-spinal neurons participating in the control of synergic eye and head movement during orienting in the cat. *Exp. Brain Res.* **66**, 355–377, with permission.

The axon and axonal collaterals in certain neurons end in a terminal arborization, i.e. numerous thin branches whose extremities, the synaptic boutons, make synaptic contacts with target cells (see **Figure 6.3**). In other neurons the axon and its collaterals have enlargements or varicosities which contact target cells along their way: these are 'boutons en passant' (see **Figures 6.14** and **6.15b**). It can be noted that both types of boutons are called *axon terminals*, although 'boutons en passant' are not the real endings of the axon.

The main characteristic of axons is their capacity to trigger sodium action potentials and to propagate them over considerable distances without any decrease in their amplitude (**Figure 1.5** left 3). Action potentials are generated at the initial segment level in response to synaptic information transmitted by the somatodendritic tree. Then they propagate along the axon and its collaterals toward the axon terminals (synaptic boutons or boutons en passant). When action potentials reach the axon terminals these trigger calcium action potentials (**Figure 1.5** left 4) which may cause the release of the neurotransmitter(s) contained in axon terminals in a specific compartment, the synaptic vesicles. This secretion is localized only at the synaptic contacts. Overall, the axon is considered as the transmitter pole of the neuron.

Chapter 4 discusses the mechanisms underlying the abrupt, large and transient depolarizations called (sodium) action potentials, and how they are triggered and propagated. Chapters 4, 5 and 7 look at how

sodium action potentials trigger calcium action potentials, the entry of calcium in synaptic terminals and the secretion of transmitter molecules.

Certain regions – such as the initial segment, nodes of Ranvier (zones between two myelinated segments; see **Figure 1.5**) and axon terminals – can also be receptive areas (a postsynaptic element) of synaptic contacts from other neurons (see Section 6.2).

1.2 NEURONS ARE HIGHLY POLARIZED CELLS WITH A DIFFERENTIAL DISTRIBUTION OF ORGANELLES AND PROTEINS

The somatodendritic tree is the neuron's receptive pole, whereas the axon and its collaterals are the neuron's transmitter pole. Neurons are highly polarized cells. Cellular morphology and accurate organelles and protein distribution lay the basis to this polarization. The organelles and cytoplasmic elements present in neurons are the same organelles found in other cell types. However, some elements such as cytoskeletal elements are more abundant in neurons. The non-homogeneous distribution of organelles in their soma and processes is one of the most distinguishing characteristics of neurons.

1.2.1 The soma is the main site of macromolecule synthesis

The soma contains the same organelles and cytoplasmic elements that exist in other cells: cellular nucleus, Golgi apparatus, mitochondria, polysomes, cytoskeletal elements and lysosomes. The soma is the main site of synthesis of macromolecules since it is the one compartment containing all the required organelles.

Compared with other types of cells, the neuron differs at the nuclear level and more specifically at the chromatin and nucleolus levels. The chromatin is light and sparsely distributed: the nucleus is in interphase. Indeed, in humans, most neurons cannot divide after birth since they are postmitotic cells. The nucleolus is the site of ribosomal synthesis and ribosomes are essential for translating messenger RNA (mRNA) into proteins. The large size of the nucleolus indicates a high level of protein synthesis in neurons.

1.2.2 The dendrites contain free ribosomes and synthesize some of their proteins

In dendrites can be found smooth endoplasmic reticulum, elongated mitochondria, free ribosomes or

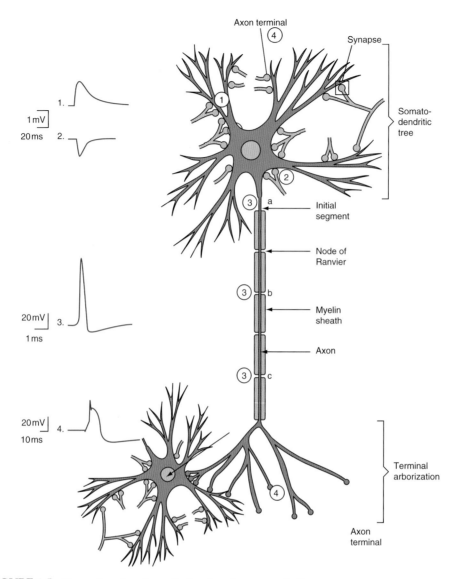

FIGURE 1.5 Comprehensive schematic drawing of neuron polarity.
The somatodendritic compartment of a neuron receives a large amount of information from other neurons that establish synapses with it. At each synapse level, the neuron generates postsynaptic potentials in response to the released neurotransmitter (1, EPSP; 2, IPSP). These postsynaptic potentials propagate and summate in the somatodendritic compartment, then they propagate to the initial segment of the axon where they generate (or not) action potential(s) (3a). The action potentials propagate along the axon (3b, 3c) and its collaterals up to the axon terminals where they evoke (or not) the entry of calcium (4) and neurotransmitter release. Note the different voltage and time calibrations.

polysomes, and cytoskeletal elements including micro-tubules which are oriented parallel to the long axis of the dendrites (but they lack neurofilaments) (**Figure 1.6**).

By using the hook procedure, microtubules have been shown to have two orientations in proximal dendrites: half of them are oriented with the plus-ends distal to the cell body, and the other half has the plus-ends proximal to the cell body. This is very different from the orientation in distal dendrites and axons (**Figure 1.7**), which is uni-form. Moreover, one microtubule-associated protein (MAP), the high-molecular-weight MAP2 protein and

more precisely the MAP2A and MAP2B, are more com-mon to dendrites than to axons. For this reason MAP2A or MAP2B antibodies coupled to fluorescent molecules are useful for labelling dendrites, particularly for den-drite identification in cell cultures.

mRNA trafficking and local protein synthesis in dendrites

The dendritic compartment contains many ribosomes whereas an axon has considerably fewer ribosomes.

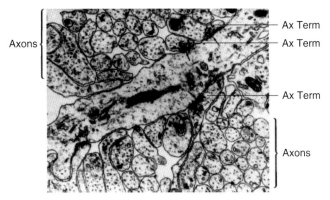

FIGURE 1.6 Photomicrograph of a tissue section of the central nervous system at the hippocampal level.
This shows the ultrastructure of a dendrite, numerous axons and their synaptic contacts (observation under the electron microscope). The apical dendrite of a pyramidal neuron contains mitochondria, microtubules, ribosomes and smooth endoplasmic reticulum. It is surrounded by fascicles of unmyelinated axons with mitochondria and microtubules but no ribosomes. The axon's trajectory is perpendicular to the section plane. Three synaptic boutons (Ax Term) with synaptic vesicles make synaptic contacts (arrows) with the dendrite. Photomicrograph by Olivier Robain.

One particular feature of dendrites, compared with axons, is the presence of synapse-associated polyribosome complexes (SPRCs); these are clusters of polyribosomes and associated membranous cisterns that are selectively localized beneath synapses (more precisely, beneath postsynaptic sites), at the base of dendritic spines when spines are present.

What is the origin of this selective distribution of ribosomes in neurons? This question is particularly important since this compartmentalization leads to different properties of dendrites and axons: dendrites can locally synthesize some of their proteins, whereas axons would synthesize very few of them, if any.

Whereas most proteins destined for dendrites and dendritic spines are conveyed from the cell body, a subset of mRNAs are transported into dendrites to support local protein synthesis. Such a local dendritic protein synthesis requires that a particular subset of mRNAs synthesized in the nucleus is transported into the dendrites up to the polysomes where they are translated.

In cultured hippocampal neurons, RNA labelled with tritiated uridin is shown to be transported at a rate of 250–500 μm per day. This transport is blocked by metabolic poisons and the RNA in transit appears to be bound to the cytoskeleton, since much of it remains following detergent extraction of the cells. Studies using video microscopy techniques and cell-permeant dyes which fluoresce on binding to nucleic acids have permitted observation of the movement of RNA-containing granules along microtubules in dendrites. These studies

(a)
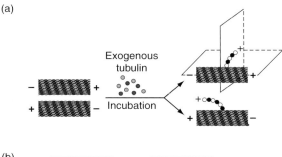

(b)
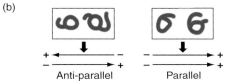

(c)
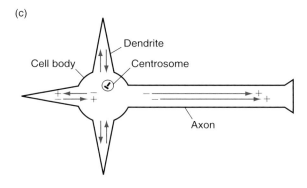

FIGURE 1.7 Microtubule polarity in neuronal processes.
(a) The polarity of microtubules is defined by the hook procedure. Neurons in culture are permeabilized in the presence of taxol to stabilize microtubules. Monomers of tubulin, purified from brain extracts, are added in the extracellular medium. Several minutes after, transversal cuts are performed at the level of dendrites or at the level of an axon. Slices are treated for electron microscopy. Hook-like structures are observed. They result from exogenous tubulin polymerization at the surface of endogenous microtubules. Hooks are always oriented toward the plus-end of microtubules. **(b)** When hooks, at the electron microscopic level, have mixed orientations (clockwise and anticlockwise), this means that endogenous microtubules are antiparallel (left). Uniformly oriented hooks (right) indicate that endogenous microtubules are parallel. **(c)** Orientation of microtubules in dendrites and the axon. Drawing (a) by Lotfi Ferhat. Drawing (b) adapted from Sharp DJ, Wenqian Yu, Ferhat L *et al.* (1997) Identification of a microtubule-associated motor protein essential for dendrite differentiation. *J. Cell. Biol.* **138**, 833–843. Drawing (c) adapted from Baas PW, Deitch JS, Black MM, Banker GA (1988) Polarity orientation of microtubules in hippocampal neurons: uniformity in the axon and nonuniformity in the dendrite. *Proc. Natl Acad. Sci. USA* **85**, 8335–8339, with permission.

suggest that mRNAs are transported as part of a larger structure. The visualized RNA particles colocalize with poly(A) mRNA, the 60S ribosomal subunit, suggesting that the granules may represent translational units or complexes (**Figure 1.8**). Therefore, this energy-dependent transport seems to be associated with the dendrite cytoskeleton as also shown by the delocalization of

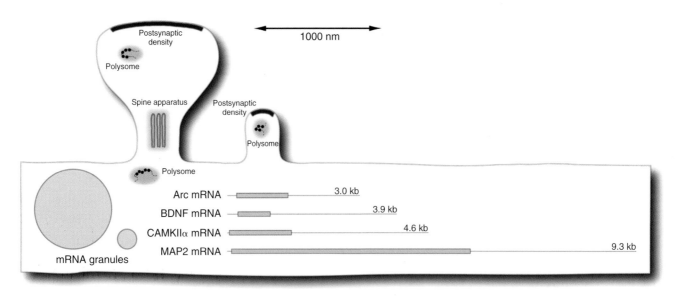

FIGURE 1.8 **Approximate sizes of representative dendritic mRNAs and translational elements at synaptic sites on dendrites.**
The drawing illustrates the approximate size range of spine synapses that would be found in rat forebrain structures such as the hippocampus and cerebral cortex. The lines represent the approximate length of representative dendritic mRNAs if they were straightened out. Shading indicates the length and position of the coding region. Adapted from Schuman EM, Dynes JL, Steward O (2006) Synaptic regulation of dendritic mRNAs. *J. Neurosci.* **26**, 7143–7146.

mRNA granules in response to colchicin (a drug which blocks microtubule polymerization).

To visualize mRNA translocation in live neurons studies used nucleic acid stains and green fluorescent protein fused to RNA-binding proteins. It showed that mRNAs are transported in the form of large granules containing mRNAs, RNA-binding proteins, ribosomes, and translational factors (RNA-containing granules) in a rapid (average speed, $0.1\,\mu m/s$), bidirectional, and microtubule dependent manner.

The exact mechanisms underlying the targeting of newly synthesized mRNAs to dendrites – which includes transport (i.e. recognition of particular mRNAs within a granule and movement along microtubules) and docking (shift from a microtubule-based transport to a cytoskeletal-based anchor) – are not yet clear. However, RNA-containing granules that are transported to dendrites bind to the C-terminal tail of the conventional kinesin KIF5 as a large detergent resistant, RNase-sensitive granule (see **Figure 1.11** and Section 1.3).

The mRNAs present in dendrites encode proteins of different functional types. Among the mRNAs detected in dendrites by *in situ* hybridization (see Appendix 6.2) are mRNAs that encode certain cytoskeletal proteins (as the high-molecular-weight MAP2), a kinase (the α subunit of calcium/calmodulin-dependent protein kinase II), an integral membrane protein of the endoplasmic reticulum (the inositol trisphosphate receptor), calcium-binding proteins, certain units of neurotransmitter

receptors (the NR1 subunit of the NMDA receptor) as well as other proteins of unknown function. Moreover, within dendrites, different mRNAs are localized in different domains and different mRNAs are localized in the dendrites of different neuron types. In summary, it has become clear that certain mRNAs are transported in dendrites in large macromolecular complexes, the granules ($>1000\,S$). These granules are transported by a kinesin KIF5, which binds RNA-binding proteins by a recognition motif in its tail domain.

The relatively large amount of RNA transported into dendrites raises the question of why neurons need this supply. Targeting of mRNAs to dendritic synthetic machinery located at the base of dendritic spines could occur, for example, in response to synaptic information and trigger local protein synthesis that would be responsible for the stability of the synaptic transmission or the modulation of it by changing, for example, the subunits or the number of receptors to the neurotransmitter in the postsynaptic membrane (this can occur during plasticity and may produce long-lasting changes in synaptic strength) (see Chapter 18).

1.2.3 The axon, to a large extent, lacks the machinery for protein synthesis

The axoplasm contains thin elongated mitochondria, numerous cytoskeletal elements and transport vesicles. It is devoid of ribosomes associated to the reticulum but may contain ribonucleoprotein complexes especially during development. Nevertheless, axons cannot restore the vast majority of the macromolecules from which they are made; neither can they ensure alone the synthesis of the neurotransmitter(s) that they release since they are unable to synthesize proteins (such as enzymes). This problem is resolved by the existence of a continuous supply of macromolecules from the cell body to the axon through anterograde axonal transport (see Section 1.3).

Another major difference between dendrites and axons is the orientation of microtubules. By using the hook procedure (see **Figure 1.7**) it has been shown that the polarity of microtubules is uniform in the axon, meaning that all their plus-ends point away from the cell body, toward the axon terminals. The polarity of the microtubules is relevant for transport properties (see Section 1.3). Moreover, one MAP, the Tau protein, is more common to axons than to dendrites. Tau antibodies coupled to fluorescent molecules are useful for labelling axons, particularly for axon identification in cell cultures.

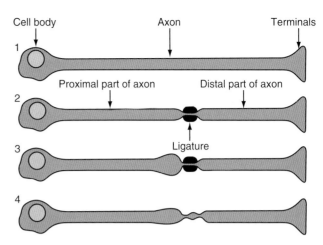

FIGURE 1.9 Experiment by Weiss and others demonstrating anterograde axonal transport.
Schematic of a chicken motoneuron (1). When a ligature is placed on the axon (2) an enlargement of the axon's diameter above the ligature is noted after several weeks (3). When this ligature is removed, the enlargement progressively disappears (4). From Weiss P, Hiscoe HB (1948) Experiments on the mechanism of nerve growth. *J. Exp. Zool.* **107**, 315–396, with permission.

1.3 AXONAL TRANSPORT ALLOWS BIDIRECTIONAL COMMUNICATION BETWEEN THE CELL BODY AND THE AXON TERMINALS

Axonal transport is the movement of subcellular structures (such as vesicles, mitochondria, etc.) and proteins (like those of the cytoskeleton) from the cell body to axonal sites (nodes of Ranvier, presynaptic release sites, etc.) and from axon terminals to the cell body.

1.3.1 Demonstration of axonal transport

Weiss and Hiscoe (1948) first demonstrated the existence of material transport in growing axons (during development) as in mature axons. Their work consisted of placing a ligature on the chicken sciatic nerve, and then examining the change in diameter of the axons over several weeks. They showed that these neurons became enlarged in their proximal part and presented degenerative signs in their distal part (**Figure 1.9**). The authors suggested that material from the cell body had accumulated above the ligature and ensured the survival of the distal part.

Later Lubinska *et al.* (1964) elaborated the concept of anterograde and retrograde transport. These authors placed two ligatures on a dog sciatic nerve, isolated part of the nerve and divided it into short segments in order to analyze their acetylcholinesterase content. This enzyme is responsible for acetylcholine degradation and was used here as a marker. They showed that it accumulates at the level of both ligatures. This result therefore suggested the existence of two types of transport: an anterograde transport (from cell body to terminals) and a retrograde transport (from terminals to cell body). Moreover, it appeared that both types of transport are distributed along the entire extent of the axon.

We presently know of three types of axonal transport: fast (anterograde and retrograde), slow (anterograde) and mitochondrial.

1.3.2 Fast anterograde axonal transport is responsible for the movement of membranous organelles from cell body towards axon terminals, and allows renewal of axonal proteins

Fast anterograde axonal transport consists in the movement of vesicles along the axonal microtubules at a rate of 200–400 mm per day (i.e. 2–5 µm/s). These transport vesicles, which are 40–60 nm in diameter, emerge from the Golgi apparatus in the cell body (**Figure 1.10a**). They transport, among other things, proteins required to renew plasma membrane and internal axonal membranes, neurotransmitter synthesis enzymes and neurotransmitter precursors when the neurotransmitter is a peptide. This transport is independent of the type of axon (central, peripheral, etc.).

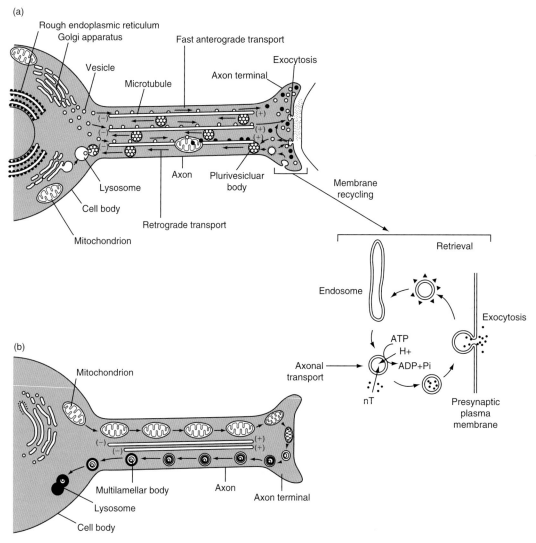

FIGURE 1.10   Fast axonal transport.
(a) Schematic of fast anterograde axonal transport (anterograde movement of vesicles) and retrograde axonal transport (retrograde movement of plurivesicular bodies). These two transports use microtubules as substrate. The detail shows recycling of small synaptic vesicles. Vesicles synthetized in the cell body and transported to the axon terminals are loaded with cytoplasmic neurotransmitter and targeted to the presynaptic plasma membrane. In response to Ca^{2+} entry, they fuse with the plasma membrane, release their content into the synaptic cleft (exocytosis); then they are recycled via an endosomal compartment. **(b)** Schematic of mitochondrial transport. Note that the neuron representation is extremely schematic since axons do not give off *one* axon terminal. Drawing (a) adapted from Allen R (1987) Les trottoirs roulants de la cellule. *Pour la Science*, April, 52–66; and Südhof TC, Jahn R (1991) Proteins of synaptic vesicles involved in exocytosis and membrane recycling. *Neuron* **6**, 665–677, with permission. Drawing (b) adapted from Lasek RJ, Katz M (1987) Mechanisms at the axon tip regulate metabolic processes critical to axonal elongation. *Prog. Brain Res.* **71**, 49–60, with permission.

The most currently used preparation

The squid's giant axon is most commonly used for these observations since its axoplasm can easily be extruded and a translucent cylinder of axoplasm devoid of its membrane is thus obtained. This living extruded axon keeps its transport properties for several hours. The absence of plasma membrane allows a precise control of the experimental conditions and entry into the axoplasm of several components that cannot usually pass through the membrane barrier *in vivo* (e.g. antibodies). The improvement of video techniques applied to light microscopy allowed the first observations of the movement of a multitude of small particles along the microtubules in a living extruded axon.

*Identification of the moving organelles and
their substrates*

Analysis of the particles that accumulate on each side
of the 1.0–1.5-mm long isolated frozen segments of the
squid axon has permitted the identification of moving
organelles in axons. Correlation between video and electron microscopy images of these axonal segments has
shown that the particles moving anterogradely on video
images are small vesicles. Indeed, when a purified fraction of small labelled vesicles (with fluorescent dyes) is
placed in an extruded axon, these vesicles and also
native vesicles are transported essentially in the anterograde direction.

Evidence demonstrating the implication of microtubules in fast anterograde transport came from experiments with antimitotic agents (colchicin, vinblastin)
which prevent the elongation of microtubules and
block this transport. Finally, video techniques have also
demonstrated that the vesicles are associated to microtubules by arms of 25–30 nm length (**Figure 1.11a**).

The role of ATP and kinesin

By analogy with actin–myosin movements in muscle
cells, scientists tried to isolate in neurons an ATPase
(the enzyme responsible for the hydrolysis of ATP)
associated with microtubules and able to generate the
movement of vesicles. To demonstrate molecular components responsible for interactions between vesicles
and microtubules, the vesicle–microtubule complex
system has been reconstituted *in vitro*: isolated vesicles
from squid giant axons are added to a preparation of
purified microtubules and placed on a glass coverslip.
These vesicles occasionally move in the presence of
ATP. If an extract of solubilized axoplasm is then added
to this system the number of transported vesicles is
considerably increased.

In order to determine the factor present in the solubilized fraction responsible for vesicle movement, a
non-hydrolyzable ATP analogue has been used: the
5'-adenylyl imidophosphate (AMP-PNP). In the presence
of AMP-PNP, the vesicles associate with the microtubules but then stop. In these conditions, vesicles are
bound to the microtubules and also, consequently,
to the transport factor. When an overdose of ATP is
added to this vesicle–microtubule complex isolated by
centrifugation, the AMP-PNP is removed and so vesicles are released and the transport factor is solubilized.
Kinesin has been thus isolated and purified. It is a soluble microtubule-associated ATPase that couples ATP
hydrolysis to unidirectional movement of vesicles
along the microtubule. As we have already seen, in
axons, all microtubules are oriented, their plus-end

being distally located from the cell body. It has been
shown that kinesin moves vesicles in one direction
only: from the minus-end toward the plus-end. All these
results show that kinesin is responsible for anterograde
transport. In mammals, kinesin is a homodimer composed of two identical heavy chains associated with
two light chains. These form a 80 nm rod-like molecule
consisting of two globular head domains (formed by
the heavy chains), a stalk domain and a tail domain
(formed by the light chains) (**Figure 1.11a,b**). Kinesin is
a microtubule-associated protein (MAP) belonging to
the family of mechanochemical ATPases. In proposed
mechanism models, the arms observed between vesicles and microtubules *in vitro* would be kinesin. The
head transiently binds to microtubules whereas the tail
would be, directly or indirectly, associated to membranous organelles. The head binds to and dissociates from a
microtubule through a cycle of ATP hydrolysis.

The effects of mutations of the kinesin heavy-chain
gene (*khc*) on the physiology and ultrastructure of
Drosophila larval neurons have been studied. Motoneuron activity and corresponding synaptic (junctional)
excitatory potentials of the muscle cells they innervate
were recorded in control and mutant larvae in response
to segmental nerve stimulation. The mutations dramatically reduced the evoked motoneuron activity and
synaptic responses. The synaptic responses were reduced
even when the terminals were directly stimulated.
However, there was no apparent effect on the number
of axons in the nerve bundle or the number of synaptic
vesicles in the nerve terminal cytoplasm. These observations show that kinesin mutations impair the function of action potential propagation and neurotransmitter
release at nerve terminals. Thus kinesin appears to be
required for axonal transport of material other than
synaptic vesicles: for example, vesicles containing ion
channels such as Na^+ channels delivered to Ranvier
nodes and Ca^{2+} channels delivered to presynaptic
membranes. These vesicles, called 'cargoes', are linked
to kinesin. The observation that mutation of kinesin
heavy chain had no effect on the number of synaptic
vesicles within nerve terminals would obviously not be
expected if conventional kinesin were the universal
anterograde axonal transport motor.

Plus-end vesicle motors

Since the original discovery of kinesin, a large
family of proteins (kinesin superfamily proteins or
KIFs) with homology to kinesin's motor domain has
been discovered. The kinesin superfamily is a large
gene family of microtubule-dependent motors with
45 members identified at present in mice and humans.
The 45 murine and human KIF genes have been classified

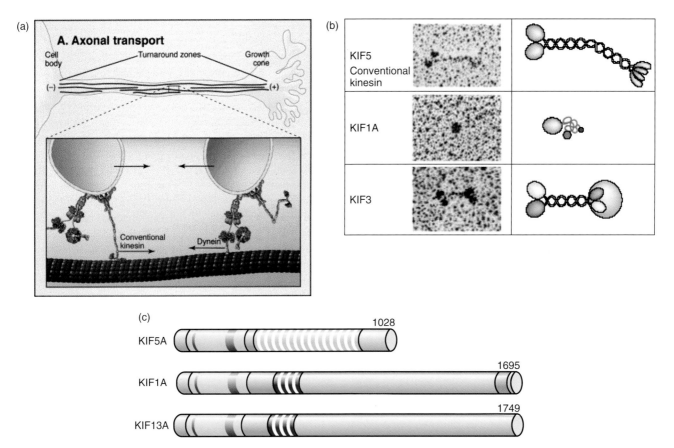

FIGURE 1.11   The motors of fast anterograde and retrograde axonal transport.
(a) Kinesin motors carry cargo (membrane organelles in the axon) along a unipolar array of microtubule towards the plus-ends. Dynein is carried along with this anterograde cargo in a repressed form, and reversals in the direction of movement are infrequent. At a 'turnaround' zone at the tip of these structures, dynein would be activated and kinesin repressed, and the processed cargo then can be transported back towards the cell body. The opposite activation/inactivation of the motors is believed to occur at the base near the cell body. **(b)** Some members of kinesin superfamily proteins (KIFs) observed by low angle rotary shadowing (left column). Diagrams, constructed on the basis of electron microscopy or predicted from the analysis of their primary structures, are shown on the right (the larger orange ovals in each diagram indicate motor domains). KIF5 (orange) forms a homodimer and kinesin light chains (blue) associate at the carboxyl C-terminus to form fanlike ends. KIF1A is monomeric and globular. KIF3 forms heterodimers. **(c)** The domain structures of the same KIFs. The motor domains are shown in orange, the ATP-binding consensus sequence by a thin purple line, the microtubule-binding consensus sequence by a thick purple line, the dimerization domains by yellow stripes, the forkhead-associated domains by red stripes and pleckstrin homology domains by orange stripes. The number of amino acids in each molecule is shown on the right. All these KIFs have their motor domains in the amino (N) terminus and are therefore N-kinesins. Part (a) adapted from Vale RD (2003) The molecular motor toolbox for intracellular transport. *Cell* **112**, 467–480, with permission. Part (b) adapted from Hirokawa N (1998) Kinesin and dynein superfamily proteins and the mechanism of organelle transport. *Science* **279**, 519–526, with permission. Part (c) adapted from Hirokawa N and Takemura R (2005) Molecular motors and mechanisms of directional transport in neurons. *Nat. Rev. Neurosci.* **6**, 201–214, with permission.

into three types on the basis of the positions of their motor domains: the amino (N)-terminal motor, middle motor and carboxy (C)-terminal motor types (referred to as N-kinesins, M-kinesins and C-kinesins, respectively). All KIFs have a globular motor domain that shows high degrees of homology and contains a microtubule-binding sequence and an ATP-binding sequence, but, outside the motor domain, each KIF has a unique sequence (**Figure 1.11b,c**). The diversity of

these cargo-binding domains explains how KIFs can transport numerous different cargoes. Kinesins have either a monomeric (KIF1A, KIF1B), a homodimeric (KIF5, KIF2) or a heterodimeric (KIF3B and KIF3C with KIF3A) structure. The 'classical kinesin' corresponds to KIF5. Many KIFs are expressed primarily in the nervous system, but KIFs are also expressed in other tissues and participate in various types of intracellular transport.

Most KIFs are plus-end motors that transport cargoes from the minus-end of microtubules toward their plus-end, i.e. from the cell body toward axon terminals. The motor domain is necessary and sufficient for ATP-driven movement along microtubules. The hypothesis is that each KIF member is targeted to a specific cargo population, allowing the trafficking of the different neuronal compartments to be regulated independently. While there is some functional redundancy among members of the kinesin superfamily, there is also a remarkable degree of cargo specialization. Many members of the kinesin superfamily have been identified as motors for specific cellular cargoes.

1.3.3 Retrograde axonal transport returns old membrane constituents, trophic factors, exogenous material to the cell body

Retrograde axonal transport allows debris elimination and could represent a feedback mechanism for controlling the metabolic activity of the soma. The vesicles or cargoes transported retrogradely are larger (100–300 nm) than those transported anterogradely. Structurally they are prelysosomal structures, multivesicular or multilamellar bodies (**Figure 1.10**). In the squid extruded axoplasm, vesicles move on to each filament in both directions and frequently cross each other without apparent collisions or interactions.

Do filaments used for the fast transport of vesicles form a complex made up of several distinct filaments where certain filaments would be implicated in fast anterograde and others in retrograde transport? By using a monoclonal antibody raised against α-tubulin (a specific component of microtubules) it has been demonstrated that all the filaments implicated in anterograde or retrograde axonal transport contain α-tubulin. Moreover, by using a toxin-binding actin (and so consequently binding microfilaments) it was shown that filaments used for fast anterograde transport or retrograde transport were devoid of actin in their structure. Thus it appeared that filaments used for the movement of vesicles in both directions are microtubules.

The minus-end motor(s)

Morphometric analysis of the arms between retrograde vesicles (pluricellular bodies) and microtubules demonstrated that these are similar to arms between anterograde vesicles and microtubules. Studies looking to find a factor different from, but homologous to, kinesin and responsible for retrograde transport were undertaken. This factor present in axoplasm homogenate might be lost during kinesin purification

procedures since no retrograde vesicles movement was observed *in vitro* with kinesin. Cytoplasmic dynein (also called MAP1C) has been thus isolated. It is a microtubule-associated protein with an ATPase activity (see **Figures 1.11a** and **1.12**).

Cytoplasmic dynein is a large and complex molecule, composed of two heavy chains, and multiple intermediate, light intermediate, and light chains to yield a 2 million Da protein complex. The two heavy chains, each around 500 kDa, fold to form globular heads on relatively flexible stalks that dimerize at their ends. Most of the other subunits of the complex are associated with the base of the molecule. Each of the globular heads forms a motor domain, while the base of the complex functions primarily to bind to an associated protein complex, dynactin, as well as participating in direct or indirect associations with cargo (**Figure 1.12**).

In vitro, cytoplasmic dynein alone is sufficient to drive microtubule gliding. However, within the cell a second protein complex, dynactin, is required for most of cytoplasmic dynein's motile functions (**Figure 1.12**). Dynactin is also a large protein complex (1 MDa) with a distinct structure, composed of 11 distinct subunits.

What mechanism regulates the direction of vesicle movement?

It can be hypothesized that kinesin and dynein are bound to only one type of vesicle, specific receptors present at their surface recognizing only one of the two motors. Or both motors might be located on the different vesicles, and by a regulation mechanism only one type is active and so transport takes place in only one direction. Anterogradely transported vesicle populations isolated from squid axoplasm have been shown to carry both motors, a kinesin and dynein. Therefore, during anterograde movement dynein would be repressed whereas kinesin would be repressed during retrograde movement (**Figure 1.11a**).

Functions of retrograde transport

The removal of misfolded or aggregated protein is a key problem in neurons. Cytoplasmic dynein's role as a retrograde motor makes it an ideal candidate for 'taking out the trash' in the cell, returning misfolded or degraded proteins from the cell periphery to the cell centre for recycling and/or degradation. Evidence for such a role has come from the analysis of aggresome formation, in which the formation of perinuclear aggregates of misfolded protein was found to be dynein-dependent. Further, dynein has been implicated as the motor driving vesicles from late endosomes to

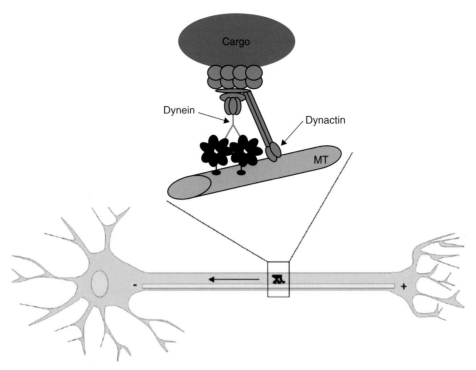

FIGURE 1.12   **Cytoplasmic dynein and dynactin drive retrograde axonal transport in motor neurons.**
Dynein and its activator dynactin are critical for the transport of neurotrophins and material targeted for
degradation from the distal regions of the neurons, including the synapse, to the cell body. This transport
occurs along microtubules (MT), which are oriented with their minus ends (-) towards the cell body and their
plus ends (+) toward the synapse. Cytoplasmic dynein and dynactin are large, multimeric protein complexes
that function together to produce minus-end-directed motility. Adapted from Levy R and Holzbaur LF (2006)
Cytoplasmic dynein/dynactin function and dysfunction in motor neurons. *Int. J. Devl. Neurosci.* **24**, 103–111,
with permission.

lysosomes, as well as driving the minus end-directed
motility of lysosomes along microtubules. Retrograde
axonal transport allows the return of membrane molecules to cell bodies, where they are degraded by acidic
hydrolases found in lysosomes.

Retrograde axonal transport is not only a means of
transporting cellular debris for their elimination, but
also a way of communicating information from the
axon terminals to the soma. The retrogradely transported
molecules would inform the cell body about activities
taking place at the axon terminal level, or they may
even have a neurotrophic action on the neuron. One
key role for cytoplasmic dynein and dynactin in neurons
is in retrograde signalling, specifically the transport
of neurotrophic factors from synapse to cell body.
Neurotrophins are a family of small molecules, such as
the nerve growth factor (NGF), the brain-derived neurotrophic factor (BDNF), and the neurotrophic factor
NT3, that are secreted by target tissues, and then bind
to receptor tyrosine kinases (Trk receptors) on the
surface of the neuron. The neurotrophin/Trk receptor

complex is then internalized (taken up by endocytosis)
transported to the cell body where it initiates signaling
cascades that regulate cell growth and survival.

Moreover, it allows the transport of tetanus toxin or
cholera toxin macromolecules that are taken up by
axon terminals and have a toxic effect on the cell body.
These toxins, as well as horseradish peroxidase (HRP),
an enzyme taken up by the axon terminals, are used in
research studies for the retrograde labelling of neuronal
pathways.

In conclusion, cargoes are transported in either the
antero- or retrograde direction, depending on whether
plus- or minus-end motors are active on their surface.
Cargoes destined for the nerve terminal, such as synaptic vesicles or their precursors, are transported by plus-
end motors; while cargoes targeted for the cell body,
such as vesicles containing neurotrophin-receptor complexes, are transported by minus-end motors. In axons,
oriented microtubules establish a 'road map' inside
the neuron to motors that are linked to particular intracellular cargoes.

1.3.4 Slow anterograde axonal transport moves cytoskeletal proteins and cytosoluble proteins

The cytoskeleton (microtubules, neurofilaments and microfilaments) and cytosolic proteins (intermediate metabolic enzymes including glycolysis enzymes) are transported anterogradely along axons at a slow rate of about 0.002–0.1 µm/s (0.17–8.6 mm/day). In the elongating axon (i.e. during development or regeneration) the function of the slow transport is to supply axoplasm required for axonal growth. In mature neurons its function is to renew continuously the total proteins present in the axon and axon terminals and to act as a substrate for the anterograde and retrograde axonal transport. To appreciate fully the structural achievement of this transport, one must put the size of cell bodies and axons into relation. The neuronal cell bodies (10–50 µm diameter) are connected by axons that can be over 1 m length (the axonal diameter is 1 to 25 µm). This is a factor of 100,000 difference. As there is relatively little protein synthesis in the axon, the proteins that comprise the microtubules, neurofilaments and microfilaments must be actively transported from the cell body into and down the length of the axon.

To understand the mechanisms involved in slow axonal transport, several questions can be raised: (i) in which state are cytoskeletal proteins transported in the axons: as soluble proteins or as polymers? (ii) in which axonal region(s) is the cytoskeleton (i.e. the complex network of filaments) assembled? The following are, in chronological order, the diverse hypotheses that have been proposed:

The different cytoskeletal elements are assembled and connected by bridges in the cell body

They then progress as a whole (a matrix) in the axon. However, studies have demonstrated that crossbridges between the different cytoskeletal elements are weak and unstable. Moreover, numerous cytoskeletal discontinuities exist along the axon as seen in the nodes of Ranvier. Thus, the hypothesis of the continuous transport of a stable matrix of assembled cytoskeletal elements explaining the ultrastructure of the axon is now known to be false.

The cytoskeletal proteins are transported in a soluble form or as isolated fibrils and assembled during their progression

Lasek and his colleagues proposed that the microtubules and other cytoskeletal elements in slow transport are moved as polymer by sliding. When they are assembled some become stationary and would be renewed onsite. This hypothesis came from pulse-labelling studies and particularly those coupled with photobleaching experiments. Purified subunits of cytoskeletal proteins (tubulin or actin) coupled to a fluorescent dye molecule are introduced into living neurons in culture by injection into their soma. The observation with fluorescent microscopy shows that these labelled subunits are gradually incorporated into the polymer pool of the corresponding cytoskeletal proteins (microtubules and microfilaments) throughout the axon. A highly focused light source is then used to extinguish or bleach the fluorescence of the molecules contained within a discrete axonal segment (about 3 µm long). The fate of the bleached zone is followed over a period of hours. The bleached zone does not move along the axon or widen and recovers a low level of fluorescence within seconds. This latter effect is ascribed to the diffusion of free fluorescent subunits from the neighbouring fluorescent regions into the bleached region. These observations suggested that microtubules and microfilaments are essentially stationary and are exchanging subunits.

The transport of microtubules and neurofilaments is bidirectional, intermittent, asynchronous, and occurs at the fast rate of known motors

However, when Wang and Brown widened the parameters of the live-cell imaging paradigm, such that a much longer bleached zone (about 30 µm in length) was created, and the zone was imaged every several seconds rather than minutes, they found that the transport of microtubules is bidirectional, intermittent, highly asynchronous, and at the fast rates of known motors (average rates of 1 µm/s) such as cytoplasmic dynein and the kinesin superfamily. These observations indicate that microtubules are propelled along axons by fast motors. The average moving microtubule length is around 3 µm.

The rapid, infrequent, and highly asynchronous nature of the movement may explain why the axonal transport of tubulin has eluded detection in so many other studies. In addition, these results offer an explanation for the slow rate of tubulin transport documented in the early kinetic studies: it reflects an average rate of fast movements and non-movements. The overall rate of microtubules movement is slow because the microtubules spend only a small proportion of their time moving.

Similarly, the initial studies of neurofilaments transport using radiographic labelling suggested a velocity of 0.25–3 mm/day which is slower than any speed produced by known molecular motors. Recent studies using green-fluorescent-protein (GFP)-tagged neurofilament subunits and real-time confocal microscopy show more accurately that the conventional fast axonal transport also applies to neurofilaments. Peak velocities of $2\,\mu m/s$ occur anterogradely and retrogradely and are interrupted by prolonged resting phases resulting in the overall slow transport originally described.

1.3.5 Axonal transport of mitochondria allows the turnover of mitochondria in axons and axon terminals

Mitochondria are prominent members of the cast of axonally transported organelles. They are essential for the function of all aerobic cells, including neurons. They produce ATP, buffer cytosolic calcium and sequester apoptotic factors. Like many other neuronal organelles, mitochondria are thought to arise mainly in the neuronal cell body, but their transport is distinctive. In postmitotic neurons, mitochondria are delivered to and remain in areas of the axon where metabolic demand is high, such as initial segments, nodes of Ranvier and synapses. How do mitochondria achieve these distributions in the axon?

The mitochondria recently formed in the cell body are transported anterogradely in axons up to axon terminals at a rate of 10–40 mm per day. A retrograde movement of mitochondria showing degenerative signs is also observed (see **Figure 1.10b**). Specific inhibition of kinesin-1 stops most mitochondrial movement in *Drosophila melanogaster* motor axons. Also time-lapse imaging of GFP-tagged mitochondria in *Drosophila* axons has shown that kinesin-1 mutations cause a profound reduction in the retrograde transport of mitochondria. In addition, the plus-end motor KIF1Bα has been shown to be associated with mitochondria with subcellular fractionation, and purified KIF1Bα can transport mitochondria along microtubules *in vitro*. Thus, kinesin-1 (KIF5A, KIF5B, and KIF5C) and KIF1Bα (**Figure 1.11b**) transport mitochondria in the anterograde direction.

1.4 NEURONS CONNECTED BY SYNAPSES FORM NETWORKS OR CIRCUITS

1.4.1 The circuit of the withdrawal medullary reflex

Sensory stimuli (including visual, auditive, tactile, gustative, olfactive, proprioceptive, and nociceptive stimuli) are detected by specific sensory receptors and transmitted to the central nervous system (encephalon and spinal cord) by networks of neurons. These stimuli are analyzed at the encephalic level. They can also evoke movements such as motor reflexes on their way to higher central structures.

Thus, when a noxious stimulus (i.e. a stimulus provoking tissue damage, for example pricking or burning) is applied to the skin of the right foot, it induces a withdrawal reflex consisting of the removal of the affected foot (contraction of flexor muscles of the right inferior limb) to protect itself against this stimulus. The noxious stimulus activates nociceptors which are the peripheral endings of primary sensory neurons whose cell bodies are located, in this case, where injury is located at the body level – in dorsal root ganglia. Action potentials are then generated (or not, if the intensity of the noxious stimulus is too small) in primary sensory neurons and propagate to the central nervous system (spinal cord). Local circuit neurons of the dorsal horn of the spinal cord (**Figure 1.13a**) relay the sensory information. Sensory information is thus transmitted to motoneurons (neurons innervating skeletal striated muscles and located in the ventral horn) through a complex network of local circuit neurons (Golgi type II neurons) which have either an excitatory or an inhibitory effect. It results on the stimulus side (ipsilateral side) in an activation of the flexor motoneurons (F) and an inhibition of the extensor motoneurons (E): the right inferior limb is being withdrawn (is in flexion). The opposite limb is extended to maintain posture.

This pathway illustrates peculiarities present in numerous other circuits.

- *Divergence of information.* Primary sensory information is distributed to several types of neurons in the medulla: local circuit neurons connected to motoneurons that innervate posterior limb muscles and also projection neurons that relay sensory informations to higher centres where they are analyzed.
- *Convergence of information.* Motoneurons receive sensory informations via local circuit neurons and also descending motor information via descending neurons whose cell bodies are located in central motor regions (motor commands elaborated at the encephalic level) (**Figure 1.13a**).
- *Anterograde inhibition* (feedforward inhibition). A neuron inhibits another neuron by the activation of an inhibitory interneuron (**Figure 1.13b**).
- *Recurrent inhibition* (feedback inhibition). A neuron inhibits itself by a recurrent collateral of its own axon which synapses on an inhibitory interneuron. The inhibitory interneuron establishes synapses on the motoneuron (**Figure 1.13c**). This recurrent inhibition allows for rapid cessation of the motoneuron's activity.

The last two circuits described are also called *microcircuits*, since they are included in a larger circuit or *macrocircuit*. In this selected example, all the neurons

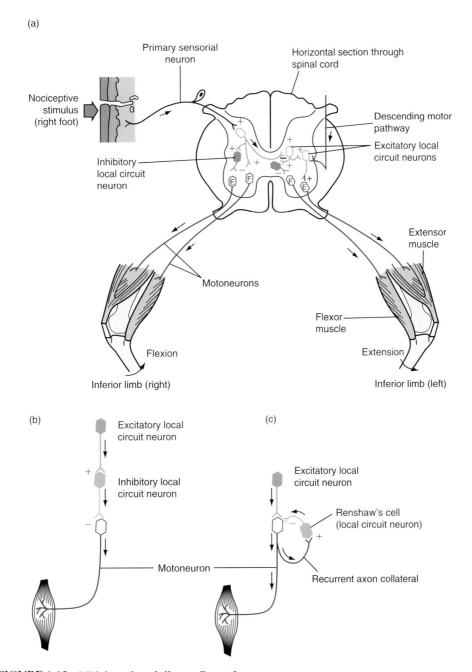

FIGURE 1.13   Withdrawal medullary reflex pathway.
(a) Schematic of a horizontal section through the spinal cord and of connections between a primary nociceptive sensory neuron, medullary local circuit neurons and ipsi- and contralateral motoneurons innervating inferior limb muscles. See text for details. **(b)** Anterograde inhibitory circuit. **(c)** Recurrent inhibitory circuit. Arrows show the direction of action potential propagation.

forming the microcircuit enable precise regulation of motoneuron activity.

1.4.2   The spinothalamic tract or anterolateral pathway is a somatosensory pathway

Noxious stimuli (temperature and sometimes touch) are detected at the skin level by free nerve endings, are transduced (or not) in action potentials and are conveyed to the somatosensory cortex via relay neurons. Information from the body reaches the dorsal horn neurons of the spinal cord, and information from the face reaches the trigeminal nuclei in the brainstem, via primary sensory neurons whose cell bodies are located in dorsal root ganglia or cranial ganglia, respectively. They relay on projection neurons located in dorsal horns or in

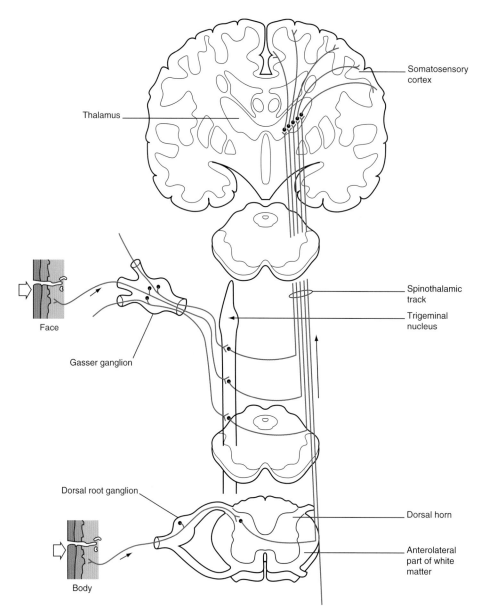

FIGURE 1.14 The spinothalamic tract or anterolateral ascending sensory pathway.
This pathway integrates and conveys sensory information such as nociception, temperature and some touch.
Bottom to top: horizontal sections through the spinal cord, the pons and frontal section through the diencephalon.
See text for explanations.

trigeminal nuclei which send axons to the thalamus. These axons cross the midline, form a tract in the anterolateral part of the white matter, and terminate in non-specific thalamic nuclei. Thalamic neurons then send the sensory information to cortical areas specializing in noxious perception (somatosensory cortex). At each level of synapses (dorsal horn or trigeminal nucleus, thalamus, cortex) the somatosensory information is not simply relayed, it is also processed through local microcircuits receiving afferent sensory information and descending information from higher centres which modulate incoming sensory information.

When superposing horizontal sections through the spinal cord (**Figures 1.13a** and **1.14**), it becomes clear that a noxious stimulus applied to the skin of the right inferior limb is transmitted to motoneurons where it can evoke a withdrawal reflex and also reaches the somatosensory cortex where it is analyzed. The reflex is evoked before the consciousness of the stimulus because of the longer distance to brain areas than to the ventral horn of the spinal cord.

1.5 SUMMARY: THE NEURON IS AN EXCITABLE AND SECRETORY CELL PRESENTING AN EXTREME FUNCTIONAL REGIONALIZATION

This chapter has described how the various functions of neurons, such as their metabolism, excitability and secretion, are localized to specific regions of the neuron. The main neuronal compartments are the dendrites (more precisely postsynaptic sites), soma, axon and axon terminals (more precisely presynaptic sites). These regions are sometimes located at great distances from each other, and so neurons have to resolve the problems of communication between these regions and harmonization of their activities.

Regionalization of metabolic functions

The essential synthesis activity of a neuron is localized in its cell body, since dendrites can synthesize only some of their proteins, and axons are able to synthesize only a few. In this cell, where the axon's volume represents up to a thousand times the volume of the cell body, the structural and functional integrity of the axon and its terminals requires an important and continuous supply of macromolecules. This supply is ensured by anterograde axonal transport. In dendrites, RNA transport from the cell body to the polysomes has been demonstrated and would allow the synthesis of some of their proteins.

The degradation of cellular metabolism debris and non-neuronal elements taken up from the external environment by endocytosis (e.g. uptake of viruses) takes place in the lysosomes of the cell body. They are transported from axon terminals to the cell body via the retrograde axonal transport. Finally, to coordinate synthesis activity in the cell body with the needs of the axon terminals, the existence of a feedback mechanism (from terminals to cell body) seems essential. This could take place through retrograde axonal transport.

Anterograde transport moves newly synthesized material outward from the cell body along the axon. Retrograde transport drives the movement of organelles, vesicles, and signalling complexes from the cell periphery and distal axon back to the cell centre. Key motors for this transport include members of the kinesin superfamily and cytoplasmic dynein.

Regionalization of functions implicated in reception and transmission of electrical signals

The neuronal regions receiving synapses are mainly the dendritic (primary segments, branches and spines of dendrites) and somatic regions, but also some axonal regions. These receptive regions, called postsynaptic elements, have a restricted surface. They contain, within their plasma membrane, proteins specialized in the recognition of neurotransmitters: the neurotransmitter receptors (receptor channels and receptors coupled to G proteins). These proteins synthesized in the cell body are then transported toward the dendritic, somatic or axonal postsynaptic membranes to be incorporated. Similarly, the proteins specialized in the generation and propagation of action potentials (voltage-dependent channels) are synthesized in the soma and have to be transported and incorporated in the axonal membrane.

Regionalization of secretory function

This function is localized in regions making synaptic contacts and more generally in presynaptic regions such as axon terminals (and sometimes in dendritic and somatic regions). At the level of presynaptic structures, the neurotransmitter is stocked in synaptic vesicles and released. The secretory function implicates the presence of specific molecules and organelles in the presynaptic region: neurotransmitter synthesis enzymes, synaptic vesicles, microtubules and associated proteins, voltage-dependent channels, etc.

In conclusion, owing to its extreme regionalization and the extreme length and volume of its processes, the neuron has the challenge to deliver the proteins synthesized in the soma at the appropriate sites (targeting) at appropriate times.

FURTHER READING

Baas PW, Nadar CV, Myers KA (2006) Axonal transport of microtubules: the long and short of it. *Traffic* **7**, 490–498.

Brady ST, Lasek RJ, Allen RD (1982) Fast axonal transport in extruded axoplasm from squid giant axon. *Science* **218**, 1129–1131.

Davies L, Burger B, Banker GA, Steward O (1990) Dendritic transport: quantitative analysis of the time course of somatodendritic transport of recently synthetized RNA. *J. Neurosci.* **10**, 3056–3068.

Gho M, McDonald K, Ganetzky B, Saxton WM (1992) Effects of kinesin mutations on neuronal functions. *Science* **258**, 313–316.

He Y, Francis F, Myers KA, Yu W, Black MM, Baas PW (2005) Role of cytoplasmic dynein in the axonal transport of microtubules and neurofilaments. *J. Cell. Biol.* **168**, 697–703.

Hirokawa N (2006) mRNA transport in dendrites: RNA granules, motors and tracks. *J. Neuroscience* **26**, 7139–7142.

Hirokawa N and Takemura R (2005) Molecular motors and mechanisms of directional transport in neurons. *Nat. Rev. Neurosci.* **6**, 201–214.

Hollenbeck PJ and Saxton WM (2005) The axonal transport of mitochondria. *J. Cell Science* **118**, 5411–5419.

Johnston JA, Illing ME, Kopito RR (2002) Cytoplasmic dynein/dynactin mediates the assembly of aggresomes. *Cell. Motil. Cytoskeleton* **53**, 26–38.

Kanai Y, Dohmae N, Hirokawa N (2004) Kinesin transports RNA: isolation and characterization of an RNA-transporting granule. *Neuron* **43**, 513–525.

Martin KC and Zukin RS (2006) RNA trafficking and local protein synthesis in dendrites: an overview. *J. Neuroscience* **26**, 7131–7134.

Mikami A, Paschal BM, Mazumdar M, Vallee R (1993) Molecular cloning of the retrograde transport motor cytoplasmic dynein (MAP1C). *Neuron* **10**, 787–796.

Muresan V, Godek CP, Reese TS, Schnapp BJ (1996) Plus-end motors override minus-end motors during transport of squid axon vesicles on microtubules. *J. Cell Biol.* **135**, 383–397.

Nangaku M, Sato-Yoshitake R, Okada Y, Noda Y, Takemura R, Yamazaki H, Hirokawa N (1994) KIF1B, a novel microtubule plus-end-directed monomeric motor protein for transport of mitochondria. *Cell* **79**, 1209–1220.

Schnapp BJ, Vale RD, Sheetz MP, Reese TS (1985) Single microtubules from squid axoplasm support bidirectional movement of organelles. *Cell* **40**, 455–462.

St Johnston D (1995) The intracellular localization of mRNAs. *Cell* **81**, 161–170.

Steward O and Levy WB (1982) Preferential localization of ribosomes under the base of dendritic spines in granule cells of the dentate gyrus. *J. Neurosci.* **2**, 284–291.

Vale RD, Reese TS, Sheetz MP (1985) Identification of a novel force-generating protein, kinesin, involved in microtubule-based mobility. *Cell* **42**, 39–50.

Wang L and Brown A (2002) Rapid movement of microtubules in axons. *Curr Biol* **12**, 1496–1501.

CHAPTER

2

Neuron–glial cell cooperation

There are roughly twice as many glial cells as there are neurons in the central nervous system. They occupy the space between neurons and neuronal processes and separate neurons from blood vessels. As a result, the extracellular space between the plasma membranes of different cells is narrow, of the order of 15–20 nm.

Virchow (1846) was the first to propose the existence of non-neuronal tissue in the central nervous system. He named it 'nevroglie' (nerve glue), because it appeared to stick the neurons together. Following this, Deiters (1865) and Golgi (1885) identified glial cells as making up the nevroglie and distinguished them from neurons.

There are several categories of glial cells. Depending on their anatomical position they are classed as follows:

- *Central glia* are found in the central nervous system, and comprise four cell types: astrocytes, oligodendrocytes, microglia (these three types are also known as interstitial glia, because they are found in interneuronal spaces) and ependymal cells which form the epithelial surface covering the walls of the cerebral ventricles and of the central canal of the spinal cord.
- *Peripheral glia* comprise a single type: Schwann cells. These cells ensheath the axons and encapsulate the cell bodies of neurons. In the latter case, they are also called satellite cells.

Glial cells, excluding microglia, have an ectodermal origin. Those of the central nervous system derive from the germinal neural epithelium (neural tube), while peripheral glia (Schwann cells) are derived from the neural crest. Microglia, in contrast, have a mesodermal origin.

Glial cells have morphological as well as functional and metabolic characteristics that distinguish them from neurons:

- They do not generate or conduct action potentials. Thus, although they extend processes, these are only of one type and are neither dendrites nor axons.
- They do not establish chemical synapses between themselves, with neurons, or any other cell type.
- Unlike most neurons in humans, glial cells are capable of division for at least several years postnatally.

Nervous tissue is made compact by glial cells and for this reason they are often ascribed the role of supporting tissue. However, as we will see in this chapter, they have additional functions. We will explain in this chapter the roles of astrocytes, oligodendrocytes and Schwann cells, only.

2.1 ASTROCYTES FORM A VAST CELLULAR NETWORK OR SYNCYTIUM BETWEEN NEURONS, BLOOD VESSELS AND THE SURFACE OF THE BRAIN

2.1.1 Astrocytes are star-shaped cells characterized by the presence of glial filaments in their cytoplasm

Astrocytes are small star-shaped cells with numerous fine, tortuous, ramified processes covered with varicosities (**Figure 2.1**). The cell body is typically 9–10 µm in diameter and the processes extend radially over 40–50 µm. These often have enlarged terminals in

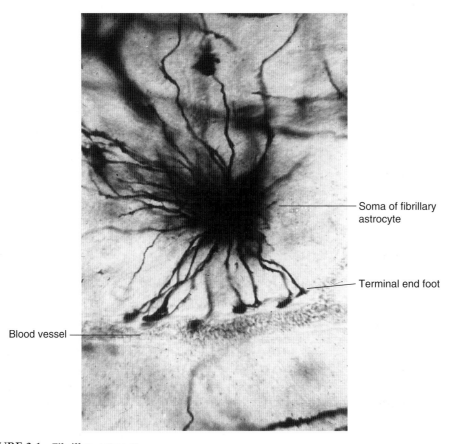

FIGURE 2.1 **Fibrillary astrocyte.**
Micrograph of a fibrillary astrocyte stained with a Golgi stain observed through an optical microscope. The processes of this astrocyte make contact with a blood vessel: these are the terminal end feet. Photograph by Olivier Robain.

contact with neurons or non-neuronal tissue (like the walls of blood vessels).

Two kinds of astrocytes are recognized. Some astrocytes contain in their cytoplasm numerous glial filaments: these are fibrillary astrocytes, principally located in the white matter (**Figure 2.1**). They have numerous, radial processes which are infrequently branched and covered with 'expansions en brindilles'. Other astrocytes contain few, if any, glial filaments: these are protoplasmic astrocytes, found normally in the grey matter. They have more delicate processes, some of which are velate (veil-like). Both types of astrocytes send out processes that end on the walls of blood vessels or beneath the pial surface of the brain and spinal cord.

The principal ultrastructural characteristics of astrocytes are the glial filaments and glycogen granules present in the cytoplasm of their somata and processes. The filaments are 'intermediate filaments' with an average diameter of 8–10 μm. They are composed of a protein specific to astrocytes, glial fibrillary acidic protein (GFAP), consisting of a single type of subunit with a molecular weight of 50 kD, different from that of

neurofilaments. This characteristic has been exploited as a method of identifying astrocytes. By using an antiserum to glial fibrillary acidic protein (anti-GFAP) linked to fluorescein, one can stain astrocytes, *in situ* or in culture, without marking either neurons or other types of glial cells.

Astrocytes, like all glial cells, do not form chemical synapses. They do, however, mutually form junctional complexes. Two types of junctions have been demonstrated: communicating junctions (or gap junctions) and desmosomes (puncta adhaerentia). Coupled to each other by numerous junctional complexes, astrocytes therefore constitute a vast cellular network, or syncytium, extending from neurons to blood vessels and the external surface of the brain.

2.1.2 Astrocytes maintain the blood–brain barrier in the adult brain

The essential characteristic of astrocytic processes is their termination on the walls of blood vessels in astrocytic end feet (**Figure 2.2**). Here the end feet are joined

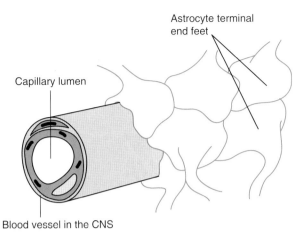

FIGURE 2.2 Diagram of the covering formed by astrocyte end feet around a capillary in the central nervous system (CNS). From Goldstein G and Betz L (1986) La barrière qui protège le cerveau. *Pour la Science*, November, 84–94, with permission.

by gap junctions and desmosomes, forming a 'palisade' between neurons and vascular endothelial cells. The space between the layer of astrocyte end feet and the endothelial cells is about 40–100 nm and is occupied by a basal lamina. Astrocytes also send processes to the external surface of the central nervous system where the astrocyte end feet, together with the basal lamina that they produce, form the 'glia limitans externa', which separates the pia mater from the nervous tissue. Astrocytes therefore constitute a barrier between neurons and the external medium (blood), preventing access of substances foreign to the central nervous system. They thus protect neurons. This barrier is not, however, totally impermeable and astrocytes are involved in selective exchange processes.

Astrocyte end feet are not the blood–brain barrier. This is formed, in most regions of the central nervous system, by vascular endothelial cells joined together by tight junctions. Even though the astrocyte end feet do not form the blood–brain barrier, they have an important role in its development and maintenance. Thus, if the layer of astrocyte end feet in the adult is destroyed, by a tumour or by allergic illnesses, for example, the capillary endothelial cells immediately take on the characteristics normally observed in capillaries outside the central nervous system: they are no longer bound by tight junctions and become 'fenestrated'. In such capillaries the blood–brain barrier no longer exists.

2.1.3 Astrocytes regulate the ionic composition of the extracellular fluid

We have seen that astrocyte end feet are involved in the formation and maintenance of the blood–brain

barrier, and that astrocytes thus contribute to regulation of the brain extracellular fluid. However, astrocytes have other important roles in controlling the composition of the extracellular fluid. We shall consider as an example the regulation of the extracellular potassium concentration.

The extracellular potassium concentration needs to be tightly regulated: if potassium increased it would depolarize neurons. This would first increase neuronal excitability and then inactivate action potential propagation. Regulation of the extracellular potassium concentration must occur in the face of large fluxes of potassium ions into the extracellular space during neuronal activity, when potassium ions leave neurons through voltage-activated potassium channels (see Sections 4.3 and 5.3). Astrocytes are thought to regulate extracellular potassium by the mechanism of 'spatial buffering'. This means that astrocytes take up potassium ions in regions where the concentration rises and eventually release through their end feet an equivalent amount of potassium ions into the vicinity of blood vessels or across the *glia limitans externa*. The details of the process are complicated, but potassium ions are thought to enter astrocytes via channels or the sodium pump and to exit at the end feet through channels. This potassium buffering role of astrocytes is likely to be of particular importance at the nodes of Ranvier, where marked accumulation of potassium ions in the restricted extracellular space can occur, due to the conduction of action potentials.

2.1.4 Astrocytes take part in the neurotransmitter cycle

After neurotransmitters are released during synaptic transmission, they need to be removed from the extracellular space to prevent the extracellular neurotransmitter concentration from rising. Steady high concentrations of transmitter would interfere with synaptic transmission, and long-lasting activation of receptors (particularly glutamate receptors) can damage neurons. Most transmitters are removed from the extracellular space by reuptake into cells (but acetylcholine is hydrolyzed; see **Figure 6.12**). Transmitters are taken up by specialized carrier molecules in the cell membrane. Although both neurons and glia express such carrier proteins, it seems that uptake into astrocytes is of particular importance. This is especially clear for the case of glutamate: astrocytes have an enormous capacity to take up this transmitter, presumably reflecting the abundance of this transmitter and the toxicity to neurons of high glutamate concentrations.

Besides their role in transmitter clearance from the synaptic cleft (by recapture), astrocytes play a role in the synthesis of transmitters and particularly glutamate and GABA. For example, thanks to the presence of glutamine synthetase in astrocytes (see **Figure 10.13**), glutamine is formed from glutamate. Glutamine is then uptaken by neurons and transformed back in glutamate.

2.2 OLIGODENDROCYTES FORM THE MYELIN SHEATHS OF AXONS IN THE CENTRAL NERVOUS SYSTEM AND ALLOW THE CLUSTERING OF NA⁺ CHANNELS AT NODES OF RANVIER

Two types of oligodendrocyte are recognized: interfascicular or myelinizing oligodendrocytes, found in the white matter where they make the sheaths of myelinated axons; and satellite oligodendrocytes which surround neuronal somata in the grey matter. We will deal with the former type in detail. Their major role is, by forming the myelin sheath, to electrically isolate segments of axons, induce the formation of clusters of Na⁺ channels at nodes of Ranvier and therefore to allow the fast propagation of Na⁺ action potentials (see Section 4.4).

2.2.1 Processes of interfascicular oligodendrocytes electrically isolate segments of central axons by forming the lipid-rich myelin sheath

The cell bodies of interfascicular oligodendrocytes are situated between bundles of axons

Interfascicular, or myelinizing, oligodendrocytes have small spherical or polyhedral cell bodies of diameter 6–8 μm and few processes. They are called interfascicular because their cell bodies are aligned between bundles (fascicles) of axons. They are distinguished from astrocytes by the sites of termination of their processes: oligodendrocyte processes enwrap axons and make no contact with blood vessels.

Observed by electron microscopy, the nucleus and perikaryon of oligodendrocytes appear dark (**Figure 2.3**), there are no glial filaments, and there are many microtubules in the somatic and dendritic cytoplasm. Because of this, oligodendrocyte processes may be confused with fine dendrites, and it is by the absence of chemical synapses that the glial processes are identified.

Oligodendrocytes can be identified by immunohistochemistry. This is done using an antigalactoceramide immune serum (anti-gal-C), galactoceramide being a glycolipid found exclusively in the membrane of processes of myelinizing oligodendrocytes.

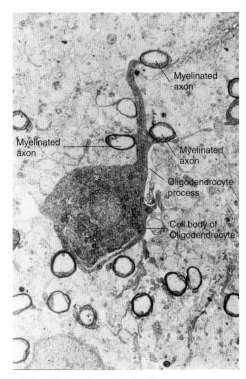

FIGURE 2.3   Myelinating oligodendrocyte.
Electron micrograph of an oligodendrocyte. The cell body and one of its processes enwrapping several axons can be seen. Section taken through the spinal cord. Photograph by Olivier Robain.

The myelin sheath is a compact roll of the plasmalemma of an oligodendrocyte process: this glial membrane is rich in lipids

Myelinated axons are surrounded by a succession of myelin segments, each about 1 mm long. The covered regions of axons alternate with short exposed lengths where the axonal membrane (axolemma) is not covered. These unmyelinated regions (of the order of a micron) are called nodes of Ranvier (**Figures 2.4** and **2.5a**).

A myelinated segment comprises the length of axon covered by an oligodendrocyte. One oligodendrocyte can form 20–70 myelin segments around different axons (**Figure 2.4**). Thus the degeneration or dysfunction of a single oligodendrocyte leads to the disappearance of myelin segments on several different axons.

Formation and ultrastructure of a myelin segment

Myelinization represents a crucial stage in the ontogenesis of the nervous system. In the human at birth, myelinization is only just beginning, and in some regions is not complete even by the end of the second year of life. The first step in the process is migration of oligodendrocytes into the bundles of axons, then the myelinization of some, but not all, axons. Once contact

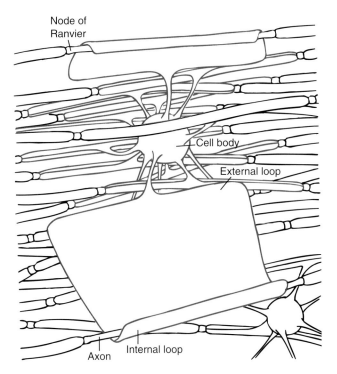

FIGURE 2.4 **Diagram of a myelinating oligodendrocyte and its numerous processes.**
Each form a segment of myelin around a different axon in the central nervous system. Two myelin segments are represented, one partially unrolled, the other completely unrolled. Drawing by Tom Prentiss. In Morell P and Norton W (1980) La myéline et la sclérose en plaques. *Pour la Science* **33**, with permission.

has been made between the oligodendrocyte and axon, the initial turn of myelin around the axon is rapidly formed. Myelin is then slowly deposited over a period which in humans can reach several months. Myelinization is responsible for a large part of the increase in weight of the central nervous system following the end of neurogenesis.

In order to form the compact spiral of myelin membrane, the oligodendrocyte process must roll itself around the axon many times (up to 40 turns) (**Figure 2.5**). It is the terminal portion of the process, called the inner loop, situated at the interior of the roll, which progressively spirals around the axon. This movement necessitates the sliding of myelin sheets which are not firmly attached. During this period, the oligodendrocyte synthesizes several times its own weight of myelin membrane each day.

Within the spiral the cytoplasm disappears entirely (except at the internal and external loops). The internal leaflets of the plasma membranes can thus adhere to each other. This adhesion is so intimate that the internal leaflets virtually fuse, forming the period, or major, dense line of thickness of 3 nm (**Figure 2.5b**). The extracellular space between the different turns of membrane

also disappears, and the external leaflets also stick to each other. This apposition is, however, less close and a small space remains between the external leaflets. The apposed external leaflets form the minor, or inter-period, dense line (**Figure 2.5b**).

Thus, a cross-section of a myelinated axon observed by electron microscopy shows alternating dark and light lines forming a spiral around the axon. The major dense line terminates where the internal leaflets separate to enclose the cytoplasm within the external loop. The interperiod dense line disappears at the surface of the sheath at the end of the spiral (**Figure 2.5b**).

In the central nervous system there is no basal lamina around myelin segments, so myelin segments of adjacent axons may adhere to each other forming an interperiod dense line.

Myelin

Myelin consists of a compact spiral (without intracellular or extracellular space) of glial plasma membrane of a very particular composition. Lipids make up about 70% of the dry weight of myelin and proteins only 30%. Compared with the membranes of other cells, this represents an inversion of the lipid:protein ratio (**Figure 2.6**).

This lipid-rich membrane is highly enriched in glycosphingolipids and cholesterol. The major glycosphingolipids in myelin are galactosylceramide and its sulfated derivative sulfatide (20% of lipid dry weight). There is also an unusually high proportion of ethanolamine phosphoglycerides in the plasmalogen form, which accounts for one-third of the phospholipids.

In myelin, a number of structural classes of proteins are present. Some proteins are extremely hydrophobic membrane-embedded polypeptides, some integral membrane proteins have a single transmembrane domain and clearly define extra- and intracellular domains, and some of the myelin proteins are cytosolic; however, they are often intimately associated with the myelin membrane.

Myelin basic protein (MBP) and the proteolipid proteins (PLP/DM20) are the two major myelin proteins in the CNS. Myelin basic proteins are found on the cytoplasmic side and play a role in the adhesion of the internal leaflets of the specialized oligodendroglial plasma membrane. Proteolipid proteins are integral membrane proteins. Though they are in high abundance (they represent around 50% of the total myelin protein in the central nervous system) their exact biological role has not yet been elucidated.

We have seen that myelin has an inverted lipid:protein ratio, while the cell body membrane of the oligodendrocyte has a ratio comparable to that of other cell

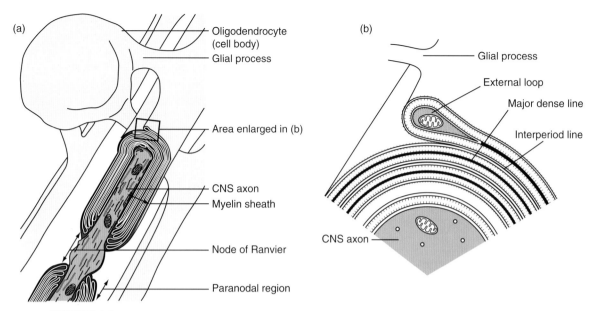

FIGURE 2.5	Myelin sheath of central axons.

(a) Three-dimensional diagram of the myelin sheath of an axon in the central nervous system (CNS). The sheath is formed by a succession of compact rolls of glial processes from different oligodendrocytes. (b) Cross-section through a myelin sheath. The dark lines, or major dense lines, and clear bands (in the middle of which are found the interperiod lines) visible with electron microscopy are accounted for by the manner in which the myelin membrane surrounds the axon, and by the composition of the membrane. The dark lines represent the adhesion of the internal leaflets of the myelin membrane while the interperiod lines represent the adhesion of the external leaflets. The lines are formed by membrane proteins while the clear bands are formed by the lipid bilayer. Drawing (a) from Bunge MB, Bunge RP, Ris H (1961) Ultrastructural study of remyelination in an experimental lesion in adult cat spinal cord. *J. Biophys. Biochem. Cytol.* **10**, 67–94, with permission of Rockerfeller University Press. Drawing (b) by Tom Prentiss. In Morell P and Norton W (1980) La myéline et la sclérose en plaques. *Pour la Science* **33**, with permission.

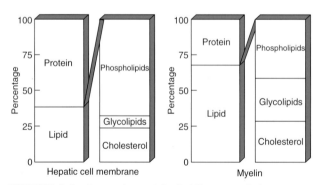

FIGURE 2.6	Comparison of the lipid content of plasma membrane and myelin.

The protein:lipid ratio is inverted between the two membranes. The proportions of the three groups of lipids are also different.

membranes. As the myelin of the oligodendrocyte process is in continuity with the plasma membrane of the cell body, it is necessary to postulate gradients in the composition of lipids and proteins (in opposite directions to each other) between the cell body and the various processes. During the active phase of myelination, each oligodendrocyte must produce as much as 5–50,000 μm^2 of myelin membrane surface area per day.

Nodes of Ranvier

In the central nervous system the nodes of Ranvier, regions between myelin segments, are relatively long (several microns) compared with those in the peripheral nervous system. Here the axolemma is exposed and an accumulation of dense material is seen on the cytoplasmic side. The myelin sheath does not terminate abruptly. Successive layers of myelin membrane terminate at regularly spaced intervals along the axon, the internal layers (close to the axon) terminating first. This staggered termination of the different layers of myelin constitutes the paranodal region (**Figure 2.5a**).

2.2.2 Myelination enables rapid conduction of action potentials for two reasons

Isolation of internode axonal segments

The high lipid content and compact structure of the myelin sheath help make it impermeable to hydrophilic substances such as ions. It prevents transmembrane ion fluxes and acts as a good electrical insulator between the intracellular (i.e. intra-axonal) and extracellular media. Between the nodes of Ranvier the axon therefore behaves as an insulated cable. This permits rapid, saltatory conduction of action potentials along the axon (see Section 5.4).

Formation of Ranvier nodes with a high density of Na$^+$ channels

Na$^+$ channels are clustered in very high density within the nodal gap whereas voltage-dependent K$^+$ channels are segregated in juxta-paranodal regions, beneath overlying myelin (see **Figure 2.5a**). To test whether oligodendrocyte contact with axon influences Na$^+$ channel distribution, nodes of Ranvier in the brain of hypomyelinating mouse *Shiverer* are examined. *Shiverer* mice have oligodendrocytes that ensheath axons but do not form compact myelin and axoglial junctions. In these mutant mice, there are far fewer Na$^+$ channel clusters than in control littermates and aberrant locations of Na$^+$ channels are observed. If Na$^+$ channel clustering depends only on the presence of oligodendrocytes and is independent of myelin and oligodendroglial contact, one would expect to find normal Na$^+$ channel distribution along axons.

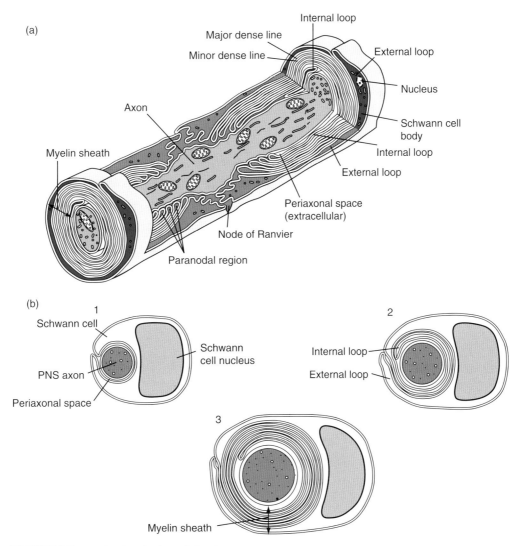

FIGURE 2.7 Myelin sheath of a peripheral axon.
(a) Three-dimensional diagram of the myelin sheath of an axon of the peripheral nervous system (PNS). The sheath is formed by successive rolled Schwann cells. **(b)** Process of myelinization. The internal loop wraps around the axon several times. During this process the axon grows and the myelin becomes compact. Contact between the Schwann cell and axon occurs only at the paranodal and nodal regions. Elsewhere an extracellular, or periaxonal, space always remains. Drawing (a) adapted from Maillet M (1977) *Le Tissu Nerveux*, Paris: Vigot, with permission. Drawing (b) by Tom Prentiss. In Morell P and Norton W (1980) La myéline et la sclérose en plaques. *Pour la Science* **33**, with permission.

2.3 SCHWANN CELLS ARE THE GLIAL CELLS OF THE PERIPHERAL NERVOUS SYSTEM; THEY FORM THE MYELIN SHEATH OF AXONS OR ENCAPSULATE NEURONS

There are three types of Schwann cell:

* those forming the myelin sheath of peripheral myelinated axons (myelinating Schwann cells);
* those encapsulating non-myelinated peripheral axons (non-myelinating Schwann cells); those that encapsulate the bodies of ganglion cells (non-myelinating Schwann cells or satellite cells).

2.3.1 Myelinating Schwann cells make the myelin sheath of peripheral axons

Along an axon, several Schwann cells form successive segments of the myelin sheath. In contrast to oligodendrocytes, it is not a process that enwraps the peripheral axon to form the segment of myelin, but the whole Schwann cell (**Figure 2.7**). Each Schwann cell therefore forms only one myelin segment.

The composition of peripheral myelin differs from that of central myelin only in the proteins it contains. The principal protein constituents of peripheral myelin are: peripheral myelin protein 2 (P2), protein zero (P0) and myelin basic proteins (MBPs). The first two proteins are specific to peripheral myelin. MBP comprises a major part of the cytosolic protein of myelin and is present both in the CNS and PNS. Protein zero is a glycoprotein that has adhesive properties and is located in the interperiod line. It functions, in part, as a homotypic adhesion molecule throughout the full thickness of the myelin sheath. It is a good marker for myelinating Schwann cells as it represents over 50% of total PNS myelin protein.

2.3.2 Non-myelinating Schwann cells encapsulate the axons and cell bodies of peripheral neurons

Non-myelinated axons are not uncovered in the peripheral nervous system as they are in the central nervous system; they are encapsulated. A single non-myelinating Schwann cell surrounds several axons (about 5–20) for a distance of 200–500 µm in man.

In addition, spinal and cranial ganglia contain a large number of Schwann cells that do not produce myelin. These Schwann cells cover the somata of the ganglionic cells, leaving an extracellular space of about 20 nm between themselves and the surface of the covered neuron.

The lipid and protein composition of the plasma membrane of non-myelinating Schwann cells is the same as that of other eukaryotic cells (30% lipid, 70% protein).

Apart from their role in the saltatory conduction of action potentials (myelinating Schwann cells), Schwann cells also play a role in the regeneration of peripheral nerve cells. It has long been known that cut peripheral nerves can, within certain limits, regrow and reinnervate deafferented regions while central axons are not capable of this. This property of regeneration is due in large part to an enabling effect of Schwann cells on axon regrowth.

FURTHER READING

Dupree JL, Mason JL, Marcus JR, *et al.* (2005) Oligodendrocytes assist in the maintenance of sodium channel clusters independent of the myelin sheath. *Neuron Glia Biol.* **1**, 1–14.

Farber K and Kettenmann H (2005) Physiology of microglial cells. *Brain Res Brain Res Rev.* **48**, 133–143.

Freeman MR and Doherty J (2006) Glial cell biology in Drosophila and vertebrates. *Trends Neurosci.* **29**, 82–90.

Greer JM and Lees MB (2002) Myelin proteolipid protein – the first 50 years. *Int J Biochem & Cell Biology* **34**, 211–215.

Montague P, McCallion AS, Davies RW, Griffiths IR (2006) Myelin-associated oligodendrocytic basic protein: a family of abundant CNS myelin proteins in search of a function. *Dev Neurosci.* **28**, 479–487.

Peles E and Salzer JL (2000) Molecular domains of myelinated axons. *Curr Opin Neurobiol.* **10**, 558–565.

Pfeiffer SE, Warrington AE, Bansal R (1993) The oligodendrocyte and its many cellular processes. *Trends Cell Biol.* **3**, 191–197.

Popko B (2000) Myelin galactolipids: mediators of axon–glial interactions? *Glia* **29**, 149–153.

Shapiro L, Doyle JP, Hensley P, Colman DR, Hendrickson WA (1996) Crystal structure of the extracellular domain from P0, the major structural protein of peripheral nerve myelin. *Neuron* **17**, 435–449.

Streit WJ (2000) Microglial response to brain injury: a brief synopsis. *Toxicol. Pathol.* **28**, 28–30.

Ionic gradients, membrane potential and ionic currents

The neuronal plasma membrane delimits the whole neuron, cell body, dendrites, dendritic spines, axon and axon terminals. It is a barrier between the intracellular and extracellular environments. The general structure of the neuronal plasma membrane is similar to that of other plasma membranes. It is made up of proteins inserted in a lipid bilayer, forming as a whole a 'fluid mosaic' (**Figure 3.1**). However, insofar as there are functions that are exclusively neuronal, the neuronal membrane differs from other plasma membranes by the nature, density and spatial distribution of the proteins of which it is composed.

The presence of a large diversity of transmembrane proteins called *ionic channels* (or simply 'channels') characterize the neuronal plasma membrane. They allow the passive movement of ions across membranes and thus

electrical signalling in the nervous system. Among the ions present in the nervous system fluids, Na^+, K^+, Ca^{2+} and Cl^- ions seem to be responsible for almost all of the action.

3.1 THERE IS AN UNEQUAL DISTRIBUTION OF IONS ACROSS NEURONAL PLASMA MEMBRANE. THE NOTION OF CONCENTRATION GRADIENT

3.1.1 The plasma membrane separates two media of different ionic composition

Regardless of the animal's environment (seawater, freshwater or air), potassium (K^+) ions are the

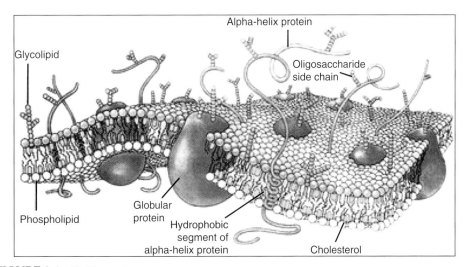

FIGURE 3.1   Fluid mosaic.
Transmembrane proteins and lipids are kept together by non-covalent interactions (ionic and hydrophobic).
From dictionary.laborlawtalk.com/Plasma_membrane.

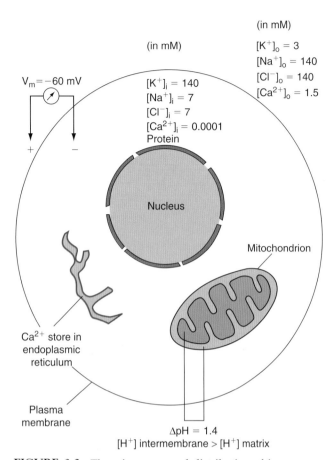

(in mM)

$[K^+]_o = 3$
$[Na^+]_o = 140$
$[Cl^-]_o = 140$
$[Ca^{2+}]_o = 1.5$

(in mM)

$[K^+]_i = 140$
$[Na^+]_i = 7$
$[Cl^-]_i = 7$
$[Ca^{2+}]_i = 0.0001$
Protein

$V_m = -60$ mV

Nucleus

Mitochondrion

Ca^{2+} store in
endoplasmic
reticulum

Plasma
membrane

ΔpH = 1.4
$[H^+]$ intermembrane > $[H^+]$ matrix

FIGURE 3.2 **There is an unequal distribution of ions across neuronal plasma membranes. Membrane potential.** Idealized nerve cell (depicted as a sphere) with relative concentrations of intra- and extracellular ions.

predominant cations in the intracellular fluid and sodium (Na^+) ions are the predominant cations in the extracellular fluid. The main anions of the intracellular fluid are organic molecules (P^-): negatively charged amino acids (glutamate and aspartate), proteins, nucleic acids, phosphates, etc… which have a large molecular weight. In the extracellular fluid the predominant anions are chloride (Cl^-) ions. A marked difference between cytosolic and extracellular Ca^{2+} concentrations is also observed (**Figure 3.2**).

Spatial distribution of Ca^{2+} ions inside the cell deserves a more detailed description. Ca^{2+} ions are present in the cytosol as 'free' Ca^{2+} ions at a very low concentration (10^{-8} to 10^{-7} M) and as bound Ca^{2+} ions (bound to Ca^{2+}-binding proteins). They are also distributed in organelles able to sequester calcium, which include endoplasmic reticulum, calciosome and mitochondria, where they constitute the intracellular Ca^{2+} stores. Free intracellular Ca^{2+} ions present in the

cytosol act as second messengers and transduce electrical activity in neurons into biochemical events such as exocytosis. Ca^{2+} ions bound to cytosolic proteins or present in organelle stores are not active Ca^{2+} ions; only 'free' Ca^{2+} ions have a role.

In spite of the unequal distribution of ions across the plasma membrane, intracellular and extracellular media are neutral ionic solutions: in each medium, the concentration of positive ions is equal to that of negative ions. According to **Figure 3.2**,

$$[Na^+]e + [K^+]e + 2[Ca^{2+}]e = 140 + 3 + (2 \times 1.5)$$
$$= 146 \, mM \text{ and } [Cl^-]e = 146 \, mM$$

$$[Na^+]_i + [K^+]_i + 2[Ca^{2+}]_i = 7 + 140 + 0.0002$$
$$= 147 \, mM \text{ but } [Cl^-]_i = 7 \, mM$$

In the intracellular compartment, other anions than chloride ions are present and compensate for the positive charges. These anions are HCO_3^-, PO_4^{2-}, aminoacids, proteins, nucleic acids, etc….. Most of these anions are organic anions that do not cross the membrane.

3.1.2 The unequal distribution of ions across the neuronal plasma membrane is kept constant by active transport of ions

A difference of concentration between two compartments is called a '*concentration gradient*'. Measurements of Na^+, K^+, Ca^{2+} and Cl^- concentrations have shown that concentration gradients for ions are constant in the external and cytosolic compartments, at the macroscopic level, during the entire neuronal life.

At least two hypotheses can explain this constancy:

- Na^+, K^+, Ca^{2+} and Cl^- ions cannot cross the plasma membrane: plasma membrane is impermeable to these inorganic ions. In that case, concentration gradients need to be established only once in the lifetime.
- Plasma membrane is permeable to Na^+, K^+, Ca^{2+} and Cl^- ions but there are mechanisms that continuously re-establish the gradients and maintain constant the unequal distribution of ions.

This has been tested experimentally by measuring ionic fluxes. When proteins are absent from a synthetic lipid bilayer, no movements of ions occur across this purely lipidic membrane. Owing to its central hydrophobic region, the lipid bilayer has a low permeability to hydrophilic substances such as ions, water and polar molecules; i.e. the lipid bilayer is a barrier for the diffusion of ions and most polar molecules.

The first demonstrations of ionic fluxes across plasma membrane by Hodgkin and Keynes (1955) were based on the use of radioisotopes of K⁺ or Na⁺ ions. Experiments were conducted on the isolated squid giant axon. When this axon is immersed in a bath containing a control concentration of radioactive *Na⁺ (^{24}Na⁺) instead of cold Na⁺ (^{22}Na⁺), *Na⁺ ions constantly appear in the cytoplasm. This *Na⁺ influx is not affected by dinitrophenol (DNP), a blocker of ATP synthesis in mitochondria. It does not require energy expenditure. This is *passive* transport. This result is in favour of the second hypothesis and leads to the following question: what are the mechanisms that maintain concentration gradients across neuronal membranes?

When the reverse experiment is conducted, the isolated squid giant axon is passively loaded with radioactive *Na⁺ by performing the above experiment, and is then transferred to a bath containing cold Na⁺. Measuring the quantity of *Na⁺ that appears in the bath per unit of time (d*Na⁺$/dt$, expressed in counts per minute) allows quantification of the efflux of *Na⁺ (**Figure 3.3a**). In the presence of dinitrophenol (DNP) this *Na⁺ efflux quickly diminishes to nearly zero. The process can be started up again by intracellular injection of ATP. Therefore the *Na⁺ efflux is *active* transport. The movement of Na⁺ from the cytosol to the outside (efflux) can be switched off reversibly by the use of metabolic inhibitors.

This experiment demonstrates that cells maintain their ionic composition in the face of continuous passive exchange of all principal ions by active transport of these ions in the reverse direction. In other words, ionic composition of cytosol and extracellular compartments are maintained at the expense of a continuous basal metabolism that provides energy (ATP) utilized to actively transport ions and thus to compensate for their passive movements (Appendix 3.1).

3.1.3 Na⁺, K⁺, Ca²⁺ and Cl⁻ ions passively cross the plasma membrane through a particular class of transmembrane proteins – the channels

Transmembrane proteins span the entire width of the lipid bilayer (see **Figure 3.1**). They have hydrophobic regions containing a high fraction of non-polar amino acids and hydrophilic regions containing a high fraction of polar amino acids. Certain hydrophobic regions organize themselves inside the bilayer as transmembrane α-helices while more hydrophilic regions are in contact with the aqueous intracellular and extracellular environments. Interaction energies are very high between hydrophobic regions of the protein and hydrophobic regions of the lipid bilayer, as well as between hydrophilic regions of the protein and the extracellular and intracellular environments. These

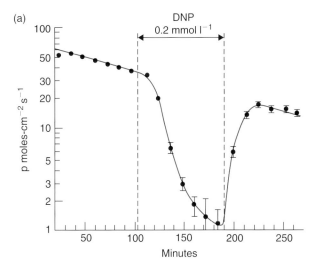

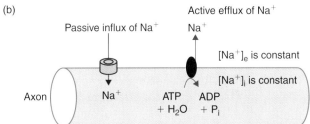

FIGURE 3.3 **Na⁺ fluxes through the membrane of giant axons of sepia.**
(a) Effect of dinitrophenol (DNP) on the outflux of *Na⁺ as a function of time. The axon is previously loaded with *Na⁺. At $t = 1$, the axon is transferred in a bath devoid of *Na⁺. The ordinate (logarithmic) axis is the quantity of *Na⁺ ions that appear in the bath (that leave the axon) as a function of time. At $t = 100$ min, DNP (0.2 mM) is added to the bath for 90 min. The efflux, which previously decreased linearly with time, is totally blocked after one hour of DNP. This blockade is reversible. **(b)** Passive and active Na⁺ fluxes are in opposite directions. Plot (a) adapted from Hodgkin AL and Keynes RD (1955) Active transport of cations in giant axons from sepia and loligo. *J. Physiol. (Lond.)* **128**, 28–60, with permission.

interactions strongly stabilize transmembrane proteins within the bilayer, thus preventing their extracellular and cytoplasmic regions from flipping back and forth.

Ionic channels have a three-dimensional structure that delimits an aqueous pore through which certain ions can pass. They provide the ions with a passage through the membrane (Appendix 3.2). Each channel may be regarded as an excitable molecule as it is specifically responsive to a stimulus and can be in at least two different states: closed and open. Channel opening, the switch from the closed to the open state, is tightly controlled (**Table 3.1**) by:

• a change in the membrane potential – these are voltage-gated channels;
• the binding of an extracellular ligand, such as a neurotransmitter – these are ligand-gated channels, also called receptor channels or ionotropic receptors;

TABLE 3.1　Examples of ionic channels

Channels	Voltage-gated	Ligand-gated			Mechanically gated
Opened by	Depolarization Hyperpolarization	Extracellular ligand	Intracellular ligand		Mechanical stimuli
Localization	Plasma membrane	Plasma membrane	Plasma membrane	Organelle membrane	Plasma membrane
Examples	Na^+ channels Ca^{2+} channels K^+ channels Cationic channels	nAChR iGluR 5-HT$_3$ GABA$_A$ GlyR	G protein-gated channels Ca^{2+}-gated channels CNG channels ATP-gated channels	IP$_3$-gated Ca^{2+} channel Ca^{2+}-gated Ca^{2+} channel	Stretch-activated channels
Closed by	Inactivation Repolarization	Desensitization Ligand recaptureior degradation			Adaptation End of stimulus
Roles	Na^+ and Ca^{2+}- dependent action potentials $[Ca^{2+}]_i$ increase	EPSP IPSP $[Ca^{2+}]_i$ increase	EPSP IPSP Action potential repolarization	$[Ca^{2+}]_i$ increase	Receptor potential

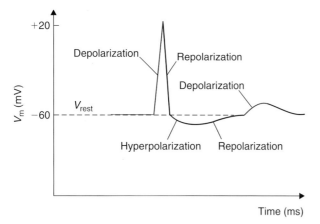

FIGURE 3.4　**Variations of the membrane potential of neurons (V_m).**
When the membrane potential is less negative than resting membrane potential (V_{rest}), the membrane is said to be depolarized. In contrast, when the membrane potential is more negative then V_{rest}, the membrane is said to be hyperpolarized. When the membrane varies from a depolarized or hyperpolarized value back to rest, the membrane repolarizes.

- the binding of an intracellular ligand such as Ca^{2+} ions or a cyclic nucleotide;
- mechanical stimuli such as stretch – these are mechanoreceptors.

The channel's response to its specific stimuli, called gating, is a simple opening or closing of the pore. The pore has the important property of selective permeability, allowing some restricted class of small ions to flow passively down their electrochemical gradients (see Section 3.3). These gated ions fluxes through pores make signals for the nervous system.

3.2 THERE IS A DIFFERENCE OF POTENTIAL BETWEEN THE TWO FACES OF THE MEMBRANE, CALLED MEMBRANE POTENTIAL (V_m)

If a fine-tipped glass pipette (usually called a microelectrode), connected via a suitable amplifier to a recording system such as an oscilloscope, is pushed through the membrane of a living nerve cell to reach its cytoplasm, a potential difference is recorded between the cytoplasm and the extracellular compartment (**Figure 3.2**). In fact, the cell interior shows a negative potential (typically between −60 and −80 mV) with respect to the outside, which is taken as the zero reference potential. Membrane potential (V_m) is by convention the difference between the potential of the internal and external faces of the membrane ($V_m = V_i − V_o$). In the absence of ongoing electrical activity, this negative potential is termed the resting membrane potential (V_{rest}) (**Figure 3.4**).

We have seen above that in the intracellular and extracellular media the concentration of positive ions is equal to that of negative ions. However, there is a very small excess of positive and negative ions accumulated on each side of the membrane. At rest, for example, a small excess of negative ions is accumulated at the internal side of the membrane whereas a small excess of positive ions is accumulated at the external side of the membrane (see Section 3.5). This creates a difference of potential between the two faces of the membrane: the external side is more positive than the internal side, which makes $V_m = V_i − V_o$, negative.

What is particular to membrane of neurons (and of all excitable cells) is that V_m varies (**Figure 3.4**). It can be more negative or hyperpolarized or less negative (depolarized) or even positive (also depolarized, the internal face is positive compared to the external face). At rest, V_m is in the range $-80/-50$ mV depending on the neuronal type. But when neurons are active, V_m varies between the extreme values -90 mV and $+30$ mV. Since nerve cells communicate through rapid (milliseconds; ms) or slow (seconds; s) changes in their membrane potential, it is important to understand V_{rest} first.

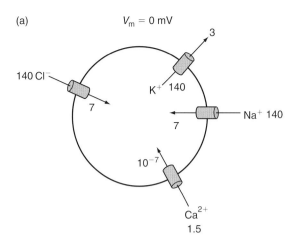

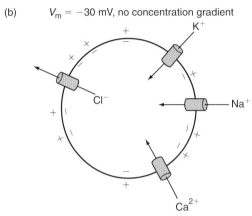

FIGURE 3.5 **Passive diffusion of ions.**
Passive diffusion of ions according to **(a)** their concentration gradient only, or **(b)** to membrane potential (electrical gradient) only ($V_m = -30$ mV).

3.3 CONCENTRATION GRADIENTS AND MEMBRANE POTENTIAL DETERMINE THE DIRECTION OF THE PASSIVE MOVEMENTS OF IONS THROUGH IONIC CHANNELS: THE ELECTROCHEMICAL GRADIENT

To predict the direction of the passive diffusion of ions through an open channel, both the concentration gradient of the ion and the membrane potential have to be known. The *resultant* of these two forces is called the electrochemical gradient. To understand what the electrochemical gradient is for a particular ion, the concentration gradient and the electrical gradient will first be explained separately.

3.3.1 Ions passively diffuse down their concentration gradient

The concentration gradient of a particular ion is the difference of concentration of this ion between the two sides of the plasma membrane. Ions passively move through open channels from the medium where their concentration is high to the medium where their concentration is lower. Suppose that membrane potential is null ($V_m = 0$ mV), there is no difference of potential between the two faces of the membrane, so ions will diffuse according to their concentration gradient only (**Figure 3.5a**). Since the extracellular concentrations of Na^+, Ca^{2+} and Cl^- are higher than the respective intracellular ones, these ions will diffuse passively towards the intracellular medium (when Na^+, Ca^{2+} or Cl^- permeable channels are open) as a result of their concentration gradient. In contrast, K^+ will move from the intracellular medium to the extracellular one (when K^+ permeable channels are open).

The force that makes ions move down their concentration gradient is *constant* for a given ion since it depends on the difference of concentration of this ion, which is itself continuously controlled to a constant value

by active transport (pumps and transporters). However, this is not always true; during intense neuronal activity, concentration of ions may change (K^+ concentration in particular) owing to the small volume of the external medium in physiological conditions. At the microscopic level this is not true also; intracellular Ca^{2+} concentration, for example, can increase locally by a factor of between 100 and 1000 but stay stable in the entire cytosol. However, these increases of ion concentration do not change the direction of the concentration gradient for this ion since ionic gradients cannot reverse by themselves.

3.3.2 Ions passively diffuse according to membrane potential

Membrane potential is a potential gradient that forces ions to passively move in one direction: positive ions are attracted by the 'negative' side of the membrane and negative ions by the 'positive' one. If we suppose that there is no concentration gradient for any ions (there is the same concentration of each ion in the extracellular and intracellular media), ions will diffuse according to membrane potential only: at a membrane potential $V_m = -30\,mV$ (**Figure 3.5b**), positively charged ions, the cations Na^+, Ca^{2+} and K^+, will move from the extracellular medium to the intracellular one according to membrane potential. In contrast, anions (Cl^-) will move from the intracellular medium to the extracellular one.

3.3.3 In physiological conditions, ions passively diffuse according to the electrochemical gradient

In physiological conditions, both the concentration gradient and membrane potential determine the direction and amplitude of ion diffusion through an open channel. Since concentration gradient is constant for each ion, the direction and amplitude of diffusion varies with membrane potential. When comparing **Figure 3.5a** and **b** it appears that at a membrane potential of $-30\,mV$, concentration gradient and membrane potential drive Na^+ and Ca^{2+} ions in the same direction, toward the intracellular medium, whereas they drive K^+ and Cl^- in reverse directions. The resultant of these two forces, concentration and potential gradients, is the electrochemical gradient. To know how to express the electrochemical gradient, the equilibrium potential must first be explained.

The equilibrium potential for a given ion, E_{ion}

All systems are moving toward equilibrium. The value of membrane potential where the concentration force that tends to move a particular ion in one direction is exactly balanced by the electrical force that tends to move the same ion in the reverse direction is called the 'equilibrium potential' of the ion (E_{ion}) or the reversal potential of the ion E_{rev}. The equilibrium potential for a particular ion is the value of V_m for which the net flux of this ion (f_{net}) through an open channel is null: when $V_m = E_{ion}$, $f_{net} = 0\,mol\,s^{-1}$.

E_{ion} can be calculated using the Nernst equation (see Appendix 3.3):

$$E_{ion} = (RT/zF)\ln([ion]_e/[ion]_i),$$

where R is the constant of an ideal gas (8.314 VCK^{-1} mol^{-1}); T is the absolute temperature in kelvin (273.16 + the temperature in °C); F is the Faraday constant (96 500 C mol^{-1}); z is the valence of the ion; and [ion] is the concentration of the ion in the extracellular (e) or intracellular (i) medium. This gives:

$$E_{ion} = (58/z)\log_{10}([ion]_e/[ion]_i), \tag{1}$$

From the equation and concentrations of **Figure 3.2**, the equilibrium potentials for each ion can be calculated:

$$E_{Na} = (58/1)\log_{10}(140/14) = +58\,mV$$

$$E_K = (58/1)\log_{10}(3/160) = -84\,mV$$

$$E_{Ca} = (58/2)\log_{10}(1/10^{-4}) = +116\,mV$$

$$E_{Cl} = (58/-1)\log_{10}(150/14) = -58\,mV.$$

These equations have the following meanings. If the channels open in a membrane where K^+ channels are the only channels open, the efflux of K^+ ions will hyperpolarize the membrane until $V_m = E_K = -84\,mV$, a potential at which the net flux of K^+ is null since K^+ ions have exactly the same tendency to diffuse towards the intracellular medium according to their concentration gradient than to move in the reverse direction according to membrane potential. At that potential the efflux of K^+ will be exactly compensated by the influx of K^+ and the membrane potential will stay stable at $V_m = E_K$ as long as K^+ channels stay open. Now, if only Na^+ channels are open, the membrane potential will move toward $V_m = +58\,mV$, the potential at which the net flux of Na^+ is null. Similarly, when $V_m = E_{Cl} = -60\,mV$, Cl^- ions have the same tendency to move down their concentration gradient than to move in the reverse direction according to membrane potential, the net flux of Cl^- is null. In contrast, when V_m is different from E_{Cl}, the net flux of Cl^- is not null. This holds true for all the other ions: when V_m is different from E_{ion} there is a net flux of this ion.

The electrochemical gradient

We have seen that when $V_m = E_{ion}$ (i.e. $V_m - E_{ion} = 0$), there is no diffusion of this particular ion ($f_{net} = 0$). In contrast, when V_m is different from E_{ion} there is a passive diffusion of this ion through an open channel. The difference ($V_m - E_{ion}$) is called the electrochemical gradient. It is the force that makes the ion move through an open channel.

3.4 THE PASSIVE DIFFUSION OF IONS THROUGH AN OPEN CHANNEL CREATES A CURRENT

To know the direction of passive diffusion of a particular ion and how many of these ions diffuse per unit of time, the direction and intensity of the net flux of ions (number of moles per second) through an open channel have to be measured. Usually the net flux (f_{net}) is not measured; the electrical counterpart of this net flux, the ionic current, is measured instead.

Passive diffusion of ions through an open channel is a movement of charges through a resistance (resistance here is a measure of the difficulty of ions moving through the channel pore). Movement of charges through a resistance is a current. Through a single channel the current is called 'single-channel current' or 'unitary current', i_{ion}. The relation between f_{net} and i_{ion} is:

$$i_{ion} = f_{net} \, zF$$

The amplitude of i_{ion} is expressed in ampères (A) which are coulombs per seconds (C s^{-1}). F is the Faraday constant (96 500 C); z is the valence of the ion (+1 for Na$^+$ and K$^+$, −1 for Cl$^-$, +2 for Ca$^+$); and f_{net} is the net flux of the ion in mol s^{-1}.

In general, currents are expressed following Ohm's Law: $U = RI$, where I is the current through a resistance R and U is the difference of potential between the two ends of the resistance. For currents carried by ions (and not by electrons as in copper wires), I is called i_{ion}, the current that passes through the resistance of the channel pore which has a resistance R (called r_{ion}). But what is U in biological systems? U is the force that makes ions move in a particular direction; it is the electrochemical gradient for the considered ion and is also called the driving force: $U = V_m - E_{ion}$ (**Figure 3.6**).

Unitary current, i_{ion}

According to Ohm's Law, the current i_{ion} through a single channel is derived from

$$(V_m - E_{ion}) = r_{ion} \cdot i_{ion}$$

So:

$$i_{ion} = (1/r_{ion})(V_m - E_{ion}) = \gamma_{ion}(V_m - E_{ion})$$

γ_{ion} is the reciprocal of resistance; it is called the *conductance* of the channel, or unitary conductance (**Figure 3.6**). It is a measure of the ease of flow of ions (flow of current) through the channel pore. Whereas resistance is expressed in ohms (Ω), conductance is expressed in siemens (S). By convention i_{ion} is negative when it represents an inward flux of positive charges (cations) and i_{ion} is positive when it represents an outward flux of positive charges (**Figure 3.5c**). It is generally of the order of pico-ampères (1 pA = 10^{-12}A). At physiological concentrations, γ_{ion} varies between 10 and 150 pico-siemens (pS), according to the channel type.

Total current, I_{ion}

In physiological conditions, several channels of the same type are open at the same time in the neuronal membrane. Suppose that only one type of channel is open in the membrane, for example Na$^+$ channels, the

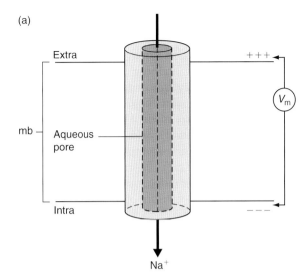

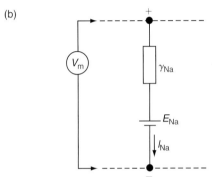

FIGURE 3.6   The Na$^+$ channel.
(a) Schematic, and **(b)** its electrical equivalent.

total current I_{Na} that crosses the membrane at time t is the sum of the unitary currents i_{Na} at time t:

$$I_{Na} = Np_o \, i_{Na}$$

where N is the number of Na$^+$ channels present in the membrane; p_o is the probability of Na$^+$ channels being open at time t (Np_o is therefore the number of open Na$^+$ channels in the membrane at time t); and i_{Na} is the unitary Na$^+$ current. More generally:

$$I_{ion} = Np_o \, i_{ion}$$

By analogy, the total conductance of the membrane for a particular ion is:

$$G_{ion} = Np_o \, \gamma_{ion}$$

and from $i_{ion} = \gamma_{ion}(V_m - E_{ion})$ above:

$$I_{ion} = G_{ion}(V_m - E_{ion})$$

I_{ion} and i_{ion} can be measured experimentally. The latter is the current measured from a patch of membrane where only one channel of a particular type is present. I_{ion} is the current measured from a whole cell membrane where N channels of the same type are present.

Roles of ionic currents

Ionic currents have two main functions:

- Ionic currents change the membrane potential: either they depolarize the membrane or repolarize it or hyperpolarize it, depending on the charge carrier. These terms are in reference to resting potential (**Figure 3.4**). Changes of membrane potential are signals. A depolarization can be an action potential (see Chapters 4 and 5) or a postsynaptic excitatory potential (EPSP; see Chapters 8 and 10). An hyperpolarization can be a postsynaptic inhibi-tory potential (IPSP; see Chapter 9). These changes of membrane potential are essential to neuronal communication.
- Ionic currents increase the concentration of a particular ion in the intracellular medium. Calcium current, for example, is always inward. It transiently and locally increases the intracellular concentration of Ca^{2+} ions and contributes to the triggering of Ca^{2+}-dependent events such as secretion or contraction.

3.5 A PARTICULAR MEMBRANE POTENTIAL, THE RESTING MEMBRANE POTENTIAL V_{rest}

In the absence of ongoing electrical activity (when the neuron is not excited or inhibited by the activation of its afferents) its membrane potential is termed the resting membrane potential (V_{rest}). For some neurons, V_{rest} is stable (silent neurons) for others it is not (pacemaker neurons for example). In this section, we will consider stable V_{rest} only. To understand unstable V_{rest} many different channels must be known that are explained later in the book (Chapter 14).

3.5.1 When most of the channels open at rest are K^+ channels V_{rest} is close to E_K

It was Julius Bernstein (1902) who pioneered the theory of V_{rest} as due to selective permeability of the membrane to one ionic species only and that nerve excitation developed when such selectivity was transiently lost. According to this theory, under resting conditions the cell membrane permeability is minimal to Na^+, Cl^- and Ca^{2+} while it is high to K^+. What is the membrane potential of a membrane permeable to K^+ ions only? This condition can be tested experimentally by measuring ionic fluxes with radioactive tracers through a plasma membrane where K^+ channels are the only open channels. K^+ moves outwards following

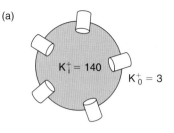

(a)

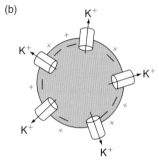

(b)

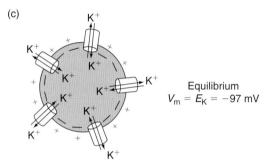

(c)

Equilibrium
$V_m = E_K = -97$ mV

FIGURE 3.7 **Establishment of V_{rest} in a cell where most of the channels open are K^+ channels.**
Suppose that at $t = 0$ and cell potential $V_m = 0$ mV (a), K^+ ions will move outwards due to their concentration gradient (b). Loss of intracellular K^+ induces a negative potential (V_m) as $V_m = E_K$ (c).

its concentration gradient (the intracellular concentration of K^+ is around 50 times higher than the extracellular one): positive charges are thus subtracted from the intracellular medium and there is an accumulation of negative charges at the intracellular side of the membrane and positive charges at the external side of the membrane. These positive charges will oppose further outward movements of K^+ until an equilibrium is reached when the concentration gradient for K^+ cancels the drive exerted by the electrical gradient. This is by definition the *equilibrium potential E_K*. Hence, at $V_m = E_K$, although K^+ keeps moving in and out of the cell, there is no net change in its concentration across the membrane (**Figure 3.7**). In a physiological situation, the exact value of E_K is unknown since the exact $[K^+]_i$ is unknown. When $V_{rest} = -80/-70$ mV, though it seems close to E_K it may not be equal to E_K.

A way to test whether $V_{rest} = E_K$ is the following. Inspection of the Nernst equation applied to K^+ indicates that a 10-fold change in the concentration ratio

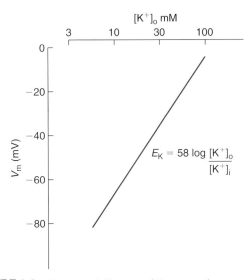

FIGURE 3.8 Theoretical diagram of E_K versus the external concentration of K^+ ions ($[K^+]o$).
$E_K = (RT/zF)2.3 \times \log([K^+]_o/[K^+]_i)$.

should alter the membrane potential of a neuron by 58 mV. This relation can be tested in experiments in which the extracellular concentration of this ion is altered and the resulting membrane potential measured with a sharp or patch microelectrode. A semilog plot of the extracellular K^+ concentration (abscissa) against the membrane potential (ordinate) should thus have a slope of 58 mV per 10-fold change in K^+ (**Figure 3.8**); this condition is rarely encountered in neurons but it seems to be more common for glial cells (which sometimes are termed K^+ electrodes because their membrane potential is linearly dependent on K^+). In the case of neurons, nonlinearity of this plot is frequently seen, particularly at low levels of extracellular K^+. These observations confirm that K^+ is a very important ion for setting the value of neuronal V_{rest} but that other ions must also play a significant role.

3.5.2 In central neurons, K^+, Cl^- and Na^+ ion movements participate in resting membrane potential and V_{rest} is different from E_K: the Goldman–Hodgkin–Katz equation

Aside from K^+, which ions play a role in V_{rest}? Since the intracellular concentration of Na^+ is not negligible, this implies that this ionic species can accumulate inside the cytoplasm, presumably because of its rather positive E_{Na} (+75 mV) versus a very negative V_{rest} creates an electrochemical gradient extremely favourable to Na^+ entry. Equally, the asymmetric distribution of Cl^- suggests its possible role in determining V_{rest}. In order to take into account various ionic species it is

useful to introduce what is commonly called the *Goldman–Hodgkin–Katz equation* (GHK), derived from the Nernst equation and named after the three physiologists responsible for its derivation:

$$V_{rest} = 58 \log \times \frac{p_K[K^+]_o + p_{Na}[Na^+]_o + p_{Cl}[Cl^-]_i}{p_K[K^+]_i + p_{Na}[Na^+]_i + p_{Cl}[Cl^-]_o} \quad (2)$$

where p is the permeability coefficient ($cm\ s^{-1}$) for each ionic species. The relative contribution of each ion species to the resting voltage is weighted by that ion's permeability.

Note that if the resting permeability to Na^+ and Cl^- is very low, the GHK equation closely resembles the Nernst equation for K^+.

In applying the GHK equation to nerve cells, the following assumptions must be made:

- The voltage gradient across the membrane is uniform in the sense that it changes linearly within the membrane. This assumption has led to the GHK equation being called the *constant field equation*.
- The overall net current flow across the membrane is zero as the currents generated by individual ionic species are balanced out.
- The membrane is in a steady state since there is no time-dependent change in ionic flux or channel density. This is obviously not applicable to non-steady state conditions of rapidly changing membrane potential as produced when a nerve cell fires action potentials.
- Any role of active transport mechanisms is ignored.
- The ionic species are monovalent cations or anions which do not interact among themselves or with water molecules. The first point does not hold true if there is a measurable permeability to divalent cations such as Ca^{2+}. Furthermore, it has been reported that ions can interact among themselves within the same channel.
- The role of membrane surface charges is ignored. This is a relatively major limitation because the cell membrane contains negative charges on its inner and outer layers (amino acid residues of membrane proteins which are typically negatively charged). The electric field generated by these charges is able to influence the kinetic properties of ionic channels (gating, activation and inactivation). Adding divalent cations such as Ca^{2+} or Mg^{2+} leads to screening of these charges and consequent changes in channel properties.
- The mobility of each ionic species and its diffusion coefficient (D) within the membrane of thickness (δ) is constant.
- The ions do not bind to specific sites in the membrane and their concentration (C) can be expressed by a

linear partition coefficient ($\beta = C_{membrane}/C_{solution}$). However, there is evidence that ions can bind to sites inside channels and influence channel kinetics.

- The ionic activities (a) can be replaced by their concentrations.

3.6 A SIMPLE EQUIVALENT ELECTRICAL CIRCUIT FOR THE MEMBRANE AT REST

Since the plasma membrane does not allow the passage of all the ions at all time, it can be equated to an insulator separating two electrically conductive media (intracellular and extracellular electrolytes): it thus plays the role of a dielectric in a *capacitor* and it can be assigned an average capacity (C_m) value of $1\ \mu F\ cm^{-2}$.

In **Figure 3.9b**, instead of three parallel current sources for K^+, Na^+ and Cl^-, we have lumped them together into only one source with driving (electromotive) force E equal to V_{rest} and an inward conductance g_m equal to the sum of the specific ionic (channel) conductances

$g_K + g_{Na} + g_{Cl}$. One may consider, instead of the absolute value of membrane potential, only its deviation from V_{rest}. In this case the equivalent electromotive force becomes equal to zero and the equivalent scheme of the cell membrane simplifies to an *RC*-circuit (**Figure 3.9c**). If one includes more channel types, then the notion of resting current still holds true. The equivalent scheme of **Figure 3.9c** is applicable only to depolarizations and hyperpolarizations characterized by linear (ohmic) current−voltage relations (**Figure 3.9d**). In standard excitable cells it means that these potential changes from V_{rest} are not activating voltage-gated currents; e.g. they are below the threshold for spike generation.

3.7 HOW TO EXPERIMENTALLY CHANGE V_{rest}

3.7.1 How to experimentally depolarize a neuronal membrane

The aim of the experiment is to lower the difference of potential between the two faces of the membrane and

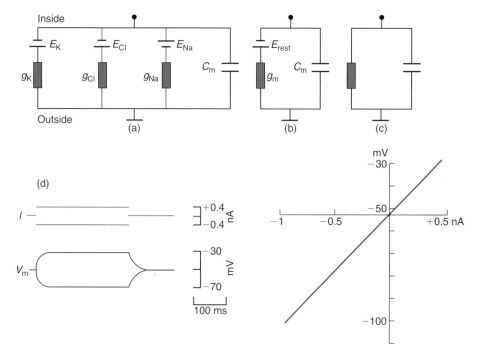

FIGURE 3.9 **Simplified equivalent scheme to account for membrane electrical characteristics near the resting potential and ohmic behaviour of the membrane potential around the resting potential.**
(a) Three main ionic current sources. Note: E_K and E_{Cl} are negative while E_{Na} is positive. **(b)** An equivalent current source for the resting potential. **(c)** Electrical scheme for below-threshold potential changes (passive de- and hyperpolarizations) relative to the resting potential. Battery symbols indicate electromotive forces, boxes represent conductances and parallel plates indicate membrane capacitors. **(d)** From top to bottom: Time-dependent responses to $\pm 0.4\ nA$ current injected for 300 ms; left: upper traces, current I; middle traces, membrane potential changes V_m; right: membrane potential at the end of the current pulse (i.e. at 300 ms) plotted against current intensity. From Adams PR, Brown DA, Constanti A (1982) M-currents and other potassium currents in bullfrog sympathetic neurones. *J. Physiol. (Lond.)* **330**, 537−572, with permission.

even to reverse it. There are at least three main ways of depolarizing a membrane: (a) by increasing the K$^+$ concentration in the external medium, (b) by applying a drug that opens cationic channels or (c) by injecting a positive current inside the neuron (**Figure 3.10**).

An *in vitro* preparation such as a neuronal culture or a brain slice is bathed in an extracellular solution of an ionic composition close to that of the extracellular medium. A recording electrode is implanted in a neuronal cell body. At rest the membrane potential is close to -70 mV. When the extracellular solution is changed to one containing a higher concentration of K$^+$ ions (30 mM instead of 3 mM) and a lower concentration of Na$^+$ ions (113 mM instead of 140 mM) to keep constant the extracellular concentration of positive ions, a depolarization is recorded. Since at rest most of the channels open are K$^+$ channels, V_m tends toward E_K which is now equal to -38 mV ($E_K = 58 \log 30/140$) instead of -97 mV. The membrane depolarizes because E_K is more depolarized than V_{rest}.

In the same preparation bathed in control extracellular medium, veratridine is applied by pressure via a pipette located close to the recorded neuron. Veratridine induces a depolarization of the recorded membrane (**Figure 3.10a**). As this drug opens Na$^+$ channels, Na$^+$ ions enter the cell and create an inward current of positive charges. The electrical circuit is closed because + charges can go out of the cell via the K$^+$ channels open at rest. Since Na$^+$ channels now represent the major population of open channels, V_m tends toward E_{Na} ($+58$ mV) and the membrane depolarizes as long as veratridine is applied.

If now a positive current is applied through the recording pipette which contains a KCl solution, K$^+$ ions are expelled from the pipette. They create a current of positive charges that depolarizes the membrane (**Figure 3.10b**). The electrical circuit is closed because K$^+$ ions can go through the membrane via the K$^+$ channels open at rest. A depolarizing current pulse is a positive current injected via an intracellular electrode. One part of the stimulating current is used to load the capacity C_m of the neuronal membrane and the other part passes through the ion channels:

$$I_{stimulus} = C_m \, dV/dt + I_{ion}$$

$$dV/dt = [-I_{ion} + I_{stimulus}]/C_m$$

A positive stimulating current applied at the inside of a neuron (cell body, dendrite, axon) will cause a depolarization of V_m according to the above equation. Inversely, a negative current will hyperpolarize the membrane (see below). Once the membrane capacity is loaded (steady state) the injected current equals the current passing through the membrane via open channels.

In the case of silver electrode inside the pipette, as a coat of AgCl is deposited on the silver metal, it provides a store of Ag$^+$ and Cl$^-$ ions and mediates between electronic conduction in the metal (Ag$^+$ + e$^-$ $\rightleftarrows$ Ag) and ionic current owing that Cl$^-$ exchanges between precipitate (AgCl) and solution.

3.7.2 How to experimentally hyperpolarize a neuronal membrane?

The aim of the experiment is to increase the difference of potential between the two faces of the membrane. There are at least two main ways of hyperpolarizing a membrane, (a) by applying a drug that opens K$^+$ channels or (b) by injecting a negative current inside the neuron.

An *in vitro* preparation such as a neuronal culture or a brain slice is bathed in a physiological saline of an ionic composition close to that of the extracellular medium. A recording electrode is implanted in a neuronal cell body. A peptide that opens K$^+$ channels is applied by pressure via a pipette located close to the recorded neuron. This induces a hyperpolarization of the membrane, due the outward flux of K$^+$ ions (**Figure 3.11a**). As this drug opens K$^+$ channels (via metabotropic receptors such as GABA$_B$ receptors, Chapter 11), K$^+$ ions exit the cell and create an outward current of positive charges. The electrical circuit is closed because ions can enter the membrane via the channels open at rest. As K$^+$ channels now represent the major population of open channels, V_m tends toward E_K (-97 mV) and the membrane hyperpolarizes as long as the peptide is applied.

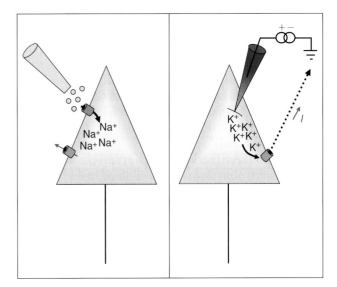

FIGURE 3.10   Vm is depolarized by applying:
(**a**) a drug in the extracellular medium that opens Na$^+$ channels (veratridine) or (**b**) a positive current via an intracellular electrode.

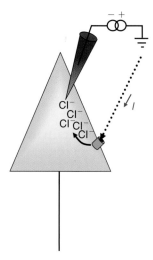

FIGURE 3.11 V_m is hyperpolarized by applying a negative current via an intracellular electrode.

If now a negative current is applied through the recording pipette which contains a KCl solution, Cl^- ions are expelled from the pipette. They hyperpolarize the membrane (**Figure 3.11**). The electrical circuit is closed because ions can go through the membrane via the channels open at rest.

3.8 SUMMARY

Passage of ions through the membrane is a regulated process and the flow of ions across the neuronal plasma membrane is not a simple and anarchic diffusion through a lipid bilayer. Instead, it is restricted through transmembrane proteins whose opening (channel proteins) or activation (pumps or transporters) is tightly controlled by different factors.

Where and how do ions passively cross the plasma membrane? (See also Appendices 3.2 and 3.3)

- Ions move passively across the plasma membrane through ionic channels that are specifically permeable to one or several ions of the same sign. They move down their electrochemical gradient. This passive movement of charges is a current that can be recorded. Through a single channel it is a unitary current i_{ion}, and through N channels it is a macroscopic current or total current I_{ion}.
- The type of ion that moves through an open channel (ionic selectivity of the channel pore) is determined by the structure of the channel itself. This ionic selectivity gives the name to the channel. For example, a Na^+ channel is permeable to Na^+ ions;

a cationic channel is permeable to cations: Na^+, K^+ and sometimes also Ca^{2+}.
- The *direction* of ion diffusion through a single channel depends on the electrochemical gradient or driving force for this particular ion ($V_m - E_{ion}$).
- The *number* of charges that diffuse through an open channel per unit of time (i_{ion}) depends on the electrochemical gradient ($V_m - E_{ion}$) but also on how easily ions move through the pore of the channel (expressed as the conductance γ_{ion} of the channel): $i_{ion} = \gamma_{ion}(V_m - E_{ion})$.

How and where do ions actively cross the plasma membrane and thus compensate for the passive movements? (see also Appendix 3.1)

Active movements of Na^+, K^+, Ca^{2+} or Cl^- ions across the membrane occur through pumps or transporters. Pumps obtain energy from the hydrolysis of ATP, whereas transporters use the energy of an ionic gradient, for example the sodium driving force. These transports require energy since they operate against the electrochemical gradient of the transported ions or molecules. They maintain ionic concentrations at constant values in the extracellular and intracellular compartments despite the continuous passive movements of ions across the membrane.

What are the roles of electrochemical gradients and passive movements of ions?

The electrochemical gradients of ions are a reserve of energy: they allow the existence of ionic currents and drive some active transports. The large asymmetries in ion distribution imply a dynamic state through which cell-to-cell signalling is made possible. Ionic currents have two main functions: (i) they evoke transient changes of membrane potential which are electrical signals of the neuron (action potentials or postsynaptic potentials or sensory potentials) essential to neuronal communication; and (ii) they locally increase the concentration of a particular ion in the intracellular medium, for example Ca^{2+} ions, and thus trigger intracellular Ca^{2+}-dependent events such as secretion or contraction.

APPENDIX 3.1
THE ACTIVE TRANSPORT OF IONS BY PUMPS AND TRANSPORTERS MAINTAIN THE UNEQUAL DISTRIBUTION OF IONS

Passive movements of Na^+, K^+, Ca^{2+} or Cl^- ions across the membrane would finally cause concentration changes in the extracellular and intracellular compartments if they were not constantly regulated during the entire life of the neuron by transport of ions in the reverse direction, against passive diffusion; i.e. against electrochemical gradients. This type of transport is described as active since it requires energy in order to oppose the electrochemical gradient of the transported ions. Ions cross the membrane *actively* through specialized proteins known as pumps or transporters. Pumps obtain energy from the hydrolysis of ATP, whereas transporters use the energy of an ionic gradient, for example the sodium driving force. The energy is needed for the conformational changes that allow the pump or the transporter to change its affinity for the ion transported during the transport: the binding site(s) must have a high affinity when facing the medium where the transported ion is at a low concentration (in order to bind it) and must change to low affinity when facing the medium where the concentration of the transported ion is high in order to release it.

Pumps are ATPases that actively transport ions

Pumps have ATPase activity (they hydrolyze ATP). This ATPase activity is generally the easiest way of identifying them. Pumps are membrane-embedded enzymes that couple the hydrolysis of ATP to active translocation of ions across the membrane. The central issue of ion motive ATPases is to couple the hydrolysis of ATP (and their auto-phosphorylation) to the translocation of ions.

The Na/K-ATPase pump

Na/K-ATPases maintain the unequal distribution of Na^+ and K^+ ions across the membrane. Na^+ and K^+ ions cross the membrane through different Na^+ and K^+ permeable channels (voltage-sensitive Na^+ and K^+ channels plus receptor channels). This pump operates continuously at a rhythm of 100 ions per second (compared with 10^6–10^8 ions per second for a channel), adjusting its activity to the electrical activity of the neuron. It actively transports three Na^+ ions towards the extracellular space for each two K^+ ions that it carries into the cell.

The energy of ATP hydrolysis is needed for the conformational changes (they are energy-dependent) that allow the pump to change its affinity for the ion transported, whether the binding sites are accessible from the cytoplasmic or the extracellular sides. For example, when the Na^+ binding sites are accessible from the cytoplasm, the protein is in a conformation with a high affinity ($K_A = 1\,mM$) for intracellular Na^+ ions, and so Na^+ ions bind to the three sites. In contrast, when the three Na^+ have been translocated to the extracellular

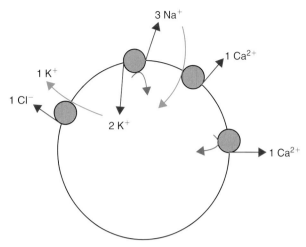

FIGURE A3.1

⟶   Transport against the electrochemical gradient of the ion

⟶   Transport along the electrochemical gradient of the ion

↻   ATP hydrolysis

side, the protein is in a conformation with a low affinity for Na^+ ions so that the three Na^+ are released in the extracellular space.

The steady unequal distribution of Na^+ and K^+ ions constitutes a reserve of energy for a cell. The neuron uses this energy to produce electric signals (action potentials, synaptic potentials) as well as to actively transport other molecules.

The Ca-ATPase pump

The function of Ca-ATPases is to maintain (with the Na–Ca transporter) the intracellular Ca^{2+} concentration at very low levels by active expulsion of Ca^{2+}. In fact, the intracellular Ca^{2+} concentration is 10,000 times lower than the extracellular concentration despite the inflow of Ca^{2+} (through receptor channels and voltage-gated Ca^{2+} channels) and the intracellular release of Ca^{2+} from intracellular stores. Maintaining a low intracellular Ca^{2+} concentration is critical since Ca^{2+} ions control several intracellular reactions and are toxic at a high concentration. Ca-ATPases are located in the plasma membrane and in the membrane of the reticulum. The former extrude Ca^{2+} from the cytoplasm whereas the latter sequester Ca^{2+} inside the reticulum (see also **Figure 7.8**).

Transporters use the energy stored in the transmembrane electrochemical gradient of Na^+, K^+, H^+ or other ions

When transporters carry Na^+, K^+ or H^+ ions (along their electrochemical gradient) in the same direction as the transported ion or molecule the process is called *symport.* When the movements occur in opposite directions the process is called *antiport.* We shall study only transporters implicated in the electrical or secretory activity of neurons.

The Na–Ca transporter

This transporter uses the energy of the Na^+ gradient to actively carry Ca^{2+} ions towards the extracellular environment. It is situated in the neuronal plasma membrane and operates in synergy with the Ca-ATPase and with transport mechanisms of the smooth sarcoplasmic reticulum to maintain the intracellular Ca^{2+} concentration at a very low level (see Section 7.2.4).

The K–Cl transporter KCC

Adult mammalian central neurons maintain a low intracellular Cl^- concentration. Cl^- extrusion is achieved by K^+-Cl^- cotransporters (KCC) fuelled by K^+. As all transporters, it does not directly consume ATP but derives its energy from ionic gradients, here the K^+ gradient generated by the Na/K/ATPase.

Neurotransmitter transporters

Inactivation of most neurotransmitters present in the synaptic cleft is achieved by rapid reuptake into the presynaptic neural element and astrocytic glial cells. This is performed by specific neurotransmitter transporters, transmembrane proteins that couple neurotransmitter transport to the movement of ions down their concentration gradient. Certain neurotransmitter precursors are also taken up by this type of active transport (glutamine and choline, for instance). Once in the cytoplasm, neurotransmitters are concentrated inside synaptic vesicles by distinct transport systems driven by the H^+ concentration gradient (maintained by the vesicular H^+-ATPase) (see Section 7.4).

APPENDIX 3.2
THE PASSIVE DIFFUSION OF IONS THROUGH AN OPEN CHANNEL

It has been stated above that a channel is said to be in a closed state (C) when its ionic pore does not allow ions to pass. In contrast, when the channel is said to be in the open state (O), ions can diffuse through the ionic pore.

$$C \rightleftarrows O$$

This diffusion of ions through an open channel is a passive transport since it does not require energy expenditure.

- Which type(s) of ions will move through a given open channel: cations, anions?
- In which direction will these ions move, from the external medium to the cytosol or the reverse?
- How many of these ions will move per unit of time?

The structure of the channel pore determines the type of ion(s) that diffuse passively through the channel

The pores of ion channels select their permeant ions. The structural basis for ion channel selectivity has been studied in a bacterial K^+ channel called the KcsA channel (it is a voltage-independent K^+ channel). All K^+ channels show a selectivity sequence $K^+ = Rb^+ > Cs^+$, whereas permeability for the smallest alkali metal ions Na^+ and Li^+ is extremely low. Potassium is at least 10,000 times more permeant than Na^+, a feature that is essential to the function of K^+ channels. Each subunit of the KcsA channel consists of an N-terminal cytoplasmic domain, followed by two transmembrane helices and a C-terminal globular domain in the cytoplasm. The P loop (P for pore) situated between transmembrane helices 1 and 2 is the region primarily responsible for ion selectivity.

The KcsA channel is overexpressed in bacteria and the three-dimensional structure of its pore investigated by the use of X-ray crystallography. The KcsA channel is a tetramer with fourfold symmetry around a central pore (**Figure A3.2**). The pore is constructed of an inverted teepee with the extracellular side corresponding to the base of the teepee. The overall length of the pore is 4.5 nm and its diameter varies along its distance. From inside the cell the pore begins as a water-filled

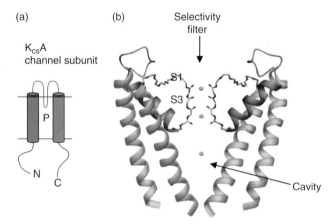

FIGURE A3.2 (**a**) membrane topology of the KcsA channel subunit showing the two transmembrane segments and the pore loop (P). (**b**) Two diametrically opposed subunits of KcsA are depicted to show the cavity in the membrane with 3 ions in cavity, S3 and S1 sites of the selectivity filter (shown in sticks). Adapted from Noskov SY and Roux B (2006) Ion selectivity in potassium channels. *Biophysical Chemistry* **124**, 279–291.

tunnel of 1.8 nm length (inner pore) surrounded by predominantly non-polar side-chains pointing to the pore axis. The diameter of this region is sufficiently wide to allow the passage of fully hydrated cations. This long entry way then opens to a wider water-filled cavity (1 nm across). Beyond this vestibule is the 1.2 nm long selectivity filter. After this, the pore opens widely to the extracellular side of the membrane.

What are the respective roles of the parts of the pore?

The pore comprises a wide, non-polar aqueous cavity on the intracellular side, leading up, on the extracellular side, to a narrow pore that is 1.2 nm long and lined exclusively by main chain carbonyl oxygens formed by the residues corresponding to the signature sequence TTVGYG common to all K^+ channels.

Electrostatic calculations show that when an ion is moved along a narrow pore through a membrane it must cross an energy barrier that is maximal at the membrane centre. A K^+ ion can move throughout the inner pore and cavity and still remain mostly hydrated, owing to the large diameter of these regions. The role of the inner pore and the cavity is to lower the electrostatic barrier. The cavity overcomes the electrostatic destabilization from the low dielectric bilayer by simply surrounding an ion with polarizable water. Another feature that contributes to the stabilization of the cation at the bilayer centre are the four pore helices which point directly at the centre of the cavity. The amino to carboxyl orientation of these helices imposes a negative electrostatic (cation attractive) potential via the helix dipole effect. These two mechanisms (large aqueous cavity and oriented helices) serve to stabilize a cation in the hydrophobic membrane interior.

The selectivity filter that follows, in contrast, is lined exclusively by polar main-chain atoms. They create a stack of sequential carbonyl oxygen rings which provide multiple closely spaced binding sites (S) for cations separated by 0.3–0.4 nm. This selectivity filter attracts K^+ ions and allows them to move.

Why are cations permeant and not anions?

As might have been anticipated for a cation channel, both the intracellular and extracellular entryways are negatively charged by acidic amino acids, that raise the local concentration of cations while lowering the concentration of anions.

Why are K^+ ions at least 10,000 times more permeant than Na^+ ions?

The selectivity filter is so narrow that a K^+ ion evidently dehydrates to enter into it and only a single K^+

ion can pass through at one time. To compensate for the energy cost of dehydration, the carbonyl oxygen atoms come in very close contact with the ion and act like surrogate water – they substitute for the hydration waters of K^+. This filter is too large to accommodate a Na^+ ion with its smaller radius (main chain oxygens are spatially inflexible and their relative distances to the centre of the pore cannot readily be changed). It is proposed that a K^+ ion fits in the filter so precisely that the energetic costs and gains are well balanced.

What drives K^+ ions to move on?

K^+ ions bind simultaneously at two binding sites 0.75 nm apart near the entry and exit point of the selectivity filter. Binding at adjacent sites may provide the repulsive force for ion flow through the selectivity filter: two K^+ ions at close proximity in the selectivity filter repel each other. The repulsion overcomes the strong interaction between ion and protein and allows rapid conduction in the setting of high selectivity. This leads to a rate of diffusion of around 10^8 ions per second.

APPENDIX 3.3
THE NERNST EQUATION

The material in this appendix is adapted from Katz B (ed.) (1966) *Nerve, Muscle and Synapse* (New York: McGraw-Hill). When $V_m = E_{ion}$, a particular ion has an equal tendency to diffuse in one direction according to its concentration gradient as to move in the reverse direction according to membrane potential. The net flux of this ion is null, so the current carried by this ion is null. $V_m = E_{ion}$ means that:

$$\text{osmotic work } (W_o) = \text{electrical work } (W_e). \qquad \text{(a)}$$

The osmotic work required to move one mole of a particular ion from a compartment where its concentration is low to a compartment when its concentration is high is equal to the electrical work needed to move one mole of this ion against the membrane potential in the opposite direction. Here, active diffusion of ions is considered instead of passive diffusion. The electrical work required to move 1 mole of an ion against a potential difference E_{ion} is:

$$W_e = zFE_{ion}, \qquad \text{(b)}$$

where z is the valence of the transported ion, equal to +1 for monovalent cations such as Na^+ or K^+, to −1 for monovalent anions such as Cl^-, and to +2 for divalent cations such as Ca^{2+}. F is the Faraday constant. F for hydrogen is the charge of one hydrogen atom: $F = Ne$.

Here N is the Avogadro number, which is 6.022×10^{23} mol^{-1} (one mole of hydrogen atoms contains 6×10^{23} protons and the same number of electrons), and e is the elementary charge of a proton, which is 1.602×10^{-19} coulombs (C). So $F = 96,500$ C mol^{-1}. Therefore zF with $z = 1$ is the charge of 1 mole of protons or 1 mole of monovalent cations (Na$^+$, K$^+$). The charge of one mole of monovalent anions (Cl$^-$) is $-F$ ($z = -1$); the charge of 1 mole of divalent cations (Ca^{2+}) is $2F$ ($z = 2$); etc.

The osmotic work required to move 1 mole of ions from a compartment where its concentration is low to a compartment where the concentration is high can be compared to the work done in compressing 1 g equivalent of an ideal gas. The gas is contained in a cylinder with a movable piston. Mechanical work to move the piston is W, calculated from force times distance of displacement of the piston (δl). The force exerted is equal to the pressure p of the gas multiplied by the surface area S of the piston. So the work δW done to displace the piston is $pS\ \delta l$, which equals $p\ \delta v$. Therefore the work done in compressing a gas from a volume v_1 to a volume v_2 is:

$$W = \int_{v_2}^{v_1} p\, dv. \qquad (c)$$

The gas law tells us that $pv = RT$ (hence $p = RT/v$), with R the constant of an ideal gas ($R = 8.314$ V C K^{-1} mol^{-1}) and T is the absolute temperature.

Equation (c) can be changed to:

$$W = RT \int_{v_2}^{v_1} (1/v) dv = RT(\ln v_1 - \ln v_2)$$
$$= RT \ln(v_1/v_2). \qquad (d)$$

By analogy the osmotic work is:

$$W_o = RT \ln([\text{ion}]_e/[\text{ion}]_i). \qquad (e)$$

From equation (a), $W_o = -W_e$, so from equations (b) and (e) the Nernst equation is obtained:

$$RT \ln([\text{ion}]_e/[\text{ion}]_i) = zFE_{\text{ion}}$$

$$E_{\text{ion}} = (RT/zF) \ln([\text{ion}]_e/[\text{ion}]_i). \qquad \text{(Nernst)}$$

At 20°C, RT/F is about 25 mV, and moving from Neperian logarithms to decimal ones a factor of 2.3 is needed. Hence:

$$E_{\text{ion}} = (58/z) \log_{10}([\text{ion}]_e/[\text{ion}]_i).$$

Of course, this description of the E_{ion} is entirely based on a physical theory of passive ion movements. Transmembrane flux of ions, however, involves active transport of ions as well. For example, the gradients for Na$^+$ and, in particular, for Ca^{2+} are regulated by complex mechanisms relying on transporters and intracellular sequestration so that the possibility of predicting the precise reversal potential of responses mediated by rises in Na$^+$ or Ca^{2+} permeability on the basis of their apparent transmembrane concentrations is limited.

CHAPTER

4

The voltage-gated channels of Na⁺ action potentials

The ionic basis for nerve excitation was first elucidated in the squid giant axon by Hodgkin and Huxley (1952) using the voltage clamp technique. They made the key observation that two separate, voltage-dependent currents underlie the action potential: an early transient inward Na⁺ current which depolarizes the membrane, and a delayed outward K⁺ current largely responsible for repolarization. This led to a series of experiments that resulted in a quantitative description of impulse generation and propagation in the squid axon.

Nearly 30 years later, Sakmann and Neher, using the patch clamp technique, recorded the activity of the voltage-gated Na⁺ and K⁺ channels responsible for action potential initiation and propagation. Taking history backwards, action potentials will be explained from the single channel level to the membrane level.

4.1 PROPERTIES OF ACTION POTENTIALS

4.1.1 The different types of action potentials

The action potential is a sudden and transient depolarization of the membrane. The cells that initiate action potentials are called 'excitable cells'. Action potentials can have different shapes, i.e. different amplitudes and durations. In neuronal somas and axons, action potentials have a large amplitude and a small duration: these are the Na⁺-dependent action potentials (**Figures 4.1** and **4.2a**). In other neuronal cell bodies, heart ventricular cells and axon terminals, the action potentials have a longer duration with a plateau following the initial peak: these are the Na⁺/Ca²⁺-dependent action potentials

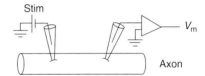

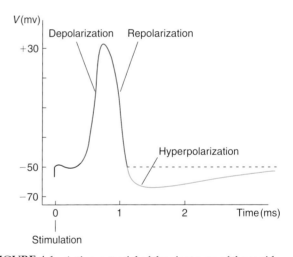

FIGURE 4.1   **Action potential of the giant axon of the squid.**
Action potential intracellularly recorded in the giant axon of the squid at resting membrane potential in response to a depolarizing current pulse (the extracellular solution is seawater). The different phases of the action potential are indicated. Adapted from Hodgkin AL and Katz B (1949) The effect of sodium ions on the electrical activity of the giant axon of the squid. *J. Physiol.* **108**, 37–77, with permission.

(**Figure 4.2b-d**). Finally, in some neuronal dendrites and some endocrine cells, action potentials have a small amplitude and a long duration: these are the Ca²⁺-dependent action potentials.

Action potentials have common properties; for example they are all initiated in response to a membrane

45

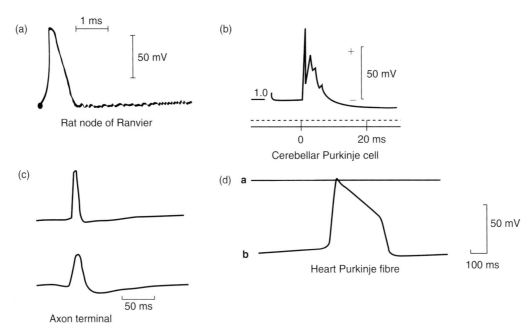

FIGURE 4.2   Different types of action potentials recorded in excitable cells.
(a) Sodium-dependent action potential intracellularly recorded in a node of Ranvier of a rat nerve fibre. Note the absence of the hyperpolarization phase flowing the action potential. **(b)–(d)** Sodium–calcium-dependent action potentials. Intracellular recording of the complex spike in a cerebellar Purkinje cell in response to climbing fibre stimulation: an initial Na⁺-dependent action potential and a later larger slow potential on which are superimposed several small Ca²⁺-dependent action potentials. The total duration of this complex spike is 5–7 ms. **(b)** Action potential recorded from axon terminals of *Xenopus* hypothalamic neurons (these axon terminals are located in the neurohypophysis) in control conditions (top) and after adding blockers of Na⁺ and K⁺ channels (TTX and TEA, bottom) in order to unmask the Ca²⁺ component of the spike (this component has a larger duration due to the blockade of some of the K⁺ channels). **(c)** Intracellular recording of an action potential from an acutely dissociated dog heart cell (Purkinje fibre). Trace 'a' is recorded when the electrode is outside the cell and represents the trace 0 mV. Trace 'b' is recorded when the electrode is inside the cell. The peak amplitude of the action potential is 75 mV and the total duration 400 ms. **(d)** All these action potentials are recorded in response to an intracellular depolarizing pulse or to the stimulation of afferents. Note the differences in their durations. Part (a) adapted from Brismar T (1980) Potential clamp analysis of membrane currents in rat myelinated nerve fibres. *J. Physiol.* **298**, 171–184, with permission. Parts (b)–(d) adapted from Coraboeuf E and Weidmann S (1949) Potentiel de repos et potentiels d'action du muscle cardiaque, mesurés à l'aide d'électrodes internes. *C. R. Soc. Biol.* **143**, 1329–1331; and Eccles JC, Llinas R, Sasaki K (1966) The excitatory synaptic action of climbing fibres on the Purkinje cells of the cerebellum. *J. Physiol.* **182**, 268–296; and Obaid AL, Flores R, Salzberg BM (1989) Calcium channels that are required for secretion from intact nerve terminals of vertebrates are sensitive to ω-conotoxin and relatively insensitive to dihydropyridines. *J. Gen. Physiol.* **93**, 715–730; with permission.

depolarization. They also have differences; for example in the type of ions involved, their amplitude, duration, etc.

4.1.2   Na⁺ and K⁺ ions participate in the action potential of axons

The activity of the giant axon of the squid is recorded with an intracellular electrode (in current clamp; see **Appendix 4.1**) in the presence of seawater as the external solution.

Na⁺ ions participate in the depolarization phase of the action potential

When the extracellular solution is changed from seawater to a Na⁺-free solution, the amplitude and

risetime of the depolarization phase of the action potential gradually and rapidly decreases, until after 8 s the current pulse can no longer evoke an action potential (**Figure 4.3**). Moreover, in control seawater, tetrodotoxin (TTX), a specific blocker of voltage-gated Na⁺ channels, completely blocks action potential initiation (**Figure 4.4a,c**), thus confirming a major role of Na⁺ ions.

K⁺ ions participate in the repolarization phase of the action potential

Application of tetraethylammonium chloride (TEA), a blocker of K⁺ channels, greatly prolongs the duration of the action potential of the squid giant axon without

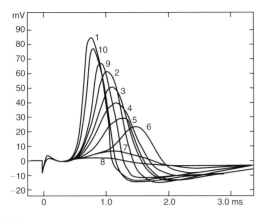

FIGURE 4.3 **The action potential of the squid giant axon is abolished in a Na$^+$-free external solution.**
(1) Control action potential recorded in sea water; (2)–(8) recordings taken at the following times after the application of a dextrose solution (Na-free solution): 2.30, 4.62, 5.86, 6.10, 7.10 and 8.11 s; (9) recording taken 9 s after reapplication of seawater; (10) recording taken at 90 and 150 s after reapplication of seawater; traces are superimposed. From Hodgkin AL and Katz B (1949) The effect of sodium ions on the electrical activity of the giant axon of the squid. *J. Physiol.* **108**, 37–77, with permission.

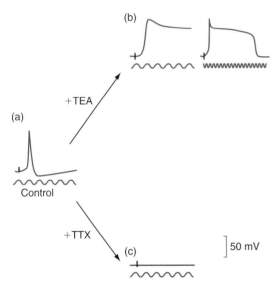

FIGURE 4.4 **Effects of tetrodotoxin (TTX) and tetraethylammonium chloride (TEA) on the action potential of the squid giant axon.** **(a)** Control action potential. **(b)** TEA application lengthens the action potential (left), which then has to be observed on a different time scale (right trace). **(c)** TTX totally abolishes the initiation of the action potential. Adapted from Tasaki I and Hagiwara S (1957) Demonstration of two stable potential states in the squid giant axon under tetraethylammonium chloride. *J. Gen. Physiol.* **40**, 859–885, with permission.

changing the resting membrane potential. The action potential treated with TEA has an initial peak followed by a plateau (**Figure 4.4a,b**) and the prolongation is sometimes 100-fold or more.

4.1.3 Na$^+$-dependent action potentials are all or none and propagate along the axon with the same amplitude

Depolarizing current pulses are applied through the intracellular recording electrode, at the level of a neuronal soma or axon. We observe that (i) to a certain level of membrane depolarization called the threshold potential, only an ohmic passive response is recorded (**Figure 4.5a,** right); (ii) when the membrane is depolarized just above threshold, an action potential is recorded. Then, increasing the intensity of the stimulating current pulse does not increase the amplitude of the action potential (**Figure 4.5a,** left). The action potential is all or none.

Once initiated, the action potential propagates along the axon with a speed varying from 1 to 100 m s^{-1} according to the type of axon. Intracellular recordings at varying distances from the soma show that the amplitude of the action potential does not attenuate: the action potential propagates without decrement (**Figure 4.5b**).

4.1.4 Questions about the Na$^+$-dependent action potential

- What are the structural and functional properties of the Na$^+$ and K$^+$ channels of the action potential? (Sections 4.2 and 4.3)
- What represents the threshold potential for action potential initiation? (Section 4.4)
- Why is the action potential, all or none? (Section 4.4)
- What are the mechanisms of action potential propagation? (Section 4.4)

4.2 THE DEPOLARIZATION PHASE OF Na$^+$-DEPENDENT ACTION POTENTIALS RESULTS FROM THE TRANSIENT ENTRY OF Na$^+$ IONS THROUGH VOLTAGE-GATED Na$^+$ CHANNELS

4.2.1 The Na$^+$ channel consists of a principal large α-subunit with four internal homologous repeats and auxiliary β-subunits

The primary structures of the *Electrophorus* electroplax Na$^+$ channel, that of the rat brain, heart and skeletal muscles have been elucidated by cloning and sequence analysis of the complementary cDNAs. The Na$^+$ channel in all these structures consists of an α-subunit of approximately 2000 amino acids (260 kDa),

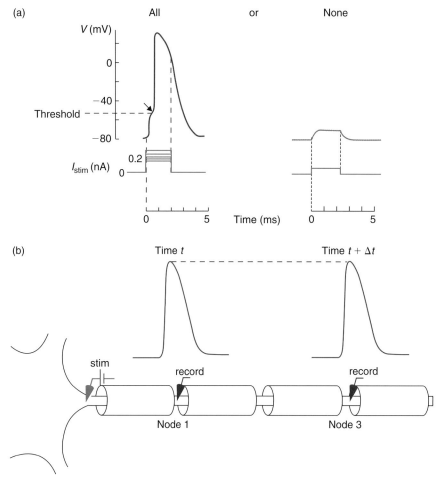

FIGURE 4.5   **Properties of the Na$^+$-dependent action potential.**
The response of the membrane to depolarizing current pulses of different amplitudes is recorded with an intra-cellular electrode. Upper traces are the voltage traces, bottom traces are the current traces. Above 0.2 nA an axon potential is initiated. Increasing the current pulse amplitude does not increase the action potential amplitude (left). With current pulses of smaller amplitudes, no action potential is initiated. **(b)** An action potential is initiated in the soma–initial segment by a depolarizing current pulse (stim). Intracellular recording electrodes inserted along the axon record the action potential at successive nodes at successive times. See text for further explanations.

composed of four homologous domains (I to IV) sepa-rated by surface loops of different lengths. Within each domain there are six segments forming six putative transmembrane α-helices (S1 to S6) and a hairpin-like P loop also called re-entrant pore loop between S5 and S6 (**Figure 4.6a**). The four homologous domains probably form a pseudo-tetrameric structure whose central part is the permeation pathway (**Figure 4.6b**). Parts of the α-subunit contributing to pore formation have been identified by site-directed mutagenesis.

Each domain contains a unique segment, the S4 seg-ment, with positively charged residues (arginine or lysine) at every third position with mostly non-polar residues intervening between them (see **Figure 4.15a**). This structure of the S4 segment is strikingly conserved in all the types of Na$^+$ channels analyzed so far and

this led to suggestions that the S4 segments serve as voltage sensors (see Section 4.2.7).

The α-subunit of mammals is associated with one or two smaller auxiliary subunits named β-subunits. They are small proteins of about 200 amino acid residues (33–36 kDa), with a substantial N-terminal domain, a single putative membrane spanning segment and a C-terminal intracellular domain.

The α-subunit mRNA isolated from rat brain or the α-subunit RNAs transcribed from cloned cDNAs from a rat brain are sufficient to direct the synthesis of func-tional Na$^+$ channels when injected into oocytes. These results establish that the protein structures necessary for voltage-gating and ion conductance are contained within the α-subunit itself. However, the properties of these channels are not identical to native Na$^+$ channels

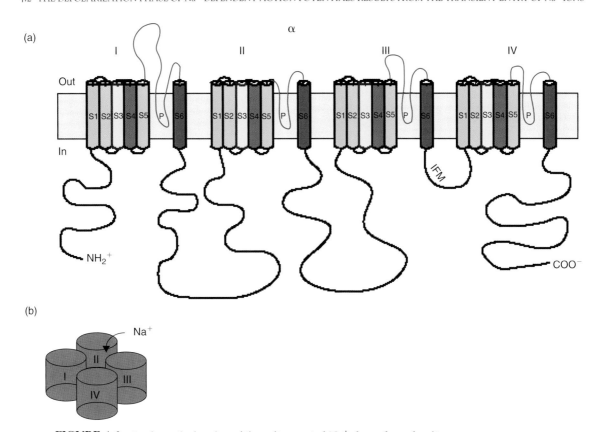

FIGURE 4.6   **A schematic drawing of the voltage-gated Na$^+$ channel α-subunit.**
(a) Cylinders represent putative membrane-spanning segments, P the P loops and IFM the critical motif of the
fast inactivation particle on the cytoplasmic linker connecting domains III and IV. The lengths that are shown
for the N and C termini and the interdomain cytoplasmic loops are consistent with the Na$_v$1.2 mammalian
Na$^+$ channel but these linkers (except for that between III and IV) vary greatly in length and sequence among
the different Na$^+$ channel isoforms. **(b)** Diagram of the pseudo-tetrameric structure whose central part is the
permeation pathway. Drawing (a) adapted from Goldin A (2002) The evolution of voltage-gated Na$^+$ channels.
J. Exp. Biol. **205**, 575–584, with permission.

as it has been shown that the auxiliary β-subunits play
a role in the targeting and stabilization of the α-subunit
in the plasma membrane, its sensitivity to voltage and
rate of inactivation.

The nomenclature of all Na$^+$ voltage-gated channels
is the following: Na$_v$ to indicate the principal permeating
ion (Na$^+$) and the principal physiological regulator
(v for voltage), followed by a number that indicates the
gene subfamily (currently Na$_v$1 is the only subfamily).
The number following the decimal point identifies the
specific channel isoform (e.g. Na$_v$1.1). At present, nine
functional isoforms have been identified.

4.2.2 Membrane depolarization favours conformational change of the Na$^+$ channel towards the open state; the Na$^+$ channel then quickly inactivates

The function of the Na$^+$ channel is to transduce *rapidly*
membrane depolarization into an entry of Na$^+$ ions.

The activity of a single Na$^+$ channel was first recorded
by Sigworth and Neher in 1980 from rat muscle cells
with the patch clamp technique (cell-attached patch;
see Appendix 4.3).

It must be explained that the experimenter does not
know, before recording it, which type of channel(s) is
in the patch of membrane isolated under the tip of the
pipette. He or she can only increase the chance of
recording a Na$^+$ channel, for example, by studying a
membrane where this type of channel is frequently
expressed and by pharmacologically blocking the other
types of channels that could be activated together
with the Na$^+$ channels (voltage-gated K$^+$ channels are
blocked by TEA). The recorded channel is then identified by its voltage dependence, reversal potential, unitary conductance, ionic permeability, mean open time,
etc. Finally, the number of Na$^+$ channels in the patch of
membrane cannot be predicted. Even when pipettes
with small tips are used, the probability of recording
more than one channel can be high because of the type

of membrane patched. For this reason, very few recordings of single native Na$^+$ channels have been performed. The number of Na$^+$ channels in a patch is estimated from recordings where the membrane is strongly depolarized in order to increase to its maximum the probability of opening the voltage-gated channels present in the patch.

Voltage-gated Na$^+$ channels of the skeletal muscle fibre

A series of recordings obtained from a single Na$^+$ channel in response to a 40 mV depolarizing step given every second is shown in **Figure 4.7a** and **c**. The holding potential is around −70 mV (remember that in the cell attached configuration, the membrane potential can only be estimated). A physiological extracellular concentration of Na$^+$ ions is present in the pipette.

At holding potential, no variations in the current traces are recorded. After the onset of the depolarizing step, unitary Na$^+$ currents of varying durations but of the same amplitude are recorded (lines 1, 2, 4, 5, 7 and 8) or not recorded (lines 3, 6 and 9). This means that six times out of nine, the Na$^+$ channel has opened in response to membrane depolarization. The Na$^+$ current has a rectangular shape and is downward. By convention, inward currents of + ions (cations) are represented as downward (inward means that + ions enter the cell; see Section 3.4). The histogram of the Na$^+$ current amplitude recorded in response to a 40 mV depolarizing step gives a mean amplitude for i_{Na} of around −1.6 pA (see Appendix 4.3).

It is interesting to note that once the channel has opened, there is a low probability that it will reopen during the depolarization period. Moreover, even when the channel does not open at the beginning of the step, the frequency of appearance of Na$^+$ currents later in the depolarization is very low; i.e. the Na$^+$ channel inactivates.

Rat brain Na$^+$ channels

The activity of rat brain Na$^+$ channels has been studied in cerebellar Purkinje cells in culture. Each trace of **Figure 4.8a** and **c** shows the unitary Na$^+$ currents (i_{Na}) recorded during a 20 ms membrane depolarization to −40 mV (test potential) from a holding potential of −90 mV. Rectangular inward currents occur most frequently at the beginning of the depolarizing step but can also be found at later times (**Figure 4.8a**, line 2). The histogram of the Na$^+$ current amplitudes recorded at −40 mV test potential gives a mean amplitude for i_{Na} of around −2 pA (**Figure 4.8d**). Events near −4 pA correspond to double openings, meaning that at least two channels are present in the patch.

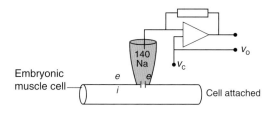

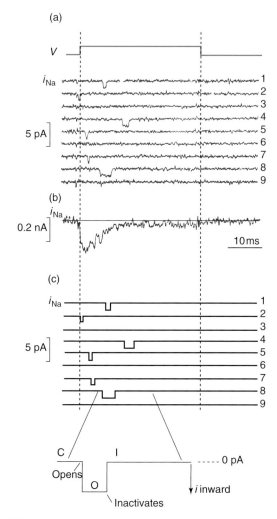

FIGURE 4.7 Single Na$^+$ channel openings in response to a depolarizing step (muscle cell).
The activity of the Na$^+$ channel is recorded in patch clamp (cell-attached patch) from an embryonic muscle cell. **(a)** Nine successive recordings of single channel openings (i_{Na}) in response to a 40 mV depolarizing pulse (V trace) given at 1 s intervals from a holding potential 10 mV more hyperpolarized than the resting membrane potential. **(b)** Averaged inward Na$^+$ current from 300 elementary Na$^+$ currents as in (a). **(c)** The same recordings as in (a) are redrawn in order to explain more clearly the different states of the channel. On the bottom line one opening is enlarged. C, closed state; O, open state; I, inactivated state. The solution bathing the extracellular side of the patch or intrapipette solution contains (in mM): 140 NaCl, 1.4 KCl, 2.0 MgCl$_2$, 1 CaCl$_2$ and 20 HEPES at pH 7.4. TEA 5 mM is added to block K$^+$ channels and bungarotoxin to block acetylcholine receptors. Adapted from Sigworth FJ and Neher E (1980) Single Na$^+$ channel currents observed in rat muscle cells. *Nature* **287**, 447–449, with permission.

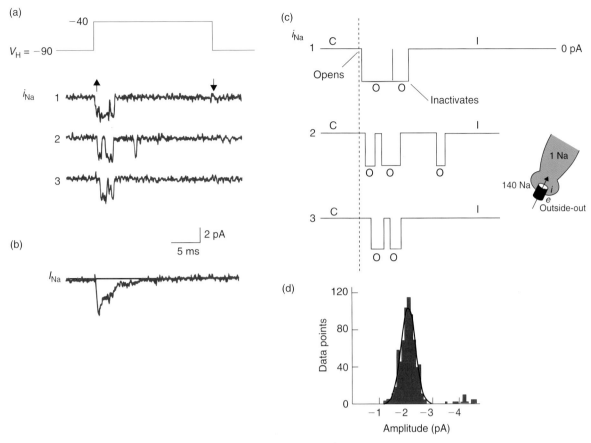

FIGURE 4.8   Single-channel activity of a voltage-gated Na⁺ channel from rat brain neurons.
The activity of a Na⁺ channel of a cerebellar Purkinje cell in culture is recorded in patch clamp (outside-out patch) in response to successive depolarizing steps to $-40\,mV$ from a holding potential $V_H = -90\,mV$. **(a)** The 20 ms step (upper trace) evokes rectangular inward unitary currents (i_{Na}). **(b)** Average current calculated from all the sweeps which had active Na⁺ channels within a set of 25 depolarizations. **(c)** Interpretative drawing on an enlarged scale of the recordings in (a). **(d)** Histogram of elementary amplitudes for recordings as in (a). The continuous line corresponds to the best fit of the data to a single Gaussian distribution. C, closed state; O, open state; I, inactivated state. The solution bathing the outside face of the patch contains (in mM): 140 NaCl, 2.5 KCl, 1 CaCl², 1 MgCl², 10 HEPES. The solution bathing the inside of the patch or intrapipette solution contains (in mM): 120 CsF, 10 CsCl, 1 NaCl, 10 EGTA-Cs⁺, 10 HEPES-Cs⁺. Cs⁺ ions are in the pipette instead of K⁺ ions in order to block K⁺ channels. Adapted from Gähwiler BH and Llano I (1989) Sodium and potassium conductances in somatic membranes of rat Purkinje cells from organotypic cerebellar cultures. *J. Physiol.* **417**, 105–122, with permission.

The unitary current has a rectangular shape

The rectangular shape of the unitary current means that when the Na⁺ channel opens, the unitary current is nearly immediately maximal. The unitary current then stays constant: the channel stays open for a time which varies; finally the unitary current goes back to zero though the membrane is still depolarized. The channel may not reopen (**Figure 4.7a,c**) as it is in an inactivated state (I) (**Figure 4.7c**, bottom trace). After being opened by a depolarization, the channel does not go back to the closed state but inactivates. In that state, the pore of the channel is closed (no Na⁺ ions flow through the pore) as in the closed state but the channel cannot reopen immediately (which differs from the closed state). The inactivated channel is refractory to opening unless the membrane repolarizes to allow it to return to the closed (resting) state.

In other recordings, such as that of **Figure 4.8a** and **c**, the Na⁺ channel seems to reopen once or twice before inactivating. This may result from the presence of two (as here) or more channels in the patch so that the unitary currents recorded do not correspond to the same channel. It may also result from a slower inactivation rate of the channel recorded, which in fact opens, closes, reopens and then inactivates.

The unitary current is carried by a few Na$^+$ ions

How many Na$^+$ ions enter through a single channel? Knowing that in the preceding example, the unitary Na$^+$ current has a mean amplitude of $-1.6\,\mathrm{pA}$ for 1 ms, the number of Na$^+$ ions flowing through one channel during 1 ms is $1.6 \times 10^{-12}/(1.6 \times 10^{-19} \times 10^3) = 10{,}000$ Na$^+$ ions (since $1\,\mathrm{pA} = 1\,\mathrm{pCs}^{-1}$ and the elementary charge of one electron is $1.6 \times 10^{-19}\,\mathrm{C}$). This number, 10^4 ions, is negligible compared with the number of Na$^+$ ions in the intracellular medium: if $[\mathrm{Na}^+]_i = 14\,\mathrm{mM}$, knowing that 1 mole represents 6×10^{23} ions, the number of Na$^+$ ions per litre is $6 \times 10^{23} \times 14 \times 10^{-3} = 10^{22}$ ions l^{-1}. In a neuronal cell body or a section of axon, the volume is of the order of 10^{-12} to 10^{-13} litres. Then the number of Na$^+$ ions is around 10^9 to 10^{10}.

The Na$^+$ channel fluctuates between the closed, open and inactivated states

where C is the channel in the closed state, O in the open state and I in the inactivated state. Both C and I states are non-conducting states. The C to O transition is triggered by membrane depolarization. The O to I transition is due to an intrinsic property of the Na$^+$ channel. The I to C transition occurs when the membrane repolarizes or is already repolarized. In summary, the Na$^+$ channel opens when the membrane is depolarized, stays open during a mean open time of less than 1 ms, and then usually inactivates.

4.2.3 The time during which the Na$^+$ channel stays open varies around an average value, τ_o, called the mean open time

In **Figures 4.7a** and **4.8a** we can observe that the periods during which the channel stays open, t_o, are variable. The mean open time of the channel, τ_o, at a given potential is obtained from the frequency histogram of the different t_o at this potential. When this distribution can be fitted by a single exponential, its time constant provides the value of τ_o (see Appendix 4.3). The functional significance of this value is the following: during a time equal to τ_o the channel has a high probability of staying open.

For example, the Na$^+$ channel of the skeletal muscle fibre stays open during a mean open time of 0.7 ms. For the rat brain Na$^+$ channel of cerebellar Purkinje cells, the distribution of the durations of the unitary currents recorded at $-32\,\mathrm{mV}$ can be fitted with a single exponential with a time constant of 0.43 ms ($\tau_o = 0.43\,\mathrm{ms}$).

4.2.4 The i_{Na}–V relation is linear: the Na$^+$ channel has a constant unitary conductance γ_{Na}

When the activity of a single Na$^+$ channel is now recorded at different test potentials, we observe that the amplitude of the inward unitary current diminishes as the membrane is further and further depolarized (see **Figure 4.11a**). In other words, the net entry of Na$^+$ ions through a single channel diminishes as the membrane depolarizes. The i_{Na}–V relation is obtained by plotting the amplitude of the unitary current (i_{Na}) versus membrane potential (V_{m}). It is linear between $-50\,\mathrm{mV}$ and $0\,\mathrm{mV}$ (**Figure 4.9a**). For membrane potentials more hyperpolarized than $-50\,\mathrm{mV}$, there are no values of i_{Na} since the channel rarely opens or does not open at all. Quantitative data for potentials more depolarized than $0\,\mathrm{mV}$ are not available.

The critical point of the current–voltage relation is the membrane potential for which the current is zero; i.e. the reversal potential of the current (E_{rev}). If only Na$^+$ ions flow through the Na$^+$ channel, the reversal potential is equal to E_{Na}. From $-50\,\mathrm{mV}$ to E_{rev}, i_{Na} is inward and its amplitude decreases. This results from the decrease of the Na$^+$ driving force ($V_{\mathrm{m}} - E_{\mathrm{Na}}$) as the membrane approaches the reversal potential for Na$^+$ ions. For membrane potentials more depolarized than E_{rev}, i_{Na} is now outward (not shown). Above E_{rev}, the amplitude of the outward Na$^+$ current progressively increases as the driving force for the exit of Na$^+$ ions increases.

The linear i_{Na}–V relation is described by the equation $i_{\mathrm{Na}} = \gamma_{\mathrm{Na}}(V_{\mathrm{m}} - E_{\mathrm{Na}})$, where V_{m} is the test potential, E_{Na} is the reversal potential of the Na$^+$ current, and γ_{Na} is the conductance of a single Na$^+$ channel (unitary conductance). The value of γ_{Na} is given by the slope of the linear i_{Na}/V curve. It has a constant value at any given membrane potential. This value varies between 5 and 18 pS depending on the preparation.

4.2.5 The probability of the Na$^+$ channel being in the open state increases with depolarization to a maximal level

An important observation at the single channel level is that the more the membrane is depolarized, the higher is the probability that the Na$^+$ channel will open. This observation can be made from two types of experiments:

- The activity of a single Na$^+$ channel is recorded in patch clamp (cell-attached patch). Each depolarizing step is repeated several times and the number of times the Na$^+$ channel opens is observed (**Figure 4.10a**). With depolarizing steps to $-70\,\mathrm{mV}$ from a holding potential of $-120\,\mathrm{mV}$, the channel very rarely

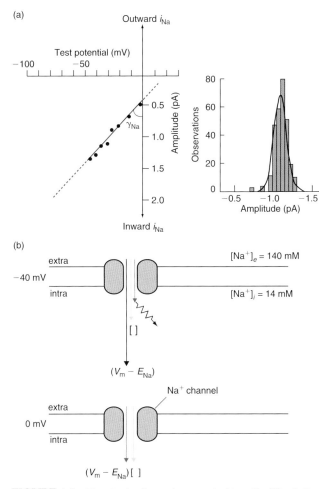

FIGURE 4.9   The single-channel current/voltage (i_{Na}/V) relation is linear.
(a) The activity of the rat type II Na⁺ channel expressed in *Xenopus* oocytes from cDNA is recorded in patch clamp (cell-attached patch). Plot of the unitary current amplitude versus test potential: each point represents the mean of 20–200 unitary current amplitudes measured at one potential (left) as shown at −32 mV (right). The relation is linear between test potentials −50 and 0 mV (holding potential = −90 mV). The slope is $\gamma_{Na} = 19$ pS. **(b)** Drawings of an open voltage-gated Na⁺ channel to explain the direction and amplitude of the net flux of Na⁺ ions at two test potentials (−40 and 0 mV). [ ], force due to the concentration gradient across the membrane; ⌇⌇→, force due to the electric gradient; $V - E_{Na}$, driving force. The solution bathing the extracellular side of the patch or intrapipette solution contains (in mM): 115 NaCl, 2.5 KCl, 1.8 CaCl2, 10 HEPES. Plot (a) adapted from Stühmer W, Methfessel C, Sakmann B *et al.* (1987) Patch clamp characterization of sodium channels expressed from rat brain cDNA. *Eur. Biophys. J.* **14**, 131–138, with permission.

opens; and if it does, the time spent in the open state is very short. In contrast, with depolarizing steps to −40 mV, the Na⁺ channels open for each trial.

- The activity of two or three Na⁺ channels is recorded in patch clamp (cell-attached patch). In response to depolarizing steps of small amplitude, Na⁺ channels do not open or only one Na⁺ channel opens at a time. With larger depolarizing steps, the

overlapping currents of two or three Na⁺ channels can be observed, meaning that this number of Na⁺ channels open with close delays in response to the step (not shown).

From the recordings of **Figure 4.10a**, we can observe that the probability of the Na⁺ channel being in the open state varies with the value of the test potential. It also varies with time during the depolarizing step: openings occur more frequently at the beginning of the step. The open probability of Na⁺ channels is voltage- and time-dependent. By averaging a large number of records obtained at each test potential, the open probability (p_t) of the Na⁺ channel recorded can be obtained at each time t of the step (**Figure 4.10b**). We observe from these curves that after 4–6 ms the probability of the Na⁺ channel being in the open state is very low, even with large depolarizing steps: the Na⁺ channel inactivates in 4–6 ms. When we compare now the open probabilities at the different test potentials, we observe that the probability of the Na⁺ channel being in the open state at time $t = 2$ ms increases with the amplitude of the depolarizing step.

4.2.6 The macroscopic Na⁺ current (I_{Na}) has a steep voltage dependence of activation and inactivates within a few milliseconds

The macroscopic Na⁺ current, I_{Na}, is the sum of the unitary currents, i_{Na}, flowing through all the open Na⁺ channels of the recorded membrane

At the axon initial segment or at nodes of Ranvier, there are N Na⁺ channels that can be activated. We have seen that the unitary Na⁺ current flowing through a single Na⁺ channel has a rectangular shape. What is the time course of the macroscopic Na⁺ current, I_{Na}?

If we assume that the Na⁺ channels in one cell are identical and function independently, the sum of many recordings from the same Na⁺ channel should show the same properties as the macroscopic Na⁺ current measured from thousands of channels with the voltage clamp technique. In **Figure 4.7b**, an average of 300 unitary Na⁺ currents elicited by a 40 mV depolarizing pulse is shown. For a given potential, the 'averaged' inward Na⁺ current has a fast rising phase and presents a peak at the time $t = 1.5$ ms. The peak corresponds to the time when most of the Na⁺ channels are opened at each trial. Then the averaged current decays with time because the Na⁺ channel has a low probability of being in the open state later in the step (owing to the inactivation of the Na⁺ channel). At each trial, the Na⁺ channel does not inactivate exactly at the same time, which explains the progressive decay of the averaged macroscopic Na⁺ current. A similar averaged Na⁺ current is shown

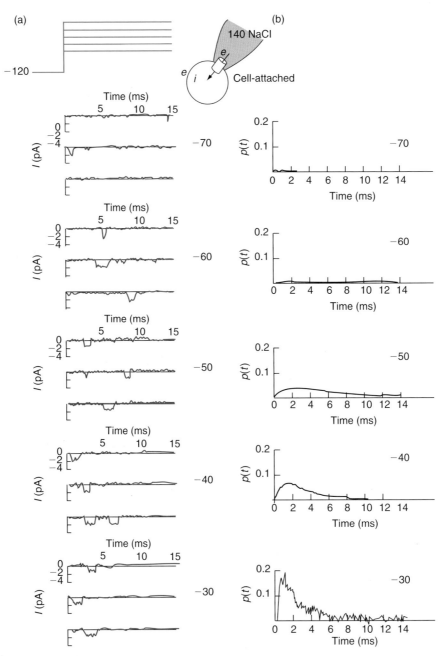

FIGURE 4.10 **The open probability of the voltage-gated Na⁺ channel is voltage- and time-dependent.**
Single Na⁺ channel activity recorded in a mammalian neuroblastoma cell in patch clamp (cell-attached patch).
(a) In response to a depolarizing step to the indicated potentials from a holding potential of −120 mV, unitary
inward currents are recorded. **(b)** Ensemble of averages of single-channel openings at the indicated voltages; 64
to 2000 traces are averaged at each voltage to obtain the time-dependent open probability of a channel ($p_{(t)}$) in
response to a depolarization. The open probability at time t is calculated according to the equation:
$p(t) = I_{Na(t)}/Ni_{Na}$, where $I_{Na(t)}$ is the average current at time t at a given voltage, N is the number of channels (i.e.
the number of averaged recordings of single channel activity) and i_{Na} is the unitary current at a given voltage. At
−30 mV the open probability is maximum. The channels inactivate in 4 ms. Adapted from Aldrich RW and
Steven CF (1987) Voltage-dependent gating of sodium channels from mammalian neuroblastoma cells. *J.
Neurosci.* **7**, 418–431, with permission.

in **Figure 4.8b**. The averaged current does not have a
rectangular shape because the Na⁺ channel does not
open with the same delay and does not inactivate at
the same time at each trial.

The *averaged* macroscopic Na⁺ current has a time
course similar to that of the *recorded* macroscopic Na⁺
current from the same type of cell at the same potential.
However, the averaged current from 300 Na⁺ channels

still presents some angles in its time course. In contrast, the macroscopic recorded Na$^+$ current is smooth. The more numerous are the Na$^+$ channels opened by the depolarizing step, the smoother is the total Na$^+$ current. The value of I_{Na} at each time t at a given potential is:

$$I_{Na} = N p_{(t)} i_{Na}$$

where N is the number of Na$^+$ channels in the recorded membrane and $p_{(t)}$ is the open probability at time t of the Na$^+$ channel; it depends on the membrane potential and on the channel opening and inactivating rate constants. i_{Na} is the unitary Na$^+$ current and $N p_{(t)}$ is the number of Na$^+$ channels open at time t.

The $I_{Na}-V$ relation is bell-shaped though the $i_{Na}-V$ relation is linear

We have seen that the amplitude of the unitary Na$^+$ current decreases linearly with depolarization (see **Figure 4.9a**). In contrast, the $I_{Na}-V$ relation is not linear. The macroscopic Na$^+$ current is recorded from a myelinated rabbit nerve with the double electrode voltage clamp technique. When the amplitude of the peak Na$^+$ current is plotted against membrane potential, it has a clear bell shape (**Figures 4.11** and **4.12a**).

Analysis of each trace from the smallest depolarizing step to the largest shows that:

- For small steps, the peak current amplitude is small (0.2 nA) and has a slow time to peak (1 ms). At these potentials the Na$^+$ driving force is strong but the Na$^+$ channels have a low probability of opening (**Figure 4.11a**). Therefore, I_{Na} is small since it represents the current through a small number of open Na$^+$ channels. Moreover, the small number of activated Na$^+$ channels open with a delay since the depolarization is just *subliminal*. This explains the slow time to peak.
- As the depolarizing steps increase in amplitude (to $-42/-35$ mV), the amplitude of I_{Na} increases to a maximum (-3 nA) and the time to peak decreases to a minimum (0.2 ms). Larger depolarizations increase the probability of the Na$^+$ channel being in the open state and shorten the delay of opening (see **Figure 4.10**). Therefore, though the amplitude of i_{Na} decreases between -63 and -35 mV, the amplitude of I_{Na} increases owing to the large increase of open Na$^+$ channels.
- After this peak, the amplitude of I_{Na} decreases to zero since the open probability does not increase enough to compensate for the decrease of i_{Na}. The reversal potential of I_{Na} is the same as that of i_{Na} since it depends only on the extracellular and intracellular concentrations of Na$^+$ ions.

- I_{Na} changes polarity for V_m more depolarized than E_{rev}: it is now an outward current whose amplitude increases with the depolarization (**Figure 4.10b**).

It is important to note that membrane potentials more depolarized than $+20$ mV are non-physiological.

Activation and inactivation curves: the threshold potential

Activation rate is the rate at which a macroscopic current turns on in response to a depolarizing voltage step. The Na$^+$ current is recorded in a voltage clamp from a node of rabbit nerve. Depolarizing steps from -70 mV to $+20$ mV are applied from a holding potential of -80 mV. When the ratio of the peak current at each test potential to the maximal peak current (I_{Na}/I_{Namax}) is plotted against test potential, the activation curve of I_{Na} can be visualized. The distribution is fitted by a sigmoidal curve (**Figure 4.12b**). In this preparation, the threshold of Na$^+$ channel activation is -60 mV. At -40 mV, I_{Na} is already maximal ($I_{Na}/I_{Namax} = 1$). This steepness of activation is a characteristic of the voltage-gated Na$^+$ channels.

Inactivation of a current is the decay of this current during a maintained depolarization. To study inactivation, the membrane is held at varying holding potentials and a depolarizing step to a fixed value is applied where I_{Na} is maximal (0 mV for example). The amplitude of the peak Na$^+$ current is plotted against the holding potential. I_{Na} begins to inactivate at -90 mV and is fully inactivated at -50 mV. Knowing that the resting membrane potential in this preparation is around -80 mV, some of the Na$^+$ channels are already inactivated at rest.

Ionic selectivity of the Na$^+$ channel

To compare the permeability of the Na$^+$ channel to several monovalent cations, the macroscopic current is recorded at different membrane potentials in the presence of external Na$^+$ ions and when all the external Na$^+$ are replaced by a test cation. Lithium is as permeant as sodium but K$^+$ ions are weakly permeant ($P_K/P_{Na} = 0.048$). Therefore, Na$^+$ channels are highly selective for Na$^+$ ions and only 4% of the current is carried by K$^+$ ions (**Figure 4.13**).

Tetrodotoxin is a selective open Na$^+$ channel blocker

A large number of biological toxins can modify the properties of the Na$^+$ channels. One of these, tetrodotoxin (TTX), which is found in the liver and ovaries of the pufferfishes, (tetrodon) totally abolishes the current through most of the Na$^+$ channels (TTX sensitive Na$^+$

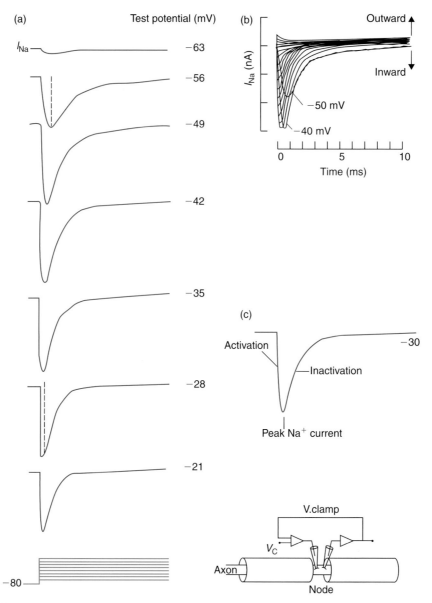

FIGURE 4.11   Voltage dependence of the macroscopic voltage-gated Na⁺ current.
The macroscopic voltage-gated Na⁺ current recorded in a node of a rabbit myelinated nerve in voltage clamp conditions. **(a)** Depolarizing steps from −70 mV to −21 mV from a holding potential of −80 mV evoke macroscopic Na⁺ currents (I_{Na}) with different time courses and peak amplitudes. The test potential is on the right. Bottom trace is the voltage trace. **(b)** The traces in (a) are superimposed and current responses to depolarizing steps from −14 to +55 mV are added. The outward current traces are recorded when the test potential is beyond the reversal potential (+30 mV in this preparation). **(c)** I_{Na} recorded at −30 mV. The rising phase of I_{Na} corresponds to activation of the Na⁺ channels and the decrease of I_{Na} corresponds to progressive inactivation of the open Na⁺ channels. The extracellular solution contains (in mM): 154 NaCl, 2.2 $CaCl_2$, 5.6 KCl; pH 7.4. Adapted from Chiu SY, Ritchie JM, Bogart RB, Stagg D (1979) A quantitative description of membrane currents from a rabbit myelinated nerve. *J. Physiol.* **292**, 149–166, with permission.

channels) (**Figure 4.4**). However, some Na⁺ channels are resistant to TTX such as those from the pufferfishes. TTX has a binding site supposed to be located near the extracellular mouth of the pore.

A single point mutation of the rat brain Na⁺ channel type II, which changes the glutamic acid residue 387 to glutamine (E387Q) in the repeat I, renders the channel insensitive to concentrations of TTX up to tens of

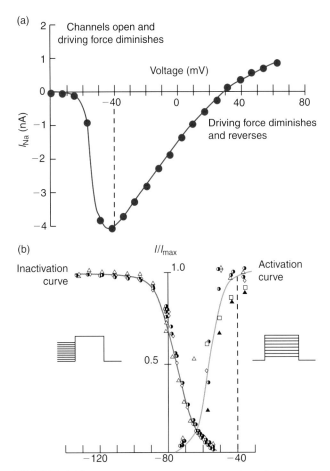

FIGURE 4.12   **Activation–inactivation properties of the macroscopic voltage-gated Na$^+$ current.**
The I_{Na}–V relation has a bell shape with a peak at -40 mV and a reversal potential at $+30$ mV (the average E_{Na} in the rabbit node is $+27$ mV). **(b)** Activation (right curve) and inactivation (left curve) curves obtained from nine different experiments. The voltage protocols used are shown in insets. In the ordinates, I/I_{max} represents the ratio of the peak Na$^+$ current (I) recorded at the tested potential of the abscissae and the maximal peak Na$^+$ current (I_{max}) recorded in this experiment. It corresponds in the activation curve to the peak current recorded at -40 mV in Figure 4.11. From Chiu SY, Ritchie JM, Bogart RB, Stagg D (1979) A quantitative description of membrane currents from a rabbit myelinated nerve. *J. Physiol.* **292**, 149–166, with permission.

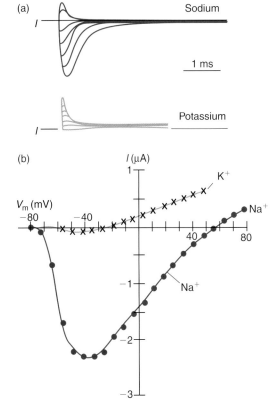

FIGURE 4.13   **Ionic selectivity of the Na$^+$ channel.**
(a) The macroscopic Na$^+$ current is recorded with the double-electrode voltage clamp technique in a mammalian skeletal muscle fibre at different test membrane potentials (from -70 to $+80$ mV) from a holding potential of -80 mV. **(a)** Inward currents in normal Na$^+$–Ringer (sodium) and in a solution where all Na$^+$ ions are replaced by K$^+$ ions (potassium). The other voltage-gated currents are blocked. **(b)** I–V relation of the currents recorded in (a). I is the amplitude of the peak current at each tested potential. Adapted from Pappone PA (1980) Voltage clamp experiments in normal and denervated mammalian skeletal muscle fibers. *J Physiol.* **306**, 377–410, with permission.

micromolars. *Xenopus* oocytes are injected with the wild-type mRNA or the mutant mRNA and the whole cell Na$^+$ currents are recorded with the double-electrode voltage clamp technique. TTX sensitivity is assessed by perfusing TTX-containing external solutions and by measuring the peak of the whole-cell inward Na$^+$ current (the peak means the maximal amplitude of the inward Na$^+$ current measured on the I_{Na}/V relation). The dose-response curves of **Figure 4.14** show that 1 μM of TTX completely abolishes the wild-type Na$^+$ current, but has no effect on the mutant Na$^+$ current. The other characteristics of the Na$^+$ channel are not

significantly affected, except for a reduction in the amplitude of the inward current at all potentials tested. All these results suggest that the link between segments S5 and S6 in repeat I of the rat brain Na$^+$ channel is in close proximity to the channel mouth.

Comparison of the predicted protein sequences of the skeletal muscle sodium channels show that pufferfish Na$^+$ channels have accumulated several unique substitutions in the otherwise highly conserved pore loop regions of the four domains. Among these substitutions, some are associated with TTX resistance as assessed with path clamp recordings. What advantages does TTX resistance offer pufferfishes? A main advantage is that the high tissue concentrations of TTX act as an effective chemical defense against predators. Also TTX resistance

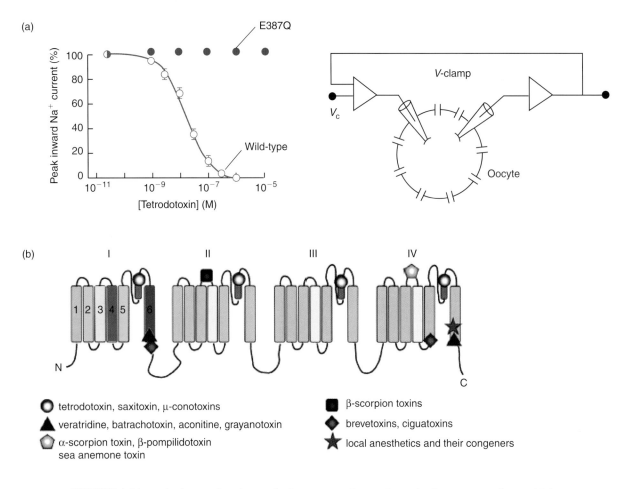

FIGURE 4.14 **A single mutation close to the S6 segment of repeat I completely suppresses the sensitivity of the Na⁺ channel to TTX.**
(a) A mutation of the glutamic acid residue 387 to glutamine (E387Q) is introduced in the rat Na⁺ channel type II. *Xenopus* oocytes are injected with either the wild-type mRNA or the mutant mRNA. The macroscopic Na⁺ currents are recorded 4–7 days later with the double-electrode voltage clamp technique. Dose-response curves for the wild-type (open circles) and the mutant E387Q (filled circles) to tetrodotoxin (TTX). TTX sensitivity is determined by perfusing TTX-containing external solutions and by measuring the macroscopic peak inward current. The TTX concentration that reduces the wild-type Na⁺ current by 50% (IC$_{50}$) is 18 nM. Data are averaged from 7–8 experiments. **(b)** Topology of drug binding sites on Na⁺ channel α-subunit. Each symbol represents the toxin or drug binding site indicated at the bottom of the figure. All these sites have been characterized by site-directed mutagenesis. Part (a) from Noda M, Suzuki H, Numa S, Stühmer W (1989) A single point mutation confers tetrodotoxin and saxitoxin insensitivity on the sodium channel II. *FEBS Lett.* **259**, 213–216, with permission. Part (b) drawing adapted from Ogata N and Ohishi Y (2002) Molecular diversity of structure and function of the voltage-gated Na⁺ channels. *Jpn. J. Pharmacol.* **88**, 365–377.

enables pufferfishes to feed on TTX-bearing organisms that are avoided by other fishes.

TTX is not the only toxin to target Na⁺ channels. Most of these toxins, except for TTX and its congeners that occlude the outer pore of the channel, bind to sites that are related to activation and inactivation processes. They fall into at least five different classes according to their corresponding receptor sites:

(1) hydrophilic toxins such as TTX, saxitoxin (STX) and α-conus toxin; (2) lipid-soluble neurotoxins such as batrachotoxin (BTX), veratridine, aconitine and grayanotoxin; (3) α-scorpion peptide toxins and sea anemone peptide toxins; (4) β-scorpion peptide toxins; (5) lipid-soluble brevetoxins and ciguatoxins. They can have opposite effects. For example, toxins (2) are activators of Na⁺ channels.

4.2.7 Segment S4, the region between segments S5 and S6, and the region between domains III and IV play a significant role in activation, ion permeation and inactivation, respectively

The major questions about a voltage-gated ionic channel and particularly the Na$^+$ channel are the following:

- How does the channel open in response to a voltage change?
- How is the permeation pathway designed to define single-channel conductance and ion selectivity?
- How does the channel inactivate?

In order to identify regions of the Na$^+$ channels involved in these functions, site-directed mutagenesis experiments were performed. The activity of each type of mutated Na$^+$ channel is analyzed with patch clamp recording techniques.

The short segments between putative membrane spanning segments S5 and S6 are membrane associated and contribute to pore formation

The Na$^+$ channels are highly selective for Na$^+$ ions. This selectivity presumably results from negatively charged amino acid residues located in the channel pore. Moreover, these amino acids must be specific to Na$^+$ channels (i.e. different from the other members of voltage-gated cationic channels such as K$^+$ and Ca^{2+} channels) to explain their weak permeability to K$^+$ or Ca^{2+} ions.

Studies using mutagenesis to alter ion channel function have shown that the region connecting the S5 and S6 segments forms part of the channel lining (see **Figure 4.6**). A single amino acid substitution in these regions in repeats III and IV alters the ion selectivity of the Na$^+$ channel to resemble that of Ca^{2+} channels. These residues would constitute part of the selectivity filter of the channel. There is now a general agreement that the selectivity filter is formed by pore loops; i.e. relatively short polypeptide segments that extend into the aqueous pore from the extracellular side of the membrane. Rather than extending completely across the lipid bilayer, a large portion of the pore loop is near the extracellular face of the channel. Only a short region extends into the membrane to form the selectivity filter. In the case of the voltage-gated Na$^+$ channel, each of the four homologous domains contributes a loop to the ion conducting pore.

The S4 segment is the voltage sensor

The S4 segments are positively charged and hydrophobic (**Figure 4.15a**). Moreover, the typical amino acid sequence of S4 is conserved among the different voltage-gated channels. These observations led to the suggestion that S4 segments have a transmembrane orientation and are voltage sensors. To test this proposed role, positively charged amino acid residues are replaced by neutral or negatively charged residues in the S4 segment of a rat brain Na$^+$ channel type II. The mutated channels are expressed in *Xenopus* oocytes. When more than three positive residues are mutated in the S4 segments of repeat I or II, no appreciable expression of the mutated channel is obtained. The replacement of only one arginine or lysine residue in segment S4 of repeat I by a glutamine residue shifts the activation curve to more positive potentials (**Figure 4.15b, c**).

It is hypothesized that the positive charges in S4 form ion pairs with negative charges in other transmembrane regions, thereby stabilizing the channel in the non-conducting closed conformation. With a change in the electric field across the membrane, these ion pairs would break as the S4 charges move and new ion pairs would form to stabilize the conducting, open conformation of the channel.

The cytoplasmic loop between domains III and IV contains the inactivation particle which, in a voltage-dependent manner, enters the mouth of the Na$^+$ channel pore and inactivates the channel

The results obtained from three different types of experiments strongly suggest that the short cytoplasmic loop connecting homologous domains III and IV, L$_{III-IV}$ loop (see **Figures 4.6a** and **4.16a**), is involved in inactivation: (i) cytoplasmic application of endopeptidases; (ii) cytoplasmic injection of antibodies directed against a peptide sequence in the region between repeats III and IV; and (iii) cleavage of the region between repeats III and IV (**Figure 4.16a–c**); all strongly reduce or block inactivation. Moreover, in some human pathology where the Na$^+$ channels poorly inactivate (as shown with single-channel recordings from biopsies), this region is mutated.

Positively charged amino acid residues of this L$_{III-IV}$ loop are not required for inactivation since only the mutation of a hydrophobic sequence, isoleucine-phenylalanine-methionine (IFM), to glutamine completely blocks inactivation. The critical residue of the IFM motif is phenylalanine since its mutation to glutamine slows inactivation 5000-fold. It is proposed that this IFM sequence is directly involved in the conformational change leading to inactivation. It would enter the mouth of the pore, thus occluding it during the process of inactivation. In order to test this hypothesis, the ability of synthetic peptides containing the IFM motif to restore fast inactivation to non-inactivating rat brain Na$^+$ channels expressed in kidney carcinoma

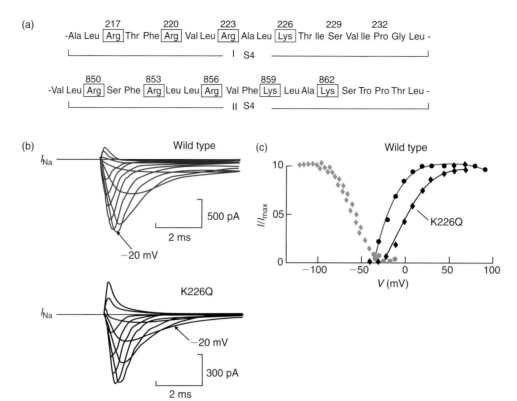

FIGURE 4.15 **Effect of mutations in the S4 segment on Na⁺ current activation.**
Oocytes are injected with the wild-type rat brain Na⁺ channel or with Na⁺ channels mutated on the S4 segment.
The activity of a population of Na⁺ channels is recorded in patch clamp (cell-attached macropatches). **(a)** Amino
acid sequences of segment S4 of the internal repeats I (I S4) and II (II S4) of the wild-type rat Na⁺ channel. Positively
charged amino acids are boxed with solid lines and the numbers of the relevant residues are given. In the mutated
channel studied here the lysine residue in position 226 is replaced by a glutamine residue (K226Q). **(b)** In response
to step depolarizations ranging from −60 to +70 mV from a holding potential of −120 mV, a family of macroscopic Na⁺ currents is recorded for each type of Na⁺ channel. The arrow indicates the response to the test potential −20 mV. Note that at −20 mV the amplitude of the Na⁺ current is at its maximum for the wild-type and less
than half maximum for the mutated channel. **(c)** Steady-state activation (right) and inactivation (left) curves for
the wild-type (circles) and the mutant (diamonds) Na⁺ channels. Adapted from Stühmer W, Conti F, Suzuki H *et al.*
(1989) Structural parts involved in activation and inactivation of the sodium channel. *Nature* **339**, 597–603, with
permission.

cells is examined. The intrinsic inactivation of Na⁺
channels is first made non-functional by a mutation of
the IFM motif. When the recording is now performed
with a patch pipette containing the synthetic peptide
with an IFM motif, the non-inactivating whole cell
Na⁺ current now inactivates. Since the restored inactivation has the rapid, voltage-dependent time course
characteristic of inactivation of the wild-type Na⁺
channels, it is proposed that the IFM motif serves as an
inactivation particle (**Figure 4.16d**).

4.2.8 Conclusion: the consequence of the opening of a population of N Na⁺ channels is a transient entry of Na⁺ ions which depolarizes the membrane above 0 mV

The function of the population of N Na⁺ channels at
the axon initial segment or at nodes of Ranvier is to

ensure a *sudden* and *brief* depolarization of the membrane above 0 mV.

*Rapid activation of Na⁺ channels makes the
depolarization phase sudden*

In response to a depolarization to the threshold
potential, the closed Na⁺ channels (**Figure 4.17a**) of the
axon initial segment begin to open (b). The flux of Na⁺
ions through the few open Na⁺ channels depolarizes
the membrane more and thus triggers the opening of
other Na⁺ channels (c). In consequence, the flux of Na⁺
ions increases, depolarizes the membrane more and
opens other Na⁺ channels until all the N Na⁺ channels
of the segment of membrane are opened (d). In (d) the
depolarization phase is at its peak. Na⁺ channels are
opened by depolarization and once opened, they contribute to the membrane depolarization and therefore
to their activation: it is a self-maintained process.

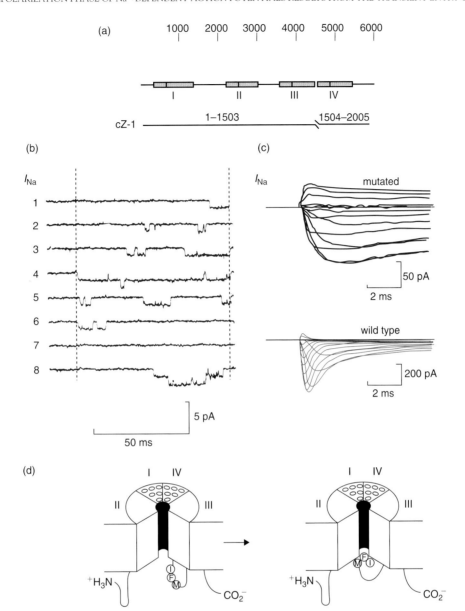

FIGURE 4.16   Effects of mutations in the region between repeats III and IV on Na⁺ current inactivation.
(a) Linear representation of the wild-type Na⁺ channel (upper trace) and the mutated Na⁺ channel (bottom trace). The mutation consists of a cut with an addition of four to eight residues at each end of the cut. An equimolar mixture of the two mRNAs encoding the adjacent fragments of the Na⁺ channel protein separated with a cut is injected in oocytes. (b) Single-channel recordings of the activity of the mutated Na⁺ channel in response to a depolarizing step to $-20\,mV$ from a holding potential of $-100\,mV$. Note that late single or double openings (line 8) are often recorded. The mean open time τ_o is 5.8 ms and the elementary conductance γ_{Na} is 17.3 pS. (c) Macroscopic Na⁺ currents recorded from the mutated (upper trace) and the wild-type (bottom trace) Na⁺ channels. (d) Model for inactivation of the voltage-gated Na⁺ channels. The region linking repeats III and IV is depicted as a hinged lid that occludes the transmembrane pore of the Na⁺ channel during inactivation. Parts (a)–(c) from Pappone PA (1980) Voltage clamp experiments in normal and denervated mammalian skeletal muscle fibers. *J Physiol.* **306**, 377–410, with permission. Drawing (d) from West JW, Patton DE, Scheuer T *et al.* (1992) A cluster of hydrophobic amino acid residues required for fast sodium channel inactivation. *Proc. Natl Acad. Sci. USA* **89**, 10910–10914, with permission.

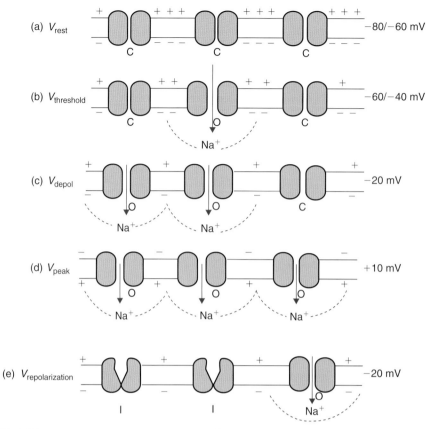

FIGURE 4.17 **Different states of voltage-gated Na$^+$ channels in relation to the different phases of the Na$^+$-dependent action potential.**
C, closed state; O, open state; I, inactivated state; →, driving force for Na$^+$ ions.

Rapid inactivation of Na$^+$ channels makes the depolarization phase brief

Once the Na$^+$ channels have opened, they begin to inactivate (e). Therefore, though the membrane is depolarized, the influx of Na$^+$ ions diminishes quickly. Therefore the Na$^+$-dependent action potential is a spike and does not present a plateau phase. Inactivation is a very important protective mechanism since it prevents potentially toxic persistent depolarization.

4.3 THE REPOLARIZATION PHASE OF THE SODIUM-DEPENDENT ACTION POTENTIAL RESULTS FROM Na$^+$ CHANNEL INACTIVATION AND PARTLY FROM K$^+$ CHANNEL ACTIVATION

The participation of a voltage-gated K$^+$ current in action potential repolarization differs from one preparation to another. For example, in the squid axon the voltage-gated K$^+$ current plays an important role in

spike repolarization, though in mammalian peripheral nerves this current is almost absent. However, the action potentials of the squid axon and that of mammalian nerves have the same duration. This is because the Na$^+$ current in mammalian axons inactivates two to three times faster than that of the frog axon. Moreover, the leak K$^+$ currents are important in mammalian axons (see below).

Voltage-gated K$^+$ channels can be classified into two major groups based on physiological properties:

- Delayed rectifiers which activate after a delay following membrane depolarization and inactivate slowly
- A-type channels which are fast activating and fast inactivating.

The first type, the delayed rectifier K$^+$ channels, plays a role in action potential repolarization. The A-types inactivate too quickly to do so. They play a role in firing patterns and are explained in Chapter 14.

This section will explain the structure and activity of the voltage-gated, delayed rectifier K$^+$ channels responsible for action potential repolarization in the squid or frog nerves. Then Section 4.4 will explain the other mode of repolarization observed in mammalian nerves, in which the delayed rectifier current does not play a significant role.

4.3.1 The delayed rectifier K⁺ channel consists of four α-subunits and auxiliary β-subunits

K⁺ channels represent an extremely diverse ion channel type. All known K⁺ channels are related members of a single protein family. Finding genes responsible for a native K⁺ current is not an easy task, because K⁺ channels have a great diversity: more than 100 of K⁺ channel subunits have been identified to date. Among strategies used to identify which genes encode a particular K⁺ channel are the single cell reverse transcriptase chain reaction (scRT-PCR) protocol combined with patch clamp recording and the injection of subfamily-specific dominant negative constructs in recorded neurons. Results of such experiments strongly suggested that α-subunits of delayed rectifiers are attributable to Kv2 and Kv3 subfamily genes.

Delayed rectifier K⁺ channels α-subunits form homo- or hetero-tetramers in the cell membrane. As for the Na⁺ channel, the P loop linking segments S5 and S6 contributes to the formation of the pore and the auxiliary small β-subunits associated with the α-subunit are considered to be intracellularly located (**Figure 4.18**).

4.3.2 Membrane depolarization favours the conformational change of the delayed rectifier channel towards the open state

The function of the delayed rectifier channel is to transduce, with a delay, membrane depolarization into an exit of K⁺ ions

Single-channel recordings were obtained by Conti and Neher in 1980 from the squid axon. We shall, however, look at recordings obtained from K⁺ channels expressed in oocytes or in mammalian cell lines from cDNA encoding a delayed rectifier channel of rat brain. Since the macroscopic currents mediated by these channels have time courses and ionic selectivity resembling those of the classical delayed outward currents described in nerve and muscle, these single-channel recordings are good examples for describing the properties of a delayed rectifier current.

Figure 4.19 shows a current trace obtained from patch clamp recordings (inside-out patch) of a rat brain K⁺ channel (RCK1) expressed in a *Xenopus* oocyte. In the presence of physiological extracellular and intracellular K⁺ concentrations, a depolarizing voltage step to 0 mV from a holding potential of −60 mV is applied. After the onset of the step, a rectangular pulse of elementary

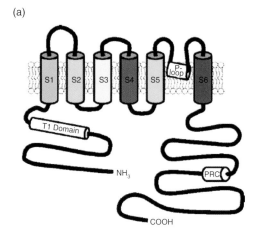

(a)

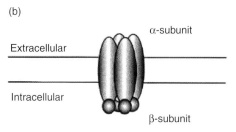

(b)

FIGURE 4.18 Putative transmembrane organization of the α-subunit of the delayed rectifier, voltage-gated K⁺ channel and its associated cytoplasmic β-subunit.

(**a**) Diagrammatic representation of the predicted membrane topology of a single Kv2.1 α-subunit. S1–S6 represent the transmembrane segments (cylinders represent putative α-helical segments), P-loop represents the amino acid residues that form the bulk of the lining of the channel pore, T1 domain represents the subfamily-specific tetramerization domain in the cytoplasmic N-terminus, and PRC represents the proximal restriction and clustering signal in the cytoplasmic C-terminus. (**b**) A schematic diagram showing the putative structure of a voltage-dependent K⁺ channel. A channel is composed of four pore-forming α-subunits, to each of which a β-subunit is associated on the cytoplasmic side. The four α-subunits can be homomers or heteromers. Part (a) from Misonou H, Mohapatra DP, Trimmer JS (2005) *NeuroToxicology* **26**, 743–752, with permission; Part (b) from Song WJ (2002) Genes responsible for native depolarization-activated K⁺ currents in neurons. *Neuroscience Research* **42**, 7–14.

current, upwardly directed, appears. It means that the current is outward; K⁺ ions leave the cell. In fact, the driving force for K⁺ ions is outward at 0 mV.

It is immediately striking that the gating behaviour of the delayed rectifier channel is different from that of the Na⁺ channel (compare **Figures 4.7a** or **4.8a** and **4.19**). Here, the rectangular pulse of current lasts the whole depolarizing step with short interruptions during which the current goes back to zero. It indicates that the delayed rectifier channel opens, closes briefly and reopens many times during the depolarizing pulse: the delayed rectifier channel does not inactivate within seconds. Another difference is that the delay of opening

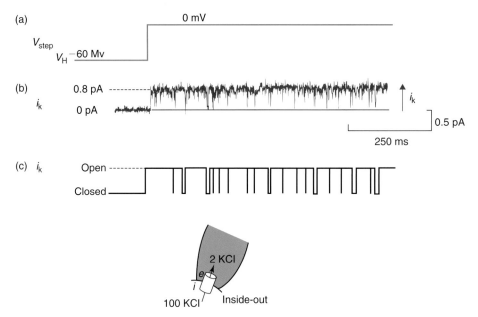

FIGURE 4.19   Single K⁺ channel openings in response to a depolarizing step.
The activity of a single delayed rectifier channel expressed from rat brain cDNA in a *Xenopus* oocyte is recorded in patch clamp (inside-out patch). A depolarizing step to 0 mV from a holding potential of −60 mV **(a)** evokes the opening of the channel **(b)**. The elementary current is outward. The channel then closes briefly and reopens several times during the depolarization, as shown in the drawing **(c)** that interprets the current trace. Bathing solution or intracellular solution (in mM): 100 KCl, 10 EGTA, 10 HEPES. Pipette solution or extracellular solution (in mM): 115 NaCl, 2 KCl, 1.8 CaCl₂, 10 HEPES. Adapted from Stühmer W, Stocker M, Sakmann B *et al.* (1988) Potassium channels expressed from rat brain cDNA have delayed rectifier properties. *FEBS Lett.* **242**, 199–206, with permission.

of the delayed rectifier is much longer than that of the Na⁺ channel, even for large membrane depolarizations (mean delay 4 ms in **Figure 4.20a**).

When the same depolarizing pulse is now applied every 1–2 s, we observe that the delay of channel opening is variable (1–10 ms) but gating properties are the same in all recordings: the channel opens, closes briefly and reopens during the entire depolarizing step (**Figure 4.20a**). Amplitude histograms collected at 0 mV membrane potential from current recordings, such as those shown in **Figure 4.20a**, give a mean amplitude of the unitary currents of +0.8 pA (**Figure 4.20c**). This means that the most frequently occurring main amplitude is +0.8 pA.

4.3.3 The open probability of the delayed rectifier channel is stable during a depolarization in the range of seconds

The average open time τ_o measured in the patch illustrated in **Figure 4.19** is 4.6 ms. The mean closed time τ_c is 1.5 ms. As seen in **Figures 4.19** and **4.20a**, during a depolarizing pulse to 0 mV the delayed rectifier channel

spends much more time in the open state than in the closed state: at 0 mV its average open probability is high ($p_o = 0.76$).

In order to test whether the delayed rectifier channels show some inactivation, long-lasting recordings are performed. Though no significant inactivation is apparent during test pulses in the range of seconds, during long test depolarizations (in the range of minutes) the channel shows steady-state inactivation at positive holding potentials (not shown). Therefore, in the range of seconds, the inactivation of the delayed rectifier channel can be omitted: the channel fluctuates between the closed and open states:

$$C \rightleftharpoons O$$

The transition from the closed (C) state to the open (O) state is triggered by membrane depolarization with a delay. The delayed rectifier channel activates in the range of milliseconds. In comparison, the Na⁺ channel activates in the range of submilliseconds. The O to C transitions of the Na⁺ channel frequently happen though the membrane is still depolarized. It also happens when membrane repolarizes.

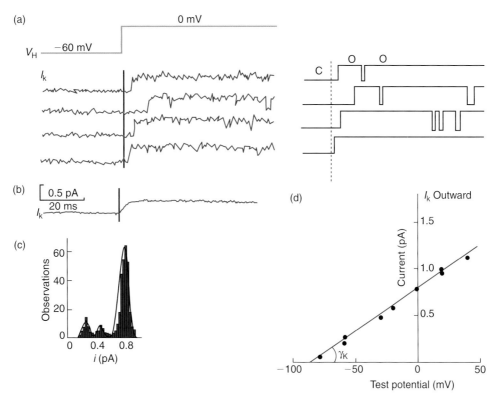

FIGURE 4.20   Characteristics of the elementary delayed rectifier current.
Same experimental design as in Figure 4.19. The patch of membrane contains a single delayed rectifier channel. **(a)** Successive sweeps of outward current responses to depolarizing steps from −60 mV to 0 mV (C for closed state, O for open state of the channel). **(b)** Averaged current from 70 elementary currents as in (a). **(c)** Amplitude histogram of the elementary outward currents recorded at test potential 0 mV. The mean elementary current amplitude observed most frequently is 0.8 pA. **(d)** Single channel current–voltage relation ($i_K − V$). Each point represents the mean amplitude of at least 20 determinations. The slope is $\gamma_K = 9.3$ pS. The reversal potential $E_{rev} = −89$ mV. From Stühmer W, Stocker M, Sakmann B *et al.* (1988) Potassium channels expressed from rat brain cDNA have delayed rectifier properties. *FEBS Lett.* **242**, 199–206, with permission.

4.3.4   The K⁺ channel has a constant unitary conductance γ_K

In **Figure 4.21a**, unitary currents are shown in response to increasing depolarizing steps from −50 to +20 mV from a holding potential of −80 mV. We observe that both the amplitude of the unitary current and the time spent by the channel in the open state increase with depolarization.

When the mean amplitude of the unitary K⁺ current is plotted versus membrane test potential, a linear i_K/V relation is obtained (**Figures 4.20d** and **4.21b**). This linear i_K/V relation (between −50 and +20 mV) is described by the equation $i_K = \gamma_K(V − E_K)$, where V is the membrane potential, E_K is the reversal potential of the K⁺ current, and γ_K is the conductance of the single delayed rectifier K⁺ channel, or unitary conductance. Linear back-extrapolation gives a reversal potential value around −90/−80 mV, a value close to E_K calculated from the Nernst equation. This means that from −80 mV

to more depolarized potentials, which correspond to the physiological conditions, the K⁺ current is outward. For more hyperpolarized potentials, the K⁺ current is inward.

The value of γ_K is given by the slope of the linear i_K/V curve. It has a constant value at any given membrane potential. This value varies between 10 and 15 pS depending on the preparation (**Figures 4.20d** and **4.21b**).

4.3.5   The macroscopic delayed rectifier K⁺ current (I_K) has a delayed voltage dependence of activation and inactivates within tens of seconds

Whole cell currents in *Xenopus* oocytes expressing delayed rectifier channels start to activate at potentials positive to −30 mV and their amplitude is clearly voltage-dependent. When unitary currents recorded from 70 successive depolarizing steps to 0 mV are averaged

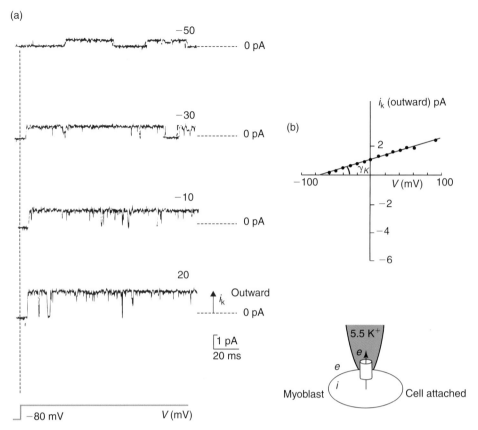

FIGURE 4.21   The single-channel current/voltage (i_K/V) relation is linear.
Delayed rectifier K$^+$ channels from rat brain are expressed in a myoblast cell line. **(a)** The activity of a single channel is recorded in patch clamp (cell-attached patch). Unitary currents are recorded at different test potentials (from -50 mV to $+20$ mV) from a holding potential at -80 mV. Bottom trace is the voltage trace. **(b)** i_K-V relation obtained by plotting the mean amplitude of i_K at the different test potentials tested. i_K reverses at $V = -75$ mV and $\gamma_K = 14$ pS. Intrapipette solution (in mM): 145 NaCl, 5.5 KCl, 2 CaCl$_2$, 2 MgCl$_2$, 10 HEPES. Adapted from Koren G, Liman ER, Logothetis DE *et al.* (1990) Gating mechanism of a cloned potassium channel expressed in frog oocytes and mammalian cells. *Neuron* **2**, 39–51, with permission.

(**Figure 4.20b**), the macroscopic outward current obtained has a slow time to peak (4 ms) and lasts the entire depolarizing step. It closely resembles the whole cell current recorded with two electrode voltage clamps in the same preparation (rat brain delayed rectifier channels expressed in oocytes; see **Figure 4.22a**). The whole-cell current amplitude at steady state (once it has reached its maximal amplitude) for a given potential is:

$$I_K = N p_o i_K$$

where N is the number of delayed rectifier channels in the membrane recorded, p_o the open probability at steady state and i_K the elementary current. The number of open channels $N p_o$ increases with depolarization (to a maximal value) and so does I_K.

The I_K/V relation shows that the whole cell current varies linearly with voltage from a threshold potential

which in this preparation is around -40 mV (**Figure 4.22b**). When the membrane is more hyperpolarized than the threshold potential, very few channels are open and I_K is equal to zero. For membrane potentials more depolarized than the threshold potential, I_K depends on p_o and the driving force state ($V - E_K$) which augments with depolarization. Once p_o is maximal, I_K augments linearly with depolarization since it depends only on the driving force.

The delayed rectifier channels are selective to K$^+$ ions

Ion substitution experiments indicate that the reversal potential of I_K depends on the external K$^+$ ions concentration as expected for a selective K$^+$ channel. The reversal potential of the whole cell current is measured as in **Figure 4.22b** in the presence of different external

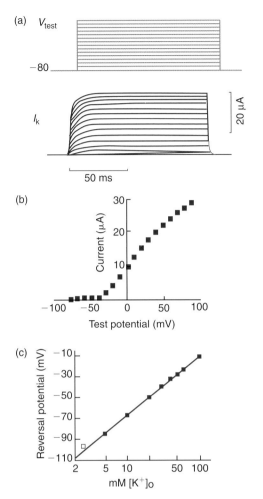

FIGURE 4.22 **Characteristics of the macroscopic delayed rectifier K+ current.**
The activity of N delayed rectifier channels expressed from rat brain cDNA in oocytes recorded in double-electrode voltage clamp. **(a)** In response to depolarizing steps of increasing amplitude (given every 2 s) from a holding potential of -80 mV (upper traces), a non-inactivating outward current of increasing amplitude is recorded (lower traces). **(b)** The amplitude of the current at steady state is plotted against test potential. The potential threshold for its activation is -40 mV. **(c)** The value of the reversal potentials of the macroscopic current is plotted against the extracellular concentration of K+ ions on a semi-logarithmic scale. The slope is -55 mV. From Stühmer W, Stocker M, Sakmann B *et al.* (1988) Potassium channels expressed from rat brain cDNA have delayed rectifier properties. *FEBS Lett.* **242**, 199–206, with permission.

concentrations of K+ ions. These experimental values are plotted against the external K+ concentration, $[K^+]_o$, on a semi-logarithmic scale. For concentrations ranging from 2.5 (normal frog Ringer) to 100 mM, a linear relation with a slope of 55 mV for a 10-fold change in $[K^+]_o$ is obtained (not shown). These data are well fitted by the Nernst equation. It indicates that the channel has a higher selectivity for K+ ions over Na+ and Cl− ions.

The delayed rectifier channels are blocked by millimolar concentrations of tetraethylammonium (TEA) and by Cs+ ions. Ammonium ions can pass through most K+ channels, whereas its quaternary derivative TEA cannot, resulting in the blockade of most of the voltage-gated K+ channels: TEA is a small open channel blocker. Amino acids in the carboxyl half of the region linking segments S5 and S6 (i.e. adjacent to S6) influence the sensitivity to pore blockers such as TEA.

4.3.6 Conclusion: during an action potential the consequence of the delayed opening of K+ channels is an exit of K+ ions, which repolarizes the membrane to resting potential

Owing to their delay of opening, delayed rectifier channels open when the membrane is already depolarized by the entry of Na+ ions through open voltage-gated Na+ channels (**Figure 4.23**). Therefore, the exit of K+ ions does not occur at the same time as the entry of Na+ ions (see also **Figure 4.24**). This allows the membrane to first depolarize in response to the entry of Na+ ions and then to repolarize as a consequence of the exit of K+ ions.

4.4 SODIUM-DEPENDENT ACTION POTENTIALS ARE INITIATED AT THE AXON INITIAL SEGMENT IN RESPONSE TO A MEMBRANE DEPOLARIZATION AND THEN ACTIVELY PROPAGATE ALONG THE AXON

Na+-dependent action potentials, because of their short duration (1–5 ms), are also named spikes. Na+ spikes, for a given cell, have a stable amplitude and duration; they all look alike, and are binary, all-or-none. The pattern of discharge (which is often different from the frequency of discharge) and not individual spikes, carries significant information.

4.4.1 Summary on the Na+-dependent action potential

The depolarization phase of Na+ spikes is due to the rapid time to peak inward Na+ current which flows into the axon initial segment or node. This depolarization is brief because the inward Na+ current inactivates in milliseconds (**Figure 4.24b**).

In the squid giant axon or frog axon, spike repolarization is associated with an outward K+ current

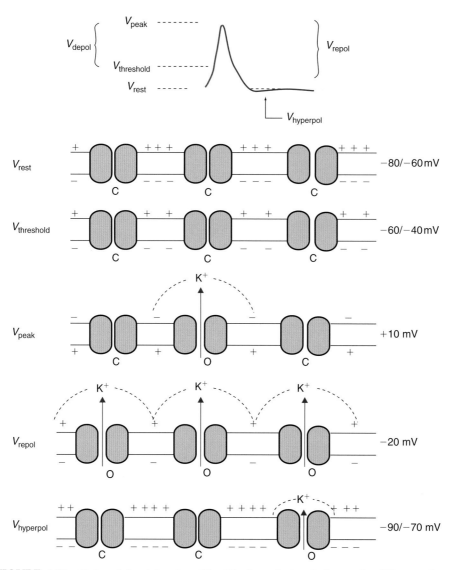

FIGURE 4.23   States of the delayed rectifier K$^+$ channels in relation to the different phases of the Na$^+$-dependent action potential.
C, closed state; O, open state; ↑, driving force for K$^+$ ions.

through delayed rectifier channels (**Figures 4.24** and **4.25**) since TEA application dramatically prolongs the action potential (see **Figure 4.4b**). As pointed out by Hodgkin and Huxley: 'The rapid rise is due almost entirely to Na$^+$ conductance, but after the peak, the K$^+$ conductance takes a progressively larger share until, by the beginning of the hyperpolarized phase, the Na$^+$ conductance has become negligible. The tail of raised conductance that falls away gradually during the positive phase is due solely to K$^+$ conductance, the small constant leak conductance being of course present throughout.'

In contrast, in rat or rabbit myelinated axons the action potential is very little affected by the application of TEA. The repolarization phase in these preparations is largely associated with a leak K$^+$ current. Voltage clamp studies confirm this observation. When the leak current is subtracted, almost no outward current is recorded in rabbit node (**Figure 4.25b**).

However, squid and rabbit nerve action potentials have the same duration (**Figure 4.25a**). In this preparation, the normal resting membrane potential is around −80 mV, which suggests the presence of a large leak K$^+$ current. Moreover, test depolarizations evoke large outward K$^+$ currents insensitive to TEA (**Figure 4.26**). How does the action potential repolarize in such preparations? First the Na$^+$ currents in the rabbit node inactivate two to three times faster than those in the frog node. Second, the large leak K$^+$ current present at depolarized membrane potentials repolarizes the membrane. The amplitude of the leak K$^+$ current augments linearly with depolarization, depending only on the K$^+$ driving force.

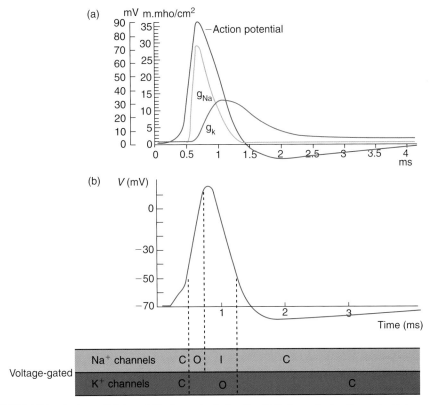

FIGURE 4.24   Gating of Na⁺ and K⁺ channels during the Na⁺-dependent action potential.
(a) Interpretation of the manner in which the conductances to Na⁺ and K⁺ contribute to the action potential.
(b) State of the Na⁺ and K⁺ voltage-gated channels during the course of the action potential. O, channels open;
I, channels inactivate; C, channels close or are closed. Trace (a) adapted from Hodgkin AL and Huxley AF
(1952) A quantitative description of membrane current and its application to conduction and excitation in
nerve. *J. Physiol.* **117**, 500–544, with permission.

4.4.2 Depolarization of the membrane to the threshold for voltage-gated Na⁺ channel activation has two origins

The inward current which depolarizes the membrane of the initial segment to the threshold potential for voltage-gated Na⁺ channel opening is one of the following:

- A depolarizing current resulting from the activity of excitatory afferent synapses (see Chapters 8 and 10) or afferent sensory stimuli. In the first case, the synaptic currents generated at postsynaptic sites in response to synaptic activity summate, and when the resulting current is inward it can depolarize the membrane to the threshold for spike initiation. In the second case, sensory stimuli are transduced in inward currents that can depolarize the membrane to the threshold for spike initiation.

- An intrinsic regenerative depolarizing current such as, for example, in heart cells or invertebrate neurons.

4.4.3 The site of initiation of Na⁺-dependent action potentials is the axon initial segment

The site of initiation was suggested long ago to occur in the axon initial segment since the threshold for spike initiation was the lowest at this level. However, this has only recently been directly demonstrated with the double-patch clamp technique. First the dendrites and soma belonging to the same Purkinje neuron of the cerebellum are visualized in a rat brain slice. Then the activity is recorded simultaneously at both these sites with two patch electrodes (whole-cell patches). To verify that somatic and dendritic recordings are made from the same cell, the Purkinje cell is filled with two differently coloured fluorescent dyes: Cascade blue at the soma and Lucifer yellow at the dendrite. To determine the site of action potential initiation during synaptic activation of Purkinje cells, action potentials are evoked by stimulation of afferent parallel fibres which make synapses on distal dendrites of Purkinje cells (see **Figures 6.8** and **6.9**).

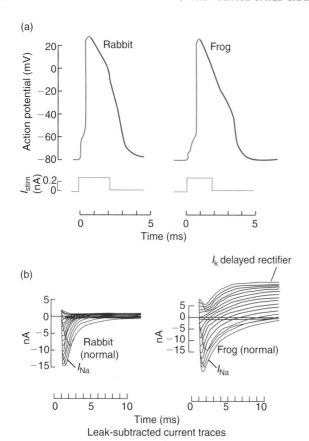

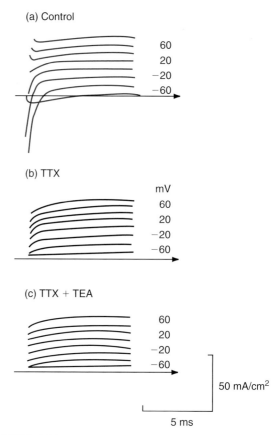

FIGURE 4.25 The currents underlying the action potentials of the rabbit and frog nerves.
(a) The action potentials are recorded intracellularly at 14°C. Bottom trace is the current of stimulation injected in order to depolarize the membrane to initiate an action potential. (b) The currents flowing through the membrane at different voltages recorded in voltage clamp. In the rabbit node, very little outward current is recorded after the large inward Na+ current. In the frog nerve, a large outward K+ current is recorded after the large inward Na+ current. Leak current is subtracted from each trace and does not appear in these recordings. Adapted from Chiu SY, Ritchie JM, Bogart RB, Stagg D (1979) A quantitative description of membrane currents in rabbit myelinated nerve. *J. Physiol.* **292**, 149–166, with permission.

FIGURE 4.26 TEA-resistant outward current in a mammalian nerve.
The currents evoked by depolarizing steps from −60 to +60 mV from a holding potential of −80 mV are recorded in voltage clamp in a node of Ranvier of an isolated rat nerve fibre. Control inward and outward currents (a), after TTX 25 nM (b), and after TTX 25 nM and TEA 5 mM (c) are added to the extracellular solution. The outward current recorded in (c) is the leak K+ current. The delayed outward K+ current is taken as the difference between the steady-state outward current in (b) and the leak current in (c). Adapted from Brismar T (1980) Potential clamp analysis of membrane currents in rat myelinated nerve fibres. *J. Physiol.* **298**, 171–184, with permission.

In all Purkinje cells tested, the evoked action potential recorded from the soma has a shorter delay and a greater amplitude than that recorded from a dendrite (**Figure 4.27a**). Moreover, the delay and the difference in amplitude between the somatic spike and the dendritic spike both augment when the distance between the two patch electrodes is increased. This suggests that the site of initiation is proximal to the soma.

Simultaneous whole-cell recordings from the soma and the axon initial segment were performed to establish whether action potential initiation is somatic or axonal in origin. Action potentials clearly occurs first in the axon initial segment (**Figure 4.27b**). These

results suggest that the actual site of Na+-dependent action potential initiation is in the axon initial segment of Purkinje cells. Experiments carried out by Sakmann *et al.* in other brain regions give the same conclusion for all the neurons tested. This may be due to a higher density of sodium channels in the membrane of the axon initial segment.

The action potential, once initiated, spreads passively back into the dendritic tree of Purkinje cells (passively means that it propagates with attenuation since it is not reinitiated in dendrites). Simultaneously it actively propagates into the axon (not shown here; see below). In some neurons, for example the pyramidal cells of the neocortex, the action potential actively

(a)

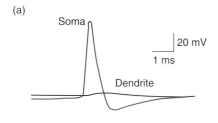

(b)

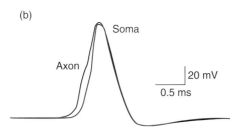

FIGURE 4.27 **The Na$^+$-dependent action potential is initiated in the axon initial segment in Purkinje cells of the cerebellum.** The activity of a Purkinje cell recorded simultaneously at the level of the soma and **(a)** 117 μm away from the soma at the level of a dendrite, or **(b)** 7 μm away from the soma at the level of the axon initial segment, with the double-patch clamp technique (whole-cell patches). Afferent parallel fibres are stimulated by applying brief voltage pulses to an extracellular patch pipette. In response to the synaptic excitation, an action potential is evoked in the Purkinje cell and recorded at the two different neuronal sites: soma and dendrite (a) or soma and axon (b). Adapted from Stuart G and Hauser M (1994) Initiation and spread of sodium action potentials in cerebellar Purkinje cells. *Neuron* **13**, 703–712, with permission.

backpropagates into the dendrites, but this is not a general rule.

4.4.4 The Na$^+$-dependent action potential actively propagates along the axon to axon terminals

Voltage-gated Na$^+$ channels are present all along the axon at a sufficient density to allow firing of axon potentials.

The propagation is active

Active means that the action potential is reinitiated at each node of Ranvier for a myelinated axon, or at each point for a non-myelinated axon. The flow of Na$^+$ ions through the open Na$^+$ voltage-gated channels of the axon initial segment creates a current that spreads passively along the length of the axon to the first node of Ranvier (**Figure 4.28**). It depolarizes the membrane of the first node to the threshold for action potential initiation. The action potential is now at the level of the

first node. The entry of Na$^+$ ions at this level will depolarize the membrane of the second node and open the closed Na$^+$ channels. The action potential is now at the level of the second node.

The propagation is unidirectional owing to Na$^+$ channel inactivation

When the axon potential is, for example, at the level of the second node, the voltage-gated Na$^+$ channels of the first node are in the inactivated state since they have just been activated or are still in the open state (**Figure 4.28**). These Na$^+$ channels cannot be reactivated. The current lines flowing from the second node will therefore activate only the voltage-gated Na$^+$ channels of the third node towards axon terminals, where the voltage-gated Na$^+$ channels are in the closed state (**Figure 4.28**). In the axon, under physiological conditions, the action potential cannot back-propagate.

The refractory periods between two action potentials

After one action potential has been initiated, there is a period of time during which a second action potential cannot be initiated or is initiated but has a smaller amplitude (**Figure 4.29**): this period is called the 'refractory period' of the membrane. It results from Na$^+$ channel inactivation. Since the Na$^+$ channels do not immediately recover from inactivation, they cannot reopen immediately. This means that once the preceding action potential has reached its maximum amplitude, Na$^+$ channels will not reopen before a certain period of time needed for their deinactivation (**Figure 4.24b**). This represents the absolute refractory period which lasts in the order of milliseconds.

Then, progressively, the Na$^+$ channels will recover from inactivation and some will reopen in response to a second depolarization: this is the relative refractory period. This period finishes when all the Na$^+$ channel at the initial axonal segment or at a node are de-inactivated. This actually protects the membrane from being depolarized all the time and enables the initiation of separate action potentials.

4.4.5 Do the Na$^+$ and K$^+$ concentrations change in the extracellular or intracellular media during firing?

Over a short timescale, the external or internal Na$^+$ or K$^+$ concentrations do not change during the emission

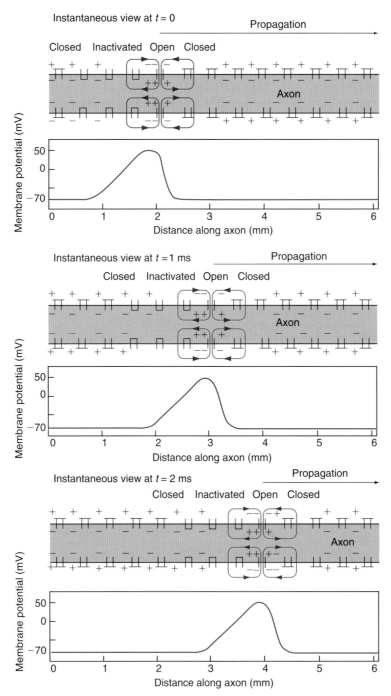

FIGURE 4.28 Active propagation of the Na⁺-dependent action potential in the axon and axon collaterals.
Scheme provided by Alberts B, Bray D, Lewis J *et al.* (1983) *Molecular Biology of the Cell*, New York: Garland Publishing.

of action potentials. A small number of ions are in fact flowing through the channels during an action potential and the Na–K pump re-establishes continuously the extracellular and intracellular Na⁺ and K⁺ concentrations at the expense of ATP hydrolysis. Over a longer timescale, during high-frequency trains of action potentials, the K⁺ concentration can significantly increase in the external medium. This is due to the very small volume of the extracellular medium surrounding neurons and the limited speed of the Na–K pump. This excess of K⁺ ions is buffered by glial cells which are highly permeable to K⁺ ions (see Section 2.1.3).

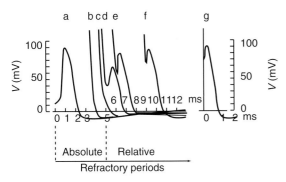

FIGURE 4.29 **The refractory periods.**
A first action potential is recorded intracellularly in the squid axon *in vitro* in response to a small depolarizing stimulus (a). Then a second stimulus with an intensity six times greater than that of the first is applied 4, 5, 6 or 9 ms after. The evoked spike is either absent (b and c; only the stimulation artifact is recorded) or has a smaller amplitude (d to f). Finally, when the membrane is back in the resting state, the evoked action potential has the control amplitude (g). Adapted from Hodgkin AL and Huxley AF (1952) A quantitative description of membrane current and its application to conduction and excitation in nerve. *J. Physiol.* **117**, 500–544, with permission.

4.4.6 Characteristics of the Na⁺-dependent action potential are explained by the properties of the voltage-gated Na⁺ channel

The *threshold* for Na$^+$-dependent action potential initiation results from the fact that voltage-gated Na$^+$ channels open in response to a depolarization positive to $-50/-40$ mV.

The Na$^+$-dependent action potential is *all or none* because voltage-gated Na$^+$ channels self-activate (see **Figure 4.17**). It propagates *without attenuation* since the density of voltage-gated Na$^+$ channels is constant along the axon or at nodes of Ranvier. It propagates *unidirectionally* because of the rapid inactivation of voltage-gated Na$^+$ channels. The instantaneous frequency of Na$^+$-dependent action potentials is limited by the *refractory periods*, which also results from voltage-gated Na$^+$ channel inactivation.

4.4.7 The role of the Na⁺-dependent action potential is to evoke neurotransmitter release

The role of the Na$^+$-dependent action potential is to propagate, without attenuation, a strong depolarization to the membrane of the axon terminals. There, this depolarization opens the high-threshold voltage-gated

Ca^{2+} channels. The resulting entry of Ca^{2+} ions into axon terminals triggers exocytosis and neurotransmitter release. The probability value of all these phenomena is not 1. This means that the action potential can fail to invade an axon terminal, the Ca^{2+} entry can fail to trigger exocytosis, etc. Neurotransmitter release is explained in Chapter 7.

FURTHER READING

Anderson PAV and Greenberg RM (2001) Phylogeny of ion channels: clues to structure and function. *Comparative Biochemistry and Physiology Part B* **129**, 17–28.

Catterall WA (2000) From ionic currents to molecular mechanisms: the structure and function of voltage-gated sodium channels. *Neuron* **26**, 13–25.

Eaholtz G, Scheuer T, Catterall WA (1994) Restoration of inactivation and block of open sodium channels by an inactivation gate peptide. *Neuron* **12**, 1041–1048.

Goldin AL, Barchi RL, Caldwell JH *et al.* (2000) Nomenclature of voltage-gated sodium channels. *Neuron* **28**, 365–368.

Hamill OP, Marty A, Neher E *et al.* (1981) Improved patch damp technique for high resolution current recording from cells and cell-free membrane patches. *Pflügers Archiv.* **391**, 85–100.

Hodgkin AL and Huxley AF (1952) A quantitative description of membrane current and its application to conduction and excitation in nerve. *J. Physiol. (Lond.)* **117**, 500–544.

Malin SA, Nerbonne JM (2002) delayed rectifier K$^+$ currents, I_K, are encoded by Kv2 α-subunits and regulate tonic firing in mammalian sympathetic neurons. *J. Neuroscience* **22**, 10094–10105.

McCormick KA, Srinivasan J, White K, Scheuer T, Caterall WA (1999) The extracellular domain of the beta1 subunit is both necessary and sufficient for beta1-like modulation of sodium channel gating. *J. Biol. Chem.* **274**, 32638–32646.

McKinnon R (2003) Potassium channels. *FEBBS Letters* **555**, 62–65.

Neher E, Sakmann B (1976) Single channel currents recorded from membrane of denervated frog muscle fibres. *Nature* **260**, 779–802.

Noda M, Ikeda T, Suzuki H *et al.* (1986) Expression of functional sodium channels from cloned cDNA. *Nature* **322**, 826–828.

Qu Y, Rogers JC, Chen SF, McCormick KA, Scheuer T, Catterall WA (1999) Functional roles of the extracellular segments of the sodium channel alpha subunit in voltage-dependent gating and modulation by beta1 subunits. *J. Biol. Chem.* **274**, 32647–32654.

Sokolov S, Scheuer T, Catterall WA (2005) Ion permeation through a voltage-sensitive gating pore in brain sodium channels having voltage sensor mutations. *Neuron* **47**, 183–189.

Stuart G and Häuser M (1994) Initiation and spread of sodium action potentials in cerebellar Purkinje cells. *Neuron* **13**, 703–712.

Vassilev PM, Scheuer T, Catterall WA (1988) Identification of an intracellular peptide segment involved in sodium channel inactivation. *Science* **241**, 1658–1661.

Venkatesh B, Lu SQ, Dandona N *et al.* (2005) Genetic basis of tetrodotoxin resistance in pufferfishes. *Current Biology* **15**, 2069–2072.

Wang SY and Wang GK (2003) Voltage-gated sodium channels as primary targets of diverse lipid soluble neurotoxins. *Cellular Signaling* **15**, 151–159.

APPENDIX 4.1 CURRENT CLAMP RECORDING

The current clamp technique, or intracellular recording in current clamp mode, is the traditional method for recording membrane potential: resting membrane potential and membrane potential changes such as action potentials and postsynaptic potentials. Membrane potential changes result from intrinsic or extrinsic currents. Intrinsic currents are synaptic or autorhythmic currents. Extrinsic currents are currents of known amplitude and duration applied by the experimenter through the intracellular recording electrode, in order to mimic currents produced by synaptic inputs.

Current clamp means that the *current applied* through the intracellular electrode is clamped to a constant value by the experimenter. It does not mean that the *current flowing through the membrane* is clamped to a constant value.

How to record membrane potential

The intracellular electrode (or the patch pipette) is connected to a unity-gain amplifier that has an input resistance many orders of magnitude greater than that of the micropipette plus the input resistance of the cell membrane ($R_p + R_m$). The output of the amplifier follows the voltage at the tip of the intracellular electrode (V_p) (**Figure A4.1**). By definition, membrane potential V_m is equal to $V_i - V_e$ (i for intracellular and e for extracellular). In **Figure A4.1**, $V_i - V_e = V_p - V_{bath} = V_p - V_{ground} = V_p - 0 = V_p$. When a current I is simultaneously passed through the electrode, $V_p = V_m$ as

long as the current I is very small in order not to cause a significant voltage drop across R_p (see the last section of this appendix).

How to inject current through the intracellular electrode

In a current injection circuit is connected to the input node, the current injected (I) flows down the electrode into the cell (**Figure A4.1**). This current source allows a constant (DC) current to be injected, either outward to depolarize the membrane or inward to hyperpolarize the membrane (**Figure A4.2**). When the recording electrode is filled with KCl, a current that expells K$^+$ ions into the cell interior depolarizes the membrane (V_m becomes less negative) (**Figure A4.2a**), whereas a current that expels Cl$^-$ ions into the cell interior hyperpolarizes the membrane (V_m becomes more negative) (**Figure A4.2b**).

Outward means that the current is flowing through the membrane from the inside of the cell to the bath; inward is the opposite

The current source can also be used to inject a short-duration pulse of current: a depolarizing current pulse above threshold to evoke action potential(s) or a low-amplitude depolarizing (**Figure A4.3**) or hyperpolarizing

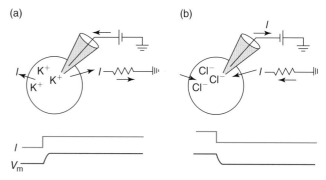

(a)   (b)

FIGURE A4.2 **(a)** When the recording electrode is filled with KCl, a current expels K$^+$ ions into the cell interior to depolarize the membrane (V_m becomes less negative). **(b)** A current expels Cl$^-$ ions into the cell interior to hyperpolarize the membrane (V_m becomes more negative).

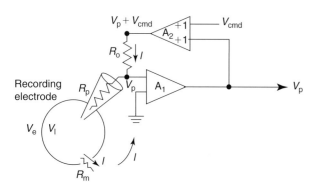

FIGURE A4.1 **A unity gain amplifier A1 and a current source made by adding a second amplifier A2.**
The micropipette voltage V_p is measured by A1. The command voltage V_{cmd} and V_p are the inputs of A2 (V_p and V_{cmd} are added). The current I applied by the experimenter in order to induce V_m changes, flows through R_o and is equal to $I = V_p/R_o$ since the voltage across the output resistor R_o is equal to V_{cmd} regardless of V_p. I flows through the micropipette into the cell then out through the cell membrane into the bath grounding electrode. I is here an outward current. Capacitances are ignored. Adapted from *The Axon Guide*, Axon Instruments Inc., 1993.

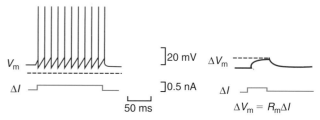

FIGURE A4.3 **Injection of a suprathreshold** (left) **and sub-threshold** (right) **depolarizing pulse.**

current pulse to measure the input membrane resistance R_m since $\Delta V_m = R_m \times \Delta I$.

How to measure the membrane potential when a current is passed down the electrode

The injected current (I) causes a corresponding voltage drop (IR_p) across the resistance of the pipette (R_p). It is therefore difficult to separate the potential at the tip of the electrode ($V_p = V_m$) from the total potential ($V_p + IR_p$). For example, if $R_p = 50\,M\Omega$ and $I = 0.5\,nA$, $IR_p = 25\,mV$, a value in the V_m range. A special compensation circuitry can be used to eliminate the micropipette voltage drop IR_p.

APPENDIX 4.2 VOLTAGE CLAMP RECORDING

The voltage clamp technique (or intracellular recording in voltage clamp mode) is a method for recording the current flowing through the cell membrane while the membrane potential is held (clamped) at a constant value by the experimenter. In contrast to the current clamp technique (see Appendix 4.1), voltage clamp does not mimic a process found in nature. However, there are several reasons for performing voltage clamp experiments:

- When studying voltage-gated channels, voltage clamp allows control of a variable (voltage) that determines the opening and closing of these channels.
- By holding the membrane potential constant, the experimenter ensures that the current flowing through the membrane is linearly proportional to the conductance G ($G = 1/R$) being studied. To study, for example, the conductance G_{Na} of the total number (N) of voltage-gated Na^+ channels present in the membrane, K^+ and Ca^{2+} voltage-gated channels are blocked by pharmacological agents, and the current I_{Na} flowing through the membrane, recorded in voltage clamp, is proportional to G_{Na}:

$$I_{Na} = V_m G_{Na} = k G_{Na}, \text{ since } V_m \text{ is constant.}$$

How to clamp the membrane potential at a known and constant value

The aim of the voltage clamp technique is to adjust continuously the membrane potential V_m to the command potential V_{cmd} fixed by the experimenter. To do so, V_m is continuously measured *and* a current I is passed through the cell membrane to keep V_m at the desired

value or command potential (V_{cmd}). Two voltage clamp techniques are commonly used. With the two-electrode voltage clamp method, one electrode is used for membrane potential measurement and the other for passing current (**Figure A4.4**). The other method uses just one electrode, in one of the following ways:

- The same electrode is used part time for membrane potential measurement and part time for current injection (also called the discontinuous single-electrode voltage clamp technique, or dSEVC). This is used for cells that are too small to be impaled with two electrodes; it will not be explained here.
- In the patch clamp technique the same electrode is used full time for simultaneously measuring membrane potential and passing current (see Appendix 4.3).

In the two-electrode voltage clamp technique, the membrane potential is recorded by a unity gain amplifier A1 connected to the voltage-recording electrode E1. The membrane potential measured, V_m (or V_p; see Appendix 4.1) is compared with the command potential V_{cmd} in a high-gain differential amplifier A2. It sends a voltage output V_o proportional to the difference between V_m and V_{cmd}. V_o forces a current I to flow through the current-passing electrode E2 in order to obtain $V_m - V_{cmd} = 0$. The current I represents the total current that flows through the membrane. It is the same at every point of the circuit.

Example of a voltage-clamp recording experiment

Two electrodes are placed intracellularly into a neuronal soma (an invertebrate neuron for example) (**Figure A4.5**). The membrane potential is first held at $-80\,mV$. In this condition an outward current flows

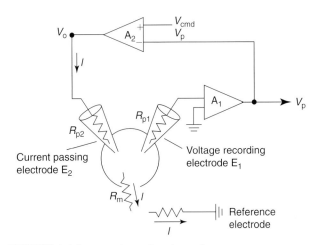

FIGURE A4.4 Two-electrode voltage clamp.
Adapted from *The Axon Guide*, Axon Instruments Inc., 1993.

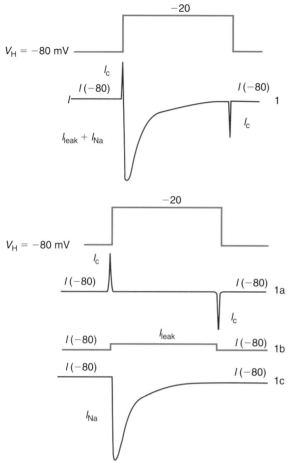

FIGURE A4.5   **Various currents.**
(1 = 1a + 1b + 1c) evoked by a voltage step to $-20\,\text{mV}$ ($V_\text{H} = -80\,\text{mV}$) in the presence of K^+ and Ca^{2+} channel blockers.

through the membrane in order to maintain the membrane potential at a value more hyperpolarized than V_rest. This stable outward current I_{-80} flows through the membrane as long as $V_\text{cmd} = -80\,\text{mV}$.

A voltage step to $-20\,\text{mV}$ is then applied for 100 ms. This depolarizing step opens voltage-gated channels. In the presence of K^+ and Ca^{2+} channel blockers, only a voltage-gated Na^+ current is recorded. To clamp the membrane at the new $V_\text{cmd} = -20\,\text{mV}$, a current $I_{(-20)}$ is sent by the amplifier A2. On the rising phase of the step this current is equal to the capacitive current I_c necessary to charge the membrane capacitance to its new value plus the leak current I_L flowing through leak channels (lines 1a and 1b). Since the depolarizing step opens Na^+ voltage-gated channels, an inward current I_{Na} flowing through open Na^+ channels will appear after a small delay (line 1c). Normally, this inward current flowing through the open Na^+ channels, I_{Na}, should depolarize the membrane but in voltage clamp experiments it does not: a current constantly equal to

I_{Na} but of opposite direction is continuously sent (in the microsecond range) in the circuit to compensate I_{Na} and to clamp the membrane to V_cmd. Therefore, once the membrane capacitance is charged, $I_{(-20)} = I_\text{L} + I_{Na}$. Usually on recordings, I_C is absent owing to the possibility of compensating for it with the voltage clamp amplifier.

Once the membrane capacitance is charged, the total current flowing through the circuit is $I = I_\text{L} + I_{Na}$ ($I_\text{c} = 0$). Therefore, in all measures of I_{Na}, the leak current I_L must be deduced. To do so, small-amplitude hyperpolarizing or depolarizing steps ($\Delta V_\text{m} = \pm 5$ to $\pm 20\,\text{mV}$) are applied at the beginning and at the end of the experiment. These voltage steps are too small to open voltage-gated channels in order to have $I_{Na} = 0$ and $I = I_\text{L}$. If we suppose that I_L is linearly proportional to ΔV_m, then I_L for a ΔV_m of $+80\,\text{mV}$ (from -80 to $0\,\text{mV}$) is eight times the value of I_L for $\Delta V_\text{m} = +10\,\text{mV}$ (see **Figure 3.9d**).

Is all the membrane surface clamped?

In small and round cells such as pituitary cells, the membrane potential is clamped on all the surface. In contrast, in neurons, because of their geometry, the voltage clamp is not achieved on all the membrane surface: the distal dendritic and axonal membranes are out of control because of their distance from the soma where the intracellular electrodes are usually placed. Such space clamp problems have to be taken into account by the experimenter in the analysis of the results. In the giant axon of the squid, this problem is overcome by inserting two long axial intracellular electrodes into a segment of axon in order to control the membrane potential all along this segment.

APPENDIX 4.3   PATCH CLAMP RECORDING

The patch clamp technique is a variation of the voltage clamp technique. It allows the recording of current flowing through the membrane: either the current flowing through all the channels open in the whole cell membrane or the current flowing through a single channel in a patch of membrane. In this technique, only one electrode is used full time for both voltage recording and passing current (it is a continuous single-electrode voltage clamp technique, or cSEVC). The patch clamp technique was developed by Neher and Sakmann. By applying very low doses of acetylcholine to a patch of muscle membrane they recorded for the first time, in 1976, the current flowing through a single

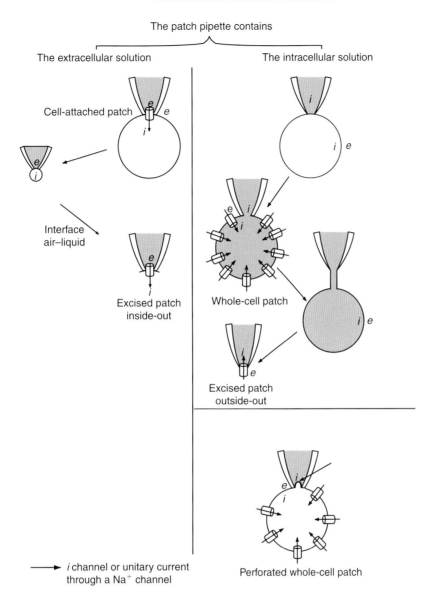

The patch pipette contains

The extracellular solution   The intracellular solution

Cell-attached patch

Interface
air–liquid

Excised patch
inside-out

Whole-cell patch

Excised patch
outside-out

⟶ *i* channel or unitary current
through a Na⁺ channel

Perforated whole-cell patch

FIGURE A4.6   Configurations of patch clamp recording.

nicotinic cholinergic receptor channel (nAChR), the unitary nicotinic current.

Some of the advantages of the patch clamp technique are that (i) with all but one configuration (cell-attached configuration) the investigator has access to the intracellular environment (**Figure A4.6**); (ii) it allows the recording of currents from cells too small to be impaled with intracellular microelectrodes; and (iii) it allows the recording of unitary currents (current through a single channel).

A4.3.1   The various patch clamp recording configurations

First a tight seal between the membrane and the tip of the pipette must be obtained. The tip of a micropipette

that has been fire polished to a diameter of about 1 μm is advanced towards a cell until it makes contact with its membrane. Under appropriate conditions, a gentle suction applied to the inside of the pipette causes the formation of a very tight seal between the membrane and the tip of the pipette. This is the cell-attached configuration (**Figure A4.6**). The resistance between the interior of the pipette and the external solution can be very large, of the order of $10\,G\Omega$ $(10^9\,\Omega)$ or more. It means that the interior of the pipette is isolated from the extracellular solution by the seal that is formed.

This very large resistance is necessary for two reasons (**Figure A4.7**):

• It allows the electrical isolation of the membrane patch under the tip of the pipette since practically

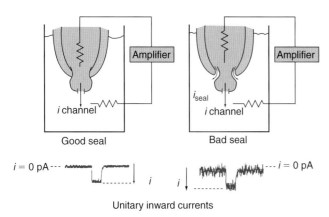

FIGURE A4.7   **Good and bad seals.**
From *The Axon Guide*, Axon Instruments Inc., 1993.

no current can flow through the seal. This is important because if a fraction of the current passing through the membrane patch leaks out through the seal, it is not measured by the electrode.

- It augments the signal-to-noise ratio since thermal movement of the charges through a bad seal is a source of additional noise in the recording. A good seal thus enables the measurement of the current flowing through one single channel (unitary current) which is of the order of picoampères.

From the 'cell-attached' configuration (the last to be explained), one can obtain other recording configurations. In total, three of them are used to record unitary currents, and one (whole-cell) to record the current flowing through all the open channels of the whole cell membrane.

Whole-cell configuration

This configuration is obtained from the cell-attached configuration. If a little suction is applied to the interior of the pipette, it may cause the rupture of the membrane patch under the pipette. Consequently, the patch pipette now records the activity of the whole cell membrane (minus the small ruptured patch of membrane). Rapidly, the intracellular solution equilibrates with that of the pipette, the volume of the latter being many times larger. This is especially true for inorganic ions.

This configuration enables the recording of the current flowing through the N channels open over the entire surface of the cell membrane. Under conditions where all the open channels are of the same type (with the opening of other channels being blocked by pharmacological agents or the voltage conditions), the total current flowing through a population of identical channels can be recorded, such that at steady state:

$$I = Np_o i,$$

where N is the number of identical channels, p_o the probability that these channels are in the open state, Np_o the number of identical channels in the open state, and i the unitary current.

The advantages of this technique over the two-electrode voltage clamp technique are: (i) the recording under voltage clamp from cell bodies too small to be impaled with two electrodes and even one; and (ii) there is a certain control over the composition of the internal environment and a better signal-to-noise ratio. The limitation of this technique is the gradual loss of intracellular components (such as second messengers), which will cause the eventual disappearance of the responses dependent on those components.

Perforated whole-cell configuration

This is a variation of the whole-cell configuration, and also allows the recording of current flowing through the N channels open in the whole membrane but avoids washout of the intracellular solution. This configuration is obtained by introducing into the recording pipette a molecule such as nystatin, amphotericin or gramicidin, which will form channels in the patch of membrane under the tip of the electrode. To record in this configuration, first the cell-attached configuration is obtained and then the experimenter waits for the nystatin channels (or amphotericin or gramicidin channels) to form without applying any suction to the electrode. The channels formed by these molecules are mainly permeable to monovalent ions and thus allow electrical access to the cell's interior. Since these channels are not permeant to molecules as large or larger than glucose, whole-cell recording can be performed without removing the intracellular environment. This is particularly useful when the modulation of ionic channels by second messengers is studied.

In order to evaluate this problem of 'washout', we can calculate the ratio between the cell body volume and the volume of solution at the very end of a pipette. For example, for a cell of $20\,\mu m$ diameter the volume is: $(4/3)\pi(10 \times 10^{-6})^3 = 4 \times 10^{-15}$ litres. If we consider $1\,mm$ of the tip of the pipette, it contains a volume of the solution approximately equal to $10^{-13}l$, which is 100 times larger than the volume of the cell body.

Excised patch configurations

If one wants to record the unitary current i flowing through a single channel and to control simultaneously the composition of the intracellular environment, the so-called excised or cell-free patch configurations have to be used. The *outside-out configuration* is obtained from the whole-cell configuration by gently pulling the pipette away from the cell. This causes the membrane patch to be torn away from the rest of the cell at the same time that its free ends reseal together. In this case the intracellular environment is that of the pipette, and the extracellular environment is that of the bath. This configuration is used when rapid changes of the extracellular solution are required to test the effects of different ions or pharmacological agents when applied to the extracellular side of the membrane.

The *inside-out configuration* is obtained from the cell-attached configuration by gently pulling the pipette away from the cell, lifting the tip of the pipette from the bath in the air and putting it back into the solution (interface of air–liquid). In this case, the intracellular environment is that of the bath and the extracellular one is that of the pipette (the pipette is filled with a pseudo-extracellular solution). This configuration is used when rapid changes in the composition of the intracellular environment are necessary to test, for example, the effects of different ions, second messengers and pharmacological agents in that environment.

Cell-attached configuration

The intracellular environment is that of the cell itself, and the extracellular environment of the recorded membrane patch is the pipette solution. This configuration enables the recording of current flowing through the channel or channels present in the patch of membrane that is under the pipette and is electrically isolated from the rest of the cell. If one channel opens at a time, then the unitary current i flowing through that channel can be recorded. The recordings in cell-attached mode present two limitations: (i) the composition of the intracellular environment is not controlled; and (ii) the value of the membrane potential is not known and can only be estimated.

Let us assume that the voltage in the interior of the patch pipette is maintained at a known value V_p (p = pipette). Since the voltage across the membrane patch is $V_m = V_i - V_e = V_i - V_p$, it will not be known unless V_i, the voltage at the internal side of the membrane, is also known. V_i cannot be measured directly. One way to estimate this value is to measure the resting potential of several identical cells under similar conditions (with intracellular or whole-cell recordings), and to calculate an average V_i from the individual values. Sometimes, however, V_i can be measured when the cell is large enough to allow a two-electrode voltage clamp recording to be made simultaneously with the patch clamp recording (with a *Xenopus* oocyte, for example). Another method consists of replacing the extracellular medium with isotonic K^+ (120–150 mM). The membrane potential under these conditions will be close to 0 mV.

To leave the intracellular composition intact while recording the activity of a single channel is particularly useful for studies of the modulation of an ionic channel by second messengers.

A4.3.2 Principles of the patch clamp recording technique

In the patch clamp technique, as in all voltage clamp techniques, the membrane potential is held constant (i.e. clamped) while the current flowing through a single open channel or many open channels (Np_o) is measured (**Figure A4.8**). In the patch clamp technique only one micropipette is used full time for both voltage clamping and current recording. How at the same time via the same pipette can the voltage of the membrane be controlled and the current flowing through the membrane be measured?

When an operational amplifier A1 is connected as shown in **Figure A4.8a** with a high megohm resistor R_f (f = feedback), a current-to-voltage converter is obtained. The patch pipette is connected to the negative input and the command voltage (V_{cmd}) to the positive one. The resistor R_f can have two values: $R_f = 1$ GΩ in the whole-cell configuration and 10 GΩ in the excised patch configurations.

How the membrane is clamped at a voltage equal to V_{cmd}

R_p represents the electrode resistance and R_m the membrane input resistance (**Figure A4.8a**). Suppose that the membrane potential is first clamped to −80 mV ($V_{cmd} = -80$ mV), then a voltage step to −20 mV is applied for 100 ms ($V_{cmd} = -20$ mV for 100 ms). The membrane potential (V_m) has to be clamped quickly to −20 mV ($V_m = V_{cmd} = -20$ mV) whatever happens to the channels in the membrane (they open or close). The operational amplifier A1 is able to minimize the voltage difference between two inputs to a very small value (0.1 µV or so). A1 compares the value of V_{cmd} (entry +) to that of V_m (entry −). It then sends a voltage output (V_o) in order to obtain $V_m = V_{cmd} = -20$ mV (**Figure A4.8b**).

What is this value of V_o? Suppose that the at time t of its peak the Na^+ current evoked by the voltage step to −20 mV is $I_{Na} = 1$ nA. V_o will force a current $I = -1$ nA to flow through $R_f = 10^9$ Ω in order to clamp the membrane potential: $V_o = R_f I = 10^9 \times 10^{-9} = 1$ V. It is said that $V_o = 1$ V/nA or 1 mV/pA.

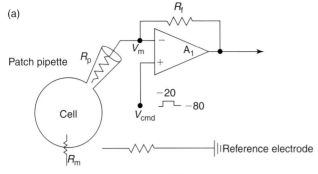

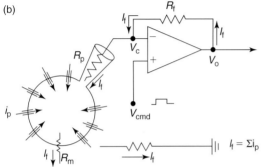

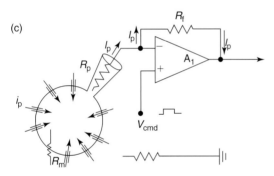

FIGURE A4.8 Example of a patch clamp recording in the whole-cell configuration.
(a) The amplifier compares V_m to the new $V_{cmd} = -20$ mV. (b) The amplifier sends V_o so that $V_m = V_{cmd} = -20$ mV. Owing to the depolarization to -20 mV, the Na⁺ channels open and unitary inward currents i_p flow through the N open channels ($Ni_p = I_p$). (c) The whole-cell current I_p flows through the circuit and is measured as a voltage change.

The limits of V_o in patch clamp amplifiers are $+15$ V and -15 V. This means that V_o cannot be bigger than these values, which is largely compatible with biological experiments where currents through the membrane do not exceed 15 nA.

The amplifier A1 compares V_m with V_{cmd} and sends V_o at a very high speed. This speed has to be very high in order to correct V_m according to V_{cmd} very quickly. The ideal clamp is obtained at the output of the circuit via R_f (black dot V_c on the scheme of **Figure A4.8b**). As in the voltage clamp technique, a capacitive current is present at the beginning and at the end of the voltage step on the current trace and a leak current during the step, but they are not re-explained here.

A4.3.3 The unitary current *i* is a rectangular step of current (see Figures 4.8a and c)

We record, for example, in the outside-out patch clamp configuration to activity of a single voltage sensitive Na⁺ channel. When a positive membrane potential step is applied to depolarize the patch of membrane from -90 mV to -40 mV, an inward current i_{Na} flowing through the open Na⁺ channel is recorded (inward current means a current that flows across the membrane from the outside to inside). By convention, inward currents are represented as downward deflections and outward currents as upward deflections.

The membrane depolarization causes activation of the voltage-dependent Na⁺ channel, and induces its transition from the closed (C) state (or conformation) to the open (O) state, a transition symbolized by:

$$C \rightleftharpoons O,$$

where C is the closed state of the channel (at -90 mV) and O is the open state of the channel (at -40 mV).

While the channel is in the O conformation (at -40 mV), Na⁺ ions flow through the channel and an inward current caused by the net influx of Na⁺ ions is recorded. This current reaches its maximum value very rapidly. Thus, the maximal net ion flux is established almost instantaneously given the timescale of the recording (of the order of microseconds). The development of the inward current thus appears as a vertical downward deflection.

A delay between the onset of the voltage step and the onset of the current *i* is observed. This delay has a duration that varies from one depolarizing test pulse to another and also according to the channel under study. This delay is due to the conformational change or changes of the protein. In fact, such changes previous to opening can be multiple:

$$C_1 \rightleftharpoons C_2 \rightleftharpoons C_3 \rightleftharpoons O$$

Notice that the opening delay does not correspond to the intrinsic duration of the process of conformational change, which is extremely short. It corresponds to the statistical nature of the equilibrium between the 2, 3, N closed and open conformations. The opening delay therefore depends on the time spent in each of the different closed states (C_1, C_2, C_3).

The return of the current value to zero corresponds to the closing of the channel. This closure is the result of the transition of the channel protein from the open state (O) to a state in which the channel no longer conducts (state in which the aqueous pore is closed). It can be either a closed state (C), an inactivated state (I) or a desensitized state (D). In the case of the Na⁺ channel, the return of the current value to zero is due mainly to the transition of the protein from the open state to the inactivated state ($O \rightarrow I$). Before closing for a long time, the channel can also flicker between the open and closed state ($C \rightleftharpoons O$):

Just as the current reaches its maximum value instantaneously during opening, it also returns instantaneously to its zero value during closing of the pore. Because of this, the unitary current i has a step-like rectangular shape.

A4.3.4 Determination of the conductance of a channel

If we repeat several times the experiment shown in **Figure 4.8a**, we observe that for a given voltage step ΔV, i varies around an average value. The current fluctuations are measured at regular intervals before, during and immediately after the depolarizing voltage pulse. The distribution of the different i values during the voltage pulse describes a Gaussian curve in which the peak corresponds to the average i value (**Figure 4.8d**). There is also a peak around $0\,pA$ (not shown on the figure) which corresponds to the different values of i when the channel is closed. Since the channel is in the closed state most of the time, where i has values around $0\,pA$, this peak is higher than the one corresponding to $i_{channel}$ (around $-2\,pA$). The width of the peak around $0\,pA$ gives the mean value of the fluctuations resulting from noise. Therefore, the two main reasons for these fluctuations of $i_{channel}$ are: the variations in the noise of the recording system and the changes in the number of ions that cross the channel during a unit of time Δt.

Knowing the average value of i and the reversal potential value of the current (E_{rev}), the average conductance value of the channel under study, γ, can be calculated: $\gamma = i/(V_m - E_{rev})$.

However, there are cases in which the distribution of i for a give membrane potential shows several peaks. Different possibilities should be considered:

- Only one channel is being recorded from but it presents several open conformational states, each one with different conductances. The peaks correspond to the current flowing through these different substrates.
- Two or more channels of the *same* type are present in the patch and their activity recorded. The peaks represent the multiples of i ($2i$, $3i$, etc.).

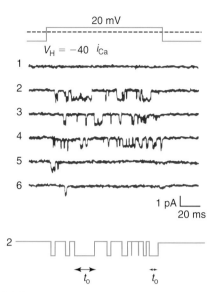

FIGURE A4.9 Example of the patch clamp recording of a single voltage-dependent Ca$^+$ channel.
In response to a voltage step to $+20\,mV$ from a holding potential of $-40\,mV$, the channel opens and closes several times during each of the six trials. Adapted from Fox AP, Nowyky MC, Tsien RW (1987) Single-channel recordings of three types of calcium channels in chick sensory neurones. *J. Physiol. (Lond.)* **394**, 173–200, with permission.

- Two or more channels of *different* types are present in the patch and their activity is simultaneously recorded. The peaks correspond to the current through different channel types.

A4.3.5 Mean open time of a channel

An ionic channel fluctuates between a closed state (C) and an open state (O):

$$C \underset{\alpha}{\overset{\beta}{\rightleftharpoons}} O$$

where α is the closing rate constant or, more exactly, the number of channel closures per unit of time spent in the open state O. β is the opening rate constant or the number of openings per unit of time spent in the closed state R (α and β are expressed in s^{-1}).

Once activated, the channel remains in the O state for a time t_o, called open time. When the channel opens, the unitary current i is recorded for a certain time t_o. t_o for a given channel studied under identical conditions varies from one recording to another (**Figure A4.9**). t_o is an aleatory variable of an observed duration. When the number of times a value of t_o (in the order of milli- or microseconds) is plotted against the values of t_o, one obtains the open time histogram; i.e. the distribution of the different values of t_o (**Figure A4.10**). This distribution declines and the shorter open times are more frequent than the longer ones.

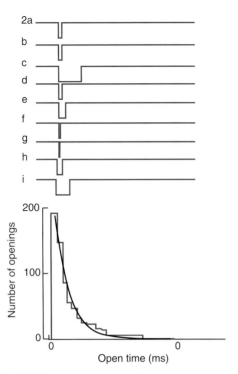

FIGURE A4.10 Determination of the mean open time of a channel.
Trial 2 of Figure A4.9 is selected and all the openings are aligned at time 0. $\tau_o = 1.2$ ms.

Why does the distribution of t_o decrease?

The histogram is constructed as follows. At time $t = 0$, all the channels are open (the delay of opening is ignored, all the openings are aligned at time 0; **Figure A4.10**). As time t increases, the number of channels that remain open can only decrease since channels progressively close. This can also be expressed as follows: the longer the observation time, the lower the probability that the channel is still in the open state. Or, alternatively, the longer the observation time, the closer the probability will be to 1 that the channel will shut (1 is the maximum value used to express a probability). It is not a Gaussian curve because the delay of opening is ignored and all the openings begin at $t = 0$.

Why is the decrementing distribution of t_o exponential?

A channel open at $t = 0$ has a probability of closing at $t + \Delta t$. It has the same probability of closing if it is still open at the beginning of any subsequent observation interval Δt. This type of probability is described mathematically as an exponential function of the observation time. Thus, when the openings of a homogeneous population of channels are studied, the decrease in the number of events is described by a single exponential.

Experimental determination of τ_0, the mean open time of a channel

The mean open time τ_o is the time during which a channel has the highest probability of being in the open state: it corresponds to the sum of all the values that t_o may take, weighted by their corresponding probability values. This value is easy to calculate if the distribution is described by a single exponential. In order to verify that the histogram is actually described by a single exponential, one has to first build the histogram by plotting the number of times a value of t_o is observed as a function of t_o; i.e. number of events = $f(t_o)$.

The exponential that describes the histogram has the form $y = y_o e^{-t/\tau 0}$, where y is the number of events observed at each time t. This curve will be linear on semi-logarithmic coordinates if it is described by a single exponential. The slope can be measured with a regression analysis. It corresponds to the mean open time τ_0 of the channel. τ_0 is the value of t_o for a number of events equal to $1/e$. It is the 'expected value' of t_o. The expected value of t_o is the sum of all the values of t_o weighted to their corresponding probabilities.

In the case of the conformational changes $C \rightleftharpoons O$, the value of τ_0 provides an estimate of the closure rate constant α, because at steady state $\tau_0 = 1/\alpha$. For example, from the open time histogram of the nicotinic receptor channel, we can determine its mean open time τ_0. Knowing that in conditions where the desensitization of the channel is negligible $\tau_0 = 1/\alpha$, we can calculated from τ_0 the closing rate constant of the channel. If $\tau_0 = 1.1$ ms, $\alpha = 900\,\text{s}^{-1}$. The channel closes 900 times for each second spent in the open state. In other words there is an average of 900 transitions of the channel to the closed state for each second spend in the open state.

5

The voltage-gated channels of Ca^{2+} action potentials: Generalization

Chapter 4 explained the Na^+-dependent action potential propagated by axons. There are two other types of action potentials: (i) the Na^+/Ca^{2+}-dependent action potential present in axon terminals or heart muscle cells (**Figure 4.2d**), for example, where it is responsible for Ca^{2+} entry and an increase of intracellular Ca^{2+} concentration, a necessary prerequisite for neurotransmitter release (secretion) or muscle fibre contraction; and (ii) the Ca^{2+}-dependent action potential (in which Na^+ ions do not participate) in dendrites of Purkinje cells of the cerebellum (see **Figure 17.9**) and in endocrine cells (**Figure 5.1a**). In Purkinje cell dendrites, it depolarizes the membrane and thus modulates neuronal integration; in endocrine cells it provides a Ca^{2+} entry to trigger hormone secretion.

5.1 PROPERTIES OF Ca^{2+}-DEPENDENT ACTION POTENTIALS

In some neuronal cell bodies, in heart ventricular muscle cells and in axon terminals, the action potentials have a longer duration than Na^+ spikes, with a plateau following the initial peak: these are the Na^+/Ca^{2+}-dependent action potentials (see **Figure 4.2b–d**). In some neuronal dendrites and some endocrine cells, action potentials have a small amplitude and a long duration: these are the Ca^{2+}-dependent action potentials (**Figure 5.1**). All action potentials are initiated in response to a membrane depolarization. Na^+, Na^+/Ca^{2+} and Ca^{2+}-dependent action potentials differ in the type of voltage-gated channels responsible for their depolarization and repolarization phases. We

will examine the properties of a Ca^{2+}-dependent action potential.

5.1.1 Ca^{2+} and K^+ ions participate in the action potential of endocrine cells

The activity of pituitary endocrine cells that release growth hormone is recorded in the perforated whole-cell configuration (current clamp mode; see Appendix 4.1). They display a spontaneous activity. When these cells are previously loaded with the Ca^{2+}-sensitive dye Fura-2, changes of intracellular Ca^{2+} concentration can be also quantified (see **Appendix 5.1**). Simultaneous recording of potential and $[Ca^{2+}]_i$ changes shows that for each action potential there is a corresponding $[Ca^{2+}]_i$ increase (**Figure 5.1a**). This strongly suggests that Ca^{2+} ions are entering the cell during action potentials.

Ca^{2+} ions participate in the depolarization phase of the action potential

When the extracellular solution is changed from control Krebs to a Ca^{2+}-free solution, or when nifedipin, an L-type Ca^{2+} channel blocker, is added to the external medium (**Figure 5.1b**), the amplitude and risetime of the depolarization phase of the action potential gradually and rapidly decreases until action potentials are no longer evoked.

K^+ ions participate in the repolarization phase of the action potential

Application of charybdotoxin (CTX) or apamin, blockers of Ca^{2+}-activated K^+ channels, increases the

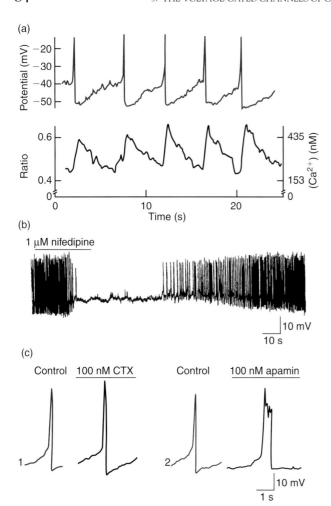

FIGURE 5.1 The Ca²⁺-dependent action potential of an endocrine cell.
Growth-hormone secreting cells of the anterior pituitary in culture are loaded with the Ca²⁺-sensitive dye Fura-2 and their activity is recorded in perforated whole-cell patch configuration (current clamp mode). **(a)** Simultaneous recordings of action potentials (top trace) and cytosolic [Ca²⁺] oscillations (bottom trace) in control conditions. **(b)** Nifedipin, an L-type Ca²⁺ channel blocker, is applied for 20 s. **(c)** Action potential in the absence and presence of blockers of Ca²⁺-activated K⁺ channels, charybdotoxin (CTX, 1) and apamin (2). Adapted from Kwiecien R, Robert C, Cannon R *et al.* (1998) Endogenous pacemaker activity of rat tumour somatotrophs. *J. Physiol.* **508**, 883–905, with permission.

peak amplitude and prolongs the duration of action potentials (**Figure 5.1c**). Note that apamin also blocks the after-spike hyperpolarization (**Figure 5.1c**, right).

5.1.2 Questions about the Ca²⁺-dependent action potential

- What are the structural and functional properties of the Ca²⁺ and K⁺ channels involved? (Sections 5.2 and 5.3)?

- What represents the threshold potential for Ca²⁺-dependent action potential initiation? Where are Ca²⁺-dependent action potentials initiated? (Section 5.4)

5.2 THE TRANSIENT ENTRY OF Ca²⁺ IONS THROUGH VOLTAGE-GATED Ca²⁺ CHANNELS IS RESPONSIBLE FOR THE DEPOLARIZING PHASE OR THE PLATEAU PHASE OF Ca²⁺-DEPENDENT ACTION POTENTIALS

The voltage-gated Ca²⁺ channels involved in these action potentials are high threshold-activated (HVA) Ca²⁺ channels. There are three main types of such channels: the L-type (L for long lasting), the N-type (N for neuronal or for neither L nor T) and the P-type (P for Purkinje cells where they have been first described).

5.2.1 The voltage-gated Ca²⁺ channels are a diverse group of multisubunit proteins

They are composed of a pore-forming α₁-subunit of about 2000 amino acid residues (190–250 kDa), with an amino acid sequence and a predicted transmembrane structure like the previously characterized pore forming α-subunit of Na⁺ channels: four repeated domains (I to IV), each of which contains six transmembrane segments (1 to 6) and a membrane-associated loop between transmembrane segments S5 and S6 of each domain (**Figure 5.2a**). It incorporates the conduction pore, the voltage sensor and gating apparatus and the known sites of channel regulation by second messengers, drugs and toxins. Auxiliary subunits can include a transmembrane disulphide-linked complex of α₂ and δ-subunits, a β-subunit and in some cases a transmembrane γ-subunit. The β-subunit has predicted α-helices but no transmembrane segments and is thought to be intracellular. They play a role in the expression and gating properties of the Ca²⁺ channels by modulating various properties of the α₁-subunit.

The pharmacological and electrophysiological diversity of Ca²⁺ channels primarily arises from the diversity of α₁-subunits. The primary structure of the different α₁-subunits has been defined by homology screening and their function characterized by expression in mammalian cells or *Xenopus* oocytes. The recent nomenclature divides the Ca²⁺ channels in three structurally and functionally related families (Ca_v1, Ca_v2, Ca_v3) to indicate the principal permeating ion (Ca) and the principal physiological regulator (v for voltage), followed by a number that indicates the gene subfamily

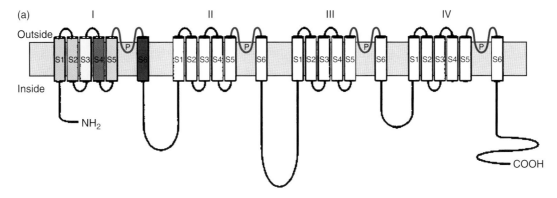

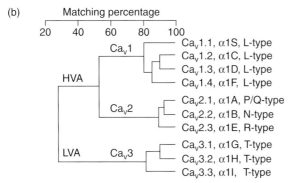

FIGURE 5.2 Subunits of voltage-gated Ca²⁺ channels.
(a) Membrane topology for the α1-subunit of a cardiac L-type Ca²⁺ channel (P: the P loops). **(b)** Evolutionary tree of voltage-gated α₁-subunit of Ca²⁺ channels. Low voltage-activated Ca²⁺ channels (LVA) appear to have diverged from an ancestral Ca²⁺ channel before the bifurcation of the high voltage-activated (HVA) channels in Ca_v1 and Ca_v2 subfamilies. (a) Adapted from Sather WA (2003) Permeation and selectivity in calcium channels. *Annu. Rev. Physiol.* **65**, 133–159. (b) Adapted from Perez-Reyes E, Cribbs LL, Daud A *et al.* (1998) Molecular characterization of a neuronal low voltage-activated T-type calcium channel. *Nature* **391**, 896–900.

(1, 2 or 3). The number following the decimal point identifies the specific channel isoform (e.g. Ca_v1.1) (**Figure 5.2b**). High-threshold Ca²⁺ channels comprise L (Ca_v1), P/Q (Ca_v2.1), N (Ca_v2.2) and R (Ca_v2.3)-type Ca²⁺ channels.

How to record the activity of Ca²⁺ channels in isolation

This needs to block the voltage-gated channels that are not permeable to Ca²⁺ ions. Different strategies can be used: in whole-cell or intracellular recordings, TTX and TEA are added to the extracellular solution and K⁺ ions are replaced by Cs⁺ in the intrapipette solution, in order to block voltage-gated Na⁺ and K⁺ channels. In cell-attached recordings the patch pipette is filled with a solution containing Ca²⁺ or Ba²⁺ ions as the charge carrier. When Ba²⁺ substitutes for Ca²⁺ in the extracellular solution, the inward currents recorded in response to a depolarizing step are Ba²⁺ currents. Ba²⁺ is often preferred to Ca²⁺ since it carries current twice as effectively as Ca²⁺ and poorly inactivates Ca²⁺ channels

(see Section 5.2.3). As a consequence, unitary Ba²⁺ currents are larger than Ca²⁺ ones and can be studied more easily.

Another challenge is to separate the various types of Ca²⁺ channels in order to record the activity of only one type (since in most of the cells they are co-expressed). These different Ca²⁺ channels are the high voltage-activated L, N and P channels (this chapter) and the low-threshold T channel. T-type Ca²⁺ channels are low threshold-activated channels, also called subliminal Ca²⁺ channels, that can be identified by their low threshold of activation and their rapid inactivation. They are studied with other subliminal channels in Section 14.2.2.

HVA Ca²⁺ channels exhibit overlapping electrophysiological profiles. It is important to separate them in order to study their characteristics and to identify their respective roles in synaptic integration (dendritic Ca²⁺ channels), in transmitter release (Ca²⁺ channels of axon terminals), hormone secretion (Ca²⁺ channels of endocrine cells), muscle contraction (Ca²⁺ channels of smooth, skeletal or cardiac muscle cells).

HVA Ca^{2+} channels can be separated by using Ca^{2+} channel blockers, which can be subdivided in three general classes: small organic blockers; peptide toxins; and inorganic blockers. Small organic blockers include the dihydropyridines (DHP) that selectively block L-type channels (these channels are also selectively opened by Bay K 8644). Peptide toxins include an ω-conotoxin of the marine snail *Conus geographicus* that selectively blocks N-type channels and a purified polyamine fraction of the funnel-web spider (*Agelenopsis aperta*) venom (FTX) or a peptide component of the same venom, ω-agatoxin IVA (ω-Aga-IVA) that selectively block P-type channels. Inorganic blockers include divalent or trivalent metal ions such as cadmium, nickel, but they are not selective and are thus not used to separate the different types of HVA channels.

5.2.2 The L, N and P-type Ca^{2+} channels open at membrane potentials positive to −20 mV; they are high-threshold Ca^{2+} channels

The L-type Ca^{2+} channel has a large conductance and inactivates very slowly with depolarization

The activity of single L-type Ca^{2+} channels is recorded in sensory neurons of the chick dorsal root ganglion in patch clamp (cell-attached patch with Ba^{2+} as the charge carrier). In response to a test depolarization to +20 mV from a *depolarized* holding potential (−40 to 0 mV), unitary inward Ba^{2+} currents are evoked and recorded throughout the duration of the depolarizing step (**Figure 5.3a**).

The voltage-dependence of activation is studied with depolarizations to various test potentials from a holding potential of −40 mV (**Figure 5.4**). With test depolarizations up to +10 mV, openings are rare and of short duration. Activation of the channel becomes significant at +10 mV: openings are more frequent and of longer duration. At all potentials tested, openings are distributed relatively evenly throughout the duration of the depolarizing step (**Figures 5.3a** and **5.4a**). At −20 mV, the mean single-channel amplitude of the L current (i_L) is around − 2 pA. i_L amplitude diminishes linearly with depolarization: the i_L/V relation is linear between −20 and +20 mV. Between these membrane potentials, the unitary conductance, γ_L, is constant and equal to 20–25 pS in 110 mM Ba^{2+} (**Figure 5.4c**).

The main characteristics of L-type channels are (i) their very slow inactivation during a depolarizing step; (ii) their sensitivity to dihydropyridines; and (iii) their loss of activity in excised patches. Bay K 8644 is a dihydropyridine compound that increases dramatically the mean open time of an L-type channel without

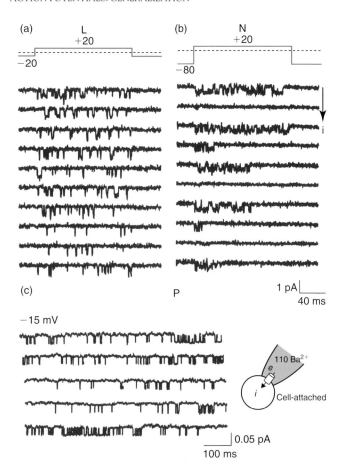

FIGURE 5.3 Single-channel recordings of the high-threshold Ca^{2+} channels: the L, N and P channels.
The activity of **(a)** single L and **(b)** N Ca^{2+} channels is recorded in patch clamp (cell-attached patches) from dorsal root ganglion cells and that of a single P channel **(c)** is recorded from a lipid bilayer in which a P channel isolated from cerebellum has been incorporated. All recordings are performed with Ba^{2+} (110 or 80 mM) as the charge carrier. In response to a test depolarizing step to +20 mV (a, b) or at a depolarized holding potential of −15 mV (c), unitary inward currents are recorded. Upper traces are voltage and the corresponding unitary current traces are the bottom traces (5–10 trials). $V_H = -20$ mV in (a), −80 mV in (b) and −15 mV in (c). In (a) and (b) the intrapipette solution contains: 110 mM BaCl$_2$, 10 mM HEPES and 200 μM TTX. The extracellular solution bathing the membrane outside the patch contains (in mM): 140 K aspartate, 10 K-EGTA, 10 HEPES, 1 MgCl$_2$ in order to zero the cell resting membrane potential. In (c) the solution bathing the extracellular side of the bilayer contains (in mM): 80 BaCl$_2$, 10 HEPES. The solution bathing the intracellular side of the bilayer in (c) contains (in mM): 120 CsCl, 1 MgCl$_2$, 10 HEPES. Parts (a) and (b) adapted from Nowycky MC, Fox AP, Tsien RW (1985) Three types of neuronal calcium channel with different calcium agonist sensitivity. *Nature* **316**, 440–443, with permission. Part (c) adapted from Llinas R, Sugimori M, Lin JW, Cherksey B (1989) Blocking and isolation of a calcium channel from neurons in mammals and cephalopods utilizing a toxin fraction (FTX) from funnel web spider poison. *Proc. Natl Acad. Sci. USA* **86**, 1689–1693, with permission.

changing its unitary conductance (**Figure 5.5**). It has no effect on the other Ca^{2+} channel types (see **Figure 5.9**). Bay K 8644 binds to a specific site on the α$_1$-subunit of L channels and changes the gating mode from brief

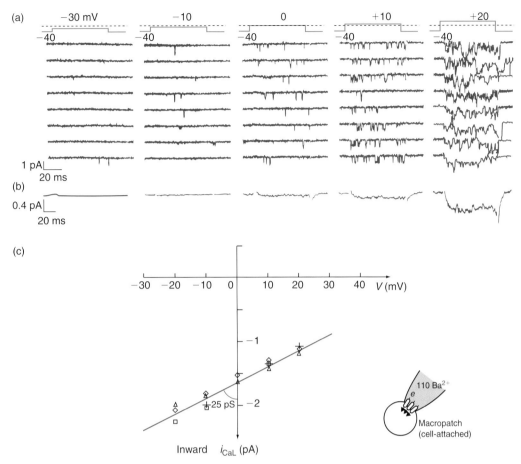

FIGURE 5.4 Voltage dependence of the unitary L-type Ca²⁺ current.

(a) The activity of L channels (the patch of membrane contains more than one L channel) is recorded in patch clamp (cell-attached patch) in a sensory dorsal root ganglion neuron. The patch is depolarized to −30, −10, 0, +10 and +20 mV from a holding potential of −40 mV. **(b)** Macroscopic current traces obtained by averaging at least 80 corresponding unitary current recordings such as those in (a). The probability of the L channels being in the open state increases with the test depolarization so that at +20 mV, openings of the 4–5 channels present in the patch overlap, leading to a sudden increase in the corresponding macroscopic current. **(c)** The unitary L current amplitude (i_L) is plotted against membrane potential (from −20 to +20 mV) in the absence (+, square) or presence (△, lozange) of Bay K 8644 in the patch pipette. The amplitude of i_L decreases linearly with depolarization between −20 and +20 mV with a slope γ_L =25 pS. The intrapipette solution contains (in mM): 110 BaCl₂, 10 HEPES. The extracellular solution bathing the extracellular side of the membrane outside of the recording pipette contains (in mM): 140 K-aspartate, 10 K-EGTA, 1 MgCl₂, 10 HEPES. A symmetric K⁺ solution is applied in order to zero the cell resting potential. Adapted from Fox AF, Nowycky MC, Tsien RW (1987) Single-channel recordings of three types of calcium channels in chick sensory neurons. *J. Physiol.* **394**, 173–200, with permission.

openings to long-lasting openings even at weakly depolarized potentials (V_{step} = −30 mV). Other dihydropyridine derivatives such as nifedipine, nimodipine and nitrendipine selectively block L channels (see **Figure 5.16**).

The loss of activity of an L channel in excised patch can be observed in outside-out patches. In response to a test depolarization to +10 mV the activity of an L channel rapidly disappears (**Figure 5.6**). To determine the nature of the cytoplasmic constituent(s) necessary to restore the activity of the L channel, inside-out patches are performed, a configuration that allows a

change of the medium bathing the intracellular side of the membrane.

The activity of a single L channel is first recorded in cell-attached configuration in response to a test depolarization to 0 mV (**Figure 5.7**). Then the membrane is pulled out in order to obtain an inside-out patch. The L-type activity rapidly disappears and is not restored by adding ATP–Mg to the intracellular solution. In contrast, when the catalytic subunit of the cAMP-dependent protein kinase (PKA) is added, the L-channel activity reappears (the catalytic subunit of PKA does not need the presence of cAMP to be active). This suggests that PKA directly

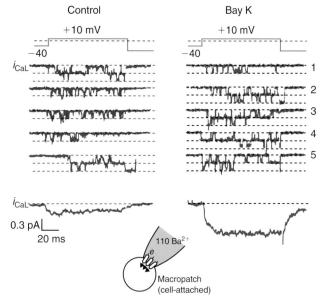

phosphorylates the L channel thus allowing its activation by the depolarization. It means that, in physiological conditions, the activity of L channels requires the activation of the following cascade: the activation of adenylate cyclase by the α-subunit of the G protein, the formation of cAMP and the subsequent activation of protein kinase A. Other kinases might also play a role. That is how neurotransmitters and a wide variety of hormones modulate L-type Ca²⁺ currents in neurons but also in endocrine cells and in smooth, skeletal and cardiac muscle.

The N-type Ca²⁺ channel inactivates with depolarization in the tens of milliseconds range and has a smaller unitary conductance than the L-type channel

The activity of single N-type channels is recorded in the same preparation in patch clamp (cell-attached patch, with Ba²⁺ as the charge carrier). In contrast to the L channels, N channels inactivate with depolarization. Therefore their activity has to be recorded in response to a test depolarization from a *hyperpolarized* holding potential (−80 to −60 mV) (**Figure 5.3b**). At holding potentials positive to −40 mV (e.g. −20 mV; **Figure 5.3a**), the N channel(s) is inactivated and its activity is absent in the recordings.

N-channel activity differs from that of the L channel in several aspects:

- N channels often open in bursts and inactivate with time and voltage (see Section 5.2.3).

FIGURE 5.5   **Bay K 8644 promotes long-lasting openings of L-type Ca²⁺ channels.**
The activity of three L channels is recorded in patch clamp (cell-attached patch). Top traces: a depolarizing step to +10 mV from a holding potential of −40 mV is applied at a low frequency. Middle traces (1 to 5): five consecutive unitary current traces recorded in the absence (left) and presence (right) of 5 μM Bay K 8644 in the bathing solution. Recordings are obtained from the same cell. Dashed line indicates the mean amplitude of the unitary current (−1.28 pA) which is unchanged in the presence of Bay K. Bottom traces: macroscopic current traces obtained by averaging at least 80 corresponding unitary current recordings. Adapted from Fox AP, Nowycky MC, Tsien RW (1987) Single-channel recordings of three types of calcium channels in chick sensory neurones. *J. Physiol.* **394**, 173–200, with permission.

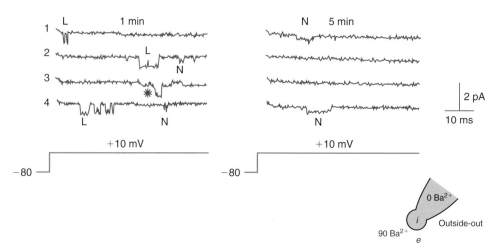

FIGURE 5.6   **In excised patches, the activity of L channels disappears within minutes.**
The activity of an L and N channel is recorded in patch clamp (outside-out patches from a pituitary cell line in culture) in response to a depolarizing pulse to +10 mV from a holding potential of −80 mV. Left: One minute after forming the excised patch, the two types of channels open one at a time or their openings overlap (line 3, *). Five minutes after, only the activity of the N-type is still present. The activity of the L-type will not reappear spontaneously. The extracellular solution contains (in mM): 90 BaCl₂, 15 TEACl, 2 × 10⁻³ TTX, 10 HEPES. The intrapipette solution contains (in mM): 120 CsCl, 40 HEPES. Adapted from Armstrong D and Eckert R (1987) Voltage-activated calcium channels that must be phosphorylated to respond to membrane depolarization. *Proc. Natl Acad. Sci. USA* **84**, 2518–2522, with permission.

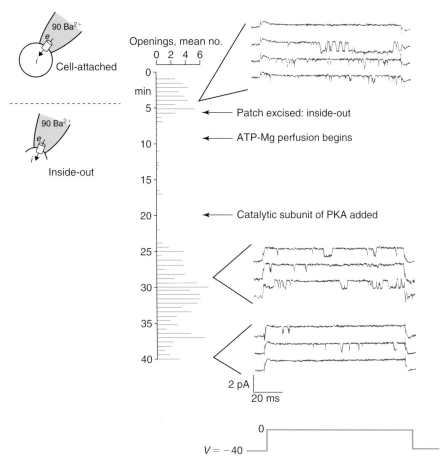

FIGURE 5.7 **Phosphorylation reverses the loss of activity of the L channels in an inside-out patch.**
The activity of an L-type channel is recorded in patch clamp (inside-out patch from a pituitary cell line in culture) in response to a depolarizing pulse to 0 mV from a holding potential of −40 mV. The horizontal traces are the unitary current traces and the vertical histogram represents the average number of channel openings per trace, determined over 30 s intervals and plotted versus time of the experiment (0–40 min). After 5 min of recording in the cell-attached configuration, the activity of the channel is recorded in the inside-out configuration. See text for further explanations. The intrapipette solution contains (in mM): 90 BaCl₂, 15 TEACl, 2×10^{-3} TTX, 10 HEPES. The solution bathing the intracellular side of the patch contains (in mM): 120 CsCl, 40 HEPES. From Armstrong D and Eckert R (1987) Voltage-activated calcium channels that must be phosphorylated to respond to membrane depolarization. *Proc. Natl Acad. Sci. USA* **84**, 2518–2522, with permission.

- Measured at the same test potential, the mean amplitude of the N unitary current is smaller than that of L (e.g. $i_N = -1.22 \pm 0.03$ pA and $i_L = -2.07 \pm 0.09$ at −20 mV; **Figure 5.3a, b**) which makes its mean unitary conductance also smaller ($\gamma_N = 13$ pS in 110 mM Ba²⁺; **Figure 5.8b**).
- N channels are insensitive to dihydropyridines but are selectively blocked by ω-conotoxin GVIA.
- N channels do not need to be phosphorylated to open (**Figure 5.6**).

The P-type Ca²⁺ channel differs from the N channel by its pharmacology

The activity of a single P-type channel is recorded from lipid bilayers in which purified P channels from

cerebellar Purkinje cells have been incorporated. Ba²⁺ ions are used as the charge carrier. The activity of the P channel is recorded at different steady holding potentials. At −15 mV, the channel opens, closes and reopens during the entire depolarization, showing little time-dependent inactivation (**Figure 5.3c**). The mean unitary conductance, γ_P, is 10–15 pS in 80 mM Ba²⁺. Recordings performed in dendrites or the soma of cerebellar Purkinje cells with patch clamp techniques (cell-attached patches) gave similar values of the unitary conductance (γ_P =9–19 pS in 110 mM Ba²⁺), but for undetermined reasons the threshold for activation is at a more depolarized potential (−15 mV) than for isolated P channels inserted in lipid bilayers (−45 mV). When the funnel web toxin fraction (FTX) is added to the recording patch pipette (the intrapipette solution bathes the extracellular side of the patch), only rare high-threshold unitary currents are recorded from Purkinje cell dendrites or soma at all potentials tested (Bay K 8644 or ω-conotoxin have no effect). These results suggest that the P channel is the predominant high-threshold Ca²⁺ channel expressed by Purkinje cells. They also show that the use of selective toxins allows the differentiation between P, N and L channels.

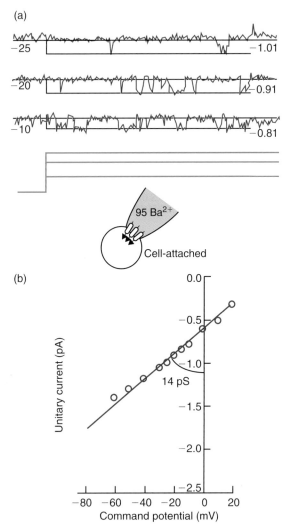

5.2.3 Macroscopic L, N and P-type Ca²⁺ currents activate at a high threshold and inactivate with different time courses

The macroscopic L, N and P-type Ca²⁺ currents (I_{Ca}), at time t during a depolarizing voltage step, are equal to: $I_{Ca} = Np_t i_{Ca}$ where N is the number of L, N or P channels in the membrane, p_t is their probability of being open at time t during the depolarizing step, Np_t is the number of open channels at time t during the depolarizing step and i_{Ca} is the unitary L, N or P current. At steady state, $I_{Ca} = Np_o i_{Ca}$, where p_o is the probability of the channel being open at steady state.

The I/V relations for L, N and P-type Ca²⁺ currents have a bell shape with a peak amplitude at positive potentials

The *I/V* relation of the different types of high threshold Ca²⁺ currents is studied in whole-cell recordings in the presence of external Ca²⁺ as the charge carrier. To separate the L, N and P currents, specific blockers are added to the external medium or the membrane potential is clamped at different holding potentials. With this last procedure, the L current can be separated from other Ca²⁺ currents since it can be evoked from depolarized holding potentials. As shown in **Figure 5.9**, the L and N currents averaged from the corresponding unitary currents recorded in 110 mM Ba²⁺ clearly differ in their time course. The averaged N current decays to zero level in 40 ms while the averaged L current remains constant during the 120 ms depolarizing step to +10 mV. As already observed (**Figures 5.3a, b**), by

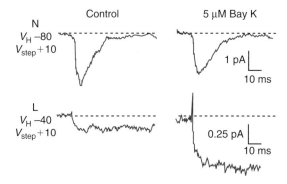

FIGURE 5.9 Averaged N- and L-type Ca²⁺ currents.
Single-channel N current averages (top traces) and L current averages (bottom traces) from cell-attached recordings of dorsal root ganglion cells with Ba²⁺ as the charge carrier (see also Figure 5.3a,b). Currents are averaged before (left) and after (right) exposure to 5 μM Bay K 8644. Voltage steps from −80 mV to +10 mV (top traces) and from −40 to +10 mV (bottom traces). From Nowycky MC, Fox AP, Tsien RW (1985) Three types of neuronal calcium channel with different calcium agonist sensitivity. *Nature* **316**, 440–443, with permission.

FIGURE 5.8 Voltage dependence of the unitary N current, i_N.
(a) The activity of an N channel is recorded in patch clamp (cell-attached patch) in a granule cell of the hippocampus. The patch is depolarized to −25, −20 and −10 mV from a holding potential of −80 mV. The amplitude of the unitary current at these voltages is indicated at the end of each recording. **(b)** The unitary N current amplitude (i_N) is plotted against membrane potential (from −60 to +20 mV). The amplitude of i_N decreases linearly with depolarization between −60 and +20 mV with a slope γ_N=14 pS ($n = 14$ patches). Adapted from Fisher RE, Gray R, Johnston D (1990) Properties and distribution of single voltage-gated calcium channels in adult hippocampal neurons. *J. Neurophysiol.* **64**, 91–104, with permission.

holding the membrane at a depolarized potential, the N current inactivates and the L current can be studied in isolation.

The macroscopic N- and L-type Ca^{2+} currents are studied in spinal motoneurons of the chick in patch clamp (whole-cell patch) in the presence of Na^{+} and K^{+} channel blockers and in the presence of a T-type Ca^{2+} channel blocker. In response to a depolarizing voltage step to $+20$ mV from a holding potential of -80 mV, a mixed N and L whole-cell current is recorded (**Figure 5.10a**). When the holding potential is depolarized to 0 mV, a voltage step to $+20$ mV now only evokes the L current

(**Figure 5.10b**). The difference current obtained by subtracting the L current from the mixed N and L current gives the N current (**Figure 5.10c**). The I/V relations of these two Ca^{2+} currents have a bell shape with a peak around $+20$ mV (**Figures 5.10d, e**). For comparison the peak amplitude of the macroscopic Na^{+} current is around -40 mV (see **Figure 4.12a**).

The macroscopic P-type Ca^{2+} current is studied in cerebellar Purkinje cells. These neurons express T, P and few L-type Ca^{2+} channels. In the presence of Na^{+} and K^{+} channel blockers and by choosing a holding potential where the low threshold T current is inactivated, the macroscopic P current can be studied. The I_P/V relation has a bell shape. The maximal amplitude is recorded around -10 mV (**Figure 5.11**).

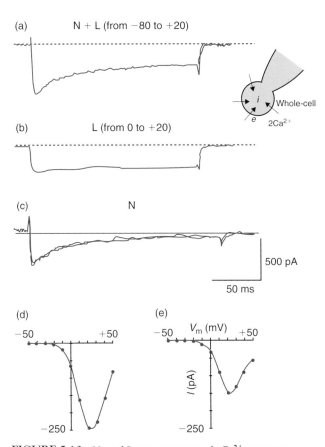

FIGURE 5.10   **N- and L-type macroscopic Ca^{2+} currents.**
(a) The mixed N and L macroscopic current is recorded with Ca^{2+} as the charge carrier (whole-cell patch) from chick limb motoneurons in culture in response to a voltage step from -80 to $+20$ mV. **(b)** The macroscopic L current is recorded in isolation by changing the holding potential to 0 mV. **(c)** The difference current obtained by subtracting the L current (b) from the N and L current (a) is the N current. **(d)** I/V relation for the L current recorded as in (b). **(e)** I/V relation for the N current obtained as the difference current. The intrapipette solution contains (in mM): 140 Cs aspartate, 5 MgCl$_2$, 10 Cs EGTA, 10 HEPES, 0.1 Li$_2$GTP, 1 MgATP. The bathing solution contains (in mM): 146 NaCl, 2 CaCl$_2$, 5 KCl, 1 MgCl$_2$, 10 HEPES. Adapted from McCobb DP, Best PM, Beam KG (1989) Development alters the expression of calcium currents in chick limb motoneurons. *Neuron* **2**, 1633–1643, with permission.

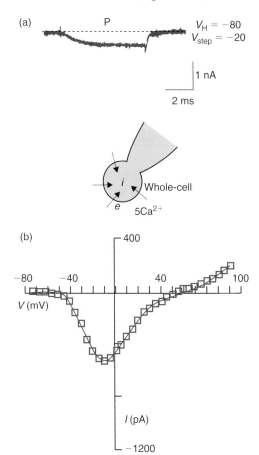

FIGURE 5.11   **P-type macroscopic Ca^{2+} current.**
The whole-cell P current recorded from acutely dissociated Purkinje cells (whole-cell patch) with Ca^{2+} as the charge carrier. **(a)** Whole-cell P current recorded in response to a depolarizing pulse to -20 mV from a holding potential of -80 mV. **(b)** I/V relation of the P current. In the recordings the low threshold T-type Ca^{2+} current was either absent, inactivated or subtracted. The intrapipette solution contains (in mM): 120 TEA glutamate, 9 EGTA, 4.5 MgCl$_2$, 9 HEPES. The bathing solution contains (in mM): 5 CaCl$_2$, 154 TEACl, 0.2 MgCl$_2$, 10 glucose, 10 HEPES. Adapted from Reagan LJ (1991) Voltage-dependent calcium currents in Purkinje cells from rat cerebellar vermis. *J. Neurosci.* **7**, 2259–2269, with permission.

The bell shape of all the I_{Ca}/V relations is explained by the gating properties of the Ca^{2+} channels and the driving force for Ca^{2+} ions. The peak amplitude of I_{Ca} increases from the threshold potential to a maximal amplitude (**Figures 5.10d,e, 5.11b** and **5.12a**) as a result of two opposite factors: the probability of opening which strongly increases with depolarization (**Figure 5.12b**) and the driving force for Ca^{2+} which linearly decreases with depolarization (i_{Ca} linearly diminishes). After a maximum, the peak amplitude of I_{Ca} decreases owing to the progressive decrease of the driving force for Ca^{2+} ions and the increase of the number of inactivated channels. Above $+30/+40\,mV$, the probability of opening (p_o) no longer plays a role since it is maximal (**Figure 5.12b**). I_{Ca} reverses polarity between $+50\,mV$ and $+100\,mV$, depending on the preparation studied. This value is well below the theoretical E_{Ca}.

This discrepancy is partly due to the strong asymmetrical concentrations of Ca^{2+} ions. To measure the reversal potential of I_{Ca}, the outward current through Ca^{2+} channels must be measured. This outward current, caused by the extremely small intracellular concentration of Ca^{2+} ions, is carried by Ca^{2+} ions but also by internal K^+ ions, which are around 10^6 times more concentrated than internal Ca^{2+} ions. This permeability

of Ca^{2+} channels to K^+ ions 'pulls down' the reversal potential of I_{Ca} towards E_K.

Activation–inactivation properties

Activation properties are analyzed by recording the macroscopic L, N or P currents in response to increasing test depolarizations from a fixed hyperpolarized holding potential ($-80\,mV$, **Figures 5.13b, 5.14b** and **5.15b**). In dorsal ganglion neurons, the L and N currents are half activated around $0\,mV$ (**Figures 5.13c** and **5.14c**) while in Purkinje cells the P current is half activated around $-20\,mV$ (**Figure 5.15c**).

Voltage-gated Ca^{2+} channels show varying degrees of inactivation

Inactivation properties are analyzed by recording the macroscopic L, N or P-type Ca^{2+} currents evoked by a voltage step to a fixed potential from various holding potentials (with Ca^{2+} as the charge carrier). The L current is half inactivated around $-40\,mV$ (**Figures 5.13a, c**), the N current around $-60\,mV$ (**Figures 5.14a, c**) and the P current around $-45\,mV$ (**Figures 5.15a, c**).

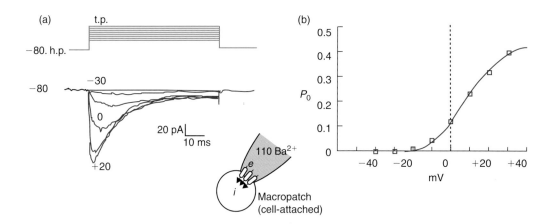

FIGURE 5.12 The peak opening probability of the N current.
The macroscopic N current is recorded in a dorsal root ganglion neuron from a cell-attached patch containing hundreds of N channels (macropatch). **(a)** Current recordings (bottom traces) in response to test potentials (t.p.) ranging from $-30\,mV$ to $+20\,mV$ from a holding potential (h.p.) of $-80\,mV$ (upper traces). **(b)** Voltage-dependence of the peak opening probability (p_o) from data obtained in (a). Values of p_o are obtained by dividing the peak current I by the unitary current i_N obtained at each test potential and by an estimate of the number of channels in the patch (599): $p_o = I/Ni_N$. N was determined by comparison with the single-channel experiment in Figure 5.3b, which shows that in response to a depolarization to $+20\,mV$ from a holding potential of $-80\,mV$, $p_o = 0.32$ and $i_N = 0.76\,pA$. I, the peak current evoked by the same voltage protocol, is $145\,pA$. $N = I/p_o\,i_N = 145/(0.32 \times 0.76) = 599$ channels. The intrapipette solution contains (in mM): $100\,CsCl$, $10\,Cs$-EGTA, $5\,MgCl_2$, $40\,HEPES$, $2\,ATP$, $0.25\,cAMP$; pH $=7.3$. The extracellular solution contains (in mM): $10\,CaCl_2$, $135\,TEACl$, $10\,HEPES$, $0.2 \times 10^{-3}\,TTX$; pH $=7.3$. From Nowycky MC, Fox AP, Tsien RW (1985) Three types of neuronal calcium channel with different calcium agonist sensitivity. *Nature* **316**, 440–443, with permission.

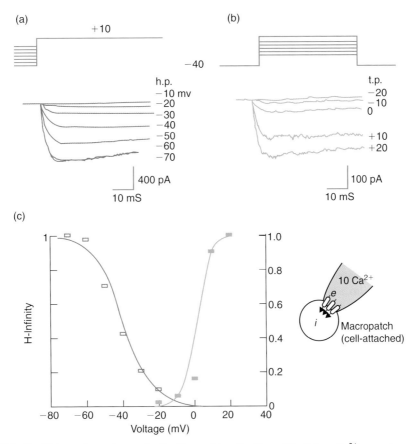

FIGURE 5.13 **Voltage dependence of activation and inactivation of the L-type Ca²⁺ current.**
The macroscopic L current is recorded in a cell with very little T or N current. **(a)** Inactivation of the L current with holding potential: a test depolarization to $+10\,\text{mV}$ is applied from holding potentials (h.p.) varying from -70 to $-10\,\text{mV}$. **(b)** Activation of the L current with depolarization: test depolarizations (t.p.) to -30, -20, -10, 0, $+10$ and $+20\,\text{mV}$ are applied from a holding potential of $-40\,\text{mV}$. **(c)** Activation–inactivation curves obtained from the data in (b) and (a), respectively. The peak Ca²⁺ current amplitudes (I) are normalized to the maximal current (Imax $= 1$) obtained in each set of experiments and plotted against the holding (inactivation curve, ■) or test potential (activation curve, •). For the activation curve, data are plotted as $I = I_{\text{max}}\{1 + \exp[(V_{1/2} - V)/k]\}^{-1}$ and for the inactivation curve as $I = I_{\text{max}}\{1 \exp[(V - V_{1/2})/k]\}^{-1}$. $V_{1/2}$ is the voltage at which the current I is half-activated ($I = I_{\text{max}}/2$ when $V_{1/2} = 2\,\text{mV}$) or half-inactivated ($I = I_{\text{max}}/2$ when $V_{1/2} = -40\,\text{mV}$). All the recordings are performed in the presence of 10 mM Ca²⁺ in the recording pipette solution which bathes the extracellular side of the channels. Adapted from Fox AP, Nowycky M, Tsien RW (1987) Kinetic and pharmacological properties distinguishing three types of calcium currents in chick sensory neurones. *J. Physiol.* **394**, 149–172, with permission.

In summary, L channels generate a large Ca²⁺ current that is activated by large depolarizations to $0/+10\,\text{mV}$ and inactivates with a very slow time course during a step. N and P channels generate smaller Ca²⁺ currents that are activated with depolarization to $-30/0\,\text{mV}$ and inactivate or not during a depolarizing step.

The inactivation process of Ca²⁺ channels can be voltage-dependent, time-dependent *and* calcium-dependent. Voltage-dependent inactivation is observed by changing the holding potential (see **Figures 5.13a, 5.14a** and **5.15a**). Time-dependent inactivation is observed during a long depolarizing step, in the presence of Ba²⁺ as the change carrier (**Figure 5.16**). Ca²⁺-dependent inactivation depends on the amount of Ca²⁺ influx through open Ca²⁺ channels. It can be considered as a negative feedback control of Ca²⁺ channels by Ca²⁺ channels.

Calcium-dependent inactivation

Several lines of evidence point to the existence of a Ca²⁺-induced inactivation of Ca²⁺ currents:

- The degree of inactivation is proportional to the amplitude and frequency of the Ca²⁺ current.
- Intracellular injection of Ca²⁺ ions into neurons produces inactivation.
- Intracellular injection of Ca²⁺ chelators such as EGTA or BAPTA reduces inactivation (**Figure 5.17**).
- Substitution of Ca²⁺ ions with Sr²⁺ or Ba²⁺ reduces inactivation.
- Very large depolarizations to near E_{Ca}, where the entry of Ca²⁺ ions is small, produce little inactivation.

Recordings of L and N channels in **Figures 5.13** and **5.14** were obtained with Ca²⁺ as the charge carrier and

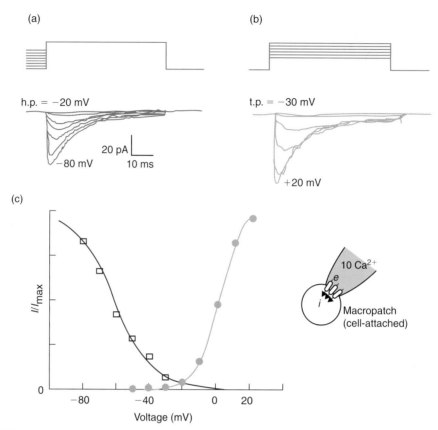

FIGURE 5.14 **Voltage dependence of activation and inactivation of the N-type Ca²⁺ current.**
The macroscopic N current is recorded in cell-attached patches containing hundreds of channels (macropatch).
(a) Inactivation of the N current with holding potential: test depolarization to +10 mV is applied from holding
potentials (h.p.) varying from −70 to −10 mV. **(b)** Activation of the N current with depolarization: test depolarizations (t.p.) to −30, −20, −10, 0, +10 and +20 mV are applied from a holding potential of −80 mV. **(c)**
Activation–inactivation curves obtained from the data in (b) and (a), respectively. The peak Ca²⁺ current amplitudes (I) are normalized to the maximal current (I_{max} =1) obtained in each set of experiments and plotted against
the holding (inactivation curve, ■) or test potential (activation curve, ●). For the activation curve, data are plotted as $I = I_{max}\{1 +\exp[(V_{1/2} - V)/k]\}^{-1}$ and for the inactivation curve as $I = I\{1 +\exp[(V - V_{1/2}/k]\}^{-1}$. $V_{1/2}$ is the
voltage at which the current I is half-activated ($I = I_{max}/2$ when $V_{1/2}$ =1.5 mV) or half-inactivated ($I = I_{max}/2$
when $V_{1/2} = -61.5$ mV). The number of channels is estimated as in Figure 5.12. Adapted from Fox AP, Nowycky
MC, Tsien RW (1987) Single-channel recordings of three types of calcium channels in chick sensory neurones. *J.
Physiol.* **394**, 173–200, with permission.

that of P channels in **Figure 5.15** with Ba²⁺ as the charge
carrier. Therefore, the inactivation seen in **Figures 5.13**
and **5.14**, results from three parameters: voltage, time
and increase of intracellular Ca²⁺ concentration. In contrast, the inactivation of the P current observed in
Figure 5.15 is a voltage- and time-dependent process.

The macroscopic Ca²⁺ current of *Aplysia* neurons is
recorded in voltage clamp. During depolarizing voltage
steps, the Ca²⁺ current increases to a peak and then
declines to a steady state Ca²⁺ current (a non-inactivating
component of current). The buffering of cytoplasmic
free Ca²⁺ ions with EGTA increases the amplitude of the
peak current and that of the steady-state current (**Figure
5.17**). This shows that the increase of intracellular Ca²⁺
ions resulting from Ca²⁺ entry through Ca²⁺ channels
causes Ca²⁺ current inactivation. It also shows that the

peak current is probably already decreased in amplitude owing to early development of inactivation.

5.3 THE REPOLARIZATION PHASE OF Ca²⁺-DEPENDENT ACTION POTENTIALS RESULTS FROM THE ACTIVATION OF K⁺ CURRENTS I_K AND I_{KCa}

The K⁺ currents involved in calcium spike repolarization are the delayed rectifier (I_K) studied in Chapter 4
and the Ca²⁺-activated K⁺ currents (I_{KCa}). Meech and
Strumwasser in 1970 were the first to describe that a
microinjection of Ca²⁺ ions into *Aplysia* neurons activates
a K⁺ conductance and hyperpolarizes the membrane.

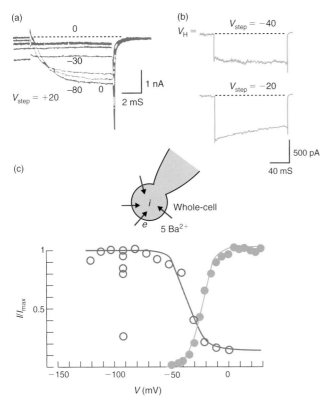

FIGURE 5.15   Voltage dependence of activation and inactivation of the P-type Ca²⁺ current.
The macroscopic P current is recorded in Purkinje cells (whole-cell patch). The T-type Ca²⁺ current present in these cells is either absent or subtracted. **(a)** Inactivation of the P current with holding potential: a test depolarization to +20 mV is applied from holding potentials varying from −80 to 0 mV. **(b)** Activation of the P current with depolarization: test depolarizations (V_{step}) to −40 and −20 mV are applied from a holding potential of −110 mV. **(c)** Activation–inactivation curves obtained from the data obtained in (b) and (a), respectively. The peak Ca²⁺ current amplitudes (I) are normalized to the maximal current ($I_{\text{max}} = 1$) obtained in each set of experiments and plotted against the holding (inactivation curve, ■) or test potential (activation curve, ●). For the activation curve, data are plotted as $I = I_{\text{max}}\{1 + \exp [(V_{1/2} - V)/k]\}^{-1}$ and for the inactivation curve as $I = I_{\text{max}}\{1 + \exp [(V - V_{1/2})/k]\}^{-1}$. $V_{1/2}$ is the voltage at which the current I is half-activated ($I = I_{\text{max}}/2$ when $V_{1/2} = -22$ mV) or half-inactivated ($I = I_{\text{max}}/2$ when $V_{1/2} = -34$ mV). In all recordings, the extracellular solution contains 5 mM Ba²⁺. Adapted from Regan L (1991) Voltage-dependent calcium currents in Purkinje cells from rat cerebellar vermis. *J. Neurosci.* **11**, 2259–2269, with permission.

On the basis of these results, the authors postulated the existence of a Ca²⁺-activated K⁺ conductance. The amount of participation of Ca²⁺-activated K⁺ currents in spike repolarization depends on the cell type.

5.3.1   The Ca²⁺-activated K⁺ currents are classified as big K (BK) channels and small K (SK) channels

Big K channels have a high conductance (100–250 pS depending on K⁺ concentrations) and are sensitive to both voltage *and* Ca²⁺ ions so that their apparent sensitivity to Ca²⁺ ions is increased when the membrane is depolarized. Their activity is blocked by TEA and charybdotoxin, a toxin from scorpion venom. Small K channels have a smaller conductance (10–80 pS depending on K⁺ concentrations) and are insensitive to TEA and charybdotoxin but sensitive to apamin, a toxin from bee venom. It is a heterogeneous class containing both voltage-dependent and voltage-independent channels. Big K and small K channels are very selective for K⁺ ions over Na⁺ ions and are activated by increases in the concentration of cytoplasmic Ca²⁺ ions.

The channels originally termed 'big' potassium (BK) channels, are also called maxi-K channels or SLO family channels, a name derived from the conserved gene that encodes this channel, which was first cloned in *Drosophila melanogaster*. Voltage-clamp recordings of currents in the flight muscles of a Drosophila mutant with a severely lethargic phenotype, named *slowpoke*, revealed that the calcium-dependent component of the outward K⁺ current was absent, implicating the *slowpoke* (*slo*) gene as the structural locus encoding the channel protein. The mammalian *slo* orthologue *Slo1* was cloned by low-stringency DNA hybridization of a mammalian cDNA library using the *Drosophila slo* cDNA. The conserved protein domains of SLO1 seem to reflect separate mechanisms for voltage and Ca²⁺ sensing. The primary sequence consists of two distinct regions. The 'core' region (which includes hydrophobic segments S0–S6) resembles a canonical voltage-gated K⁺ channel except for the inclusion of the additional S0 segment (**Figure 5.20a**). The distal part of the carboxyl region (containing S9–S10), termed the tail, includes the region that is most highly conserved among SLO1 proteins from different species, the calcium bowl. The gating of SLO1 channels by both voltage and the binding of intracellular Ca²⁺ suggests that the two independent sensing mechanisms converge near the gates of the pore.

The genes that encode the SK channels belong to the KCNN gene family. SK channels have a similar topology to members of the voltage-gated (Kv) K⁺ channel superfamily. They consist of six transmembrane segments (S1–S6), with the pore located between S5 and S6. The S4 segment, which confers voltage sensitivity to the Kv channel, shows in SK channels a reduced number and a disrupted array of positively charged amino acids. The SK channels retain only two of the seven positively charged amino acids that are found in the S4 segment of Kv channels, and only one of these residues corresponds to the four arginine residues that carry the gating charges in Kv channels. These differences in the primary sequence could represent the molecular framework for the observed voltage independence of SK channels.

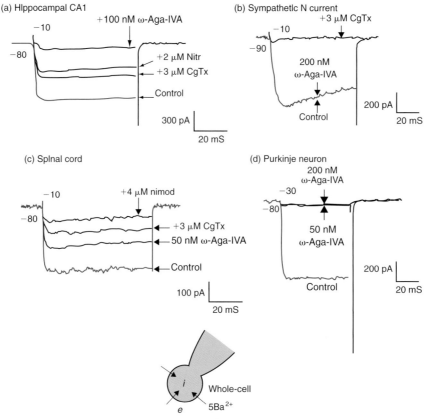

FIGURE 5.16 Pharmacology of L-, N- and P-type Ca²⁺ channels.
The macroscopic mixed Ca²⁺ currents are recorded in different neurons with Ba²⁺ as the charge carrier (whole-cell patch). High-threshold Ca²⁺ currents are evoked by depolarizations to -30 or -10 mV from a holding potential of -90 or -80 mV. Various blockers or toxins are applied in order to block selectively one type of high-threshold Ca²⁺ current at a time: ω-conotoxin (CgTx, 3 μM) selectively blocks N current, nitrendipine or nimodipine (nitr., nimod., 2–4 μM) selectively blocks L current, and ω-agatoxin (ω-Aga-IVA, 50–200 nM) selectively blocks P current. In hippocampal cells of the CA1 region and in spinal cord interneurons, the high-threshold Ca²⁺ current is a mixed N, L and P current. In sympathetic neurons it is almost exclusively N and in Purkinje cells almost exclusively P. The intrapipette solution contains (in mM): 108 Cs methanesulphonate, 4 MgCl₂, 9 EGTA, 9 HEPES, 4 MgATP, 14 creatine phosphate, 1 GTP; pH = 7.4. The extracellular solution contains (in mM): 5 BaCl₂, 160 TEACl, 0.1 EGTA, 10 HEPES; pH = 7.4. Adapted from Mintz IM, Adams ME, Bean B (1992) P-type calcium channels in rat central and peripheral neurons. *Neuron* **9**, 85–95, with permission.

5.3.2 Ca²⁺ entering during the depolarization or the plateau phase of Ca²⁺-dependent action potentials activates K_Ca channels

To study Ca²⁺-activated K⁺ channels from rat brain neurons, plasma membrane vesicle preparation is incorporated into planar lipid bilayers. In such conditions, the activity of four distinct types of Ca²⁺-activated K⁺ channels is recorded. We will look at one example of a big K and one example of a small K channel. This preparation allows the recording of single-channel activity (**Figure 5.18**).

The current–voltage relations obtained in the presence of two different extracellular K⁺ concentrations show that the current reverses at E_K, the theoretical reversal potential for K⁺ ions as expected for a purely K⁺-selective channel. The Ca²⁺-dependence is studied by raising the intracellular Ca²⁺ concentration in the range of 0.1–10 μM. Channels are activated by micromolar concentrations of Ca²⁺. The open probabilities of the big K and small K channels are largely increased when the medium bathing the intracellular side of the membrane contains 0.4 μM Ca²⁺ instead of 0.1 μM (**Figures 5.19a, b**). For comparison the Ca²⁺-sensitivity of big K channels from cultured rat skeletal muscle is shown in **Figure 5.19c**. The rat brain big K channels are sensitive to nanomolar concentrations of charybdotoxin (CTX) and millimolar concentrations of extracellular TEA ions (**Figure 5.20**).

The macroscopic Ca²⁺-activated K⁺ currents are recorded from a bullfrog sympathetic neuron in single-electrode voltage clamp mode ($V_H = -28$ mV). The iontophoretic injection of Ca²⁺ ions via the recording

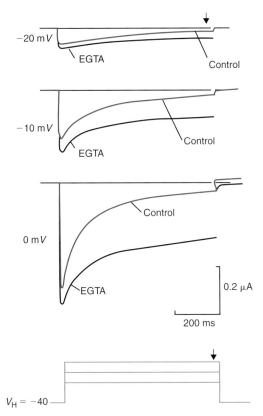

FIGURE 5.17   Intracellular EGTA slows Ca^{2+}-dependent inactivation of Ca^{2+} channels.

The macroscopic Ca^{2+} current is recorded in axotomized *Aplysia* neurons in double-electrode voltage clamp (axotomy is performed in order to improve space clamp). Control Ca^{2+} currents are recorded in response to step depolarizations to -20, -10 and $0\,mV$ from a holding potential of $-40\,mV$ (control traces). Iontophoretic ejection of EGTA (300–$500\,nA$ for 4–$8\,min$) increases the peak amplitude of the Ca^{2+} current and slows its inactivation at all potentials tested (EGTA traces). The amplitude of the non-inactivating component of the current is measured at the end of the steps (arrow). Adapted from Chad J, Eckert R, Ewald D (1984) Kinetics of calcium-dependent inactivation of calcium current in voltage-clamped neurones in *Aplysia californica. J. Physiol. (Lond.)* **347**, 279–300, with permission.

electrode triggers an outward current (**Figure 5.21a**). Its amplitude increases when the iontophoretic current is increased; i.e. when the amount of Ca^{2+} ions injected is increased. To study the voltage-dependence and the kinetics of activation of this Ca^{2+}-activated outward current, depolarizing steps from a holding potential of $-50\,mV$ are applied in the presence of $2\,mM$ of Ca^{2+} in the extracellular medium (**Figure 5.21b**, 2 Ca). Suppression of Ca^{2+} entry by removal of Ca^{2+} ions from the extracellular medium (0 Ca) eliminates an early Ca^{2+}-activated outward current. In the Ca-free medium, only the sigmoidal delayed rectifier K$^+$ current I_K is recorded. In the presence of external Ca^{2+}

ions, both I_K and a $I_{K(Ca)}$ are recorded (**Figure 5.21b**, right). The recorded $I_{K(Ca)}$ corresponds to a big K current also called I_C in some preparations. It has activation kinetics sufficiently rapid to play a role in spike repolarization (**Figure 5.22**).

In nerve terminals at the motor end plate, big K channels are co-localized with voltage-dependent Ca^{2+} channels. They play an important role in repolarizing the plasma membrane following each action potential. This repolarization resulting from the increased activity of Ca^{2+}-activated K$^+$ channels closes voltage-dependent Ca^{2+} channels and constitutes an important feedback mechanism for the regulation of voltage-dependent Ca^{2+} entry. K$_{Ca}$ current thereby lowers intracellular Ca^{2+} concentration and dampens neurotransmitter secretion. Conversely when it is strongly reduced by TEA or apamin, transmitter release is increased.

5.4 CALCIUM-DEPENDENT ACTION POTENTIALS ARE INITIATED IN AXON TERMINALS AND IN DENDRITES

5.4.1 Depolarization of the membrane to the threshold for the activation of L-, N- and P-type Ca$^+$ channels has two origins

L-, N- and P-type Ca^{2+} channels are high-threshold Ca^{2+} channels. This means that they are activated in response to a relatively large membrane depolarization. In cells (e.g. neurons, heart muscle cells) where the resting membrane potential is around $-80/-60\,mV$, a 40–$60\,mV$ depolarization is therefore needed to activate the high-threshold Ca^{2+} channels. Such a membrane depolarization is too large to result directly from the summation of excitatory postsynaptic potentials (EPSPs). It usually results from a Na$^+$ spike. In heart Purkinje cells, Na$^+$ entry during the sudden depolarization phase of the action potential depolarizes the membrane to the threshold for L-type Ca^{2+} channel activation: the Na$^+$-dependent depolarization phase is immediately followed by a Ca^{2+}-dependent plateau (see **Figure 4.2d**). In axon terminals, the situation is similar: the Na$^+$-dependent action potential actively propagates to axon terminals where it depolarizes the membrane to the threshold potential for N- or P-type Ca^{2+} channel activation: a Na$^+$/Ca^{2+}-dependent action potential is initiated (see **Figure 4.2c**).

In cerebellar Purkinje neurons the situation is somehow different: dendritic P-type Ca^{2+} channels are opened by the large EPSP resulting from climbing fibre

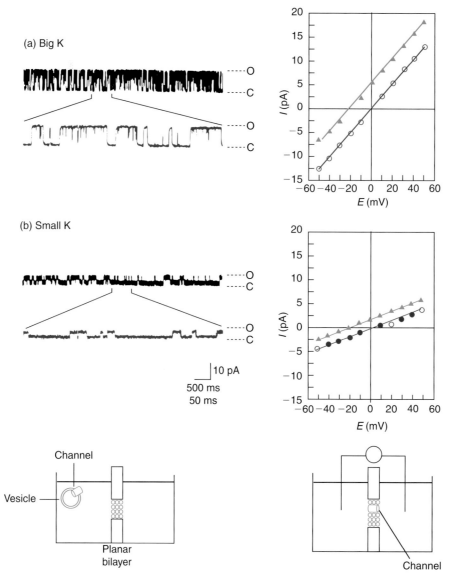

FIGURE 5.18   Two types of rat brain Ca²⁺-activated K⁺ channels incorporated into lipid bilayers.
(a, b) Left: Single-channel recordings in symmetrical K⁺ (the extracellular and intracellular solutions contain 150 mM KCl) at $V_H = 40$ mV. For all traces channel openings correspond to upward deflections. The recording length of upper traces is 6.4 s and each lower trace is expanded to show a 640 ms recording. Right: I/V relations for the big K channel and the small K channel in symmetrical K⁺ (150 mM, circles) and 150 mM KCl inside, 50 mM KCl outside (triangles). The slope conductance for each of these channels in symmetrical 150 mM KCl is 232 pS (big K channel) and 77 pS (small K channel). All the recordings are performed in the presence of 1.05 mM CaCl₂ in the intracellular solution. Adapted from Reinhart PH, Chung S, Levitan IB (1989) A family of calcium-dependent potassium channels from rat brain. *Neuron* **2**, 1031–1041, with permission. (a) Big K (rat brain neurons).

EPSP. As a result, Ca²⁺-dependent action potentials are initiated and actively propagate in dendrites (see **Figure 4.2b**; also see Sections 16.2 and 17.3).

The cells that do not express voltage-gated Na⁺ channels and initiate Ca²⁺-dependent action potentials (endocrine cells for example; see **Figure 5.1**) usually present a depolarized resting membrane potential

($-50/-40$ mV) close to the threshold for L-type Ca²⁺ channel activation. In such cells, the activation of high-threshold Ca²⁺ channels results from a depolarizing current generated by receptor activation or from an intrinsic pacemaker current (for example, activation of the T-type ²⁺ current – see Section 14.2.2 – or the turning off of a leak K⁺ current).

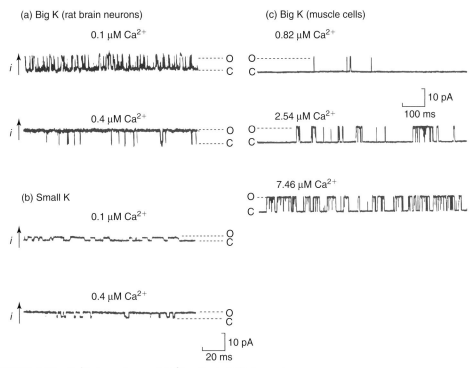

FIGURE 5.19   Ca²⁺-dependence of Ca²⁺-activated K⁺ channels.
(a, b) Single-channel activity of Ca²⁺-activated channels from the rat brain. The activity of the 232 pS big K channel and that of the 77 pS small K channel is recorded in the presence of 0.1 μM Ca²⁺ (upper traces) and 0.4 μM Ca²⁺ (lower traces) in symmetrical 150 mM KCl (V_H = + 20 mV). **(c)** Single-channel activity of a big K channel from rat skeletal muscle recorded at three different Ca²⁺ concentrations in symmetrical 140 mM KCl (V_H = + 30 mV). O, open state; C, closed state. Part (a) from Chad J, Eckert R, Ewald D (1984) Kinetics of calcium-dependent inactivation of calcium current in voltage-clamped neurones in *Aplysia californica*. *J. Physiol. (Lond.)* **347**, 279–300, with permission. Part (b) adapted from McManus OB and Magleby KL (1991) Accounting for the calcium-dependent kinetics of single large-conductance Ca²⁺-activated K⁺ channels in rat skeletal muscle. *J. Physiol.* **443**: 739–777, with permission.

5.4.2 The role of the calcium-dependent action potentials is to provide a local and transient increase of [Ca²⁺]ᵢ to trigger secretion, contraction and other Ca²⁺-gated processes

In some neurons, Ca²⁺ entry through high-threshold Ca²⁺ channels participates in the generation of various forms of electrical activity such as dendritic Ca²⁺ spikes (Purkinje cell dendrites) and activation of Ca²⁺-sensitive channels such as Ca²⁺-activated K⁺ or Cl⁻ channels. However, the general role of Ca²⁺-dependent action potentials is to provide a local and transient increase of intracellular Ca²⁺ concentration. Under normal conditions, the intracellular Ca²⁺ concentration is very low, less than 10^{-7} M. The entry of Ca²⁺ ions through Ca²⁺ channels locally and transiently increases the intracellular Ca²⁺ concentration up to 10^{-4} M. This local [Ca²⁺]ᵢ increase can trigger Ca²⁺-dependent intracellular events such as exocytosis of synaptic vesicles, granules or sliding of the myofilaments actin and myosin. It thus couples action potentials (excitation) to secretion (neurons and other excitable secretory cells, see Chapter 7) or it couples action potentials to contraction (heart muscle cells). The influx of Ca²⁺ also couples neuronal activity to metabolic processes and induces long-term changes in neuronal and synaptic activity. During development, Ca²⁺ entry regulates outgrowth of axons and dendrites and the retraction of axonal branches during synapse elimination and neuronal cell death.

5.5 A NOTE ON VOLTAGE-GATED CHANNELS AND ACTION POTENTIALS

Voltage-gated Na⁺, K⁺ and Ca²⁺ channels of action potentials share a similar structure and are all activated by membrane depolarization. The Na⁺, Na⁺/Ca²⁺ and Ca²⁺ action potentials have a similar pattern: the depolarization phase results from the influx of cations, Na⁺ and/or Ca²⁺, and the repolarization phase results from

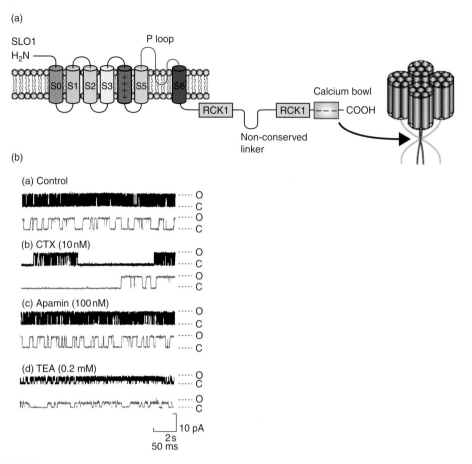

FIGURE 5.20   The big K channel.
(a) Schematic representation of SLO1 α-subunit. SLO1 subunit has an additional S0 membrane-spanning domain compared to delayed rectifier K⁺ channels and also includes an extensive cytosolic carboxy-terminal extension containing sites that sense cytosolic factors, such as the calcium bowl. RCK, regulators of conductance of K⁺ domain. **(b)** Single-channel activity of the big K channel (232 pS channel) in symmetrical 150 mM KCl at two different time bases ($V_H = + 40$ mV). From top to bottom: Control conditions. In the presence in the extracellular solution of, respectively, 10 nM charybdotoxin (CTX), 100 nM apamin, and 0.2 mM tetraethylammonium chloride (TEA). All the recordings are performed in the presence of 1.05 mM Ca2+ in the intracellular solution. Part (a) from Salkoff L, Ferreira G, Santi C, Wei A (2006) High-conductance potassium channels of the SLO family Nature reviews. *Neuroscience* **5**, 921–931. Part (b) from Chad J, Eckert R, Ewald D (1984) Kinetics of calcium-dependent inactivation of calcium current in voltage-clamped neurones in Aplysia californica. *J. Physiol. (Lond.)* **347**, 279–300, with permission.

the inactivation of Na⁺ or Ca²⁺ channels together with the efflux of K⁺ ions. However, these action potentials have at least one important difference. The Na⁺-dependent action potential is all-or-none. In contrast the Ca²⁺-dependent action potential is gradual. This reflects different functions. The Na⁺-dependent action potential propagates over long distances *without attenuation* in order to transmit information from soma-initial segment to axon terminals where they trigger Ca²⁺-dependent action potentials. Ca²⁺-dependent action potentials have

the general role of providing a local, *gradual* and transient Ca²⁺ entry.

FURTHER READING

Catterall WA (2000). Structure and regulation of voltage-gated Ca²⁺ channels. *Annu. Rev. Cell Dev. Biol.* **16**, 521–555.
Coetzee WA, Amarillo Y, Chiu J *et al.* (1999) Molecular diversity of K⁺ channels. *Ann. NY Acad. Sci.* **868**, 233–285.

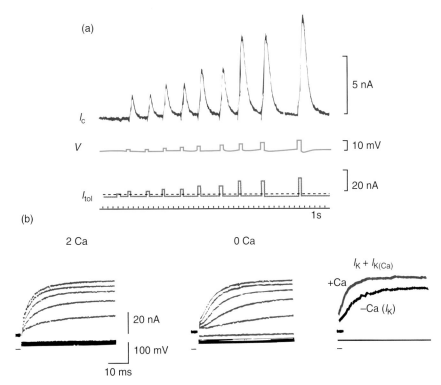

FIGURE 5.21   The macroscopic Ca^{2+}-activated K^+ current of bullfrog sympathetic neurons.
(a) Outward currents recorded in single-electrode voltage clamp at a holding potential of $-28\,mV$. In response to increasing $0.4\,s$ intracellular iontophoretic injections of Ca^{2+} from a microelectrode containing $200\,mM$ $CaCl_2$, increasing outward currents are recorded. (b) Outward currents recorded during voltage steps to -20, -10, 0 and $+20\,mV$ from a holding potential of $-50\,mV$ in the presence of $2\,mM$ external Ca^{2+} ($2\,Ca$) and a Ca-free external medium ($0\,Ca$). The leak current is subtracted. The two superimposed current traces recorded at the same potential in the presence ($+Ca$) or absence ($-Ca$) of external Ca^{2+} ions show that an early component of the outward current is present ($I_{K(Ca)}$) in the presence of Ca^{2+} ions. Adapted from Brown DA, Constanti A, Adams PR (1983) Ca^{2+}-activated potassium current in vertebrate sympathetic neurons. *Cell Calcium* **4**, 407–420, with permission.

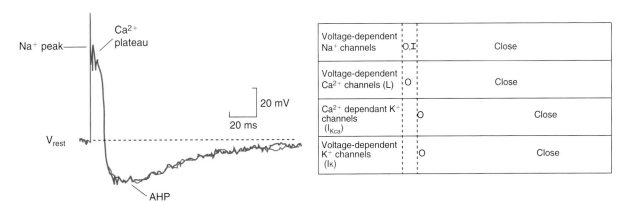

FIGURE 5.22   States of voltage-gates Na^+, Ca^{2+} and K^+ channels.
Different states in relation to the various phases of the Na^+/Ca^{2+}-dependent action potential. Example of the action potential recorded in olivary neurons of the cerebellum.

De Leon M, Wang Y, Jones J *et al*. (1995) Essential Ca²⁺-binding motif for Ca²⁺-sensitive inactivation of L-type Ca²⁺ channels. *Science* **270**, 1502–1506.

Denk W, Piston DW, Webb WW (1995) Two-photon molecular excitation in laser-scanning microscopy. In Pawley JB (ed.) *Handbook of Biological Microscopy*. New York: Plenum Press.

Elkins T, Ganetzky B, Wu CF (1986) A Drosophila mutation that eliminates a calcium-dependent potassium current. *Proc. Natl Acad. Sci. USA* **83**, 8415–8419.

Ertel EA, Campbell KP, Harpold MM *et al* (2000) Nomenclature of voltage-gated calcium channels. *Neuron* **25**, 533–535.

Grynkiewicz G, Poenie M, Tsien RY (1985) A new generation of calcium indicators with greatly improved fluorescence properties. *J. Biol. Chem.* **260**, 3440–3448.

Hodgkin AL and Huxley AF (1952) A quantitative description of membrane current and its application to conduction and excitation in nerve. *J. Physiol. (Lond.)* **117**, 500–544.

Miller C (1995) The charybdotoxin family of K⁺ channel-blocking peptides. *Neuron* **15**, 5–10.

Neher E and Augustine GJ (1992) Calcium gradients and buffers in bovine chromaffin cells. *J. Physiol. (Lond.)* **450**, 273–301.

Schreiber M, Yuan A, Salkoff L (1999) Transplantable sites confer calcium sensitivity to BK channels. *Nature Neurosci.* **2**, 416–421.

Stocker M (2004) Ca²⁺-activated K⁺ channels: molecular determinants and function of the SK family. *Nature Reviews Neuroscience* **5**, 758–770.

Stotz SC and Zamponi GW (2001) Structural determinants of fast inactivation of high voltage-activated Ca²⁺ channels. *Trends in Neurosciences* **24**, 176–181.

Tan YP, Llano I, Hopt A, Wuerrihausen F, Neher E (1999) Fast scanning and efficient photodetection in a simple two-photon microscope. *J. Neurosci. Meth.* **92**, 123–135.

Tsien RY (1989) Fluorescent probes of cell signalling. *Ann. Rev. Neurobiol.* **12**, 221–253.

Zamponi GW, Bourinet E, Nelson D, Nargeot J, Snutch TP (1997) Crosstalk between G proteins and protein kinase C mediated by the calcium channel alpha₁ subunit. *Nature* **385**, 442–446.

Zhong H, Li B, Scheuer T, Catterall WA (2001) Control of gating mode by a single amino acid residue in transmembrane segment IS3 of the N-type Ca²⁺ channel. *PNAS* **98**, 4705–4709.

APPENDIX 5.1 FLUORESCENCE MEASUREMENTS OF INTRACELLULAR Ca²⁺ CONCENTRATION

A5.1.1 The physical basis of fluorescence

The interaction of light with matter

Light is electromagnetic radiation that oscillates both in space and time, and has electric and magnetic field components that are perpendicular to each other. If for the sake of simplicity one focuses only on the electromagnetic component, it can be seen that the molecule, which is much smaller than the wavelength of light, will be perturbed by light because its electronic charge distribution will be altered by the oscillating electric field component of the light. Without resorting to complicated quantum mechanical calculations we can say that light will interact with matter via a resonance

phenomenon; i.e. the matter will absorb light only if the energy of the incoming photon is exactly equal to the difference between the potential energy of the lowest vibrational level of the ground state and that of one of the vibrational levels of the first excited state (**Figure A5.1**). The absorption of light therefore occurs in discrete amounts termed quanta. The energy E in a quantum of light (a photon) is given by:

$$E = h\nu = hc/\lambda \,,$$

where h is Planck's constant, ν and λ are the frequency and wavelength of the incoming light, and c is the speed of light in a vacuum. When a quantum of light is absorbed by a molecule, a valence electron will be boosted into a higher energy orbit, called *the excited state*. This phenomenon will take place in 10^{-15} s, resulting in conservation of the molecular coordinates. For the sake of simplicity the rotational energy levels are not taken into account and it is assumed that at room temperature the electrons will be at their lowest vibrational energy level.

The difference of energy between the vibrational levels being typically in the order of 10 kcal mol⁻¹, there is not enough thermal energy to excite a transition to higher vibrational levels at room temperature. One might thus assume that most of the electrons will lie at the lowest vibrational level of the ground state $S_v = 0$ (S for singlet, as the electrons spins are antiparallel). Because absorptive transitions occur to one of the vibrational levels of the excited state, had there been no interaction with the solvent molecules one could have measured the energy difference between the ground state and each of the vibrational levels of the excited

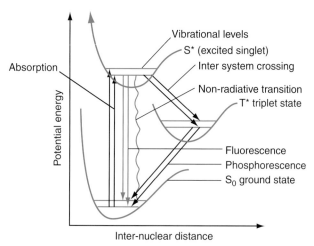

FIGURE A5.1 Pathways to excitation and de-excitation of an electron.

The rotational levels between the vibrational levels and higher excited states are not shown for the sake of simplicity.

state. This type of spectra can only be obtained for chemical compounds in the gaseous state. The absorption spectra under those circumstances would resemble narrowly separated bands; however, the interaction of the orbital electrons with solvent molecules will broaden those peaks, producing the absorption spectra of the more familiar form.

The return from the excited state

The electrons that have been promoted to one of the vibrational levels of the excited state will lose their vibrational energy through interaction with solvent molecules by a process known as *vibrational relaxation*. This process has a timescale much shorter than the lifetime of the electrons in the excited state (10^{-9} to 10^{-7} s for aromatic molecules). The electrons that have been promoted to the excited state will return to the ground state from the lowest-lying excited vibrational state, by one of the following ways.

Fluorescence emission

Some of the electrons in the excited state will return to one of the vibrational levels of the ground state by a radiative transition, whose frequency will be a function of the energy difference separating these levels. If one simply assumes that the energy spacing the vibrational levels of the excited and ground states are similar, one expects the fluorescence emission spectrum to be a mirror image of the absorption spectrum (**Figure A5.2**).

A further expectation will be that the $S_{v=0}$ to $S^*_{v=0}$ absorption will be at the same frequency as the $S^*_{v=0}$ to $S_{v=0}$ emission; however, this is rarely the case, as the absorption process takes place in about 10^{-15} s. The orientation of the solvent molecules with respect to the electronic states will be conserved as well as the quantum coordinates of the molecule; however, as the

excited level lifetimes are rather long, the solvent molecules will reorient favourably about the electronic levels, resulting in a difference in the zero-zero frequencies. This difference between $S_{v=0}$ to $S^*_{v=0}$ absorption and $S^*_{v=0}$ to $S_{v=0}$ emission is termed the *Stoke's shift*.

Non-radiative transition

In this process the excitation energy will be lost mainly by interactions with solvent molecules, resulting in some of the electrons of the excited state returning to the ground state with a non-radiative transition. This process is favoured by an increase in temperature, and can explain why increasing the temperature causes a decrease in fluorescence intensities.

Quenching of the excited state

The excitation energy might be lost through interactions, in the form of collisions of quenchers with the electrons in the excited orbital. Typical quenchers such as O_2, I^- and Mn^{2+} ions will quench every time they collide with an excited singlet.

Intersystem crossing

Intersystem crossing is a mechanically forbidden quantum process that occurs by a spin exchange of the electron of the excited singlet state, resulting in an excited triplet state T^* (**Figure A5.1**). As this process involves a forbidden transition its probability of occurrence will be extremely low; nevertheless it will occur because the potential energy of the excited triplet is usually lower than that of the excited singlet state. The electron in the excited triplet state can then become de-excited by a non-radiative transition, quenching, or by a radiative transition called *phosphorescence* (the light emitted will be of longer wavelength than fluorescence because of the lower potential energy of the excited triplet). One should note that the return to the ground state necessitates a novel forbidden transition $T^*_{v=0}$ to $S_{v=x}$ (x for any vibrational level of the ground state). The probability of this transition will be extremely low, for the same reasons given above, resulting in a long lifetime of the excited triplet state (seconds to days). This long-lived triplet state will result in a very weak intensity of radiation, will be prone to quenching by collisions with quenchers, and the non-radiative processes will compete well with the phosphorescence. Phosphorescence in solution will rarely be observed. In order to observe phosphorescence at all, one must rigorously remove oxygen from the medium, and should use rigid glasses at very low temperatures, in order to minimize the competing non-radiative processes.

Some of the electrons that have undergone intersystem crossing, and therefore are in the $T^*_{v=0}$ state, may

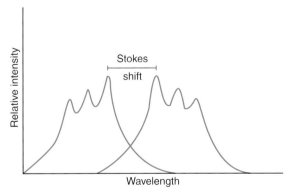

FIGURE A5.2 **Excitation (left) and emission (right) spectra of a hypothetical molecule.**
The excitation spectrum has the same peaks as the absorption spectrum; the separation between the individual peaks reflects the potential energy differences between the vibrational levels.

undergo a novel intersystem crossing to the S* level by the thermal energy provided by the solution, provided the energy difference between the T* and S* states is small; the return from the $S^*_{v=0}$ to $S^*_{v=x}$ level by fluorescence emission is called *delayed fluorescence* and has the effect of lengthening the fluorescence lifetime of the molecule beyond what is expected in normal fluorescence emissions.

Photolysis: bleaching and toxicity

The molecules in the excited state undergo certain chemical reactions resulting in the loss of fluorescence; this is called *photobleaching*. It is estimated that a good organic fluorophore can be excited about 10^4 to 10^5 times before it bleaches. Some of the reaction products might be damaging for the cell, resulting in phototoxicity. One of the important ingredients in bleaching is the interaction between the triplet state of the fluorophore (**Figure A5.1**) and molecular oxygen (O_2). The triplet state can transfer its energy to oxygen and bring it to its singlet excited state. Singlet oxygen is a reactive molecule that participates in many kinds of chemical reactions with organic molecules. As a result, the fluorophore looses its ability to fluoresce (it bleaches). In addition, the singlet oxygen can interact with other organic molecules causing phototoxicity for living cells. The minimal intensity of excitation and the minimal exposure time must be used in order to keep photobleaching and phototoxicity to a minimum.

A5.1.2 Fluorescence measurements: general points

Advantages

In the absence of fluorophore, provided there is no background fluorescence, the level of the signal is zero, so that even a very small change of fluorescence of the fluorophore is detected. This might need a large amplification, itself limited by the noise level of the amplifier chain.

Observation of fluorescence emission

The best fluorimeter should maximize collection of the fluorescence emission and minimize collection of excitation light. This is usually achieved by selecting a band of excitation wavelengths located outside the emission spectrum using filters (interference or combination filters), or monochromator on the excitation side and highpass or bandpass filters on the emission side. The emission-side filters pass wavelengths longer than the excitation wavelengths (remember the Stoke's shift).

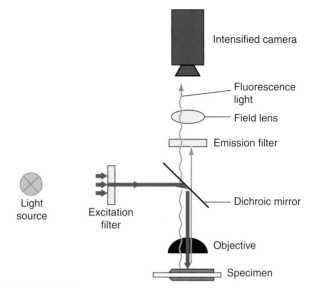

FIGURE A5.3   Epi-illumination microscopy.

For the measurement of fluorescence from individual cells, epi-illuminated fluorescence microscopes are used. The epiluminescence technique means that both the excitation and emission light have a common optical path through the objective. The key element of epi-illumination is the dichroic mirror; an interference mirror formed by successive depositions of dielectric layers on a transparent substrate. The dichroic mirror reflects the wavelengths below its cutoff frequency and transmits those that are above the cutoff. This cutoff frequency is chosen so that it reflects all of the excitation wavelengths, and transmits most of the emission wavelengths. The Stoke's shift is an aid in this respect. It is also possible to find polychroic mirrors that allow the simultaneous detection of many chromophores (**Figure A5.3**).

Confocal and multiphoton microscopy

The principle of confocal imaging was developed by Marvin Minsky in the mid-1950s. In a conventional (i.e., wide-field) fluorescence microscope, the entire specimen is flooded in light from a light source. All parts of the specimen throughout the optical path are excited and the fluorescence detected by a photodetector or a camera. The confocal microscope uses a laser to provide the excitation light (in order to get very high intensity of excitation and high resolution). The laser beam reflected from the dichroic beam splitter hits two mirrors which are mounted on motors and scan the sample. Dye in the sample fluoresces and the emitted light captured by the objective lens (epi-fluorescence) gets de-scanned by the same moving mirrors. The

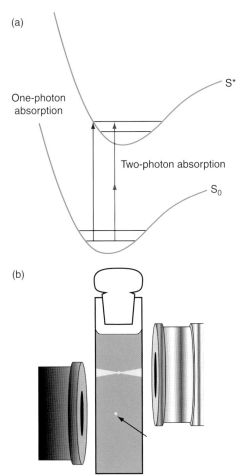

(a)

One-photon
absorption

S*

Two-photon absorption

S₀

(b)

FIGURE A5.4 Comparison of one-photon and two-photon absorption.
(a) Two photons in the red (right) combine their energies to get absorbed as a one blue photon (left). The energies of the photons can be thought as equal to the amplitude of the vectors. The two photons that get absorbed need not have equal energies. (b) Fluorescence emission profile produced by one-photon absorption occurs throughout the laser beam focused in a fluorescent solution by the objective on the right. With the two-photon scheme, excitation is limited to the focal point of the objective on the left (shown by the arrow) providing inherent 3-dimensional resolution.

fluorescence passes through the dichroic and is focused onto the pinhole in an optically conjugate plane. The pinhole in front of the detector eliminates out-of-focus light. Only the fluorescence within the focal plane is detected by the detector, i.e. a photomultiplier tube. In fact, there is not a complete image of the sample at any given instant; only one point of the sample is observed (laser scanning confocal microscopy: LSCM). The detector is connected to a computer which builds up the image, one pixel at a time. In practice, this can be done around twice per second for a 512×512 pixel image. The time limitation is due to the scanning mirrors. Several new features have been

recently developed to improve speed performance (spinning nipkow disc, multibeam excitation, linned 'pinhole' etc).

The resolution improvement due to confocal microscopy, either in xy or in z dimension, and the general use of laser as light source, represent a real revolution in imaging biological material. Despite these benefits, the confocal approach presents some limitations for specific applications such as deep imaging or long duration experiments in living tissue. Indeed, biological tissues strongly scatter visible light and prevent more than $100\,\mu m$ depth images by defect of excitation light. On the other hand, increasing the laser power is not the solution because thermal and photolysis effects will definitively affect the sample. Despite the fact that laser beam excites only the fluorophores in its path, scanning the sample induces a large photobleaching effect in the entire light cone (**Figure A5.4b**). One has to deal with the laser power for the preservation of the sample.

In order to alleviate some of these limitations Denk *et al*. in 1990 took advantage of an old physical theoretic prediction and of new powerful pulsed lasers to develop multiphoton microscopy. In 1931, Maria Goeppert-Mayer predicted the possibility of simultaneous absorption by a molecule of two photons of long wavelength, combining their energies to cause the transition of the molecule to the excited state (remember that the energy of a photon is inversely proportional to its wavelength $\lambda : E = hc/\lambda$). This can be viewed as two IR (near infrared) photons being absorbed simultaneously by a molecule normally excited by UV (**Figure A5.4a**). This technique, however, did not find practical use until the advent of very short pulse-width lasers for the following reasons.

The probability of two-photon absorption is $\sim 10^{31}$ times lower than the probability of one-photon absorption, and therefore does not occur under normal illumination conditions. Typical cross-sections for one-photon absorption are of the order of 10^{-16} cm²; for the two-photon case they are 10^{-48} cm⁴ s⁻¹. The two-photon cross-sections are cited in GM (Goeppert-Mayer) units, with 1 GM being 10^{-50} cm⁴ s⁻¹. In order for this absorption to occur, very high density photon fluxes confined to a small volume are needed. This was made possible by the advent of pulsed lasers, which typically generate pulses of 70 to 100 femtoseconds (1 fs = 10^{-15} s) width, each at power levels of 500 kW (1–3 W in average) and repetition rates of 80 MHz. For fluorescence measurements of biological samples, 10 mW laser intensity on the specimen plane are typically used. Under these conditions the excitation is confined to the focal volume only, as the necessary photon flux can only be reached at this plane. This has two very important implications: (i) as the excitation is limited to the focal plane, the emission is also limited to this plane (**Figure A5.4b**), resulting in an intrinsic 'confocal' image. At this time, image reconstruction does not need any pinhole and allows the use of 'non-descanned' detectors placed just after the condenser or the objective (transmission or epifluorescence). This short-length optical tract is several times more efficient than the confocal one to collect emitted fluorescence and greatly improves recorded signals; (ii) other chromophores in the light cone are not excited, so photodamage and phototoxicity resulting from photolysis of the chromophore are greatly reduced. Light at long wavelengths is less prone to scattering and better penetrates biological tissue, enabling researchers to measure Ca^{2+} dynamics in a non-invasive way and from deeper locations. It is possible to measure Ca^{2+} signals from rat brain at a depth of $800\,\mu m$ from the surface. Absorption is not limited to two photons only. Triple-photon absorption by UV dyes (DAPI…) or by nucleic acids with cross-sections as large as $10^{-75}\,cm^6\,s^{-2}$ has been reported. This technique currently has the drawback of requiring rather expensive lasers, but one can expect prices to come down in the future, allowing their routine use.

A5.1.3 Measurement of ion concentration by fluorescence techniques

The main requirement for an indicator to report the concentration of an ion is a change in its optical properties, and at the same time it must be highly specific for the ion in question, at physiological pH values. Furthermore, its binding and release from the ion must be faster than the kinetics of the intracellular ionic changes. Its affinity to that ion should be compatible with physiological conditions and prevent buffering effect. One can therefore envisage the production of probes that will change their absorption, bioluminescence (such as aequorin) or fluorescence properties as a function of ion concentration. Fluorescence is the technique of choice because of its higher sensitivity. In fluorescence measurements, the change in optical property sought to report an ionic concentration might be a change in quantum yield, excitation spectra or emission spectra.

Indicators for Ca^{2+}

Richard Tsien and colleagues have developed many probes sensitive to the free Ca^{2+} ions concentration. The common property of these probes is that they are all fluorescent derivatives of the calcium chelator BAPTA, which in turn is an aromatic analogue of the commonly used calcium chelator (EGTA, ethyleneglycol bis (13-aminoether) -N,N,N',N' tetra-acetic acid) (**Figure A5.5**). The probes form an octahedral complex,

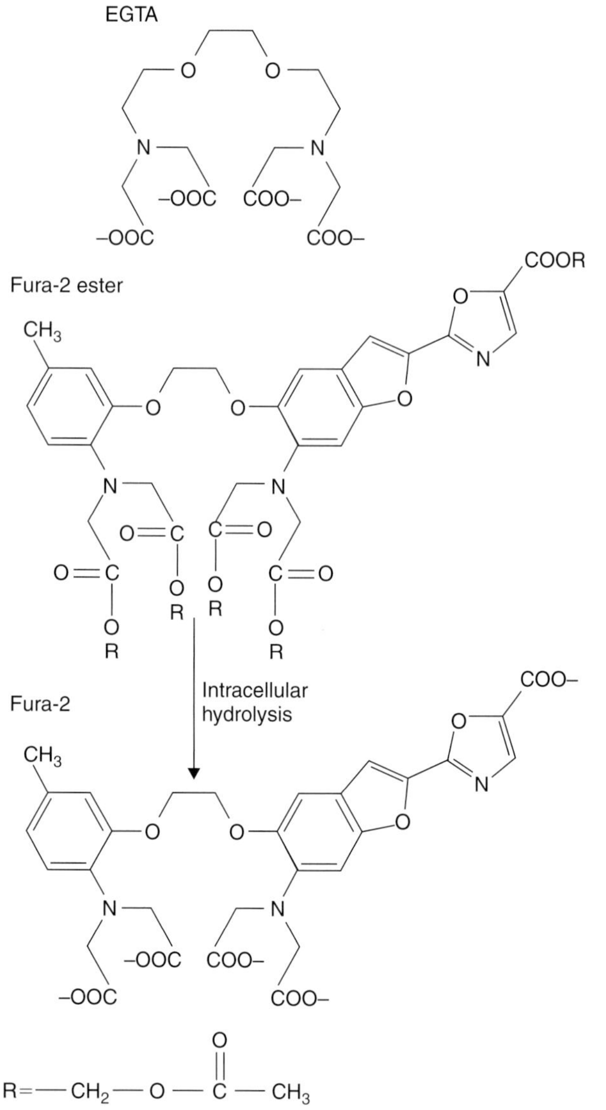

FIGURE A5.5 Chemical structure of Fura-2.
Note the similarities between Fura-2 and the acetoxymethylester variety Fura-2AM and EGTA. The AM variety is membrane permeant, and is de-esterified by intracellular esterases, liberating Fura-2, formaldehyde and acetate ions.

with the calcium ion at the centre of the plane formed by the COO⁻ groups of the carboxylic acid. The binding and unbinding of the ion induces a strain or relaxation on the electron cloud of the aromatic groups, which in turn results in changes of the spectral properties of the reporter chromophore.

Once these probes synthesized, several strategies were developed to make them penetrate in the cell, including direct injection of the chromophore salt through the patch clamp pipette allowing Ca^{2+} measurement combined with electrophysiological monitoring from a single neuron. Another way is to neutralize carboxylic groups with ester residues to form acetoxy-methylester variety (AM). This neutral molecule diffuses through the plasma membrane and is de-esterified enzymatically to produce functional chromophore (**Figure A5.5**), this is convenient to load a population of cell either *in vivo* or *in vitro*.

Three such reporter chromophores have found much use in the measurement of intracellular free Ca^{2+} concentrations, namely INDO, Fura-2, and FLUO-3. Each of these probes has a certain number of advantages over the others, depending on the measurement technique sought. INDO and Fura-2 are ratiometric probes; i.e. the change in spectral properties occur at two different wavelengths, and by measuring the fluorescence intensities at these two wavelengths and taking their ratio; one can calculate the absolute value of the free Ca^{2+} ion concentration within the cytosol, given by the following formula (see Grynkiewicz *et al.*, 1985):

$$[Ca^{2+}] = K_i\{(R - R_{min})/(R_{max} - R)\},$$

where R_{min} is the ratio at two wavelengths at zero ion concentration, R_{max} is the ratio at 'infinite' ion concentration, R is the ratio of the measurements, and K_i is constant unifying instrumental parameters together with the K_D of the chromophore for calcium. The major advantage of ratiometric probes is the fact that they are insensitive to the intensity of the emitted light, which changes from the centre to the periphery of most of the cells. This is because of differences in thickness at the centre and towards the edges, so there are more chromophores in the centre than at the edges.

INDO's emission properties at 405 nm and 480 nm change upon binding to Ca^{2+} ($\lambda_{exc} = 350$ nm). The two emission intensities can easily be measured by using a beam splitter, two interference filters and two photomultipliers; it is fairly difficult to envisage the use of two intensified cameras to form an image unless one uses a specifically split CCD array. Therefore INDO has been applied in processes that require either rapid determination of the free calcium concentration (i.e. cell sorting), or where the kinetics of the free calcium change are fast.

Fura-2, upon binding to calcium, undergoes a change in its absorption spectrum and therefore in its excitation spectrum; namely the emission intensity (collected at $\lambda_{em} = 510$ nm and higher) increases at $\lambda_{exc} = 340$ nm and decreases at $\lambda_{exc} = 380$ nm (**Figure A5.6**). A typical property of all the indicators that undergo either an excitation or emission shift is the presence of an 'isosbestic point', namely the presence of a 'unique point' in the spectrum when the parameter concentration is changed (Ca^{2+} in the case of Fura-2). The isosbestic point is only present when two species are in equilibrium (in our case calcium-bound and free forms of Fura-2 or of INDO). The

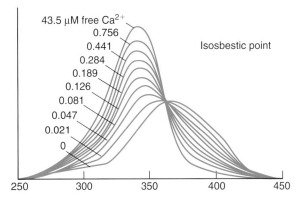

FIGURE A5.6 **Excitation spectral changes of Fura-2 as a function of Ca²⁺ concentration**.
Each curve represents the intensity of fluorescence emitted by Fura-2 (at $\lambda = 510$ nm) as a function of the wavelength of excitation (from 250 to 450 nm) and for a given Ca^{2+} concentration (from 0 to 43.5 μM). Knowing that Fura-2 + $Ca^{2+} \rightleftharpoons$ Fura-2-Ca, the curve obtained in the presence of the maximal Ca^{2+} concentration (43.5 μM) represents the excitation spectrum of the bound form of Fura (Fura-2-Ca). In contrast, the curve obtained for the minimal CA^{2+} concentration (0 μM) represents the excitation spectrum of the free form of Fura (Fura-2). It appears clearly that measured at $\lambda = 340/350$ nm and at $\lambda = 380$ nm, the intensity of fluorescence emitted by Fura-2 varies with the ratio free/bound forms of Fura; i.e., with Ca^{2+} concentration.

absence of this point can be taken as an indicator for the contamination by an other ion. This point appears at 360 nm for Fura-2 (**Figure A5.6**). As Fura-2 undergoes a change in its absorption properties, alternating the excitation filters at the two chosen wavelengths, mostly at 340 and 380 nm, and collecting the emission above 510 nm with an intensified camera, one can construct the free calcium image, or the time series of the changing free Ca^{2+} concentration in a living cell, by calculating the free Ca^{2+} concentration at each pixel (picture element). One is not limited to these two wavelengths; it might even be advantageous to take the images at longer wavelengths than 340 nm as most of the optical path of old fluorescence microscopes is opaque to this wavelength.

With the advent of confocal microscopy, an indicator with absorption properties in the visible part of the spectrum was needed (confocal microscopes use laser scanning, and ultraviolet lasers have been recently developed but are still too expensive). FLUO-3, Calcium Green and Oregon Green were developed to respond to this need. The main disadvantage of these probes is that their quantum efficiency changes at one wavelength only (~530 nm, when excited at the 488 nm line of the argon laser) upon binding to Ca^{2+} ions. It is therefore not possible to measure absolute values of Ca^{2+} concentrations directly; nevertheless, if the resting level of the free Ca^{2+} concentration in the cell is known, the values obtained before stimulation can be used to calculate the approximate value of the free Ca^{2+} concentration under stimulation.

Experience shows that fluorescence intensity always seems too low, and it is tempting to increase the concentration of the reporter molecule inside the cell to overcome this problem. In the case of Ca^{2+} measurements this will have the adverse effect of buffering free Ca^{2+} ions and to prevent its rise (BAPTA-like backbone). It is necessary to find a compromise between the signal-to-noise level and the buffering of the ion in general, the best approach being the use of the least amount of indicator required for the job.

Indicators for Mg²⁺ and other divalents

Mag-Fura and Magnesium Green are Mg^{2+} indicators. They are designed around the same EGTA chelator structure as for Ca^{2+} indicators. Mg^{2+} indicators are designed to respond maximally to the Mg^{2+} concentrations commonly found in cells – typically 0.1–6 mM. They also bind Ca^{2+} with a low affinity. Typical physiological Ca^{2+} concentrations (10 nM–1 μM) do not usually interfere with Mg^{2+} measurements. Although Ca^{2+} binding to Mg^{2+} indicators can be a complicating factor in Mg^{2+} measurements, this property can also be exploited for measuring high Ca^{2+} concentrations (1–100 μM) such as those seen in the mitochondria. Mag-Fura and Magnesium Green do have similar spectral properties as their calcium counterparts.

Most of the reporter molecules synthesized for Ca^{2+} or Mg^{2+} also interact with other polyvalent ions such as Tb^{3+}, Cd^{2+}, Hg^{2+}, Ni^{2+} and Ba^{2+}, and in some cases with better quantum yields. This annoying property can be turned into advantage, when one wants to measure changes in the concentrations of those ions.

Indicators for Na⁺ and K⁺

SBFI and PBFI are designed around a crown ether chelator to which benzofuranyl chromophores are linked, conferring to those molecules the same spectroscopic properties as Fura. Hence the same filter sets can be used as for Fura. The cavity size of the crown ether is the factor which determines the specificity of the molecule for Na^+ or K^+. The specificities of both SBFI and PBFI for their respective ions is much smaller than that of Fura for Ca^{2+}, and the K_D changes as a function of the concentration of the other ion, the ionic strength, pH and temperature.

Sodium Green is designed around a crown ether chelator to which two dichlorofluorescein chromophores are linked, resulting in similar spectroscopic properties as Calcium Green (i.e. excited at 488 nm). The cavity size of the crown ether results in a greater selectivity for Na^+ over K^+ compared with SBFI – 41 versus

18 times, respectively. The spectral properties, however, result in emission changes at one wavelength only, so ratiometric measurements with this reporter molecule are not possible.

All of the cation reporter molecules suffer K_D changes as a result of intracellular interactions as mentioned above, so they need to be calibrated *in situ* using pore-forming antibiotics like gramicidin and loading the cells with known ionic conditions. Another point which must be borne in mind is the fact that protein dye interactions might dampen or completely eliminate the signals.

Indicators for Cl⁻

All of the chloride indicators are based on methoxyquinolinium derivatives and report the chloride by the diffusion-limited collisional quenching of the chromophore in the excited state interacting with the halide ion. The quenching is not accompanied by spectral shifts, so ratiometric measurements are not possible. As the quenching depends on collisional encounters of the halide ion, it is very sensitive to intracellular viscosity and temperature. The quenching efficiency is greater for the other halides such as Br^- and I^-.

Voltage-sensitive dyes

Voltage-sensitive dyes enable us to measure membrane potential in cells or organelles that are too small for microelectrode impalement. Moreover, these probes can map variations of membrane potential with spatial resolution and sampling frequency that are difficult to achieve using microelectrodes, such as cells microdomains or full network studies. Potentiometric probes include many chemical structures (styryl, carbocyanines and rhodamines, oxonols derived … as examples), the existence of numerous dyes analogs reflects the observation that no single dye provides the optimal response under all experimental conditions. Selecting the best voltage-sensitive probe might be empirical and the choice of the class of dye is determined by different factors such as accumulation in cells, response mechanisms, toxicity or kinetics of the electrical events observed. Voltage-sensitive dyes are divided into two categories concerning this last parameter: (i) fast-response probes that are sufficiently fast to detect transient (millisecond) potential changes in excitable cells. However, the magnitude of their potential-dependent fluorescence change is often small; they typically show a 2–10% fluorescence change per 100 mV; (ii) slow-response probes; the magnitude of their optical responses is much larger than that of fast-response probes (typically a 1% fluorescence change per mV). Slow-response probes are suitable for detecting changes in average membrane potentials of non-excitable cells caused by respiratory activity, ion-channel permeability, drug binding and other factors.

FURTHER READING

Denk W, Strickler JH, Webb WW (1990) Two-photon laser scanning fluorescence microscopy. *Science* **248**, 73–76.

Tsien RY (1989) Fluorescent probes of cell signaling. *Annu. Rev. Neurosci.* **12**, 227–253.

http://www.molecularexpressions.com/primer/techniques/ fluorescence/filters.html

http://www.molecularexpressions.com/index.html

http://probes.invitrogen.com/handbook/

APPENDIX 5.2 TAIL CURRENTS

Tail currents are observed in voltage or patch clamp experiments. 'Tail' means that the voltage-gated current is observed at the end of a depolarizing voltage step, upon sudden removal of the depolarization of the membrane. Tail currents do not exist in physiological conditions; they are 'experimental artifacts'. However, there are several reasons for studying tail currents: they are tools for determining characteristics of currents such as reversal potential and inactivation rate constants. Tail currents were first described by Hodgkin and Huxley (1952) in the squid giant axon.

Single-channel tail current

In patch clamp recordings of the activity of a single voltage-gated channel, a unitary current of much larger amplitude is occasionally observed at the end of the voltage step (**Figure A5.7**). It corresponds to the current flowing through a channel that is not yet closed at the end of the depolarizing step. Therefore tail currents are recorded for voltage-gated channels that do not rapidly close or inactivate during a depolarizing step, such as delayed rectifier K$^+$ or L-type Ca^{2+} channels.

The activity of an L-type Ca^{2+} channel is recorded in patch clamp (cell-attached patch) in the presence of the selective agonist Bay K 8644. On stepping back the membrane to the holding potential, the L-type Ca^{2+} channel opened by the preceding depolarization does

not immediately close since the transition O $\rightleftharpoons$ C is not immediate. The inward unitary Ca^{2+} current recorded at this moment is larger (**Figure A5.7**) because of the larger driving force upon removal of depolarization than during the depolarizing step: during the depolarizing step to 0 mV, $i_{Ca} = \gamma_{|Ca} (V_m - E_{Ca}) = \gamma_{|Ca} (0 - 50) = -50\gamma_{|Ca}$; upon removal of depolarization $i_{Ca} = \gamma_{|Ca} (V_m - E_{Ca}) = \gamma_{|Ca}(-60 - 50) = -110\gamma_{|Ca}$.

Then, after a few milliseconds, owing to closing of the channel, the tail current returns to zero (the voltage-gated channel closes in response to the repolarization of the membrane).

Whole-cell tail current

In voltage or whole-cell patch clamp recordings (in the presence of Na$^+$ and K$^+$ channel blockers), a voltage step to 0 mV from a holding potential of −40 mV activates a number N of L-type Ca^{2+} current is recorded. At the end of the voltage step a Ca^{2+} current of larger amplitude and small duration is always recorded: the tail Ca^{2+} current (**Figure A5.8**). Then the amplitude of this tail current progressively diminishes. The peak of the whole-cell tail current has a larger

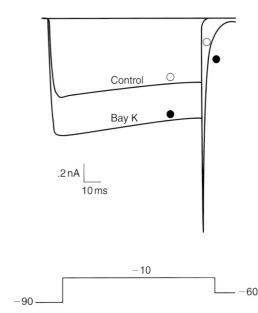

FIGURE A5.8 Activity of a dorsal root ganglion neuron.
Recorder in single-electrode voltage clamp in the presence of Na$^+$ and K$^+$ channel blockets and 2 mM external Ba^{2+}. A depolarization to −10 mV followed by a repolarization to −60 mV is applied to the membrane from a holding potential V_H = −90 mV. The depolarizing step evokes an inward Ba^{2+} current followed by an inward Ba^{2+} tail current (control, open circle). The presence of 1 μm Bay K 8644 increases the amplitude of the Ba^{2+} current during the step. It also prolongs the Ba^{2+} tail current (black circle). Adapted from Carbone E, Formenti A, Pollo A (1990) Multiple actions of Bay K 8644 on high-threshold Ca channels in adult rat sensory neurons. *Neurosci. Lett.* **111**, 315–320, with permission.

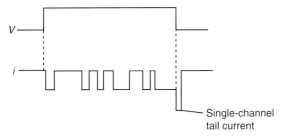

FIGURE A5.7 Activity of an L-type Ca^{2+} channel.
Recorded in patch clamp (cell-attached patch) in the presence of 110 mM external Ba^{2+}. In response to a depolarizing step in the presence of 5 μM Bay K 8644, a single-channel current of larger amplitude is recorded upon repolarization. It is a single-channel Ca^{2+} tail current.

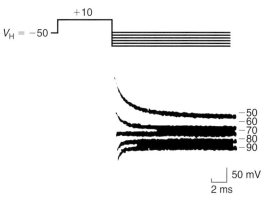

FIGURE A5.9   Activity of a chick dorsal root ganglion cell.
Recorded in single-electrode voltage clamp, in the presence of 1 μm TTX and 10 mM CO^{2+} to block, respectively, Ca^{2+} and Na^+ channels. Depolarizations to +10 mV from a holding potential of −50 mV followed by successive repolarizations to −50, −60, −70, −80 and −90 mV are applied (V traces). Bottom I traces show the K^+ tail currents at the corresponding membrane potentials (the outward K^+ current during the step is not shown). The reversal potential of the K^+ tail current is −70 mV. It indicates the value of E_K in these cells. To separate the ionic tail current from the capacitive current, the latter was subtracted from the total current by digital summation of the currents elicited with identical depolarizing and hyperpolarizing test pulses. Adapted from Dunlap K, Fischbach GD (1981) Neurotransmitters decrease the calcium conductance activated by depolarization of embryonic chick sensory neurones. *J. Phsiol.* **317**, 519–535, with permission.

amplitude than that of the whole-cell current recorded during the voltage step since the driving force for Ca^{2+} ions is larger upon removal of depolarization than during the depolarization, as explained above.

The tail current diminishes progressively owing to the progressive closure of the N open Ca^{2+} channels: the channels do not all close at the same time once the membrane is repolarized. The whole-cell tail current of **Figure A5.8** represents the summation of hundreds to thousands of recordings of single-channel tail currents.

In **Figures A5.7** and **A5.8**, the tail currents are inward. The direction of a tail current (as for any type of current) depends on the sign of the driving force; i.e. the value of membrane potential upon repolarization (V_H) and that of the reversal potential of the current (E_{rev}) which depends on the ions flowing through the open channels. By varying the voltage at the end of the depolarizing step the tail current varies in amplitude and direction (inward to outward or the reverse) and it is possible to determine the reversal potential of the tail current under study: when $V_H = E_{rev}$ the tail current is equal to zero (**Figure A5.9**). This values of E_{rev} is the same for the tail

current and the current recorded during the voltage step since it concerns the same channels.

The voltage protocol of **Figure A5.9** allows the determination of E_{rev} and consequently identification of the type of ions that carry the current E_{rev} can also be determined directly by changing the voltage-step value. However, for K^+ channels for example, E_{rev} is near −100 mV, a membrane potential where the open probability of voltage-gated channels is very low. By using tail currents, this problem is overcome.

FURTHER READING

Bertolino M and Llinas RR (1992) The central role of voltage-activated and receptor-operated calcium channels in neuronal cells. *Ann. Rev. Pharmacol. Toxicol.* **32**, 399–421.

Caterall WA (1998) Structure and function of neuronal Ca^{2+} channels and their role in transmitter release. *Cell Calcium* **24**, 307–323.

De Leon M, Wang Y, Jones J et al. (1995) Essential CA^{2+}-binding motif for Ca^{2+}-sensitive inactivation of L-type CA^{2+} channels. *Science* **270**, 1502–1506.

Denk W, Piston DW, Webb WW (1995) Two-Photon molecular excitation in laser-scanning microscopy. In Pawley JB (ed.) *Handbook of Biological Microscopy.* New York: Plenum Press.

Felix R (1999) Voltage-dependent Ca^{2+} channel alpha$_2$-delta auxiliary subunit: structure, function and regulation. *Recept. Chann.* **6**, 351–362.

Grynkiewicz G, Poenie M, Tsien RY (1985) A new generation of calcium indicators with greatly improved fluorescence properties. *J. Biol. Chem.* **260**, 3440–3448.

Hodgkin AL and Huxley AF (1952) A quantitative description of membrane current and its application to conduction and excitation in nerve. *J. Physiol. (Lond.)* **117**, 500–544.

Hofmann F, Lacinova L, Klugbauer N (1999) Voltage-dependent calcium channels: from structure to function. *Reo. Physiol. Biochem. Pharmacol.* **139**, 33–87.

Miller C (1995) The charybdotoxin family of K^+ channel-blocking peptides. Neurons **15**, 5–10.

Neher E and Augustine GJ (1992) Calcium gradients and buffers in bovine chromaffin cells. *J. Physiol.(Lond.)* **450**, 273–301.

Randall A and Benham CD (1999) Recent advances in the molecular understanding of voltage-gated Ca^{2+} channels. *Cell Neurosci.* **14**, 255–272.

Stea A, Soong TW, Snuth TP (1995) Determinants of PKC-dependent modulation of a family of neuronal calcium channels. *Neuron* **15**, 929–940.

Tan YP, Llano I, Hopt A, Wuerrihausen F, Neher E (1999) Fast scanning and efficient photodetection in a simple two-photon microscope. *J. Neurosci. Meth.* **92**, 123–135.

Tsien RY (1989) Fluorescent probes of cell signalling. *Ann. Rev. Neurobiol.* **12**, 221–253.

Varadi G, Mori Y, Mikala G, Schwartz A (1995) Molecular determinants of Ca^{2+} channels function and drug action. *Trend. Pharmacol. Sci.* **16**, 43–49.

Zamponi GW, Bourinet E, Nelson D, Nargeot J, Snutch TP (1997) Crosstalk between G proteins and protein kinase C mediated by the calcium channel alpha$_1$ subunit. *Nature* **385**, 442–446.

6

The chemical synapses

In 1888, Ramon y Cajal suggested that the contacts between the axon terminals of a neuron and the dendrites or the perikaryon of another neuron are the points at which information flows from one neuron to the other: 'Les articulations ou contacts utiles et efficaces entre neurones ne s'effectuent qu'entre cylindre-axiles, collatérales ou terminales d'un neurone et les prolongements ou le corps cellulaire d'un autre neurone.' The term *synapse* was introduced by Sherrington (1897) to describe these zones of contact between neurons, specialized in the transmission of information.

In fact, the term 'synapse' is not used exclusively to describe connections between neurons (interneuronal connections) but also those between neurons and effector cells such as muscular and glandular cells (neuroeffector synapses) and those between receptive cells and neurons (**Figure 6.1**). These contacts are the points where the information is transmitted from one cell to the other: synaptic transmission.

According to morphological and functional criteria, there are various types of synapses, including chemical, electrical and mixed types.

Chemical synapses

These are characterized morphologically by the existence of a space between the plasma membranes of the connected cells. These spaces are called *synaptic clefts*. In this case, a molecule – the neurotransmitter – conveys information between the presynaptic cell and the postsynaptic cell. Chemical synapses will be described in this chapter (**Figure 6.2a**). Some of the chemical synapses have particular characteristics:

- *Reciprocal synapses* are formed by the juxtaposition of two chemical synapses oriented in the reverse direction to each other (**Figure 6.2d**).
- *Glomeruli* are formed by a group of chemical synapses. In some cases a group of dendrites form

chemical synapses with the axon they surround (**Figure 6.2e**). In other cases, numerous axon terminals form synapses with the dendrite they surround.

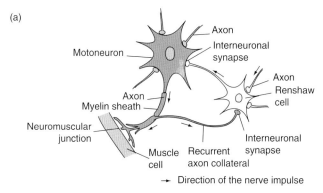

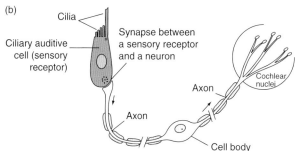

FIGURE 6.1   Types of cells connected by chemical synapses.
(a) Interneuronal synapses and neuromuscular junction. Example of synapses between a motoneuron (Golgi type I neuron that innervates striated muscle fibres) and a Renshaw cell (a Golgi type II neuron in the spinal cord) and between a motoneuron and a striated muscle cell. **(b)** Synapse between a sensory receptor and a neuron. Example of synapses between an auditory receptive cell (ciliary cell in the cochlea) and a primary sensory neuron whose cell body is located in the spiral ganglion. This neuron is free of dendrites and has a T-shaped axon that drives sensory information from the periphery to the central nervous system. Drawing (a) from Eckert R, Randall D, Augustine G (1988) *Animal Physiology*, New York: W. A. Freeman, with permission.

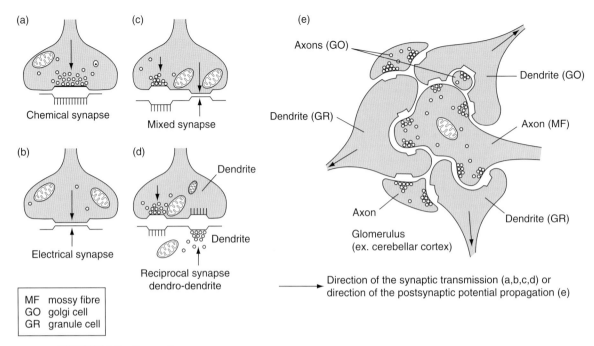

FIGURE 6.2 **Types of synapses.**
See text for explanations. MF, mossy fibre; GO, Golgi cell; GR, granule cell. Parts **(a)–(d)** from Bodian D (1972)
Neuronal junctions: a revolutionary decade. *Anat. Rec.*, **174**, 73–82, with permission. Drawing **(e)** from Steiger
U (1967) Uber den Feinbau des Neuropils im Corpus pedunculatum des Waldaneise. *Z. Zelforsch.* **81**, 511–536,
with permission.

- *Electrical synapses* or gap junctions are characterized by the apposition of the plasma membranes of the connected cells. In this case the ions flow directly from one cell to the other without the use of a chemical transmitter. Gap junctions also allow the exchange of small-diameter intracellular molecules such as second messengers and metabolites (**Figure 6.2b**). These synapses are common between glial cells in the mammalian central nervous system.

- *Mixed synapses* are formed by the juxtaposition of a chemical synapse and a gap junction (**Figure 6.2c**). In mammals, between neurons, these synapses are more common than the electrical synapses.

6.1 THE SYNAPTIC COMPLEX'S THREE COMPONENTS: PRESYNAPTIC ELEMENT, SYNAPTIC CLEFT AND POSTSYNAPTIC ELEMENT

This section takes as an example the interneuronal chemical synapses. Under an electron microscope, a section of brain tissue taken from a region of the central nervous system rich in cell bodies and dendrites (grey matter) reveals many synaptic contacts at the surface of a dendritic shaft or of a dendritic spine (**Figure 6.3**, arrows). One of these synaptic contacts represents a synaptic complex (**Figure 6.4a**). The synaptic complex

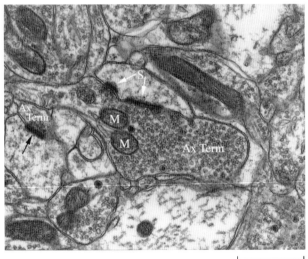

1μm

FIGURE 6.3 **Axo-spinous synapses.**
One of which (centre) is a 'perforated' synapse. Microphotography of a section of the hippocampus (molecular layer of the fascia dentata) observed under the electron microscope. Two synaptic boutons (Ax Term.) filled with synaptic vesicles and forming one or two asymmetric synaptic contacts (arrows) with dendritic spines (S) of pyramidal neurons can be visualized. M: mitochondries, Bar: 1 micrometre (Microphotography Alfonso Represa).

includes three components: the presynaptic element, the synaptic cleft and the postsynaptic element. The synaptic complex is *the non-reducible basic unit* of each

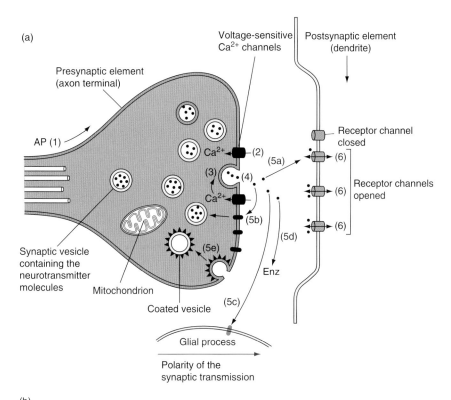

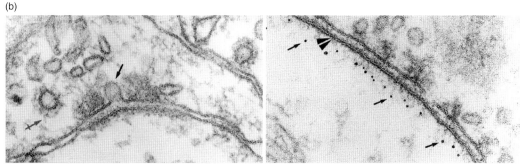

FIGURE 6.4   **Pre- and postsynaptic specializations.**
(a) Schematic of the synaptic transmission (AP, action potential; see text for explanation). **(b)** Electron photomicrographs of transverse sections at the level of synaptic complexes. Left: A figure of exocytosis (long arrow) between two dense presynaptic projections. A coated vesicle (crossed arrow) characteristic of the recycling of the membrane is also seen (inhibitory synapse afferent to the Mauthner cell in the fish). Right: Postsynaptic localization of the glycine receptors, visualized with gold particles associated to a specific monoclonal antibody (single arrow). These are lined up at distance from the membrane; this space originates mainly from the aggregation of the antibodies used to label indirectly the receptors (inhibitory synapse afferent to a motoneuron in the spinal cord of the rat). Part (b, left) from Triller A and Korn H (1985) Activity-dependent deformations of presynaptic grids at central synapses. *J. Neurocytol.* **14**, 177–192, with permission. Part (b, right) from Triller A, Cluzeaud F, Pfeiffer F, Korn H (1986) Distribution and transmembrane organization of glycine receptor at central synapses: an immunocytochemical touch. In Levi-Montalcini R *et al.* (eds) *Molecular Aspects of Neurobiology*, Berlin: Springer Verlag, with permission.

chemical synapse since it includes the minimal requirement for efficient chemical synaptic transmission.

6.1.1   The pre- and postsynaptic elements are morphologically and functionally specialized

The presynaptic element is characterized by the presence of numerous mitochondria and synaptic vesicles

which store the neurotransmitter (**Figures 6.3** and **6.4**). Two types of synaptic vesicles are described: the clear vesicles (40–50 nm in diameter) and the dense-core vesicles or dense granules, which have an electron-dense core (40–60 nm in diameter). Occasionally, under the presynaptic membrane can be seen an electron-dense zone with a geometry more or less distinguishable, the presynaptic grid (see **Figure 6.10**). It corresponds to a particular organization of the cytoskeleton which is related to the exocytotic machinery.

The postsynaptic element in the interneuronal synapses is characterized by a submembranous electron-dense zone, which most probably corresponds to the region where the postsynaptic receptors are anchored. In cases where the postsynaptic element is non-neuronal, we shall see that various other postsynaptic specializations exist.

The synaptic complex displays a particular asymmetric structure, the synaptic vesicles being present only in the presynaptic element. This structural asymmetry suggests a functional asymmetry.

6.1.2 General functional model of the synaptic complex

A general functional model of chemical synaptic transmission is as follows. The newly synthesized neurotransmitter molecules are stored in the synaptic vesicles present in the presynaptic element. In a non-depolarized presynaptic element, the voltage-sensitive Ca^{2+} channels are closed and Ca^{2+} ions cannot enter the intracellular space. The exocytosis of synaptic vesicles is normally triggered by an increase of the intracellular Ca^{2+} concentration. Then, while the presynaptic membrane is at rest (i.e. as long as it is not depolarized by the arrival of an action potential), the probability of exocytosis of a synaptic vesicle and the release of its content into the synaptic cleft is very low: the neurotransmitter molecules are not released in significant quantities into the synaptic cleft. There is no synaptic transmission.

Now, when Na^+-dependent action potentials (AP) propagate to axon terminals, they induce a depolarization of the presynaptic membrane (**Figure 6.4a**, 1). This results in opening of the voltage-sensitive Ca^{2+} channels present in the presynaptic membrane (2). Ca^{2+} entry through the opened channels evokes an increase of the intracellular Ca^{2+} concentration ($[Ca^{2+}]_i$) a factor required for triggering exocytosis of synaptic vesicles.

Thus, the probability of exocytosis of synaptic vesicles is strongly increased. This results in fusion of a docked vesicle(s) with the presynaptic plasma membrane (3) and release of the neurotransmitter molecules in the synaptic cleft (extracellular medium; 4). Once released into the synaptic cleft, neurotransmitter molecules bind to an ensemble of receptors: postsynaptic receptors (5a; receptor-channels and G-protein coupled receptors), presynaptic receptors (G-protein coupled receptors, 5b neurotransmitter transporters), glial receptors (5c; neurotransmitter transporters), and in some synapses enzymes (5d) that degrade the neurotransmitter molecules present in the synaptic cleft.

All these receptors are proteins that bear specific receptor sites to the neurotransmitter. By binding to postsynaptic receptor-channels (5a) or to postsynaptic receptors coupled to G-proteins, the neurotransmitter will induce the movement of ions through postsynaptic channels (6) and a postsynaptic current. At that stage, the synaptic transmission is completed. By binding to transporters present in the neuronal and glial membranes or in the cleft, the neurotransmitter is rapidly eliminated from the synaptic cleft. In the presynaptic element, the neurotransmitter is taken back into vesicles or degraded. The membrane is recycled by an endocytotic process (5e). All the events here called (5) are simultaneous.

To refill the synaptic vesicles, the neurotransmitter has also to be synthesized *de novo*. Neurotransmitters are generally synthesized in axon terminals from a precursor present in the axon terminals or taken up from blood. The enzymes necessary for its synthesis are synthesized in the soma and carried via the anterograde axonal transport to the axon terminals (see Section 1.3.2). However, neurotransmitter peptides are synthesized as an inactive precursor form in the neuronal soma, and are carried to the axonal terminals via anterograde axonal transport (see **Figure A6.1**).

This general scheme is of course oversimplified. For example, the presynaptic element can contain more than a single neurotransmitter; the intracellular concentration of Ca^{2+} ions can be increased also by the release of such ions from intracellular stores; and the role of presynaptic receptors has been ignored, for which there is evidence in the majority of synapses.

As we have seen, the *presynaptic element* contains the machinery for the synthesis, storage, release and inactivation of neurotransmitter(s). The presynaptic active zone is the complex formed by the synaptic vesicles and the region of the presynaptic membrane where exocytosis occurs (**Figure 6.4b**). Various methods can be used to characterize the neurotransmitter(s) present in a presynaptic element: immunohistochemical methods that identify the synthesis enzyme of the neurotransmitter in various parts of the neuron, and the *in situ* hybridization technique that identifies the mRNA coding for the synthesis enzyme of the neurotransmitter (see **Appendix 6.2**). However, identification of a substance as a neurotransmitter requires experimental proof of a number of other criteria (see **Appendix 6.1**). If these criteria have not been satisfied, the substance is called a *putative* neurotransmitter.

The *postsynaptic element* is specialized to receive information. Its plasma membrane contains proteins that are receptors for the neurotransmitter: receptor channels (**Figure 6.4c**) and G-protein linked receptors. Various methods can be used to characterize the receptors present in the postsynaptic membrane: radioautographic techniques with monoclonal antibodies, or more rarely anti-idiotype antibodies.

In most cases synaptic transmission is unidirectional (or polarized): it propagates only from the presynaptic element, which contains the neurotransmitter, to the postsynaptic element at the surface of which are receptors for the neurotransmitter (**Figure 6.4a**). In the case of dendro-dendritic synapses (olfactory bulb of the rat), we recognize two juxtaposed synaptic complexes that work in opposite polarities; these are the reciprocal synapses (**Figure 6.2d**). However, it is worth noting that, here also, the synaptic transmission is polarized in each of the synaptic complexes.

6.1.3 Complementarity between the neurotransmitter stored and released by the presynaptic element and the nature of receptors in the postsynaptic membrane

In all synapses, receptors present in the postsynaptic membrane are those that specifically recognize the neurotransmitter released from the corresponding presynaptic element. For example, in glutamatergic synapses, glutamate receptors are found highly concentrated in the membrane of the corresponding postsynaptic element. Efficient synaptic transmission requires, in fact, specific localization of receptors on the postsynaptic membrane apposed to the transmitter release site. This is the case even in synapses where the pre- and postsynaptic cells have different embryonic origin (as in a nerve–muscle junction, for example). This pre–post complementarity requires, at least, the following steps to be completed: targeting, anchoring and clustering of postsynaptic receptors.

Targeting of receptors to a specific postsynaptic membrane

How do neurons target specific proteins to specialized neuronal subdomains? We shall take the example of metabotropic glutamate receptors (G-protein linked receptors of glutamate). They are a homologous family of differentially targeted receptors. Among mGluRs (see Chapter 12), mGluR1a and mGluR2 are targeted to dendrites and excluded from axons, whereas mGluR7 is targeted to dendrites and axons. In order to study the peptide sequence that could be responsible for this differential targeting, native or chimeric mGluRs are expressed, one at a time, in cultured hippocampal neurons, from viral vectors. The distribution of these expressed mGluRs is then checked by labelling mGluR with a specific antibody coupled to a fluorescent marker (Texas Red) and by labelling axon with a tau antibody (green) or dendrites with a MAP2 antibody (green).

First, the distribution of expressed mGluRs is checked. The selective distribution of endogenous mGluRs is reproduced for expressed mGluRs: axon exclusion of mGluR1a and mGluR2 and axon targeting for mGluR7 (all three are also targeted to dendrites). What mediates the axon exclusion of mGluR1a and mGluR2? The working hypothesis is that, since the C-terminal cytoplasmic domain of these receptors is the most divergent region of the primary sequence of mGluRs, it is involved in targeting. To answer this question, the distribution of chimeric mGluRs, such as mGluR2tail7 and mGluR7tail2 constructs, is studied (**Figure 6.5a**). To be sure that tails are intact in the constructs, antibodies against tail2 or tail7 are tested. As they still recognize the tails, the C-terminal epitope of chimeric mGluR forms correctly from the transfected chimeric cDNA.

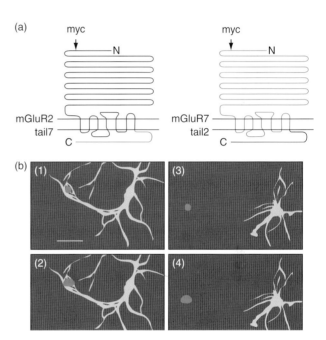

FIGURE 6.5 Differential targeting of mGluR chimera.
(a) Schematics of mGluR chimera (mGluR2tail7 and mGluR7tail2) primary structures with mGluR2 structure shown in black and mGluR7 structure shown in green. The *myc* ten aminoacid epitope tag (arrow) is inserted in the N-terminal extracellular domains, three amino acids past the signal sequence. **(b)** Left: Expressed *myc*-mGluR2tail7 recognized by surface labelling with the *myc* antibody (1) and labelling after permeabilization with an antibody against the C-terminus of mGluR7 (2). Both epitopes give the same distribution pattern, with labelling the full extent of the transfected neuron on the right and the axon of the transfected neuron in contact with a non-transfected neuron on the left. Right: expressed *myc* mGluR7tail2 was recognized by surface labelling with the *myc* antibody (3) and labelling after permeabilization with an antibody against the C-terminus of mGluR2 (4). Again, both epitopes give the same distribution pattern, with labelling of the somatodendritic domain of the transfected neuron on the right. Scale bar, 50 mm. From Nash Stowell J and Craig AM (1999) Axon/dendrite targeting of metabotropic glutamate receptors by their cytoplasmic carboxy-terminal domains. *Neuron* **22**, 525–536, with permission.

Analysis of these chimeric constructs reveals that the C-terminal cytoplasmic domain of mGluRs contains the axon/dendrite targeting information. The mGluR2tail7– containing the backbone of mGluR2 up to and including the seventh transmembrane domain followed by the C-terminal 65-amino-acid domain of mGluR7 – is targeted to axons (**Figure 6.5b**, left column). The reciprocal chimera mGluR7tail2 is present in dendrites but excluded from axons (**Figure 6.5b**, right column). To more narrowly define the targeting signal only the 30-amino-acid distal part of the C-terminal domain between mGluR2 and mGluR7 is swapped. These constructs are not as efficiently targeted as the first ones, showing that axon exclusion of mGluR7 versus mGluR2 is dependent on the 65-amino-acid C-terminal sequences primarily and not exclusively on the more distal amino acids. The mGluR2 C-terminus is required for axon exclusion and the mGluR7 C-terminus is required for axon targeting of the native proteins. These are 'axon exclusion' and 'axon targeting' signals. The mGluR targeting signals may function at any stage in targeting: sorting into specific vesicles from the *trans*-Golgi network, transport by association with specific motors, selection of plasma membrane addition sites, etc. When mGluR1a and mGluR2 are not detected at the surface of the axon, they are also not detected in the axoplasm. Therefore, the mGluR targeting signals such as 'axon exclusion' may act at an early stage, such as sorting out into vesicles directly targeted to dendrites.

Anchoring and clustering of receptors in the postsynaptic membrane

A single neuron may receive input from thousands of synaptic connections on its cell body and dendrites. To integrate these signals rapidly and specifically, the neuron anchors a high concentration of receptors at postsynaptic sites, matching the correct receptor with the neurotransmitter released from the presynaptic terminal. The mechanism of site-specific receptor clustering has been most thoroughly investigated at the neuromuscular junction (a cholinergic synapse; see Section 6.3). Rapsyn is believed to be one of the molecules responsible for nicotinic cholinergic receptors (nAChR) clustering. For glycine receptors (GlyR), postsynaptic clustering is dependent on gephyrin, a 93 kD channel associated protein, that is totally unrelated to rapsyn. We shall take here the example of the identification of anchor proteins for an ionotropic glutamatergic receptor, the NMDA (*N*-methyl-D-aspartate) receptor (see Chapter 10).

NMDARs are composed of NR1 and NR2 subunits, which cotranslationally assemble in the endoplasmic reticulum to form functional channels. A conspicuous feature of the NR2 subunits is their extended, intracellular C-terminal sequence distal to the last transmembrane

region. The working hypothesis is that this region participates in anchoring. In an attempt to identify molecules that can mediate the association between NMDA receptors and cytoskeleton, the yeast two-hybrid system is used to search for such gene products that bind to the intracellular C-terminal tails of NR2 subunits at synapses.

The NR2 subunits are found to interact specifically with a family of membrane-associated synaptic proteins. In mammals, this family includes PSD-95, PSD-93 and PSD-95-synapse-associated-proteins (SAP), a subfamily of membrane-associated guanylate kinase (MAGUK). In their N-terminal half, this family of proteins is characterized by the presence of three domains with a length of approximately 90 amino acids, termed PDZ domains; they are therefore called PDZ-containing proteins (P for PSD-95, D for *dlg* and Z for ZO-1, the first proteins to be identified with these domains) (**Figure 6.6a**). PDZ repeats are protein-binding sites that recognize a short consensus peptide sequence of NR2 subunits (**Figure 6.6b**). PDZ domains also bind to intracellular proteins (**Figure 6.6b**).

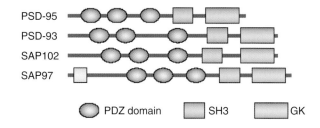

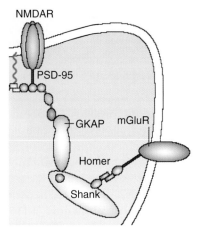

FIGURE 6.6 PDZ-containing proteins and anchoring of glutamate receptors.

(a) Schematic diagram of PDZ proteins. PDZ domains are often found in scaffold proteins as multiple tandem arrays and/or linked to other kinds of modular protein-interaction domain. PDZ domains are shown as purple ellipses. GK, guanylate-kinase-like domain; SH3, Src homology 3 domain. **(b)** Schematic of possible interactions in the postsynaptic element between the NMDA receptor (NMDAR) and anchor proteins: PSD-95, GKAP guanylate kinase-associated protein, Shank SH3 and ankyrin repeat-containing protein. The metabotropic glutamate receptor (mGluR) is anchored via Homer. From Kim E and Sheng M (2004) PDZ domain proteins of synapses. *Nature Rev. Neurosci.* **5**, 771–781.

Synaptic NMDARs are thus concentrated in postsynaptic densities (PSDs), where they are structurally organized (and spatially restricted) in a large macromolecular signalling complex composed of kinases, phosphatases, and adaptor and scaffolding proteins. Scaffolding proteins serve to structurally organize and localize proteins to the PSD and physically link receptors such as NMDARs in close proximity to protein kinases, phosphatases, and other downstream signalling proteins.

6.2 THE INTERNEURONAL SYNAPSES

6.2.1 In the CNS the most common synapses are those where an axon terminal is the presynaptic element

As described in Section 1.1.2, the axon terminals are either *terminal boutons* (**Figure 6.3a**), which are terminals

of axonal branches, or *boutons en passant* (**Figure 6.3b**), which appear as swellings located along the non-myelinated axons and at the nodes of Ranvier along myelinated axons. These two types of axon terminals form synaptic contacts with various neuronal postsynaptic elements: a dendrite (axo-dendritic synapse), a soma (axo-somatic synapse) or an axon (axo-axonic synapse) (**Figure 6.7**). More rarely, there are synapses in which the presynaptic element is a dendrite (dendro-dendritic synapse; see **Figure 6.2d**) or a soma (soma-somatic or soma-dendritic synapses).

6.2.2 At low magnification, the axo-dendritic synaptic contacts display features implying various functions

We will consider as an example the cerebellar cortex, a layered structure in which the cells and their afferents are well characterized. The Purkinje cells are the

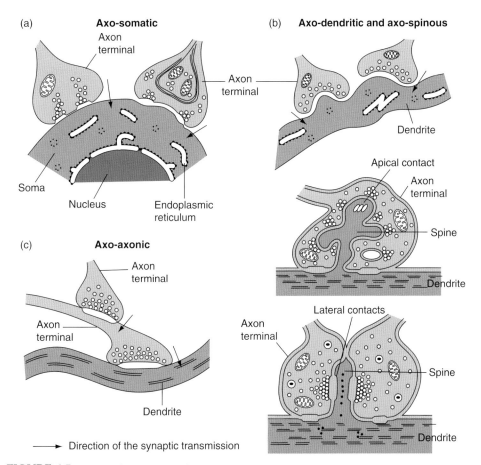

FIGURE 6.7 **Types of interneuronal synapses.**
(a) Terminal boutons forming axo-somatic synapses. (b) Top to bottom: Synapses between terminal boutons and a smooth dendritic branch (axo-dendritic synapse) and two examples of indented synapses between terminal boutons and a dendritic spine (axo-spinous synapses). (c) Synapse between an axon terminal and a terminal axon collateral (axo-axonic synapse). The 'postsynaptic' axon terminal is itself 'presynaptic' to a dendrite. From Hamlyn LH (1972) The fine structure of the mossy fiber endings in the hippocampus of the rabbit. *J. Anat.* **96**, 112–120, with permission.

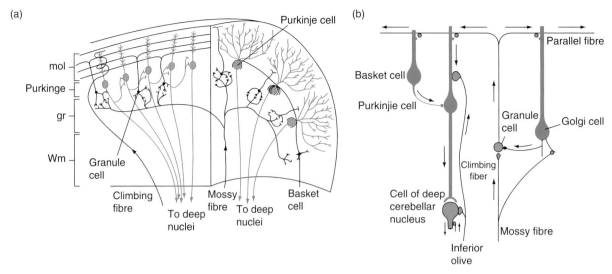

FIGURE 6.8   Synaptic connections in the cerebellar cortex.
(a) Diagram of cells in a folium of the cerebellum. The layers are depicted: from the surface (pia-mater) to the depth, the molecular layer (mol), the layer of Purkinje cells, the granular layer (gr) and the white matter (wm). By comparing the drawings of the Purkinje cells in the sagittal plane with those in the transverse plane, notice that the dendritic tree of the Purkinje cells is planar. A climbing fibre which arborizes along the dendrites of one of the Purkinje cells is shown. It synapses directly with one Purkinje cell. Mossy fibres synapse with many granule cells. The axons of the granule cells enter the molecular layer and bifurcate in a T to form the parallel fibres running lengthwise in the folium. Granule cell axons synapse with Purkinje cells and basket cells. **(b)** Schematic representation of the principal synaptic connections within the cerebellar cortex. Inhibitory GABAergic Purkinje cells are in green; inhibitory GABAergic interneurons are in black; excitatory neurons are stippled. Drawing (a) from Gardner E (1975) *Fundamentals of Neurology*, Philadelphia: W. B. Saunders. Drawing (b) from Eccles JC (1973) *J. Physiol. (Lond.)* **299**, 1–3, with permission.

single 'output' cells of the cerebellar cortex (Golgi type I neurons, **Figure 6.8a**) which send their axons to the deep cerebellar nuclei. They have a cell body with a large diameter (20–30 μm) from which emerges a single dendritic trunk that gives rise to numerous spiny dendritic branches which arborize in the molecular layer. The dendritic tree is planar, and the dendritic branches extend mainly in the transverse plane. The neurotransmitter of Purkinje cells is γ-aminobutyric acid (GABA).

Purkinje cells receive two types of excitatory afferents: the climbing fibres (axons of the neurons in the inferior olivary nucleus) and the parallel fibres (axons of the granule cells in the cerebellar cortex). The inhibitory afferents arise mainly from the numerous local circuit neurons in this structure: the basket cells, the stellate cells and the Golgi cells (**Figure 6.8b**).

A single climbing fibre innervates each single Purkinje cell. The climbing fibre gives rise to numerous axon collaterals that 'fit' the shape of the postsynaptic dendritic tree: the axon collaterals 'climb' along the dendrites (**Figures 6.8a** and **6.9a,b**) forming numerous synaptic contacts with the soma and the dendrites of the Purkinje cell. These contacts are axodendritic or axospinous, the presynaptic element being a terminal bouton. Such a synaptic organization implies that this

excitatory afferent is very efficient: a single action potential along the climbing fibre can in fact induce a response in the Purkinje cell.

The axons of the granule cells form very different synaptic contacts with the Purkinje cells. The axons enter the molecular layer where they bifurcate and extend for 2 mm in a plane perpendicular to the plane of the dendritic tree of the Purkinje cell and form what are called parallel fibres (**Figure 6.8**). Parallel fibres form a few 'en passant' synapses (axospinous synapses between axonal varicosities and distal dendritic spines), with the numerous Purkinje cells (about 50). Therefore, each Purkinje cell receives synaptic contacts from about 200,000 parallel fibres. The consequence of such a synaptic organization is as follows. Activation of a parallel fibre cannot induce a Purkinje cell response since the activation of one or a few of these excitatory synapses cannot trigger postsynaptic action potentials; numerous parallel fibres converging on to a single Purkinje cell must be activated to induce a response in this cell.

The basket cells are local circuit neurons (Golgi type II neurons) which inhibit the activity of Purkinje cells. The axons of these neurons project to a large number of Purkinje cells and give rise to numerous axon collaterals which form 'baskets' around the soma of Purkinje

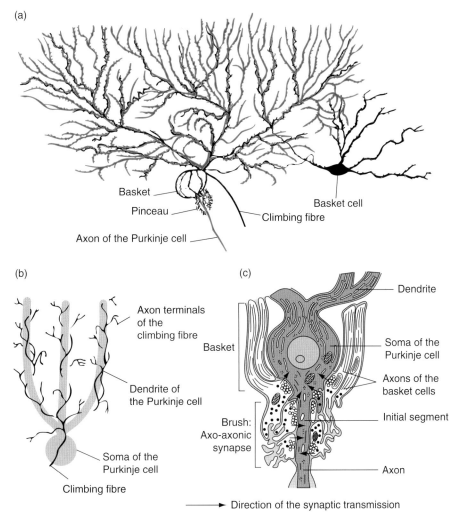

FIGURE 6.9 **Varieties of synaptic arrangements at the level of a Purkinje cell.**
Representation on a single drawing **(a)** and in two separated schematic drawings **(b** and **c)** of the synaptic arrangements between a climbing fibre and a Purkinje cell (a and b) and between a basket cell and a Purkinje cell (a and c). Part (a) from Chan-Palay V and Palay S (1974) *Cerebellar Cortex: Cytology and Organization*, Berlin: Springer Verlag, with permission. Part (b) from Scheibel ME and Scheibel AB (1958) *Electroenceph. Clin. Neurophysiol.* **Suppl. 10**, 43–50, with permission. Part (c) from Hamori J and Szentagothai J (1965) The Purkinje cell baskets: ultrastructure of an inhibitory synapse. *Acad. Biol. Hung.* **15**, 465–479, with permission.

cells. The axonal branches extend further and terminate 'en pinceau' around the initial segment of the Purkinje cells' axon (**Figures 6.8** and **6.9a,c**). Such an organization allows inhibition of the Purkinje cells at a strategic point where the sodium action potentials arise. This represents an efficient way of counteracting the excitatory potentials propagating along the dendritic branches to the initial axonal segment.

6.2.3 Interneuronal synapses display ultrastructural characteristics that vary between two extremes: types 1 and 2

A classification of the synapses on the basis of the form of their synaptic complex was proposed by Gray (1959). This author described two types of synaptic complexes in the cerebral cortex which he named types 1 and 2 (**Figure 6.10**).

- Type 1 synapses are asymmetrical because they have a prominent accumulation of electron-dense material on the postsynaptic side. These synapses are found more often on dendritic spines or distal dendritic branches. The presynaptic element contains round vesicles and the synaptic cleft is about 30 nm wide.

- Type 2 synapses are symmetrical because they have electron-dense zones of the same size in both the pre- and postsynaptic elements. The presynaptic element contains oval-shaped vesicles and the

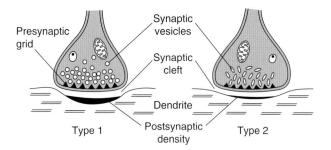

FIGURE 6.10 Schematic representation of type 1 (asymmetric) and type 2 (symmetric) synapses according to Gray.

synaptic cleft is narrow. These synapses are more commonly found at the surface of dendritic trunks and soma.

On the basis of correlations between physiological and morphological data obtained in the cerebellar cortex, Gray proposed that type 1 synapses are excitatory whereas type 2 synapses are inhibitory.

In the central nervous system (CNS), types 1 and 2 synapses are the extremes of a morphological continuum since synaptic complexes may have intermediate forms and display features that characterize both types of synapse; e.g. a large synaptic cleft (type 1) and a narrow postsynaptic density (type 2). In addition, it has been shown that the form of the synaptic vesicles is dependent on the fixation technique used.

6.3 THE NEUROMUSCULAR JUNCTION IS THE GROUP OF SYNAPTIC CONTACTS BETWEEN THE TERMINAL ARBORIZATION OF A MOTOR AXON AND A STRIATED MUSCLE FIBRE

The motoneurons or motor neurons have their cell body located in motor nuclei of the brainstem or in the ventral horn of the spinal cord. The axons of these neurons are myelinated and form the cranial and spinal nerves that innervate the skeletal striated muscles (see Section 1.4.1). In general, a single striated muscle fibre is innervated by one motoneuron but a single motoneuron can innervate many muscle fibres. The myelin sheath of each axon is interrupted at the zone where the axon arborizes at the surface of the muscle fibre. At this point, the thin non-myelinated axonal branches possess numerous varicosities which are located in the depression at the surface of the muscle fibre: the synaptic gutter. The axon terminals are covered by the non-myelinating Schwann cells (**Figure 6.11**; see also Section 2.3.2).

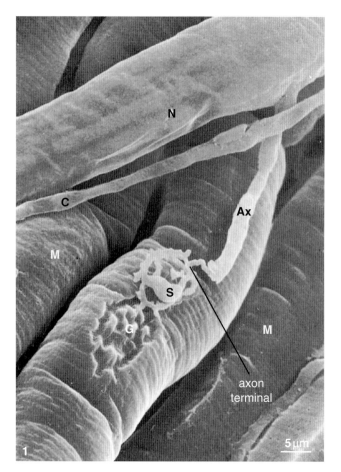

FIGURE 6.11 **The neuromuscular junction.**
Photograph of a rat neuromuscular junction observed under the scanning electron microscope. The terminal part of the axon (ax) is detached from the muscle cell (M) in order to show the synaptic gutter (G); c, capillary; N, motor nerve; S, nucleus of a Schwann cell. From Matsuda Y *et al.* (1988) Scanning electron microscopic study of denervated and reinnervated neuromuscular junction. *Muscle Nerve* **11**, 1266–1271, with permission.

6.3.1 In the axon terminals, the synaptic vesicles are concentrated at the level of the electron-dense bars; they contain acetylcholine

The neuromuscular junction is formed by the juxtaposition of the terminals of a motor axon and the corresponding sub-synaptic domains of a striated muscle fibre, these two elements being separated by a 50–100 nm-wide cleft. In a transverse section of a neuromuscular junction observed under electron microscopy (**Figure 6.12a**) the vesicles in the presynaptic element are small (40–60 nm diameter), clear and contain acetylcholine, the neurotransmitter of all the neuromuscular junctions. Larger vesicles (80–120 nm diameter) that contain an electron-dense material are also present but in a much lower proportion (1% of the total population). The vesicles are aggregated in the presynaptic

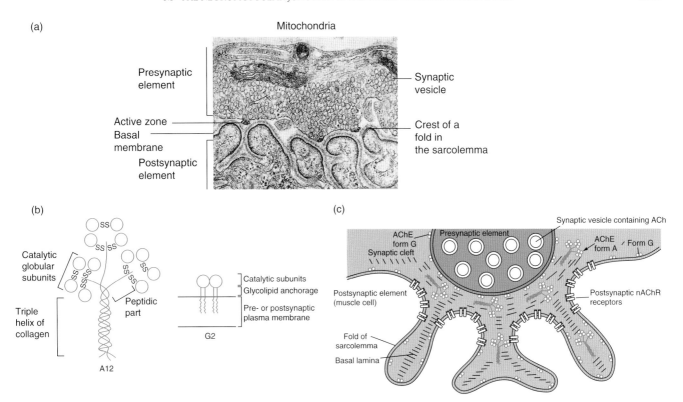

FIGURE 6.12 **Ultrastructure of a neuromuscular junction and the location of acetylcholinesterases.**
(a) Microphotography of the neuromuscular junction of a batrachian observed under the electron microscope. In the axon terminal can be seen mitochondria and numerous vesicles. The axonal plasma membrane displays signs of exocytosis (active zones). The basal lamina is located in the synaptic cleft. The postsynaptic muscle cell membrane has numerous folds. **(b)** Schematic of the asymmetric (A12) and globular (G2) forms of acetylcholinesterase (AChE) (top). The index number of A or G indicates the number of catalytic subunits. The asymmetric forms consist of a collagen tail, three peptide parts and catalytic subunits. The globular forms consist of one or more catalytic subunits (hydrophilic domain) and a glycolipid part (hydrophobic domain) which permits their insertion in the lipid bilayer. **(c)** Location of acetylcholinesterase in the neuromuscular junction. The A forms are synthesized in the motoneurons and secreted into the synaptic cleft where they are associated with the basal lamina. The globular forms are synthesized in the motoneurons and inserted into the presynaptic plasma membrane or secreted into the synaptic cleft. Microphotograph (a) by Pécot-Dechavassine. Drawings (b) and (c) from Berkaloff A, Naquet R, Demaille J (eds) (1987) *Biologie 1990: Enjeux et Problématiques*, Paris: CNRS, with permission.

zones where an electron-dense material is present, the dense bars. These dense bars are functionally homologous to the presynaptic grid of interneuronal synapses. They are 100 nm wide and are located perpendicularly to the largest axis of each axonal branch. The vesicles are aligned along each side of these bars. The complex of dense bars and synaptic vesicles forms a presynaptic active zone (Couteaux 1960). There are many active zones per varicosity. They are located opposite to the folds of the postsynaptic plasma membrane. Each active zone with the folds of the sarcolemma in front of them forms a synaptic complex. Therefore, the neuromuscular junction contains numerous synaptic complexes.

Synthesis of acetylcholine takes place in the cytoplasm of the presynaptic element from two precursors: choline and acetylcoenzyme A (acetyl CoA). The reaction is catalyzed by choline acetyltransferase (CAT). Acetylcholine is transported actively into synaptic vesicles where it is stored (see **Figure 8.20a**). The protein responsible for this active transport is a transporter which uses the energy of the proton (H^+) gradient. This gradient of protons is established by active transport of H^+ ions from the cytoplasm towards the interior of the vesicles by a H^+-ATPase pump.

6.3.2 The synaptic cleft is narrow and occupied by a basal lamina which contains acetylcholinesterase

The postsynaptic muscular membrane (sarcolemma) is covered, on the extracellular surface, with a layer of electron-dense material, the basal lamina

(**Figure 6.12a, c**). This lamina, which follows the folds of the sarcolemma, is a conjunctive tissue secreted by the non-myelinating Schwann cells covering the axon terminals. It contains, *inter alia,* collagene, proteoglycans and laminin.

Acetylcholinesterases are glycoproteins synthesized in the soma and carried to the terminals via anterograde axonal transport. They are inserted into the presynaptic membrane and the basal lamina. They display an important structural polymorphism (**Figure 6.12b**): they have a globular form (G) or an asymmetric form (A). These different forms have distinct localizations. Globular forms (G) are anchored in the pre- or postsynaptic membrane (these are ectoenzymes) and are secreted as a soluble protein into the synaptic cleft. Asymmetric forms (A) are anchored in the basal lamina (**Figure 6.12c**). The molecules of acetylcholine, released in the synaptic cleft when the neuromuscular junction is activated, cross the basal lamina through its loose stitches. But a part of the acetylcholine molecules is also degraded before being fixed to postsynaptic receptors, by the acetylcholinesterase inserted in the basal lamina. The other part is quickly degraded after its fixation. Acetylcholinesterases hydrolyze acetylcholine into acetic acid and choline. Choline is taken up by presynaptic terminals for the synthesis of new molecules of acetylcholine. This degradation system of acetylcholine is a very efficient system for inactivation of a neurotransmitter.

6.3.3 Nicotinic receptors for acetylcholine are abundant in the crests of the folds in the postsynaptic membrane

The plasma membrane of muscle cells, the sarcolemma, presents numerous folds in mammalian neuromuscular junctions. By using a radioactive ligand for a type of acetylcholine nicotinic receptor, α-bungarotoxin labelled with a radioactive isotope or a fluorescent molecule, it has been shown that the radioactive material accumulates predominantly in the crests of the folds in the sarcolemma. Immunocytochemical techniques produce similar results. Other studies have shown that they are anchored to the underlying cytoskeleton (see the following section).

The nicotinic receptor is a transmembrane glycoprotein comprising four homologous subunits assembled into a heterologous $\alpha_2 \beta \gamma \delta$ pentamer. It is a receptor channel permeable to cations whose activation results in the net entry of positively charged ions and in depolarization of the postsynaptic membrane. The structure and functional characteristics of the muscular nicotinic receptors are given in Chapter 8.

6.3.4 Mechanisms involved in the accumulation of postsynaptic receptors in the folds of the postsynaptic muscular membrane

The acetylcholine nicotinic receptors are, in the adult neuromuscular junction, present in high density (about 10,000 molecules per μm) in the postsynaptic regions and occur in a much lower density in the nonsynaptic membrane (extrajunctional membrane). Under the nerve terminal, the muscle cell is free of the myofimanents actin and myosin. At this level, four to eight cell nuclei are found, the fundamental nuclei (Ranvier 1875). The myonuclei located outside the postsynaptic region (extrasynaptic) are the sarcoplasmic nuclei. The formation of this well organized subsynaptic domain – which concerns not only the nicotinic receptors but also the Golgi apparatus and the cytoskeleton (it also comprises the organization of the basal lamina and the distribution of the asymmetric form of acetylcholinesterase in the synaptic cleft) – occurs in numerous steps during maturation of the neuromuscular junction (**Figure 6.13a**):

- There is an increase in the number of nicotinic receptors (1 and 2) during fusion of the myoblasts to form myotubes, owing to the neosynthesis of these receptors. They have an even distribution over the membrane surface. This phenomenon is independent of the neuromuscular activity since it is not affected by the injection *in ovo* of nicotinic antagonists such as curare.

- There is formation of aggregates of nicotinic receptors under the nerve terminal (3–5) and disappearance of extrajunctional receptors (5). Upon innervation, nAChR rapidly accumulates under the nerve endings. *In situ* hybridization experiments with a genomic coding probe (see **Appendix 6.2**) have shown that in innervated 15-day-old chick muscle, the nAChR α-subunit mRNAs accumulate under the nerve endings. More precisely, accumulation of the mRNAs increases around the sub-synaptic (fundamental) nuclei and decreases around the sarcoplasmic nuclei. This can be interpreted as a differential expression of the nAChR α-subunit gene in the fundamental and sarcoplasmic nuclei. The presence of motor nerve and muscle activity are both crucial for the regulation of nAChR mRNA levels in the developing fibre.

- Distribution of the Golgi apparatus, studied by using a monoclonal antibody directed against it, shows a similar evolution. In cultured myotubes, the Golgi apparatus is associated with every nucleus. Conversely, in 15-day-old innervated chick muscle, the Golgi apparatus is now restricted to discrete, highly focused regions that appear to co-distribute

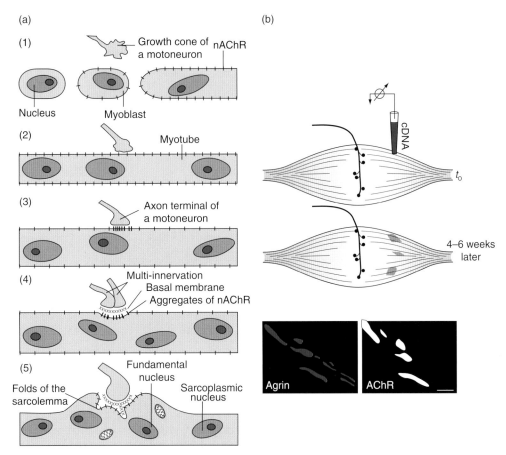

FIGURE 6.13 Postsynaptic differentiation at the neuromuscular junction.
(a) The different steps in postsynaptic differentiation. The black thin bars indicate the nicotinic receptor (nAChR). (1, 2) Fusion of myoblasts to form myotubes and approach of the axon growth cone; (3) the growth cone forms contact with the myotube and induces the clustering of nicotinic receptors at this level; (4) numerous motor terminals converge towards a single aggregate of nicotinic receptors; but (5) a single terminal stabilizes and the folds of the sarcolemma develop. **(b)** Injection of expression constructs that encode full-length neural agrin into single muscle fibres *in vivo* (t_o). After 4–6 weeks, muscle fibres have a deposit of chick neural agrin in muscle basal lamina at ectopic sites. Staining of the muscle anti-chick agrin antibodies reveals the depositing of neural agrin as visualized in optical longitudinal sections through ectopic sites (Agrin). In this area, ectopic neural agrin induces aggregation of nAChRs (AChR) as visualized by *in situ* hybridization. Scale bar, 15 μm. Drawing (a) from Laufer R and Changeux JP (1989) Activity-dependent regulation of gene expression in muscle and neuronal cells. *Mol. Neurobiol.* **3**, 1–53, with permission. Part (b) from Ruegg MA and Bixby JL (1998) Agrin orchestrates synaptic differentiation at the vertebrate neuromuscular junction. *Trend. Neurosci.* **21**, 22–27, with permission.

with endplates (revealed by fluorescein isothiocyanate conjugated α-bungarotoxin, a labelled ligand of nAChR).

- There is stabilization of nicotinic receptors in the postsynaptic membrane (5).

These observations raise questions about the nature of the signalling pathways which underlie such reorganization. Is there activation of second messengers by anterograde signals from the nerve endings that would lead to positive regulation of the expression of the nicotinic receptor in the junctional regions and negative regulation in the extrajunctional regions? Are there retrograde signals too?

Aggregation of proteins at the nerve–muscle contact depends, in fact, on instructive signals that are released by the motor axon. More than 20 years ago, McMahan and colleagues identified the basal lamina as the carrier of the information necessary to induce pre- and postsynaptic specializations during neuromuscular regeneration. A protein, agrin, was purified from basal lamina extracts of the cholinergic synapse of the electric organ of *Torpedo californica* (see **Figure 8.1**). When added to cultured myotubes, soluble agrin induces the aggregation of acetylcholine receptors. This led McMahan to formulate the following hypothesis: 'Agrin is released from motor neurons, binds to a receptor on the muscle cell surface and induces postsynaptic specializations. Subsequent binding of agrin to synaptic basal lamina will then immobilize agrin.'

The fact that neural agrin is necessary and sufficient for postsynaptic differentiation is confirmed by the following experiment. Agrin is a protein of 225 kD, consisting of domains found in other basal lamina proteins. The region of agrin necessary and sufficient to bind to the basal lamina (it binds in fact to laminins) maps to the amino-terminus end of the molecule. The most carboxy-terminus is necessary and sufficient for its nAChR-aggregating activity. Agrin mRNA undergoes alternative splicing at several sites, two of which modulate agrin's ability to induce nAChR clustering in cultured muscle cells. In innervated adult rat muscle, injection of expression constructs that encode full-length chick neural agrin is sufficient to induce postsynaptic specializations: after 4–6 weeks, staining with anti-chick agrin fluorescent antibodies reveals the deposit of neural agrin in basal lamina at the ectopic site of injection (**Figure 6.13b**). In this area, aggregation of nAChRs is observed with immunocytochemistry. Therefore, ectopic expression of recombinant agrin in adult muscle *in vivo* induces the formation of postsynaptic-like structures that closely resemble the muscle endplate.

Which molecule is the agrin receptor(s) and what are the second messengers activated by agrin and responsible for postsynaptic differentiation? Agrin initiates a signalling cascade which is still under study. Interestingly, voltage-dependent Na^+ channels are also concentrated at the synapse where they are restricted to the depth of postjunctional folds. This clustering pathway also involves agrin.

6.4 THE SYNAPSE BETWEEN THE VEGETATIVE POSTGANGLIONIC NEURON AND THE SMOOTH MUSCLE CELL

Smooth muscle cells are present in most of the visceral organs (digestive system, uterus, bladder, etc.) but also in the wall of blood vessels and around the hair follicles. They are innervated by postganglionic neurons of the autonomic nervous system (orthosympathetic neurons and parasympathetic neurons) (**Figure 6.14**).

6.4.1 The presynaptic element is a varicosity of the postganglionic axon

The axons of the postganglionic neurons are not myelinated. Before contacting the smooth muscles, the axons divide into numerous thin filaments 0.1–0.5 μm in diameter which travel alone or in fascicles over long distances along the smooth muscle cells (**Figures 6.14**

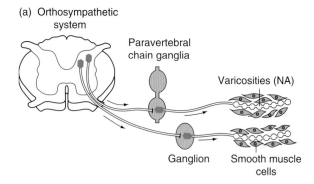

(a) Orthosympathetic system

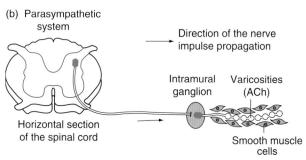

(b) Parasympathetic system

FIGURE 6.14 **The orthosympathetic (a) and parasympathetic systems (b).**
The synapses between postganglionic neurons and smooth muscle fibres are shown. The axon terminals of postganglionic neurons are varicosities and the synaptic contacts are of the 'boutons en passant' type. ACh, acetylcholine; NA, noradrenaline; arrows show the direction of action potential propagation.

and **6.15a**). Each of these filaments has swellings or varicosities 0.5–2.0 μm in diameter spaced 3–5 μm apart. The varicosities contain mitochondria and numerous synaptic vesicles whereas the intervaricose segments contain many elements of the cytoskeleton (**Figure 6.15b**). The varicosities are the presynaptic elements. There are no electron-dense regions in the presynaptic membrane, which suggests the absence of a preferential zone for exocytosis (active zone) in these synapses.

The axonal varicosities contain a large number of small granular synaptic vesicles with an electron-dense central core (30–50 nm diameter), but also some large granular vesicles (60–120 nm diameter) and small agranular vesicles. The neurotransmitter of the orthosympathetic postganglionic neurons is noradrenaline (NA) (**Figure 6.14a**). It is stored in small and large granular vesicles. Noradrenaline is a catecholamine (as dopamine and adrenaline) synthesized from the amino acid tyrosine.

The noradrenaline receptors present in the postsynaptic membrane of smooth muscle cells are G-protein-coupled receptors. The inactivation of noradrenaline released from nerve terminals is, to a large extent, achieved by reuptake by the catecholaminergic neurons or the nearby glial cells. It is recycled into the synaptic

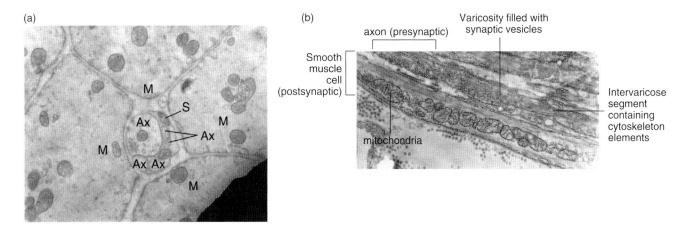

FIGURE 6.15   **A nerve–smooth muscle synapse.**
(a) Microphotograph of smooth muscle cells of the intestine (M) and postganglionic axon fascicles (parasympathetic nervous system, Ax) which are half-covered by a Schwann cell (S), sectioned in the transverse plane and observed under the electron microscope. Note the width of the synaptic cleft. **(b)** Longitudinal section of a postganglionic axon showing a varicosity filled with vesicles and an intervaricose region. The postsynaptic smooth muscle cell contains numerous mitochondria. Microphotography by Jacques Taxi.

vesicles or degraded by specific enzymes such as the monoamine oxidase (MAO). Some of the catecholamines are degraded in the synaptic cleft by the catechol-O-methyl-transferase (COMT).

Acetylcholine (ACh) is the neurotransmitter of the parasympathetic postganglionic neurons (**Figure 6.14b**). The varicosities of these axons contain mainly small agranular vesicles but also large granular vesicles. Acetylcholine is stored in small agranular vesicles. The acetylcholine receptors present in the postsynaptic membrane of smooth muscle cells are cholinergic muscarinic receptors (mAChR); they are protein-G coupled receptors.

6.4.2 The width of the synaptic cleft is very variable

Where the synaptic cleft is narrowest, in the vas deferens or in the pupil for example, it measures between 15 and 20 nm. However, in the wall of blood vessels, the closest contacts are spaced 50–100 nm apart.

6.4.3 The autonomous postganglionic synapse is specialized to ensure a widespread effect of the neurotransmitter

The large width of the synaptic cleft results in a widespread effect of the neurotransmitter on the postsynaptic membrane compared with the neuromuscular junction or central synapses, where secretion of the neurotransmitter is focused on a small postsynaptic region. Moreover, in some autonomous synapses, there is no distinguishable specialization of the presynaptic

membrane, which suggests that the vesicles have no preferential site for exocytosis. Formation of a dense plexus by the postganglionic axons also contributes to the extended diffusion of presynaptic messages. Therefore, activation of a postganglionic neuron results in activation of numerous postsynaptic cells. Finally, the presence of numerous gap junctions which connect smooth muscle cells permits spread of the synaptic response to neighbouring muscle cells, even to those not innervated.

6.5 EXAMPLE OF A NEUROGLANDULAR SYNAPSE

This section considers the synapse between an orthosympathetic preganglionic neuron and the chromaffin cell of the adrenal medulla (**Figure 6.16**).

The adrenal medulla is the central part of the adrenal gland, the endocrine gland located above each kidney. It is formed by secretory cells, which are called chromaffin cells since they are coloured by chromium salts. The adrenal medulla is innervated by orthosympathetic preganglionic neurons which have axons that form the splanchnic nerve. When this nerve is stimulated, the chromaffin cells secrete essentially adrenaline but also noradrenaline and enkephalins (endogenous opioid peptides). These hormones are then transported via the blood to numerous target tissues and mainly the heart and blood vessels.

The *presynaptic element* of this synapse is the axon terminal of the splanchnic orthosympathetic preganglionic

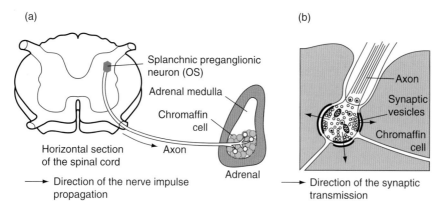

FIGURE 6.16 **Synapses between an orthosympathetic preganglionic neuron and chromaffin cells in the adrenal medulla.**
(a) The cell bodies of preganglionic neurons are localized in the spinal cord and their axons form the orthosympathetic splanchnic nerve. These neurons innervate the chromaffin cells. **(b)** Postganglionic axon terminal forming numerous contacts with chromaffin cells. Drawing (b) from Coupland RE (1965) Electron microscopic observations on the structure of the rat adrenal medulla: II. Normal innervation *J. Anat. (Lond.)* **99**, 255–272, with permission.

neurons. Their cell bodies are located in the intermediate horn of the spinal cord. Most of the axons are non-myelinated and are surrounded by the extensions of the non-myelinating Schwann cells. This glial sheath is present until the axon collaterals penetrate the junctional space. The axon terminals are mainly terminal boutons with a diameter ranging between 1 and 3 μm. They contain clear vesicles (10–60 nm diameter) as well as some dense-core and granular vesicles (25–115 nm diameter). Acetylcholine is the neurotransmitter of this synapse. The terminal boutons form a large variety of synaptic contacts with the chromaffin cells. They are characterized by a narrow synaptic cleft (15–20 nm) and the presence of electron-dense pre- and postsynaptic zones similar to those observed at the level of the central interneuronal synapses.

In the *postsynaptic region*, the cytoplasm of the chromaffin cells is free of chromaffin granules, organelles that store hormones in the adrenal medulla. The acetylcholine receptors present in the postsynaptic membrane are cholinergic nicotinic receptor channels. In the rest of the chromaffin cell cytoplasm there are numerous chromaffin granules. These granules are coloured by chromium salts which react with adrenaline to form a yellow–brown precipitate.

6.6 SUMMARY

Chemical synapses are connections between two neurons or between a neuron and a non-neuronal cell (muscle cell, glandular cell, sensory cell). The synaptic

complex is the non-reducible basic unit of each chemical synapse as it represents the minimal requirement for an efficient chemical synaptic transmission. It includes three elements: the presynaptic element (such as an axon terminal), a synaptic cleft, and a postsynaptic element (such as a dendritic spine).

The *presynaptic element* is characterized by (i) an active zone (i.e. a specialized presynaptic membrane area) where the density of Ca^{2+} channels is high and where occurs the fusion of synaptic vesicles (exocytosis), and (ii) a nearby cytoplasmic region where the synaptic vesicles are found close to the presynaptic membrane, with a particular cytoskeletal arrangement. The regulated release of neurotransmitter occurs at active zones. However, the active zone is not a characteristic of all synapses. A few monoaminergic synapses and peptidergic synapses do not have discernible active zones. The presence of an active zone would be a clue to focal neurotransmitter release.

The *postsynaptic element* is characterized, in interneuronal synapses, by a sub-membranous electron-dense zone (postsynaptic density), which most probably corresponds to the region where the postsynaptic receptors are anchored. There is a strict complementarity between the neurotransmitter released by the presynaptic element and the postsynaptic receptors inserted in the postsynaptic membrane. This includes specific targeting, anchoring and clustering of postsynaptic receptors.

The ultrastructure of chemical synapses is asymmetric, synaptic vesicles that contain the neurotransmitter(s) being present only in the presynaptic element. Synaptic transmission is unidirectional – it always occurs from the presynaptic element to the postsynaptic one.

APPENDIX 6.1 NEUROTRANSMITTERS, AGONISTS AND ANTAGONISTS

Neurotransmitters are molecules of varied nature: quaternary amines, amino acids, catecholamines or peptides, which are released by neurons at chemical synapses. They transmit a message from a neuron to another neuron, or to an effector cell, or a message from a sensory cell to a neuron.

A6.1.1 Criteria to be satisfied before a molecule can be identified as a neurotransmitter

Identification of a substance as a neurotransmitter requires the experimental proof of several criteria. If these are not satisfied, the term *putative* neurotransmitter is used. The criteria are:

- The putative neurotransmitter must be present in the presynaptic element.
- The precursors and enzymes necessary for *synthesis* of the putative neurotransmitter must be present in the presynaptic neuron.
- The putative neurotransmitter must be released in response to activation of the presynaptic neuron and in a quantity sufficient to produce a postsynaptic response. This release should be dependent on Ca^{2+} ions.
- There should be *binding to specific postsynaptic receptors*: (i) specific receptors of the neurotransmitter are present in the postsynaptic membrane; (ii) application of the substance at the level of the postsynaptic element reproduces the response obtained by stimulation of the presynaptic neuron; and (iii) drugs, which specifically block or potentiate the postsynaptic response, have the same effects on the response induced by the application of the putative neurotransmitter.
- The elements of the synaptic nervous tissue (pre- or postsynaptic elements, glial cells, basal membrane) must possess one or several mechanisms for *inactivation* of the putative neurotransmitter.

Currently, few molecules have satisfied all these criteria to be firmly identified as a neurotransmitter at a particular synapse. In most cases there is no more than fragmentary evidence owing to technical limitations.

A6.1.2 Types of neurotransmitter

Acetylcholine: a quaternary amine

In the peripheral nervous system, acetylcholine is the neurotransmitter of all the synapses between motoneurons and striated muscle cells, of all the synapses between preganglionic and postganglionic neurons of the para- and orthosympathetic systems, and of all the synapses between parasympathetic postganglionic neurons and effector cells (see **Figures 6.12, 6.14** and **6.16**). It is also a neurotransmitter in the central nervous system. Choline acetyltransferase (CAT), the enzyme required for acetylcholine synthesis, is a specific marker of cholinergic neurons. Using immunocytochemical or *in situ* hybridization techniques (see **Appendix 6.2**), one can visualize cholinergic neuronal pathways by labelling choline acetyltransferase or its mRNA. At the same time, the acetylcholine receptors can be localized.

Amino acids: glutamate, GABA (γ-aminobutyric acid) and glycine

In contrast to other neurotransmitters, glutamate also plays an important role in cellular metabolism (in intermediary metabolism, in the synthesis of proteins and as precursor of GABA). It is, therefore, present in all neurons and its identification as a neurotransmitter poses several problems.

In fact, evidence for the enzymes of its synthesis or degradation cannot represent a valid criterion for the identification of glutamatergic neurons. These difficulties can be overcome as glutamate is present in much higher concentrations in neurons where it plays a neurotransmitter role. In addition, these neurons have the property of recapturing selectively glutamate with the help of a high-affinity transport system, and localization of this transport system can be used to identify glutamate neurons. Glutamic acid decarboxylase, the enzyme required for GABA synthesis, is a good marker of GABAergic neurons. In the CNS, two isoforms of glutamic acid decarboxylase (GAD67 and GAD65), each encoded by a different gene and highly conserved among vertebrates, are co-expressed in most GABA neurons. Thus GABAergic neurons (cell bodies, axon terminals and fibres) can be visualized by the localization of these two forms of GADs or their mRNAs by immunocytochemistry and *in situ* hybridization, respectively, or by immunohistochemical detection of GABA itself.

Monoamines

These are classified as catecholamines (adrenaline, noradrenaline, dopamine), indolamine (serotonin) and imidazole (histamine). Adrenaline, noradrenaline and dopamine are all catecholamines. Their structure has a common part, the catechol nucleus (a benzene ring with two adjacent substituted hydroxyl groups). They are synthesized from a common precursor, tyrosine.

Serotonin or 5-hydroxytryptamine (5HT) is synthesized from tryptophan, a neutral amino acid.

Neuropeptides

Peptides that are present in neurons with a supposed role in synaptic transmission are called neuropeptides. They are, for example, opioid peptides (enkephalins, dynorphin, β-endorphin) or they are peptides that have already been identified in the gastrointestinal tract (substance P, cholecystokinin, vasoactive intestinal peptide (VIP)) or in the hypothalamo-hypophyseal complex (luteinizing hormone releasing hormone (LHRH), somatostatin, adrenocorticotropic hormone (ACTH), vasopressin) or they are also circulating hormones (corticotropin or ACTH, insulin), before they were suggested as neurotransmitters in the central nervous system. It seems reasonable that other neuropeptides await discovery. These peptides were proposed to be neurotransmitters on the basis of their presence and synthesis in the neurons as well as by their release from axonal terminals by a Ca^{2+}-dependent mechanism. For some peptides other criteria have also been demonstrated.

The differences in the chemical nature of neurotransmitters have a fundamental consequence. Non-peptidic neurotransmitters are synthesized in axonal terminals: a precursor (or precursors) synthesized by the neuron or taken up from the extracellular medium is transformed into a neurotransmitter via an enzymatic reaction in axon terminals. The synthesis enzyme(s) is (are) synthesized in the cell body and transported to axon terminals via axonal transport. The newly synthesized neurotransmitter is then actively transported inside the synaptic vesicles (**Figure A6.1a**). Peptidic neurotransmitters are synthesized in the cell body since axon terminals, being deprived of the organelles responsible for protein synthesis, cannot themselves synthesize neuropeptides (**Figure A6.lb**). These are synthesized in cell bodies in the form of larger peptides called *precursors*. These precursors are then transported to the axon terminals by fast axonal transport. Cleavage of the precursors into neuroactive peptides is carried out by vesicular peptidases during anterograde axonal transport. Since these precursors have no biological activity, regulation of peptidase activity seems to be an important factor in the regulation of the synthesis of peptidic neurotransmitters.

Concerning their mode of inactivation, most of the neurotransmitters are taken up by axon terminals or glial cells via specific transporters. The major exception is acetylcholine, which is degraded in the synaptic cleft by acetylcholinesterases. Since enzymatic degradation is more rapid than a transporter reaction, acetylcholine is much more rapidly inactivated than the other neurotransmitters.

A6.1.3 Agonists and antagonists of a receptor

An *agonist* is a molecule (drug, neurotransmitter, hormone) that binds to a specific receptor, activates the receptor and thus elicits a physiological response:

$$A + R \underset{k_{-1}}{\overset{k_{+1}}{\rightleftharpoons}} AR \; \rightleftharpoons \; AR^* \; ----\rightarrow \; \text{physiological} \atop \text{response}$$

where A is the agonist, R is the free receptor, AR is the agonist–receptor complex and AR^* is the activated state of the receptor bound to the agonist. k_{+1} and k_{-1} measure the rate at which association and dissociation occur. An agonist (and an antagonist) is defined in relation to a receptor and not to a neurotransmitter. For example, an agonist of nicotinic acetylcholine receptors such as nicotine is not an agonist of muscarinic acetylcholine receptors though both receptors are activated by acetylcholine.

An *antagonist* is a molecule that prevents the effect of the agonist. A *competitive antagonist* (C) is a receptor antagonist that acts by binding reversibly to agonist receptor site (R). It does not activate the receptor and thus does not elicit a physiological response:

$$B + R \underset{k_{-1B}}{\overset{k_{+1B}}{\rightleftharpoons}} BR \; ----\rightarrow \; \text{no physiological} \atop \text{response}$$

The effect of a *reversible competitive antagonist* can be reversed when the agonist concentration is increased since the agonist (A) and the reversible competitive antagonist (B) compete for the same receptor site (R). Competition means that the receptor can bind only one molecule (A or B) at a time. An *irreversible competitive antagonist* is a receptor antagonist that dissociates from the receptor slowly or not at all. For this reason its effect cannot be reversed when the agonist concentration is increased.

APPENDIX 6.2 IDENTIFICATION AND LOCALIZATION OF NEUROTRANSMITTERS AND THEIR RECEPTORS

A6.2.1 Immunocytochemistry

Principle and definitions

Immunocytochemical techniques are based on the high specificity of the antigen–antibody reaction. They consist of the detection of an antigen present in histological or cellular structures by application on tissues or cells of ifs specific antibody or antiserum. The complex antigen–antibody formed is then visualized under

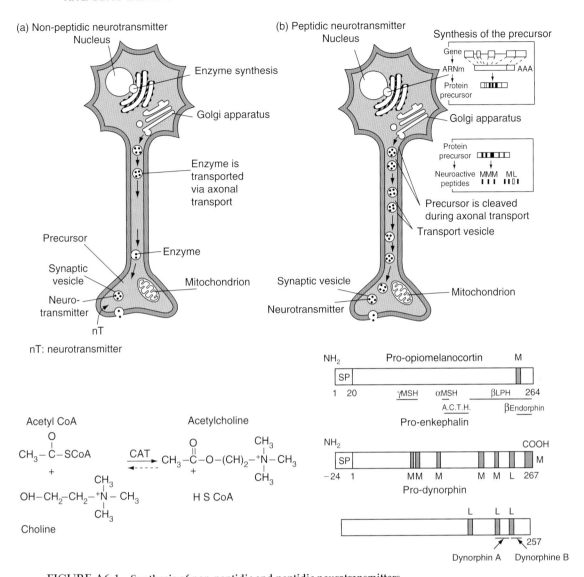

FIGURE A6.1 **Synthesis of non-peptidic and peptidic neurotransmitters.**
(a) Non-peptidic type (example, acetylcholine). **(b)** Peptidic type (example, opioid peptides). Bottom left: The synthesis reaction of acetylcholine from acetylcoenzyme A. Bottom right: Precursors of endomorphines are pro-opiomelanocortin, pro-enkephalin A and pro-dynorphin. The peptides they contain are shown. The numbers indicate the position of the peptides along the protein. CAT, choline acetyltransferase; HSCoA, coenzyme A; MSH, melanocyte stimulating hormone; ACTH, corticotropin; enk, enkephalin; L, leu-enkephalin; M, met-enkephalin; LPH, lipotropin.

light or electron microscopy by means of various methods of detection described below.

The antigen is an endogenous molecule able to induce the formation of antibodies when injected into a foreign body. The antigen will be recognized specifically by these antibodies. Antigens are endogenous particles such as synthesis enzymes or receptors for neurotransmitters. Neuroactive peptides or amino acid neurotransmitters can become antigenic after being conjugated to a carrier protein or a polysaccharide.

The antibodies are immunoglobulins of type G (IgG) or type M (IgM), Y-shaped molecules that display a minimum of two binding sites for the antigen. These two binding sites recognize a very short amino acid sequence of the antigen. This sequence is called an antigenic determinant. The term 'hapten' is used to describe an amino acid sequence that binds specifically to the binding site of the antibody but cannot induce on its own an immune response (example: an amino acid neurotransmitter like GABA).

Two families of antibodies are commonly used in immunocytochemistry: polyclonal antibodies (or antiserum) and monoclonal antibodies. A *monoclonal antibody* is the product of a single B lymphocyte clone

(**Figure A6.2a**). It is made of a population of identical antibody molecules, each of them recognizing the same antigenic determinant (or hapten) on the antigen. A *polyclonal antibody* consists of a heterogeneous family of antibodies that recognize different antigenic determinants on the same antigen. They are generated in a host animal (usually a rabbit) after its immunization by injection of the antigen. The antibody (polyclonal or monoclonal) used to recognize an antigen into the tissues is called the *primary antibody*.

Antibodies are themselves antigenic, so it is possible to produce antibodies that will recognize antigenic determinants on various regions of an antibody. Antibodies directed against 'primary antibodies' are called *secondary antibodies*. They are anti-IgG or anti-IgM antibodies. These secondary antibodies can be labelled and are then used to detect the antigen/primary-antibody complex.

Among the different antigenic determinants of an antibody those that are associated with the antigenbinding site are called *idiotypes*. Secondary antibodies directed against the specific antigen-binding sites (idiotypes) of a primary antibody are called *anti-idiotype antibodies*. These anti-idiotype antibodies are also useful tools for the localization of receptors.

Applications

Localization of neurons synthesizing a specific neurotransmitter

If we want to localize, for example, the cholinergic neurons in a section of brain tissue, the approach is to reveal the neurons that contain the synthesis enzyme for acetylcholine, choline acetyltransferase (ChAT). Sections of brain tissue are incubated with a primary antibody directed against ChAT. The antibodies will bind specifically to the antigen into sections and a stable complex antigen–antibody will be formed only in the neurons that contain ChAT. After washing the sections to remove the antibodies that did not link with the antigen, the complex antigen–antibody is detected according to one of the methods described below (detection).

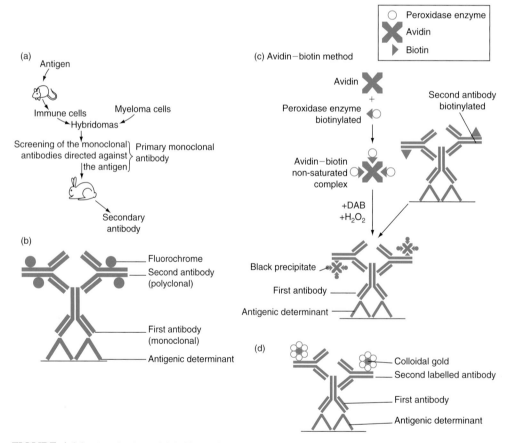

FIGURE A6.2 **Synthesis and labelling of secondary antibodies raised against monoclonal antibodies specific to a neuronal antigen.**

Localization of receptors of a neurotransmitter

Primary antibodies directed against a purified receptor can be used to localize a specific type of receptor on sections of brain tissue, similarly to the localization of synthesis enzymes of neurons.

A second method for localization of receptors use the anti-idiotype antibodies (**Figure A6.3**). Anti-idiotype antibodies are generated against primary antibodies specific to the ligand of the receptor. For example, for the localization of substance P receptors, anti-idiotype antibodies are generated against the antibody specific to substance P. Anti-idiotype antibodies are used since their antigen-binding sites have structural similarities with the ligand itself, of which they constitute a sort of 'molecular image'. This property allows them to bind the biological receptor. This method displays the advantage to enable receptor antibodies to be obtained without a pre-purification of the receptor. The receptor can therefore be localized and its stereospecificity studied.

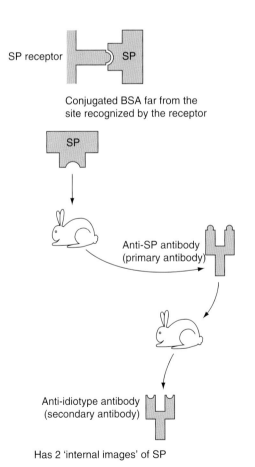

FIGURE A6.3 **Synthesis of anti-idiotype antibodies for substance P (SP).**

Detection of the antigen–antibody complex

Most of the detection methods use markers (labels). The marker is bound to the secondary antibody to obtain *labelled secondary antibodies* which, by reacting with the *antigen–primary antibody complex*, will allow the visualization of the antigen under light or electron microscopy. The markers used for light microscopy are (i) fluorochromes that can be detected with a microscope equipped with an epifluorescence system, or (ii) enzymes (such as peroxidase) that will induce a chromogen reaction in presence of its substrate (such as the diaminobenzide). For electron microscopy, electron dense compounds such as colloidal gold particles are used (**Figure A6.2d**).

Why use labelling of secondary antibodies and not primary ones?

Labelling of antibodies (markers are directly conjugated to the primary antibodies) displays many disadvantages. The labelling of the antibody reduces significantly its capacity to recognize the antigen, thus its specificity. Moreover there is only one molecule of marker for a single antigen–antibody complex. Therefore this technique is poorly sensitive and cannot be used to localize antigens that are present in small quantity in neurons. In addition, for each antigen studied, it is necessary to label the corresponding antibody. For these reasons indirect labelling methods have been developed to increase the sensitivity of detection by amplification of the labelling. All of them use as a first step an unlabelled primary antibody that binds to its antigen into the section. In the following step the primary antibody is recognized by a secondary antibody raised in another species. The secondary antibody is a serum anti-heterologous IgG (or IgM) that binds to many antigenic sites on the primary antibody.

Three main methods commonly used

In the first method (**Figure A6.2b, d**), the secondary antibody is labelled with one of the markers previously described. Since many labelled secondary antibodies bind to a single primary antibody molecule, this technique allows an increase in the labelling and consequently a better visualization of the antigen.

In the peroxidase anti-peroxidase (PAP) detection method, the molecules of secondary antibody are unlabelled and applied in excess, allowing one of their antigen binding sites to be left free. A third step consists in incubating the tissue in a solution containing the peroxidase–anti-peroxidase complex formed by several molecules of peroxidase and antibodies directed against those molecules. The antibodies of the PAP complex are raised in the same species as the primary antibody and thus are recognized by the free antigen-binding sites of the secondary antibody.

The avidin (or streptavidin) biotin method (**Figure A6.2c**) uses the very high affinity of a little hydrosoluble vitamin, biotin, for the protein avidin. The tissue is incubated in the presence of the secondary antibody previously conjugated to biotin molecules (biotinylated secondary antibodies), then in the presence of the preformed complex consisting of molecules of biotin and avidin covalently bound with markers such as peroxidase or fluorochromes.

Both PAP and avidin–biotin methods allow a strong amplification of the labelling and are the most sensitive immunocytochemical techniques.

A6.2.2 *In situ* hybridization

Principle

The aim of these techniques is the detection of a specific nucleic acid sequence in cells on histological sections or in cultured cells. *In situ* hybridization techniques are based on the capacity of all nucleic or ribonucleic acids sequences (ADN or ARN) to bind to a complementary sequence. The sequence of nucleic acids to recognize may correspond to chromosomal DNA, called 'hybridization on chromosomes'. But in general the term '*in situ* hybridization' relates to the detection of messenger RNA (mRNA). This detection or recognition is made possible by the use of a probe that corresponds to a sequence of nucleic acids complementary to the DNA or RNA that is to be detected. The probe is labelled with a marker. When they are in the presence of each other, the specific labelled probe and the endogenous RNA (or DNA) recognized by the probe, hybridize (because of the complementary sequences). The hybrids thus formed are detected by means of the marker linked to the probe.

Application

The *in situ* hybridization technique allows visualization of gene transcripts and, therefore, localization of the potential site of synthesis for the protein or the peptide coded by this gene. In the case of nerve cells, this technique can localize neurons that express a gene coding for a neurotransmitter (if it is a peptide), for a synthesis or degradative enzyme of a specific neurotransmitter, or for a receptor for a neurotransmitter. For example, it has been used to localize neurons that contain the mRNA coding for the precursors of the enkephalins. Moreover, the role of various factors that regulate the expression of these peptides can be studied because this technique can be used to analyze variations in the level of transcribed mRNA in relation to the activity of the neuron, the presence of hormonal factors, etc.

Probes

The most commonly used types are double-stranded DNA, single-stranded DNA, single-stranded RNA and oligodeoxyribonucleotides. The latter three, which include only complementary strands (antisense) to the targeted sequence (cellular mRNA), provide the highest sensitivity. The labelling of the probe is performed in general during probe synthesis by incorporation of markers into the probe.

Double-stranded DNA probes are cDNAs (complementary DNA to cellular mRNA). They are obtained by reverse transcription of cellular RNAs by means of a reverse transcriptase, an enzyme which is able to make complementary single-stranded DNA chains from RNA templates. This is followed by second-strand synthesis using a DNA polymerase. cDNAs are inserted into plasmid vectors, this cDNA library is then screened to identify and isolate the cDNA of interest. This cDNA is then amplified in bacteria or by the polymerase chain reaction (PCR) using oligoprimers on opposite strands. The double-stranded DNA probe can be labelled by different techniques: the nick-translation method consists to induce cuts in the double strand DNA with a DNAse and to repair these cuts by incorporation of labelled and unlabelled desoxynucleotides in the presence of DNA polymerase; the random primed method use, after denaturation of the two strands of cDNA, the ability of the fragment of DNA polymerase to copy single-stranded DNA templates primed with random hexa-nucleotide mixture. Finally, the probes can be labelled during the PCR reaction in the presence of labelled nucleotides. Random priming and PCR give the highest efficiency of labelling.

Double-stranded DNA probes need to be denatured before use for hybridization. These probes are less sensitive than the single-stranded type because many of the two strands can reappariate during the hybridization reaction instead of hybridizing with the target.

Single-stranded DNA probes are preferentially obtained by PCR-based methods using a specific primer from the complementary strand of the RNA transcript and a mixture of labelled and unlabelled desoxynucleotides. Probes which do not reappariate with themselves are thus more sensitive and produce less background noise than double-stranded DNA probes.

Single-stranded RNA probes are produced by *in vitro* transcription of specific cDNAs sub-cloned into a transcription vector (plasmid containing the appropriate polymerase initiation site), by means of an RNAse polymerase. Before this transcription step, the plasmid is linearized with a restriction enzyme to avoid the transcription of plasmid sequences that will cause high backgrounds. The labelling is performed during the transcription by incorporation of labelled nucleotides. The labelled transcript is in general hydrolysed to obtain probes of approximately 150–200 nucleotides in length.

Among the different types of probe, RNA probes are the most sensitive. Since RNA–RNA hybrids are more stable than DNA–RNA ones, strong specific staining with low background can be achieved with RNA probes by using post-hybridization treatment with RNAse and high-temperature washes.

Oligonucleotides are obtained through automated chemical synthesis. They are small (typically 20–30 bases in length) single-stranded DNA probes. Labelling of the probes is performed during synthesis or by adding a tail of labelled nucleotides. Their small size gives them good access to the targeted nucleic acid sequence but limits their sensitivity. However, they are useful when target abundance is high, when gene-specific probes cannot be obtained otherwise, or when only protein sequence information is available.

Markers

Two major families of markers are used for probe labelling: radioactive labels which are detected by autoradiography, and non-radioactive labels which are detected by immunocytochemistry. For many years radioisotopes have been the only markers available to label nucleotides. Among the different radioisotopes used to label probes, ^{35}S is the most commonly used for radioactive *in situ* hybridization. ^{35}S-labelled probes give a resolution of about one cell diameter and relatively rapid results (the time exposure for detection of hybrids is about one week). Despite their high sensitivity, radiolabelled probes have disadvantages such as safety measures required during experimental procedures, limited utilization time, and limited spatial resolution due to scattering of emitted radiation.

More recently non-radioactive *in situ* hybridization techniques have been introduced with the development of haptenlabelled nucleotides that can be well incorporated during probe synthesis. Biotin- or dioxigenin-labelled probes are the most commonly used for cellular mRNA detection. Fluorescent labelling is successfully used for chromosomal *in situ* hybridization. Non-radioactive labelled probes are stable, give rapid results

and display high levels of cellular resolution comparable to that obtained with immunocytochemistry. Furthermore, they open up new opportunities with the possibility of using different labels for simultaneous detection of different sequences in the same tissue.

Detection of the hybrids

For radioactive probes, a low-resolution signal can be obtained by placing the tissue or cells mounted on slides in contact with X-ray film for overnight exposure. This step allows one to control the efficiency of the reaction. If satisfactory, a greater resolution is obtained by dipping the slides in a liquid photographic emulsion, exposed for one or several weeks and developed. In general, a nuclear stain of cells is performed (e.g. with Toluidine Blue) before observation under a microscope with brightfield or darkfield illumination.

Detection of hybrids labelled with non-radioactive probes is performed by immunocytochemistry using specific antibody conjugated with enzyme such as peroxidase or alkaline phosphatase that will give a colour precipitate in the presence of their substrates.

FURTHER READING

Betz H (1999) Structure and functions of inhibitory and excitatory glycine receptors. *Ann. NY Acad. Sci.* **868**, 667.

Brenman JE, Topinka JR, Cooper EC *et al.* (1998) Localization of postsynaptic density-93 to dendritic microtubules and interaction with microtubule-associated protein 1A. *J. Neurosci.* **18**, 8805–8813.

Cohen I, Rimer M, Lomo T, McMahan UJ (1997) Agrin-induced postsynaptic-like apparatus in skeletal muscle fibers in vivo. *Mol. Cell. Neurosci.* **9**, 237–253.

Couteaux R (1998) Early days in the research to localize skeletal muscle actylcholinesterases. *J. Physiol. (Paris)* **92**, 59–62.

Massoulie J, Anselmet A, Bon S *et al.* (1998) Acetylcholinesterase: C-terminal domains, molecular forms and functional localization. *J. Physiol. (Paris)* **92**, 183–190.

Niethammer M and Sheng M (1998) Identification of ion channel-associated proteins using the yeast two-hybrid system. *Meth. Enzymol.* **293**, 104–122.

Nitkin RM, Smith MA, Magill C *et al.* (1987) Identification of agrin, a synaptic organizing protein from Torpedo electric organ. *J. Cell Biol.* **105**, 2471–2478.

Sanes JR (1998) Agrin receptors at the skeletal neuromuscular junction. *Ann NY Acad. Sci.* **841**, 1–13.

Wang ZZ, Mathias A, Gautam M, Hall ZW (1999) Metabolic stabilization of muscle nicotinic acetylcholine receptor by rapsyn. *J. Neurosci.* **19**, 1998–2007.

CHAPTER

7

Neurotransmitter release

The neuron is a secretory cell. The secretory product, the neurotransmitter, is released at the level of chemical synapses (see Chapter 6 and Appendix 6.1). Neurotransmitters achieve the transmission of information at the level of chemical synapses between neurons, neurons and muscle cells, neurons and glandular cells, and sensory receptors and neurons.

Neurotransmitters synthesized by the neuron are stored in the presynaptic element, inside the synaptic vesicles. In the absence of presynaptic activity, the probability of a neurotransmitter being released in the synaptic cleft is very low. This probability increases strongly when the presynaptic element is depolarized by an action potential. The vesicle hypothesis of neurotransmitter release, first formulated by Del Castillo and Katz (1954), is the generally accepted theory of neurotransmitter release (see Appendix 7.1). It states that the neurotransmitter molecules released in the synaptic cleft are those stored in synaptic vesicles (see **Figure 6.4a**). Many recent studies have confirmed the existence of vesicular release, such as data obtained with combined capacitance measurements and amperometry or optical analysis of labelled synaptic vesicles.

Many presynaptic elements contain small synaptic vesicles as well as large dense-core vesicles (see Section 6.1.1). Demonstration at the frog neuromuscular junction that exocytosis of small and large dense-core vesicles can be dissociated pharmacologically strongly suggests the existence of differences in the mechanisms that regulate exocytosis of the two types of secretory vesicles. However, the two systems share at least some common mechanisms for final fusion since both are sensitive to botulinum toxins. This chapter focuses on Ca^{2+}-regulated release of neurotransmitter from small synaptic vesicles.

The work of B. Katz and his collaborators led also to the formulation of the hypothesis that neurotransmitter release from presynaptic vesicles is triggered by elevations

of the intracellular Ca^{2+} concentration ($[Ca^{2+}]_i$). Then, this local rise of $[Ca^{2+}]_i$ has been clearly established as one of the prerequisites for neurotransmitter release. The molecular mechanisms responsible for the coupling between Ca^{2+} ion influx and exocytosis are being elucidated. This includes the identification of the proteins involved in exocytosis, the steps regulating exocytosis and their order of appearance in the phenomenon. It is noteworthy that even when all the proper conditions come together (a presynaptic spike, opening of Ca^{2+} channels, Ca^{2+} entry), the existence of exocytosis of a synaptic vesicle is still not guaranteed (see Appendix 7.2).

The examples in this chapter will be taken mainly from studies on the glutamatergic synapses of the mammalian central nervous system (**Figure 7.1a, b**), and on other synapses that have been examined owing to the large diameter of their presynaptic element, such as the squid giant synapse (**Figure 7.1c**), the neuromuscular junction (see **Figures 6.11** and **6.12a**) and the central synapse, the calyx of head (**Figure 7.1b**).

7.1 OBSERVATIONS AND QUESTIONS

7.1.1 Quantitative data on synapse morphology and synaptic transmission

The regulated release of neurotransmitter occurs at the active zone of a synaptic complex. Synapses of the mammalian central nervous system generally exhibit one or two active zones (or release sites) per bouton (**Figures 7.2a, b**); but there are exceptions, such as the perforated synapse of the hippocampus (see **Figure 6.3**) or the calyx of Held in the medial trapezoid body of the brainstem (**Figure 7.1b**) that contain more than one active zone (**Figure 7.2c**) (around 4–5 for the former and at least 200 for the latter). Giant synapses such

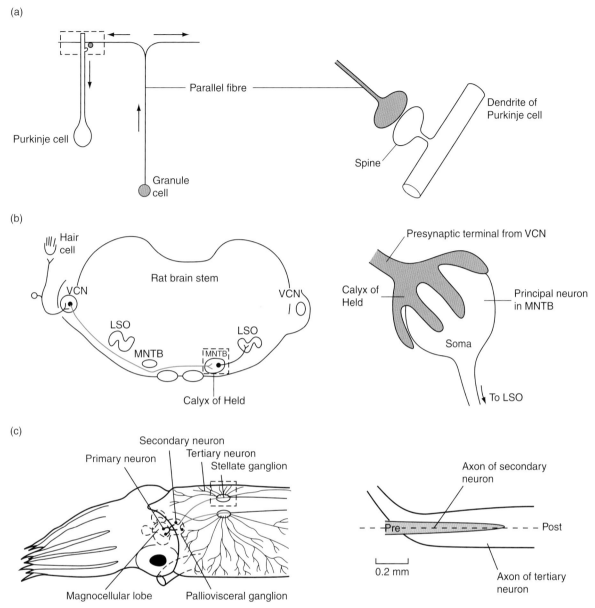

FIGURE 7.1   Three examples of preparations in which synaptic transmission has been studied.
(a) *The cerebellar cortex* Left: Schematic showing the connections between a granular cell axon (parallel fibre) and a Purkinje cell. These synapses are axo-spinous (right). **(b)** *The calyx of Held* Left: Frontal section of the brainstem drawn at the level of the 8th nerve. The axon collaterals of the globular cells of the ventral cochlear nucleus (VCN) project to the neurons of the contralateral medial nucleus of the trapezoid body (MNTB). This synapse is axosomatic and each MNTB neuron receives only one axon terminal that forms the calyx of Held (right). **(c)** *Squid giant synapse.* Secondary neurons, that receive sensory information from the primary ones (left), establish giant axo-axonic synapses with tertiary neurons (right). The tertiary neurons are responsible for contraction of the mantle muscles thus permitting expulsion of water and propelling the animal out of the danger zone. The dotted square indicates the region enlarged on the right of the figures. LSO, lateral superior olive. Drawing (b) adapted from Forsythe ID, Barnes-Davies M, Brew HM (1995) The calyx of Held: a model for transmission at mammalian glutamatergic synapses. In: *Excitatory Aminoacids and Synaptic Transmission*, 2nd edn, New York: Academic Press. Drawing (c) from Llinas R (1982) Calcium in synaptic transmission *Sci. Amer* **247** , 56–65, with permission.

as in the squid (**Figure 7.1c**) comprise around 4400 active zones and the neuromuscular junction (see **Figure 6.12a**) from 300 to 1000 active zones. Synapses of the mammalian central nervous system have small dimensions. In the hippocampus, synaptic terminals are rarely more than 1–5 μm wide, but there are exceptions. Vesicles have diameters of 25–60 nm. Cleft diameter ranges from 0.1 to 1 μm and pre- and postsynaptic

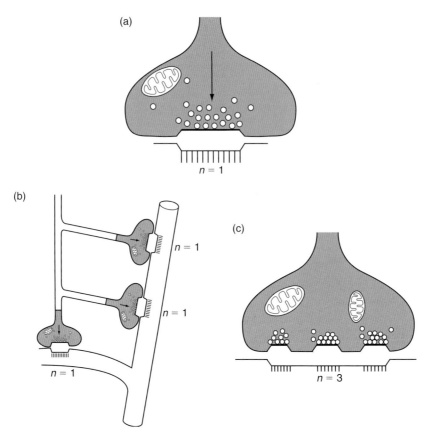

FIGURE 7.2 **Number *n* of active zones per presynaptic terminal.**
Drawings of synaptic boutons of the central nervous system with presynaptic active zone(s) and postsynaptic membranes. The active zones are represented as a black bar with adjacent synaptic vesicles. The number of active zones per bouton is *n* = 1 in **(a)** and **(b)** but *n* = 3 in **(c)**. The synaptic connections are such that a single presynaptic spike will activate a maximum of one active zone in (a), and a maximum of three in (b) and (c).

elements are distant by 10–30 nm (cleft width). Postsynaptic density areas range from 0.01 to 0.5 μm². In three dimensions, the surfaces of presynaptic and postsynaptic membranes have the shape of two plates facing one another.

Postsynaptic responses are either a depolarization or a hyperpolarization of the postsynaptic membrane. The former is called the *excitatory postsynaptic potential* (EPSP; **Figure 7.3a**) since it brings the membrane potential closer to the spike threshold, and the latter is called the *inhibitory postsynaptic potential* (IPSP) since it does the opposite. These responses, being voltage changes, are recorded in current clamp. The underlying currents can be recorded in voltage clamp; they are the *excitatory or inward* (EPSC; **Figure 7.3b**) and the *inhibitory or outward* (IPSC) postsynaptic currents. In this chapter, only excitatory responses are studied.

In response to a single spike in the presynaptic axon, a small-amplitude EPSP (or EPSC) is recorded; it is called 'single-spike EPSP' (or EPSC). It has the following characteristics: the synaptic delay ranges from 200 μs to 1 ms (**Figure 7.3b**); the amplitude of the

single-spike EPSP ranges from 200 μV to 1 mV (**Figure 7.3a**) depending on the number of boutons and active zones activated by the presynaptic spike (**Figure 7.2**). The half-duration of a single-spike EPSP is in the order of milliseconds to tens of milliseconds. Each single presynaptic spike does not necessarily evoke a postsynaptic potential – there are failures of synaptic transmission (**Figure 7.3a**).

In summary, the synapse is an electrochemical unit specialized to function on a distance scale of micrometres and a timescale of submilliseconds.

7.1.2 Ways of estimating neurotransmitter release in central mammalian synapses

Recording of the postsynaptic response

If one considers the schematic representation of synaptic transmission in **Figure 6.4a**, neurotransmitter release corresponds to steps 2, 3 and 4. One way to estimate neurotransmitter release is to measure the postsynaptic response that it evokes. It is an *indirect*

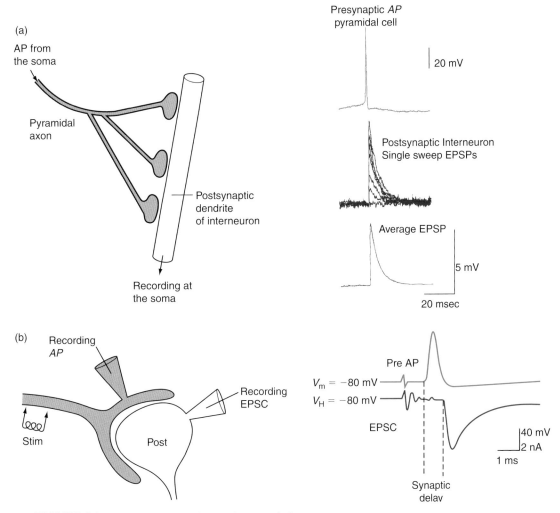

FIGURE 7.3 Basic properties of synaptic transmission.
(a) In the neocortex, pyramidal neurons are Golgi type I neurons, whose main axons leave the cortex after giving off collaterals. Left: These collaterals establish excitatory synapses with dendrites of local interneurons (Golgi type II neurons). Three synapses are represented but their exact number in the experiment performed at right were not determined. Right: In response to each presynaptic action potential (AP) evoked in the pyramidal neuron (top trace), a single-spike EPSP is recorded in the postsynaptic interneuron: it fluctuates in amplitude from 0 mV (failure) to 5 mV (middle traces); EPSPs recorded in response to four successive presynaptic spikes are superimposed (bottom traces). $V_m = -75$ mV in middle and bottom traces. **(b)** The synaptic delay between a presynaptic action potential (AP) evoked by stimulation of the afferent axon and recorded in the calyx of Held, and the postsynaptic response (excitatory postsynaptic current, EPSC) recorded in the postsynaptic soma varies from 500 µs to 1 ms. Drawing (a) from Thomson AM personal communication. Drawing (b) from Borst JGG, Helmchen F, Sakmann B (1995) Pre- and postsynaptic whole-cell recordings in the medial nucleus of the trapezoid body of the rat. *J. Physiol* **489**, 825–840, with permission.

measure since it includes events following release, such as neurotransmitter diffusion from the pre- to the postsynaptic element, binding of neurotransmitter molecules to postsynaptic receptors (step 5a), and induction of the postsynaptic current (step 6). In addition, to be a reliable detector of release events the postsynaptic responses (EPSP or IPSP) should not activate postsynaptic voltage-dependent currents that would amplify or decrease them. Most of the data on neurotransmitter release

explained here have been obtained from recordings of postsynaptic responses to a single presynaptic action potential.

Other techniques

Other techniques can be used to monitor transmitter release from peripheral synapses, large synaptosomes or endocrine cells; they are the patch clamp technique

7. NEUROTRANSMITTER RELEASE

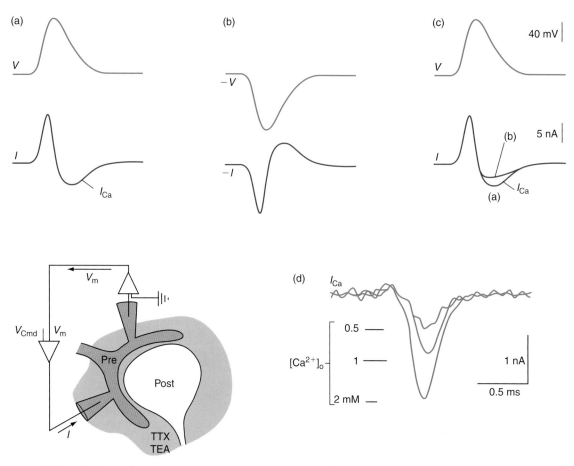

FIGURE 7.4 Ca²⁺ **current flows into presynaptic terminal during the repolarizing phase of the presynaptic spike.**
Two whole-cell electrodes are positioned in the calyx of Held; one measures the membrane potential (V) and the other one injects current (I). It is a two-electrode voltage clamp configuration. The preparation is bathed in 2 mM external Ca²⁺ with TTX and TEA to block the voltage-gated Na⁺ and K⁺ currents. **(a)** The voltage clamp command is an action potential waveform (V) from a holding potential of −80 mV. The current that flows through the membrane is shown on the bottom trace (I). **(b)** The action potential waveform command is inverted ($-V$) and so is the current that flows through the membrane ($-I$). In **(c)** both signals are superimposed (those in (b) have been inverted for superimposition). It shows that the current needed for the action potential is similar in (a) and (b) during the depolarization phase of the action potential. In contrast, the (a) current is larger during the repolarization phase of the action potential. **(d)** The difference between the (a) and (b) currents is an inward current (I_{Ca}). This current is reduced in 1 or 0.5 mM external Ca²⁺ ($[Ca^{2+}]_o$). To perform the experiments in (b), the voltage command (i.e. the action potential waveform injected) is reduced to avoid a large signal in the negative direction (from −80 mV it would hyperpolarize the membrane to −200 mV). Then the currents are scaled up. Adapted from Borst JGG and Sakmann B (1996) Calcium influx and transmitter release in a fast CNS synapse. *Nature* **383**, 431–434, with permission.

and amperometry. The first consists of measuring the membrane capacitance that is directly proportional to the membrane surface area. Upon fusion of a secretory vesicle, membrane area and therefore capacitance increases stepwise by an amount equal to the vesicle or granule membrane area. Amperometry monitors the release of some secretory products by measuring the oxidation of electroactive substances (serotonin, catecholamines, dopamine) with a carbon fibre microelectrode placed near the cell.

7.1.3 Questions

Considering the general functional model of synaptic transmission which states that exocytosis of synaptic vesicles is triggered by an increase of the intracellular Ca²⁺ concentration in the presynaptic element ($[Ca^{2+}]_i$) (in **Figure 6.4a**), the following questions can be asked:

- Is $[Ca^{2+}]_i$ increase a prerequisite for transmitter release? Which type of Ca²⁺ channels are present in the presynaptic membrane and in response to which signal do they open (step 2)?

- Is the presynaptic [Ca^{2+}]$_i$ increase local and transient in response to a presynaptic spike (step 3)?
- How does [Ca^{2+}]$_i$ increase trigger exocytosis? Why is it so rapid (step 3)?
- How do vesicles fuse with the presynaptic membrane in response to a single presynaptic spike (step 4)?
- How much neurotransmitter is released into the synaptic cleft from a single vesicle (step 4)? What is the neurotransmitter lifetime in the cleft?
- What is the mechanism underlying the clearance of the transmitter from the synaptic cleft (steps 5)?

To answer the above questions, we will study the processes of transmitter release in chronological order, from depolarization of the presynaptic membrane by an action potential to the release of transmitter from synaptic vesicles (presynaptic processes I and II; Sections 7.2 and 7.3). Processes occurring in the synaptic cleft, just after transmitter release, are studied in Section 7.4. Details on the quantal and probabilistic nature of neurotransmitter release are given in Appendices 7.1 and 7.2.

7.2 PRESYNAPTIC PROCESSES I: FROM PRESYNAPTIC SPIKE TO [Ca^{2+}]$_i$ INCREASE

7.2.1 The presynaptic Na$^+$-dependent spike depolarizes the presynaptic membrane, opens presynaptic Ca^{2+}channels and triggers Ca^{2+}entry

In a resting presynaptic element, Ca^{2+} ions are present at a very low concentration, 10^{-8} to 10^{-7}M. This intracellular Ca^{2+} concentration, [Ca^{2+}]$_i$, is at least 10,000 times smaller than the extracellular one (see **Figure 3.2**). It is maintained at this resting level by various Ca^{2+} clearance mechanisms (see Section 7.2.4). In response to a presynaptic action potential [Ca^{2+}]$_i$ suddenly increases in the presynaptic element (**Figure 6.4a**, step 1, and **Figure 7.6**).

What is the origin of [Ca^{2+}]$_i$ increase in the presynaptic element? Is [Ca^{2+}]$_i$ increase a prerequisite for transmitter release?

In an extracellular medium deprived of Ca^{2+} ions or containing Ca^{2+}channel blockers such as Co^{2+} or Cd^{2+} ions, presynaptic [Ca^{2+}]$_i$ increase and postsynaptic response are no longer observed although presynaptic

action potentials are unchanged. This suggests that external Ca^{2+} ions enter the presynaptic element through voltage-gated Ca^{2+} channels (that are blocked by Co^{2+}and Cd^{2+}; see Section 5.1). It also shows that presynaptic [Ca^{2+}]$_i$ increase is a prerequisite for synaptic transmission. This has been shown to be valid for all chemical synapses that have been studied.

What triggers the opening of voltage-gated Ca^{2+} channels?

The following hypothesis has been proposed. The brief membrane depolarization that occurs in the ascending phase of each Na$^+$-dependent presynaptic spike triggers the opening of voltage-dependent Ca^{2+}channels and allows subsequent Ca^{2+}influx into the presynaptic element.

In order to check that a presynaptic depolarization can trigger Ca^{2+}entry, Llinas and co-workers (1966) performed the following experiment in the squid giant synapse. They introduced two microelectrodes, one in the presynaptic element to inject a depolarizing current and one in the postsynaptic element to record its activity, and blocked Na$^+$-dependent spikes with TTX. In such conditions, direct depolarization of the presynaptic membrane, though it fails to evoke a presynaptic spike (because of TTX), evokes a presynaptic [Ca^{2+}]$_i$ increase and a postsynaptic response. A presynaptic membrane depolarization can thus trigger Ca^{2+} channel opening and neurotransmitter release.

7.2.2 Ca^{2+}enters the presynaptic bouton during the time course of the presynaptic spike through high-voltage-activated Ca^{2+}channels (N- and P/Q-types)

The calyx of Held is an axo-somatic synapse located in the rat brainstem, in the medial nucleus of the trapezoid body (MNTB), a nucleus that participates in sound localization. It is the largest synapse in the mammalian central nervous system. The presynaptic axon originates in the contralateral cochlear nucleus and each MNTB postsynaptic neuron receives only one calyx (see **Figure 7.1b**). Synaptic transmission is glutamatergic.

When does Ca^{2+}enter the presynaptic element in response to a presynaptic spike?

In order to record the presynaptic Ca^{2+} current, the presynaptic membrane is clamped at $V_H = -80$mV by means of two whole-cell electrodes. To isolate the presynaptic Ca^{2+}current, voltage-dependent Na$^+$ and K$^+$ currents are blocked with TTX and TEA, respectively.

An action potential waveform is injected into the presynaptic terminal through one of the whole-cell electrodes. It evokes a presynaptic Ca^{2+} current that is recorded by the second whole-cell electrode. Recordings show that Ca^{2+} influx is tightly associated with the repolarizing phase of the action potential (**Figure 7.4**): it is essentially a tail current (see Appendix 5.2) that activates shortly after the peak of the action potential and ends before repolarization is complete. It has a peak amplitude of $2.6 \pm 0.2\,nA$ and a half-width of about $350\,ms$. The delay between the beginning of the action potential and that of Ca^{2+} current is about $500\,\mu s$ at $23-24°C$.

Which types of Ca^{2+} channels are involved in transmitter release?

This has been first investigated in the frog neuromuscular junction and then in different other preparations. Antibodies against ω-conotoxin GVIA that selectively bind to N-type Ca^{2+} channels were seen to label active zones on the terminals of motoneurons. In central synapses, to examine the Ca^{2+} channels responsible for Ca^{2+} influx and transmitter release, pharmacological agents that selectively block a type of Ca^{2+} channel have been tested on the amplitude of the presynaptic Ca^{2+} increase and the postsynaptic response.

Consider the example of the glutamatergic synapse between the axons of granule cells (parallel fibres) and Purkinje cells in the rat cerebellum (see **Figures 6.8** and **7.1a**). The presynaptic Ca^{2+} concentration is determined with the Ca^{2+}-sensitive dye magfura, a low-affinity Ca^{2+}-sensitive dye, that emits light in the presence of free Ca^{2+} with a sensitivity of $10^{-4}M$ (see Appendix 5.1). The transmitter release is estimated from the postsynaptic excitatory current recorded in voltage clamp (whole-cell configuration) from the Purkinje soma. Parallel fibres are excited by a stimulating electrode placed in the molecular layer (**Figure 7.5a**). Ca^{2+} entry in response to presynaptic stimulation is measured as a fluorescence signal.

A single stimulus produces an abrupt change in fluorescence (a fluorescence transient) which returns to resting levels within a few hundreds of milliseconds. At saturating concentration, ω-conotoxin GVIA ($0.5\,\mu M$) inhibits by $27.0 \pm 1.7\%$ the fluorescence transient elicited by the stimulation (**Figure 7.5b**, top traces). In comparison, ω-agatoxin IVA ($200\,nM$) reduces the amplitude of the transient by $50.1 \pm 0.9\%$ (**Figure 7.5c**, top traces) and nimodipine ($5\,\mu M$), an L-type channel blocker, has no effect. Simultaneous application of the two toxins has an additive effect and inhibits Ca^{2+} influx by $77 \pm 3\%$ (**Figure 7.5d**). In conclusion, at this cerebellar synapse, the ω-conotoxin-sensitive N-type and the ω-agatoxin-sensitive P/Q-type Ca^{2+} channels are both present in the presynaptic membrane and

allow around 80% of Ca^{2+} entry in response to a presynaptic spike.

For all synapses studied so far, Ca^{2+} enters presynaptic terminals mainly through N- and/or P/Q-type Ca^{2+} channels. The functional properties of these channels impose at least one constraint: since N and P/Q-type Ca^{2+} channels are high-voltage-activated channels, Ca^{2+} enters presynaptic terminals only in response to a *large* membrane depolarization (such as the depolarizing phase of the Na^{+} action potential).

7.2.3 Presynaptic $[Ca^{2+}]$ increase is transient and restricted to micro- or nanodomains close to docked vesicles

How does Ca^{2+} rise in a presynaptic terminal, uniformly or in domains?

This has been studied in the squid giant synapse (see **Figure 7.1c**) with Ca^{2+} imaging techniques. A low-affinity Ca^{2+}-sensitive dye (*n*-aequorin J, a protein that emits light in the presence of free Ca^{2+} ions with a minimum sensitivity of $10^{-4}M$; see Appendix 5.1) is injected into the presynaptic terminal in order to visualize only zones where Ca^{2+} concentration is high. Then the presynaptic axon is stimulated at $10\,Hz$. Multiple fluorescent domains, equally spaced, with a mean size of 0.250 to $0.375\,\mu m^2$ (mean $0.313\,\mu m^2$) are seen in the presynaptic terminal (**Figure 7.6**). Their number 4500 is quite close to the average number of release sites in this terminal (4400). These microdomains are located at active zones of the presynaptic plasma membrane. This observation is evidence for the fact that voltage-dependent Ca^{2+} channels are clustered at active zones. Since the presynaptic stimulation used in this experiment is not an action potential, it does not allow to apprehend the physiological Ca^{2+} concentration in presynaptic microdomains.

Where are presynaptic Ca^{2+} channels located?

The rapidity with which neurotransmitter release can be triggered after Ca^{2+} influx (within $200\,\mu s$) makes it likely that Ca^{2+} ions act at a very short distance from the Ca^{2+} channels. This suggested that: (i) Ca^{2+} channels and release sites are located at a close distance; and (ii) there exists a stable complex between synaptic vesicles and plasma membrane, preassembled in the resting state, before $[Ca^{2+}]_i$ increases. Ca^{2+} diffusion is too restricted and delay of exocytosis too short to allow for vesicle movement before fusion with the plasma membrane (Ca^{2+} ions diffuse no more than a few vesicle diameters into the cytoplasm). In other words, presumably synaptic vesicle available for rapid transmitter release must be predocked in the vicinity of Ca^{2+} channels.

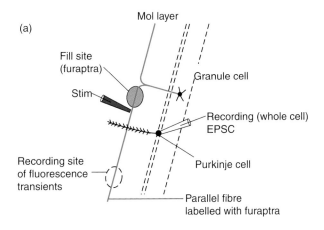

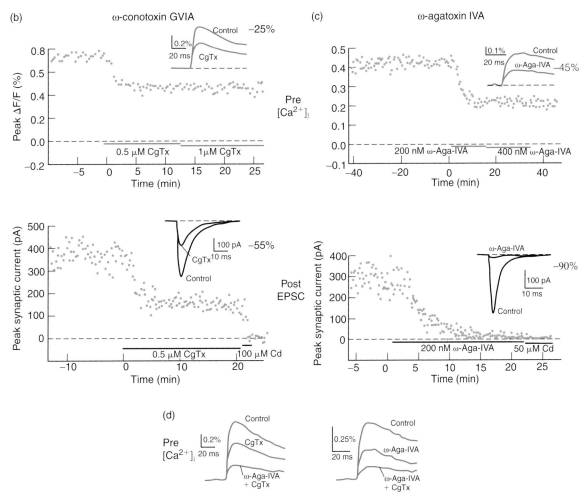

FIGURE 7.5 **N- and P/Q-type Ca^{2+} channel blockers reduce presynaptic Ca^{2+} influx and synaptic transmission at a cerebellar synapse.**

(a) In a transverse cerebellar slice, the relative locations of labelling with the dye furaptra (fill site), stimulus electrode and recording sites (whole-cell recording of EPSC from a Purkinje soma and recording of presynaptic fluorescence transients in the molecular layer). **(b)** Amplitude of furaptra fluorescence transients ($\Delta F/F$) in the presence of increasing concentrations of ω-conotoxin GVIA (CgTx) to block the N-type Ca^{2+} current (top traces). Concomitant recording of the postsynaptic current is shown in the bottom trace. Each furaptra transient is elicited by a single stimulus of the parallel fibre tract and is a measure of the presynaptic Ca^{2+} influx. The inset shows superimposed fluorescence transients in control conditions and after addition of 0.5 and 1 μM CgTx. **(c)** The same experiment as in (b) but in the presence of increasing concentrations of ω-agatoxin IVA (ω-Aga IVA) to block the P/Q-type Ca^{2+} current. The inset shows superimposed fluorescence transients in control conditions and after addition of 200 and 400 nM ω-Aga IVA. Concomitant recording of the postsynaptic current is shown in the bottom trace. **(d)** Additive effects of the sequential application of saturating concentrations of the toxins on the furaptra transients. Adapted from Mintz I, Sabatini BL, Regehr WG (1995) Calcium control of transmitter release at a cerebellar synapse. *Neuron* **15**, 675–688, with permission.

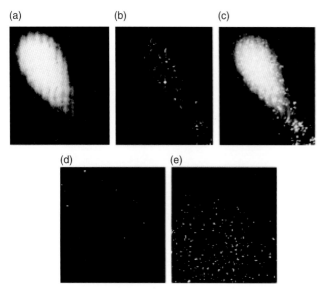

FIGURE 7.6   Microdomains of Ca^{2+} increase in the presynaptic terminal of the squid giant synapse.
(a) Fluorescence image of a presynaptic terminal injected with n-aequorin-J. When the presynaptic fibre is fully loaded, it is continuously stimulated at 10 Hz for 10 s. **(b)** The acquisition during these 10 s of tetanic stimulation reveals stable quantum emission domains that appear as white spots. The background fluorescence shown in (a) disappears in (b) due to subtraction. **(c)** Superposition of the fluorescent images in (a) and (b) reveals that the distribution of the microdomains of high calcium coincides with the presynaptic terminal. Emission domains in an unstimulated terminal **(d)** and in the same terminal during tetanic stimulation **(e)** are shown at high magnification. Adapted from Llinas R, Sugimori M, Silver RB (1992) Microdomains of high calcium concentration in a presynaptic terminal. *Science* **256**, 677–679, with permission.

The localization of Ca^{2+} channels relative to the position of transmitter release sites was first investigated with imaging (**Figure 7.7a**) and immunocytochemical (**Figure 7.7b**) techniques. In the frog neuromuscular junction, the presynaptic nerve terminal is a long structure (several hundreds of micrometres) characterized by the presence of neurotransmitter release sites or active zones spaced at regular 1 μm intervals. Directly across the synaptic cleft just facing active zones are clusters of nicotinic acetylcholine receptors (nAChR) located on the edge of the postjunctional folds (see **Figure 6.12a**). The preparation is double-labelled to disclose both postsynaptic nAChR and presynaptic N-type Ca^{2+} channels. The idea is that the localization of postsynaptic nAChRs indicates exactly the localization of presynaptic active zones. Ca^{2+} channels are labelled with biotinylated ω-conotoxin, a specific and irreversible blocker of N-type Ca^{2+} channels and of synaptic transmission at the frog neuromuscular junction. To reveal ω-conotoxin-sensitive Ca^{2+} channel labelling, preparations are then incubated with streptavidin

Texas Red (see Appendix 6.2) which fluoresces red. When nerve terminals are removed by pulling off branches of the motor nerve, the Ca^{2+} channel labelling totally disappears, indicating that ω-conotoxin binding sites are strictly located on presynaptic terminals. Postsynaptic nicotinic receptors nAChR are labelled with α-bungarotoxin coupled to boron dipyrromethane difluoride which fluoresces green. Under the light microscope, each fluorescent band of presynaptic Ca^{2+} channels is matched by a fluorescent stain of postsynaptic nAChR (**Figure 7.7b**). Bands of labelled Ca^{2+} channels and labelled nAChRs are thus almost perfectly aligned, suggesting that Ca^{2+} channels are clustered in the membrane of presynaptic active zones, opposite the postjunctional folds.

The clustering of presynaptic N- and P/Q-type Ca^{2+} channels (Ca$_v$2.2 and Ca$_v$2.1) at active zones has been confirmed by the discovery of a physical link between these Ca^{2+} channels and syntaxin, an integral protein of the presynaptic plasma membrane involved in vesicle docking at the presynaptic active zones (see Section 7.3.2). This was shown (i) by co-precipitation of N and P/Q channels (labelled with their specific toxins) with syntaxin (labelled with a specific antibody), (ii) by the identification of a syntaxin-binding domain on N- and P/Q-type α$_1$-subunits, and (iii) by inhibition of synaptic transmission after injection of peptide inhibitors into presynaptic neurons. N- and P/Q-type Ca^{2+} channels interact directly with SNARE proteins through the *synaptic protein interaction* (synprint) site, which resides in the N-terminal half of the L$_{II–III}$ cytoplasmic loop connecting domains II and III of their α$_1$-subunit (**Figure 7.7c**). The synprint site of Ca$_v$2.2 binds the plasma membrane proteins syntaxin and SNAP-25 as well as the vesicle protein synaptotagmin (**Figures 7.7d** and **7.12**). Injection of peptide inhibitors of this interaction into presynaptic neurons inhibits synaptic transmission, consistent with the conclusion that this interaction is required to position docked synaptic vesicles near Ca^{2+} channels for effective fast exocytosis. These results define a second functional activity of the presynaptic Ca^{2+} channel: targeting docked synaptic vesicles to a source of Ca^{2+} for effective transmitter release (**Figure 7.7d**) and making release as fast as possible.

These results exemplify some general principles of rapid Ca^{2+} signalling in neurotransmitter release:

- Ca^{2+} entry into the presynaptic element occurs in close proximity to the exocytotic apparatus.
- Clustering of Ca^{2+} channels close to release sites ensures that a Ca^{2+} signal is rapidly available (in the hundreds of microseconds timescale) to the nearby Ca^{2+}-sensitive proteins which initiate transmitter release.

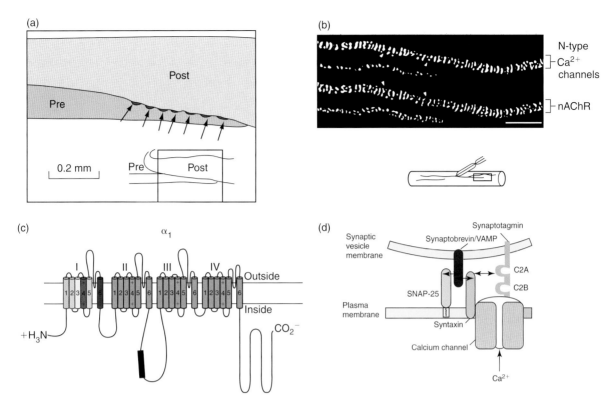

FIGURE 7.7 Presynaptic N-type Ca^{2+} channels are clustered at active zones.
(a) In the squid giant synapse, presynaptic zones of $[Ca^{2+}]_i$ increase in response to a train of brief presynaptic stimuli (0.5 s at 80 Hz) are visualized with the FURA-2 technique. They are localized at active zones. The diagram below illustrates the synapse and the box indicates the region studied. **(b)** In the frog neuromuscular junction, N-type Ca^{2+} channels and nicotinic acetylcholine receptors (nAChR) are labelled with two different selective toxins coupled to different fluorescent dyes. The preparation is viewed with a confocal laser microscope. The diagram below illustrates the structure of the neuromuscular junction and the box indicates the region scanned by the microscope. The images showing the distribution of presynaptic Ca^{2+} channels (top) and postsynaptic nAChRs (bottom) are separated for clarity but they are in fact superimposed. **(c)** Predicted topological structure of the α_1 subunit of N-type Ca^{2+} channel. Rectangle indicates the synprint site. **(d)** Theoretical model of interaction of presynaptic $Ca_v2.2$ channels (N- and P/Q-type Ca^{2+} channels) with SNARE proteins (syntaxin, SNAP-25 and synaptotagmin) at the presynaptic plasma membrane. Part (b) from Robitaille R, Adler EM, Charlton MP (1990) Strategic location of calcium channels at transmitter release sites of frog neuromuscular synapses. *Neuron* **5**, 773–779, with permission. Part (c) from Sheng ZH, Rettig J, Takahashi M, Catterall WA (1994) Identification of a syntaxin-binding site on N-type calcium channels. *Neuron* **13**, 1303–1313, with permission. Part (d) Catterall WA (2000) Structure and regulation of voltage-gated Ca^{2+} channels. *Annu. Rev. Cell Dev. Biol* **16**, 521–555.

7.2.4 Ca^{2+} clearance makes presynaptic $[Ca^{2+}]_i$ increase transient: it shapes its amplitude and duration

Ca^{2+} clearance is the removal of Ca^{2+} ions (in excess compared to the resting state) from the presynaptic terminal. The aim of Ca^{2+} clearance mechanisms is to rapidly re-establish the resting level of $[Ca^{2+}]_i$. Ca^{2+} clearance is achieved by proteins that extrude Ca^{2+} ions toward the extracellular space or toward organelles such as the endoplasmic reticulum or mitochondria: these are Ca^{2+} pumps (Ca^{2+}-ATPases) and Ca^{2+}-transporters (Na–Ca exchanger) (see **Figure A3.1** and **Figure 7.8**). Ca^{2+} ions that enter the presynaptic terminal will also rapidly bind to cytosolic proteins (Ca^{2+}-buffers, Ca-B).

Because such binding confiscates Ca^{2+} ions, it can rapidly diminish freely diffusing Ca^{2+} ions.

But such buffering is not a real clearance since Ca^{2+} ions will unbind from these proteins; it is a temporary clearance. Unbound Ca^{2+} ions will have then to be extruded by pumps and transporters. Therefore, the more numerous the number of Ca^{2+} pumps and transporters, the more efficient and rapid is Ca^{2+} clearance.

Extrusion of Ca^{2+} to the extracellular medium by the plasma membrane Ca-ATPase (PMCA) pump and by the Na–Ca exchanger

The former uses the hydrolysis of ATP as a source of energy and is independent of the extracellular Na^+

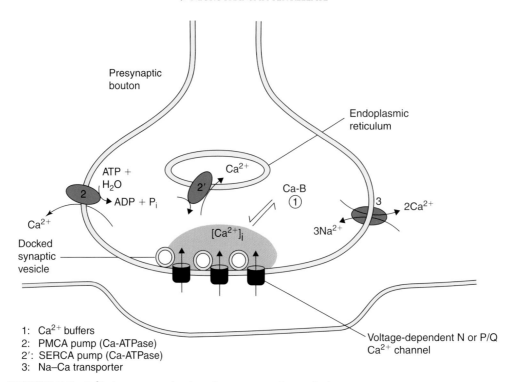

1: Ca^{2+} buffers
2: PMCA pump (Ca-ATPase)
2': SERCA pump (Ca-ATPase)
3: Na–Ca transporter

FIGURE 7.8 Ca^{2+} clearance mechanisms in a presynaptic terminal.
While Ca^{2+} ions enter at the level of the presynaptic active zone through high-voltage-activated Ca^{2+} channels, they are rapidly buffered by cytoplasmic Ca^{2+}-binding proteins (Ca-B) (1). Ca^{2+} ions are also actively cleared from the intracellular medium towards the extracellular medium via Ca-ATPases of the plasma membrane (PMCA pumps) (2) and the Na–Ca exchangers (3). They are also cleared by active transport toward endoplasmic reticulum via another type of Ca-ATPase (SERCA pumps) (2^1). This clearing has a time constant of the order of tens of milliseconds to seconds.

concentration. The latter is driven by the Na$^+$ electrochemical gradient across plasma membrane and is thus sensitive to extracellular Na$^+$ concentration. The Ca-ATPase pump is proposed to be a low-capacity high-affinity system (K$_D$ = 0.2–0.3 µM) whereas the Na–Ca exchanger would have a high-capacity low-affinity system (K$_D$ = 0.5–1.0 µM). The Ca-ATPase pump would thus be the most efficient system in the presence of a low presynaptic activity, and the two systems would act in synergy to regulate the intracellular Ca^{2+} concentration after a train of action potentials.

Sequestration of Ca^{2+} ions in smooth endoplasmic reticulum and mitochondria

This is achieved by sarco-endoplasmic Ca-ATPase (SERCA) pumps present in the membrane of these organelles. The smooth endoplasmic reticulum is a Ca^{2+} storage compartment. In the different cell types studied, the smooth endoplasmic reticulum Ca-ATPase pump has a better affinity for Ca^{2+} than that of mitochondrion. This latter would function in rare situations in cases of massive Ca^{2+} entry. Noteworthy, following an appropriate signal (such as the formation of inositol trisphosphate, IP$_3$), is that the Ca^{2+} ions stored in these compartments can be released in the cytoplasm through Ca^{2+}-permeable channels.

Ca^{2+} buffering by cytosolic proteins

Different cytosolic proteins have the ability to bind Ca^{2+} with a high affinity. These proteins have, in general, a low molecular weight and act primarily as Ca^{2+} buffers (such as parvalbumin and calbindin) or subserve messenger functions (calmodulin). Parvalbumin is found in great amount in most GABAergic neurons in the mammalian central nervous system (the co-localization of GABA and parvalbumin has been shown immunohistochemically using highly specific antibodies to GABA and parvalbumin); whereas in other neurons the concentration of parvalbumin is much lower. The high concentration of parvalbumin might have a consequence on a neuron's ability to rapidly buffer Ca^{2+}. This is especially important for neurons that have a high tonic activity since trains of action potentials trigger repetitive [Ca^{2+}]$_i$ increases in their synaptic terminals.

Calbindin is a protein that was originally found in the gut, where it binds Ca^{2+} and is vitamin D-dependent. Its presence has been shown in neurons of the mammalian central nervous system, notably the Purkinje cells of the cerebellar cortex and the dopaminergic neurons of the substantia nigra. Calmodulin has a high affinity for Ca^{2+} and a role of intracellular messenger. The buffering of free intracellular Ca^{2+} ions by cytoplasmic calcium-binding proteins is a very efficient system responsible for the rapid disappearance of free Ca^{2+} ions.

The relative contribution of the clearance systems: example of Purkinje cells

Cerebellar Purkinje cells are GABAergic neurons that have powerful systems to control $[Ca^{2+}]_i$. Immunocytochemical studies demonstrate considerable amounts of cytosolic Ca^{2+}-binding proteins, particularly calbindin D_{28K} and parvalbumin. There are also numerous Ca^{2+} pumps localized in the endoplasmic reticulum (SERCA pumps) or the plasma membrane (PMCA pumps). In order to understand the respective roles of these clearance systems, Purkinje cells are loaded with the fluorescent Ca^{2+} dye FURA-2 (see Appendix 5.1), $[Ca^{2+}]_i$ transients are evoked by direct membrane depolarization and measured by microfluorometry, and clearance systems are pharmacologically inhibited one at a time. Since all these experiments are achieved with the use of a whole-cell electrode, the duration of the study is limited to 25 minutes in order to avoid the washing out of intracellular constituents that would give an artefactual diminution of $[Ca^{2+}]_i$.

The contribution of SERCA pumps on the amplitude and decay phase of $[Ca^{2+}]_i$ transients is studied by applying cyclopiazonic acid (CPA) or thapsigargin, specific inhibitors of this ATPase. For blocking PMCA pumps, 5,6-succinimidyl carboxyeosin (CE) is applied; and for blocking the Na–Ca exchanger, external Na^+ is replaced with Li^+, choline or N-methyl-D-glucamine, cations that cannot substitute for Na^+ in the exchange reaction. The rate of decay of $[Ca^{2+}]_i$ transients with similar peak values are compared in control and experimental conditions in order to calculate the rate of clearance. All these inhibitors do not affect resting $[Ca^{2+}]_i$ levels, indicating that the passive leak of Ca^{2+} into the somata is small. For low-intensity $[Ca^{2+}]_i$ transients (0.5 μM at the peak), the proportion of intracellular Ca^{2+} removed by SERCA pumps, PMCA pumps and the Na–Ca exchanger is balanced (**Figure 7.9**). They equally remove 78% of the intracellular Ca^{2+}.

7.3 PRESYNAPTIC PROCESSES II: FROM [Ca²⁺] INCREASE TO SYNAPTIC VESICLE FUSION

In a resting nerve terminal the small synaptic vesicles loaded with the neurotransmitter are either localized in the cytoplasm of the active zone (linked to cytoskeletal elements) or docked to the presynaptic plasma membrane. Depolarization of the presynaptic membrane by an action potential leads to an influx of Ca^{2+} ions that *may* trigger fusion (exocytosis) of a docked vesicle.

7.3.1 Overview of the hypothetical vesicle cycle in presynaptic terminals

From observations and experiments described in the following sections, synaptic vesicle traffic in nerve terminals is considered to involve several hypothetical stages (**Figure 7.10**).

- *Targeting and docking.* After they fill with neurotransmitter by active transport, synaptic vesicles dock at morphologically defined sites of the presynaptic plasma membrane.
- *Fusion.* The local increase of $[Ca^{2+}]_i$ triggers exocytosis of docked vesicles; i.e. fusion of synaptic vesicle membrane with the presynaptic plasma membrane and release of the vesicular content through a fusion pore.
- *Retrieval and recycling.* Empty vesicles form coated pits that undergo endocytosis.

Comparison of neurotransmitter release with other secretory systems shows that targeting, docking and fusion of vesicles involve common mechanisms. Synaptic transmission makes use of a mechanism that is common to biology.

7.3.2 Docking: a subpopulation of synaptic vesicles is docked to the active zone close to Ca²⁺ channels by means of specific pairing of vesicular and plasma membrane proteins

How do synaptic vesicles recognize the presynaptic plasma membrane for docking?

Selective targeting of a vesicle to its correct destination has been proposed by G. Palade in 1970 to result from specific recognition sites between vesicle and plasma membranes. Later, the molecular basis of this specific interaction was studied by J. Rothman and colleagues in a cell-free preparation of Golgi stacks that reconstitutes vesicle-mediated transport between Golgi cisternae, a model of constitutive vesicle fusion with a membrane. The experiments described briefly here led to the discovery and purification of several proteins crucial for exocytosis.

Everything began with the discovery that N-ethyl maleimide (NEM) blocks the fusion of Golgi vesicles with Golgi stacks: the vesicles still bud off from cisternae but the released vesicles no longer fuse with the next stack

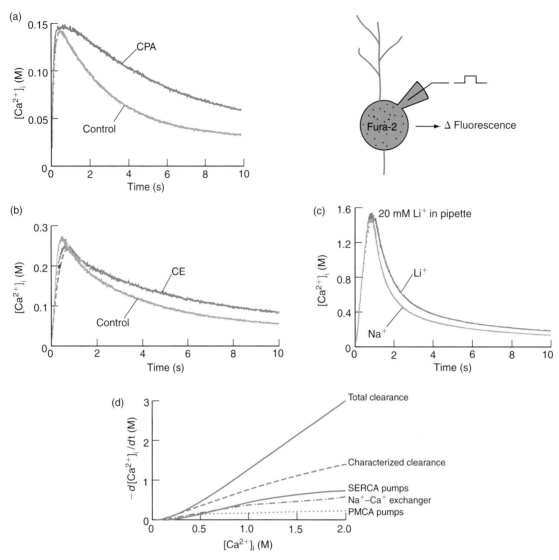

FIGURE 7.9   Relative contribution of the different mechanisms of Ca²⁺ clearance in Purkinje somata.
$[Ca^{2+}]_i$ transients are evoked in FURA-2-loaded Purkinje cells by a depolarizing current pulse of varying duration (60–250 ms). Effects on $[Ca^{2+}]_i$ transients of **(a)** cyclopiazonic acid (CPA), a blocker of SERCA pumps, **(b)** 5,6-succinimidyl carboxyeosin (CE), a blocker of PMCA pumps, and **(c)** Li^{2+} saline, a blocker of Na–Ca exchanger. **(d)** Total Ca^{2+} clearance rate is presented in comparison with the rate of the different components characterized in the above experiments. Clearance rate is plotted as a function of the $[Ca^{2+}]_i$ in the range between 50 nM and 2 μM. The clearance rate is calculated as follows: (i) The decay phase of each transient is fitted by a single or double exponential function and the derivative function $(d[Ca^{2+}]_i/dt)$ is calculated from the fit. (ii) $-d[Ca^{2+}]_i/dt$ is then plotted as a function of the $[Ca^{2+}]_i$ values obtained from the experimental fit. (iii) The plots from transients with equal peak $[Ca^{2+}]_i$ in each condition (control *versus* inhibitor) are pooled and fitted with a polynomial function of fifth to seventh order. Adapted from Fierro L, DiPolo R, Llano I (1998) Intracellular calcium clearance in Purkinje cell somata from rat cerebellar slices. *J. Physiol.* **510**, 499–512, with permission.

membrane; they accumulate docked to the target membrane. This suggested the involvement of a NEM-sensitive fusion protein (NSF) in the fusion step. That protein was purified according to its ability to restore fusion after NEM inactivation. NSF is a 76 kDa protein, a water-soluble ATPase with two distinct ATP-binding sites. Adaptor proteins required for NSF function were subsequently purified and named soluble NSF-attachment proteins or SNAPs. There are three forms α, β, and γ-SNAPs. The existence of such a membrane bound form of NSF + SNAPs suggested that these proteins recognized specific receptors situated in the membrane.

The Rothman group used immobilized α-SNAP and NSF to isolate the SNAP Receptors (SNAREs) (**Figure 7.11**). Since the complex NSF–SNAP is attached to membranes via SNAREs in the absence of hydrolyzable ATP,

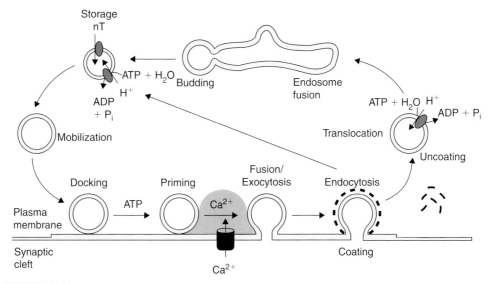

FIGURE 7.10   Diagram of the hypothetical synaptic vesicle cycle in a presynaptic terminal.
The same synaptic vesicle is shown at different stages. Sites of docking, priming and fusion have been separated for clarity. NT, neurotransmitter. Adapted from Südhof TC (1995) The synaptic vesicle cycle: a cascade of protein–protein interactions. *Nature* **375**, 645–653, with permission.

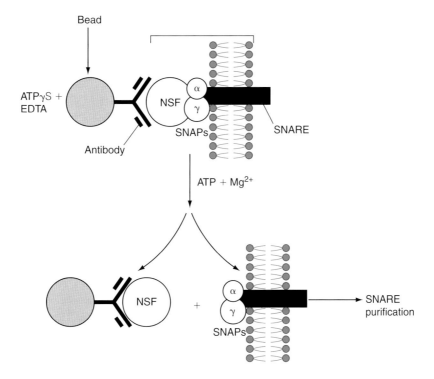

FIGURE 7.11   The SNARE discovery.
Scheme of the experiment that identified the integral membrane proteins (SNAREs) of the vesicle and presynaptic plasma membranes of brain synapses. A stable complex between the NSF protein, the SNAPs and the membranes can be isolated in the presence of a non-hydrolyzable analogue of ATP (ATPγS). Inversely, the membrane-bound form of the NSF protein is released from the membranes to the cytoplasm by ATP hydrolysis (i.e. in the presence of ATP and Mg²⁺). Solubilized brain membranes and NSF–SNAPs are immobilized on beads via a specific anti-*myc* antibody (NSF is tagged with the marker *myc*), in the presence of the non-hydrolyzable analogue of ATP, ATPγS, and in the absence of Mg²⁺. The stable complex [NSF–SNAP–membrane proteins] is thus captured. It is then dissociated in the presence of ATP and Mg²⁺ in NSF on the one hand and the complex SNAPs-membrane proteins on the other hand. Membranes are collected and SNAREs are characterized. Rothman JE (1994) Intracellular membrane fusion. *Adv. Second Messenger Phosphoprotein Res.* **29**, 81–96.

(a)

Example of SNARE subfamilies

Example of the corresponding proteins

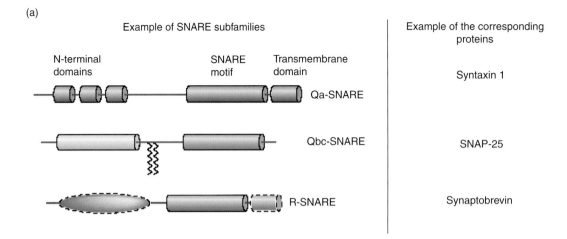

Syntaxin 1

SNAP-25

Synaptobrevin

(b)

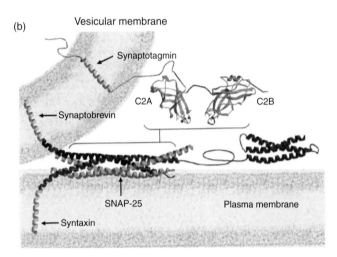

FIGURE 7.12   The structures of three SNARE subfamilies and model of a trans-SNARE complex.
(a) Qa-SNAREs have N-terminal antiparallel three-helix bundles. Qbc-SNAREs represent a small subfamily of
SNAREs – the SNAP-25 (25-kDa synaptosome-associated protein) subfamily – that contain two SNARE
motifs connected by a linker that is frequently palmitoylated (zig-zag lines), and most of the members of this
subfamily function in constitutive or regulated exocytosis. The various N-terminal domains of R-SNAREs are
represented by a basic oval shape. Dashed domain borders highlight domains that are missing in some subfamily members. (b) Model of a *trans* -SNARE complex and synaptotagmin. The regions that interact are indicated with *brackets*. Part (a) adapted from Jahn R and Scheller RH (2006) SNAREs-engines for membrane
fusion. *Nature Rev. Mol. Cell Biol.* 7, 631–643. Part (b) adapted from Littleton JT, Bai J, Vyas B *et al.* (2001)
Synaptotagmin mutants revel essential functions for the C2B domain in Ca-triggered fusion and recycling of
synaptic vesicles *in vivo. J. Neurosci.* **21**, 1421–1433, with permission.

this property was utilized to purify SNAREs of the
vesicle membrane (v-SNARE; v for vesicle) and SNAREs
of the target presynaptic plasma membrane (t-SNARE;
t for target). In the vesicle membrane, synaptobrevin (VAMP/synaptobrevin) was thus identified as a
v-SNARE. In the presynaptic membrane, syntaxin and
SNAP-25 (SyNaptosome-Associated Protein 25, a protein of 25 kDa which has no relation to the similarly
named SNAPs) were thus identified as t-SNAREs
(**Figure 7.12a**). The cytoplasmic domains of the SNARE
proteins assemble into a SNARE complex. The associat-
ing segments are called SNARE motifs and contain

60–70 residues. The SNARE complex is composed of
four α-helices. In neurons SNAP-25 contributes two
helices to the SNARE complex while syntaxin and
synaptobrevin each contribute one helix. These three
SNAREs, syntaxin, SNAP-25 and VAMP/synapto-
brevin, form a stable heterotrimeric complex (**Figure
7.12b**) that has been proposed to mediate the docking of
synaptic vesicles at the presynaptic plasma membrane.

The crucial function of SNARE proteins in numerous
membrane trafficking pathways has been established
in various organisms. Each transport vesicle has its
own specific v-SNARE that pairs up in a unique match

with a cognate t-SNARE found only at the intended target membrane. The SNAREs differ from one secretion system to the other, while NSF proteins and soluble NSF attachment proteins (SNAPs) are very general cytoplasmic proteins.

In summary, at the docking stage, the membrane of synaptic vesicles and the target plasma membrane are tight together via the binding of v-SNAREs (synaptobrevin) and t-SNAREs (syntaxin, SNAP-25) that are respectively vesicular and plasma membrane proteins. The cytoplasmic domains of these proteins assemble into a four-helix bundle that pulls the vesicle and target membranes together.

7.3.3 Multiple Ca^{2+} ions must bind to Ca^{2+} receptor(s) to initiate vesicle fusion (exocytosis)

SNAREs constitute the core of a conserved fusion machine, but additional accessory proteins must serve to regulate the fusion reaction, in particular at least one protein that senses Ca^{2+}, since in Ca^{2+}-regulated exocytosis as in neuronal synapses, the docked vesicles do not fuse with a high probability until a significant [Ca^{2+}]$_i$ rise occurs. The delay between [Ca^{2+}]$_i$ rise and the postsynaptic response can be as short as 60 to 200 µs, placing strong kinetic constraints on the transduction pathway that ends up in exocytosis and release of neurotransmitter by the presynaptic terminal.

What is the local Ca^{2+} concentration required to trigger vesicle fusion?

The Ca^{2+} sensor (also called Ca^{2+} receptor) should have a high affinity for Ca^{2+} ions, in the order of tens or hundreds of micromolars depending on the nerve terminal that is under study.

Why is Ca^{2+} entry close to docked vesicles and release sites?

As explained in Section 7.2.3, SNARES are tightly coupled to either N- or P/Q-type Ca^{2+} channels. This makes Ca^{2+} entry close to docked vesicles and release sites (see **Figure 7.7**).

What is the identity of the Ca^{2+} receptors that transduce the [Ca^{2+}]$_i$ rise in secretory trigger?

The search for presynaptic Ca^{2+} receptors was performed with antibodies raised against synaptic junctional complexes. This led to the isolation of a 65 kDa antigen, p65, which once cloned and sequenced was renamed synaptotagmin I. It represents a candidate protein for the role of Ca^{2+} receptor as its large cytoplasmic domain is composed of tandem Ca^{2+}-binding motifs called C2 domains (C2A and C2B) (**Figures 7.7d** and

7.12b). It spans the vesicle membrane once, near its N terminus, and possesses a short intravesicular domain. Synaptotagmin I seems to bind a total of five Ca^{2+} ions, three by C2A and two by C2B. Since the initial cloning, more than a dozen of synaptotagmin isoforms have been identified.

Synaptotagmin co-immunoprecipitates with syntaxin from rat brain extracts and binds to isolated t-SNAREs. Binding occurs at the base of t-SNAREs in the region that assembles into SNARE complexes (**Figure 7.12b**). This binding site suggests that synaptotagmin regulates SNARE function.

7.3.4 Fusion: from Ca^{2+} binding to exocytosis

Fusion is a lipid–lipid interaction. Fusion between synaptic vesicles and presynaptic plasma membrane is called exocytosis: it is a regulated fusion (as opposed to constitutive fusion which is Ca^{2+}-independent). In synapses, synaptic vesicles fuse upon receipt of a signal, the increase of [Ca^{2+}]$_i$.

How does binding of Ca^{2+} to its receptor(s) initiate exocytosis?

One model (**Figure 7.13**) proposes that in the absence of Ca^{2+}, v- and t-SNAREs weakly preassemble into a ring-like structure. In response to [Ca^{2+}]$_i$ rise, synaptotagmin binds Ca^{2+} ions and by changing conformation, drives complete assembly of trans-SNARE complexes, bringing the membranes that are destined to fuse into close proximity such as in the centre of the ring, so the hydratation barrier at the surface of the membranes is overcome and the membranes touch.

Other interpretations have been proposed. For example, synaptotagmin would bind to t-SNAREs and block SNARE assembly as long as [Ca^{2+}]$_i$ does not rise. Then upon binding of Ca^{2+} ions, synatotagmin would change conformation and let SNAREs assemble. In this model Ca^{2+}-free synaptotagmin acts as a fusion clamp.

Whichever the mechanism, one must keep in mind that even when all the proper conditions come together (a presynaptic spike, opening of Ca^{2+} channels, Ca^{2+} entry, Ca^{2+} binding), the average probability of the exocytosis of a synaptic vesicle still remains below 1 (see Appendix 7.2). In fact, the mean release probability is between 0.05 and 0.5 according to the synapse studied. In other words, Ca^{2+}-triggered exocytosis is *inefficient*: only one out of two to twenty action potentials leads to exocytosis.

Exocytosis involves the formation of a fusion pore

The earliest attempts to characterize the fusion pore came from the ultrastructural studies of Heuser and

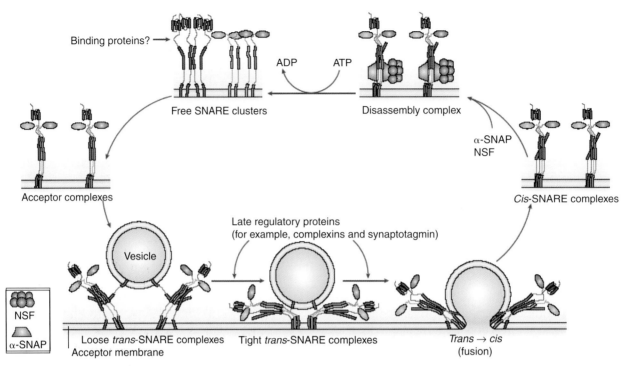

FIGURE 7.13 Hypothetical SNARE conformational cycle during vesicle docking and fusion.
As an example, we consider three Q-SNAREs (Q-soluble *N*-ethylmaleimide-sensitive factor attachment protein receptors) on an acceptor membrane (red and green) and an R-SNARE on a vesicle (blue). Q-SNAREs, which are organized in clusters (top left), assemble into acceptor complexes. Acceptor complexes interact with the vesicular R-SNAREs through the N-terminal end of the SNARE motifs, and this nucleates the formation of a four-helical *trans*-complex. *Trans*-complexes proceed from a loose state (in which only the N-terminal portion of the SNARE motifs are 'zipped up') to a tight state (in which the zippering process is mostly completed), and this is followed by the opening of the fusion pore. In regulated exocytosis, these transition states are controlled by late regulatory proteins that include complexins (small proteins that bind to the surface of SNARE complexes) and synaptotagmin (which is activated by an influx of calcium). During fusion, the strained *trans*-complex relaxes into a *cis*-configuration. *Cis*-complexes are disassembled by the AAA+ (ATPases associated with various cellular activities) protein NSF (*N*-ethylmaleimide-sensitive factor) together with SNAPs (soluble NSF attachment proteins) that function as cofactors. The R- and Q-SNAREs are then separated by sorting (e.g. by endocytosis). Adapted from Jahn R, Scheller RH (2006) SNAREs- engines for membrane fusion. *Nature Rev. Mol. Cell Biol.* **7**, 631–643.

Reese in 1981. When exocytosis has been captured by rapid freezing (in the presence of pharmacological agents to prolong the excitation of the nerve terminal, see Section 7.3.5), images suggested the presence of a narrow fusion pore. Later, the development of patch-clamp capacitance techniques demonstrated the formation of a fusion pore during exocytosis. When chromaffin cells that release adrenalin are excited, small step-like increases of membrane capacitance of about 1 fF are recorded. These capacitance steps correspond to exocytosis of large dense core vesicles, and for each exocytosis to the formation of a fusion pore: an increase of capacitance corresponds to an increase of membrane surface that occurs if the vesicle fuses, even transiently, with the plasma membrane. After formation, the pore would expand in order to allow rapid transmitter release in the cleft. Theoretical models propose that the fusion pore expands at a rate approaching $100\,nm\,ms^{-1}$ within a few tens of microseconds to achieve the observed transmitter time course.

7.3.5 Pharmacology of neurotransmitter release

Agents blocking K^+ currents potentiate neurotransmitter release

K^+ channels present at the presynaptic membrane are voltage-gated K^+ channels of the Na^+ action potential (see Section 4.3) and Ca^{2+}-activated K^+ channels ($K_{(Ca)}$; Section 5.3). The outflow of K^+ ions through these channels repolarizes the presynaptic membrane and limits Ca^{2+} entry (in amplitude and duration). At the neuromuscular junction, simultaneous labelling with specific fluorescent toxins of presynaptic $K_{(Ca)}$ and Ca^{2+} channels showed that they are localized close to one another, at presynaptic active zones. This organization

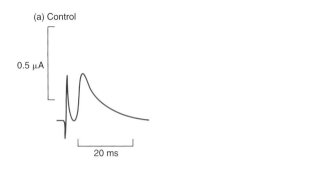

(a) Control

0.5 μA

20 ms

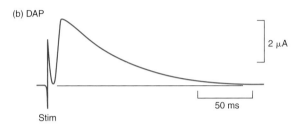

(b) DAP

2 μA

50 ms

Stim

FIGURE 7.14   Diaminopyridine (DAP) increases the duration and amplitude of motor endplate current.
The postsynaptic endplate current of a frog sartorius muscle cell is evoked by stimulation (2 μA intensity, 5 ms duration) of the motor nerve (V_H = −90 mV). **(a)** The average amplitude of the evoked postsynaptic current is 0.5 μA in control Ringer solution. **(b)** The postsynaptic current evoked by the same stimulation in the presence of DAP (1 mM) is always greater than 1 μA and can reach 3.3 μA. These inward currents are represented upwardly, which is unusual. Adapted from Katz B and Miledi R (1979) Estimates of quantal content during chemical potentialization of transmitter release. *Proc. R. Soc. Lond. B* **205**, 369–378, with permission.

ensures a rapid activation of these K⁺ channels during Ca²⁺ entry triggered by an action potential.

When these K⁺ channels are blocked, Ca²⁺ entry is increased and so is transmitter release. At the neuromuscular junction, potassium channel blockers such as TEA and 3,4-diaminopyridine, when used in conjunction with TTX (to block voltage-dependent Na⁺ channels), potentiate the postsynaptic response (**Figure 7.14**). In the presence of K⁺ blocking agents, the endplate current has an average time of decline much longer than that of the miniature current. This is due in part to the fact that, in this case, the release of vesicles is asynchronous and there is a temporal spread of quantal release during the entire presynaptic calcium spike. On the other hand, the massive release possibly saturates the enzyme acetylcholinesterase and as a result there is a repeated activation of postsynaptic nicotinic receptors.

Example of botulinum toxins

Botulinum toxin blocks synaptic transmission at all peripheral cholinergic synapses. The transmission is

blocked for several months. The patient usually dies from asphyxia (paralysis of respiratory muscles) and, in cases where the patient survives, he suffers from muscle atrophy owing to the non-functioning of muscle cells. Botulinum toxins derived from the microorganism *Clostridium botulinum* is a powerful neurotoxin made up of seven different toxins (A to G). Botulinum toxins are proteases, each of which cleaves a single target at a single site. These proteins have a heavy chain (100 kDa) joined by a single disulphide bond to a light chain (50 kDa). The heavy chain is responsible for the selective binding of the toxin to neuronal cells and for the penetration of the light chain into neurons; the light chain bears the activity. It contains a consensus sequence of the catalytic site of metallopeptidases: it has a Zn²⁺-dependent endopeptidase activity.

Botulinum toxins enter nerve terminals and decrease the number of acetylcholine vesicles released in the synaptic cleft without affecting acetylcholine synthesis or action potential conduction in the motor nerve (it blocks synaptic transmission). In cases where the decrease is not total, the effect of the toxin can be reversed by increasing Ca²⁺ concentration in the extracellular medium or by adding TEA or 3,4-diaminopyridine, agents that potentiate Ca²⁺ influx. Aminopyridines are used to cure patients poisoned by the botulinum toxins contained, for example, in damaged preserves.

Knowing these facts, it was hypothesized that botulinum toxin affected either Ca²⁺ entry or the coupling between intracellular Ca²⁺ concentration increase and exocytosis of synaptic vesicles. To verify the first proposition, Ca²⁺ entry was recorded in terminals 'paralyzed' by botulinum toxins. The results showed that presynaptic Ca²⁺ current is not significantly changed (**Figure 7.15a**). The botulinum toxin would therefore act at the presynaptic level to decrease acetylcholine release, after the entry of Ca²⁺ ions.

In order to identify the intracellular target of botulinum toxins, synaptic vesicles from rat cerebral cortex were purified. Of the many proteins detected in these purified synaptic vesicles, one protein band was altered by incubation of the vesicles with botulinum toxin. The electrophoretic mobility of this band corresponds to that of synaptobrevins, the v-SNARE which plays a role in vesicle docking. Syntaxins and SNAP-25 are also targets for botulinum toxin. Botulinum toxins B, D, F and G are specific for synaptobrevin, botulinum toxin C cleaves syntaxins, and botulinum toxins A and E are specific for SNAP-25 (**Figure 7.15b**).

Botulinum toxins specifically bind to nerve terminals and deliver their zinc-endopetitidase N-terminal domain inside the cytosol, where it specifically cleaves a SNARE protein at a single site within its cytosolic

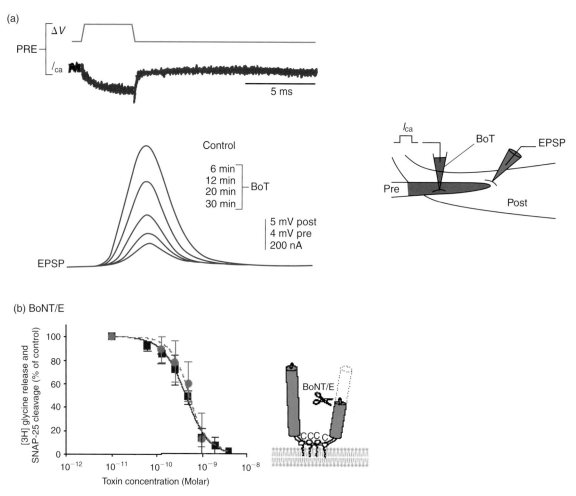

FIGURE 7.15   Botulinum toxins strongly decrease synaptic transmission without affecting presynaptic Ca²⁺ current.

(a) In the squid giant synapse, at the stellate ganglion, presynaptic and postsynaptic intracellular electrodes are implanted to allow simultaneous recording of the presynaptic Ca^{2+} current (voltage clamp mode) and the post-synaptic response (EPSP, current clamp mode). A presynaptic voltage step (ΔV) evokes a presynaptic Ca^{2+} current (I_{Ca}) and after a delay, a postsynaptic response (EPSP, control). After injection of botulinum toxin (BoT) through the presynaptic electrode, the EPSP decreases with time. Note that in the same time the presynaptic Ca^{2+} current is unchanged. **(b)** Dose-response curve of the effect of botulinum neurotoxin E (BoNT/E) on SNAP-25 proteolysis (purple squares, continuous line) and glycine release (green circles, broken line) in cultured neurons. The right-hand panel shows a schel of SNAP-25 and the position of BoNT/E cleavage. Part (a) adapted from Marsal J, Ruiz-Montasell B, Blasi J *et al.* (1997) Block of transmitter release by botulinum C1 action on syntaxin at the squid giant synapse. *Proc. Natl Acad. Sci. USA* **94**, 14871–14876, with permission. Part (b) adapted from Keller JE *et al.* (2001) Uptake of botulinum neurotoxin into cultured neurons. *Biochemistry* **43**, 526–532, with permission.

domain. The cleavage of one of the SNAREs greatly reduces the probability of neurotransmitter release.

7.4 PROCESSES IN THE SYNAPTIC CLEFT: FROM TRANSMITTER RELEASE IN THE CLEFT TO TRANSMITTER CLEARANCE FROM THE CLEFT

The time course of a transmitter in the cleft depends on the balance between the amount of transmitter released per unit of time and the efficacy of clearance mechanisms that clear transmitter molecules from the cleft.

7.4.1 The amount of neurotransmitter released in the synaptic cleft

The amount of transmitter released per vesicle and per unit of time in the synaptic cleft depends on (i) the concentration of the transmitter in the exocytotic vesicle,

(ii) the volume of the vesicle, and (iii) the rate of transmitter release through the vesicle fusion pore into the cleft (i.e. the dimension of the fusion pore).

What is the concentration of neurotransmitter in a synaptic vesicle?

Consider the example of synaptic vesicles that contain glutamate as a neurotransmitter. To isolate synaptic vesicles, antibodies against a vesicular protein (such as synaptophysin) are immobilized on the surface of non-porous methacrylate microbeads. Using these immunobeads, synaptic vesicles are isolated. To avoid the loss of glutamate, the vesicular H^+ gradient (responsible for the transport of neurotransmitter molecules into vesicles; see Appendix 3.1) is preserved by adding an ATP-regenerating system. Under these conditions, high levels of glutamate are found in vesicles: 0.8 µmoles of glutamate per milligram of synaptophysin. Knowing that synaptophysin represents 7% of total vesicle protein, it gives 60 nmoles of glutamate per milligram of protein. This gives an intravesicular concentration of 60 mM, assuming an internal volume of $1 \, \mu l \, mg^{-1}$ of protein. The concentration of glutamate in synaptic vesicles is estimated at 60–210 mM depending on the preparation studied. The concentrations of other transmitters such as GABA and glycine are not known.

Is the vesicle content stable? If yes, what mechanisms regulate the amount of neurotransmitter per vesicle?

Miniature spontaneous postsynaptic responses that correspond to the exocytosis of a very small number of vesicles (say one, two or three; see Appendix 7.1) have a quite stable amplitude. This suggests that the vesicle content is relatively stable. Therefore, at a given synapse, synaptic vesicles would contain the same amount of transmitter. Active transport into synaptic vesicles is achieved by proteins that couple the uptake of transmitter to the movement of H^+ in the opposite direction (along its electrochemical gradient). A vesicular H^+-ATPase provides the H^+ electrochemical gradient that drives transmitter uptake. Thus, to achieve a stable vesicular content the number of synaptic transporters in the vesicle membrane, the activity of the H^+-ATPase and the cytoplasmic concentration of transmitter must be stable.

Another hypothesis is that the amount of transmitter in a vesicle is variable but very high so that it saturates postsynaptic receptors; in such a condition, variations in the vesicle content would not be 'seen' by the postsynaptic ligand-gated channels and the size of miniature potentials would show minimal variation.

How does the transmitter diffuse from the vesicle into the cleft?

The detailed nature of the vesicle fusion process remains unclear. In particular, the formation of the fusion pore, its opening rate, as well as how molecules diffuse out from the vesicle are not known.

What is the peak concentration of neurotransmitter in the cleft?

The time course of neurotransmitter concentration can be evaluated experimentally. The technique utilizes the non-equilibrium displacement of a competitive antagonist following the synaptic release of transmitter. A specific antagonist of the postsynaptic ionotropic receptors that mediate the transmission is applied. Attenuation of synaptic response amplitude is measured at one or more concentrations of a rapidly dissociating competitive antagonist and a dose–response curve is constructed for the inhibition of synaptic transmission. Transmitter peak concentration in the cleft would be around 1 mM for glutamate and achieved in around 20 µs.

7.4.2 Transmitter time course in the synaptic cleft is brief and depends mainly on a transmitter binding to target proteins

The speed of transmitter clearance from the synaptic cleft is a fundamental parameter influencing many aspects of synaptic function. The amount of time the neurotransmitter stays in the cleft depends on (i) the amount of transmitter released, (ii) the transmitter diffusion coefficient, the geometry of the cleft and adjacent extra synaptic space, (iii) the distribution and affinity of transmitter binding sites, and (iv) the transporters uptake rate and/or degradative enzymes turnover rate. Therefore the transmitter time course varies significantly from synapse to synapse.

Theoretical models predict that within 50 µs the transmitter is evenly distributed throughout the cleft and by 500 µs the cleft is clear of transmitter. Only transporters, degradative enzymes and diffusion achieve the real removal of neurotransmitter molecules from the cleft. Binding to pre- and postsynaptic receptors is a temporary clearance (buffering) since transmitter molecules will be back in the cleft as soon as they unbind from receptors (as already seen for Ca^{2+} clearance; see Section 7.2.4). The turnover rate for known neurotransmitter transporters is in the range $1–15 \, s^{-1}$: a single transporter requires at least 60 ms to complete its cycle. This is extremely slow compared with the turnover rate for AChEsterase which is in the order of $10^4 s^{-1}$. It shows how highly efficient is this enzyme to remove acetylcholine from the cleft at nicotinic synapses (e.g. the neuromuscular junction). Therefore, for glutamate and GABA, which are not enzymatically degraded in the cleft, uptake transporters are too slow to achieve a rapid disappearance of transmitter molecules from the cleft.

In central synapses, rapid buffering of transmitter molecules from the cleft arises from the binding of these molecules to receptor proteins: postsynaptic receptor channels (ionotropic receptors), and pre- and postsynaptic G protein-linked receptors (metabotropic receptors). These receptors bind the transmitter molecules tightly enough (i.e. with a high affinity) to prevent release from binding sites for tens of milliseconds, sufficient time for transporters to become less saturated.

In other words, each transporter stands ready to bind a transmitter molecule as soon as it is released from a receptor. Otherwise, neurotransmitter molecules would be released a second time in the cleft and evoke a second postsynaptic response or amplify the duration of the first one.

7.5 SUMMARY (Figures 7.16 and 7.17)

The answers to the questions raised in Section 7.1.3 are as follows:

- The opening of presynaptic high-voltage-activated^{2+} channels (N- and P/Q-type Ca^{2+} channels) is triggered by membrane depolarization that occurs

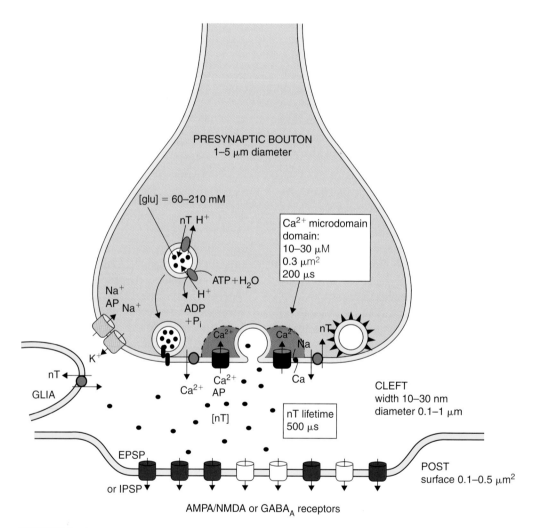

FIGURE 7.16 **Cascade of events leading to neurotransmitter release and its clearance from the synaptic cleft.** Schematic of some steps of synaptic transmission (between the presynaptic action potential and the postsynaptic response, EPSP or IPSP). In the presynaptic and glial plasma membranes only one example of each channel, pump or transporter, is represented owing to the lack of space. In the postsynaptic membrane many examples of ionotropic glutamatergic (AMPA and NMDA) or GABAergic (GABA$_A$) channels are represented. nT, neurotransmitter.

during the depolarizing phase of presynaptic action potentials.

- Presynaptic $[Ca^{2+}]_i$ increase is local in response to a presynaptic spike due to the clustering of N- and P/Q-type Ca^{2+} channels close to docked vesicles, at release sites (active zones). $[Ca^{2+}]_i$ increase is transient since Ca^{2+} channels open and close quickly and Ca^{2+} ions are cleared from the cytoplasm by binding to receptor proteins (cytoplasmic Ca^{2+}-binding proteins and transmembrane proteins such as Ca/ATPase or Na–Ca exchanger).

- Exocytosis of a docked vesicle is triggered when local $[Ca^{2+}]_i$ increase is around 0.5–40 μM. The first step includes binding of multiple Ca^{2+} ions to Ca^{2+}-binding protein(s) (Ca^{2+}-sensor(s)) but the exact mechanism of vesicle fusion is not yet known. Exocytosis is rapidly triggered since Ca^{2+} entry is close to release sites and each docked vesicle is ready to fuse.

- The transmitter is released in the synaptic cleft through a fusion pore.

- The neurotransmitter molecules present in the cleft bind to specific receptors such as pre- and postsynaptic ionotropic and metabotropic receptors. Binding to postsynaptic ionotropic receptors results in an EPSC or IPSC.

- The clearance of the neurotransmitter molecules (total disappearance from the cleft) is achieved by their binding to neuronal and glial transporters and/or to degradative enzymes and by their diffusion out of the cleft. Theoretical models predict that by 500 μs the cleft is clear of neurotransmitter molecules.

Neurotransmitter release, including steps from Ca^{2+} entry into the presynaptic terminal to vesicle exocytosis, is achieved by a cascade of chemically-gated events (**Table 7.1**). It differs fundamentally from electrical events such as action potentials that are achieved by a cascade of voltage-gated events. Once presynaptic Ca^{2+} channels open, synaptic transmission is determined by the different affinity constants of the reactions between ligands and receptors that underlie synaptic transmission: this includes binding of Ca^{2+} to Ca^{2+} sensor(s), and binding of neurotransmitter molecules to pre- and postsynaptic receptors (receptor channels, G-protein coupled receptors, uptake transporters). Transmitter release from presynaptic elements is a Ca^{2+}-regulated, multiprotein process.

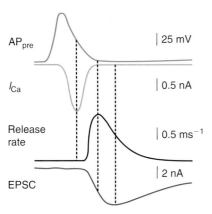

FIGURE 7.17 Summary (example of the calyx of Held).
Time course of the signalling cascade, showing (top to bottom) the presynaptic action potential (AP) waveform and resulting Ca^{2+} current (I_{Ca}), (inferred) release rate and postsynaptic EPSC. Adapted from Meinrenken CJ, Borst JGG, Sakmann B (2003) Local routes revisited: the space and time-dependence of the Ca^{2+} signal for phasic transmitter release at the rat calyx of Held. *J. Physiol.* **547**, 665–689 with permission.

TABLE 7.1 Some of the reactions (voltage-gated and ligand-gated) from the opening of N- and P/Q-type Ca^{2+} channels to transmitter release into the synaptic cleft and its clearance from the synaptic cleft. The numbers indicate the values of K_D or EC_{50}. All the vesicle steps have been omitted

Location of reactions	Type of reaction	Steps of Fig 6.4a	Reactions			Effect
PRE	Voltage-gated	Step 2	N channels closed P/Q	action potential ⇌	N channels open P/Q	Ca^{2+} entry
	Ligand gated {	Step 3a	$_{2-5}Ca^2$ + sensors	10–30 μM ⇌	$_{2-4}$Ca-sensor	Exocytosis
		Step 3b	Ca^{2+} + pumps Ca^{2+} + transporters	0.25–1.50 mM ⇌	Ca-pumps Ca-transporters	Ca^{2+} clearance
CLEFT	Ligand-gated	Step 5a/6	2Glu + AMPAR	0.25–1.50 mM ⇌	Glu_2-AMPAR	EPSP
			2Glu + NMDAR	1 μM ⇌	Glu_2-NMDAR	
		Step 5a/6	2Gaba + $GABA_A$	8–40 μM ⇌	$Gaba_2$-$GABA_A$	IPSP
		Step 5b/5c	nT + transporter	high affinity ⇌	nT + transporter	nT uptake

APPENDIX 7.1 QUANTAL NATURE OF NEUROTRANSMITTER RELEASE

A7.1.1 Spontaneous release of acetylcholine at the neuromuscular junction evokes miniature endplate potentials: the notion of quanta

At the neuromuscular junction (also called the motor endplate), when an intracellular electrode is implanted in a muscle fibre at the level of the postsynaptic membrane in the presence of TTX and in the absence of extracellular Ca^{2+}, spontaneous postsynaptic potentials of very small amplitude (0.5–1.0 mV on average) are recorded, though presynaptic spikes and synaptic transmission are blocked (**Figure A7.1a**). They occur randomly at a low frequency (about 1 s^{-1}). They have been called 'miniature' endplate potentials (mEPP) by Fatt and Katz (1952). As explained by Katz (1966): 'Except for their spontaneous occurrence and their small size, the miniatures are indistinguishable from the EPSPs evoked by presynaptic nerve stimulation'. For example, curare suppresses them and acetylcholinesterase inhibitors enhance their amplitude and

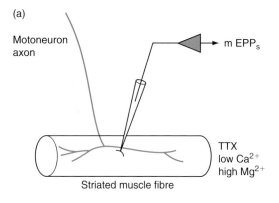

(a)

Motoneuron axon

m EPP$_s$

TTX
low Ca^{2+}
high Mg^{2+}

Striated muscle fibre

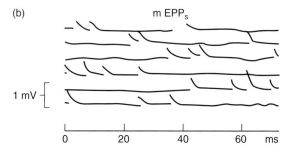

(b)

m EPP$_s$

1 mV

0   20   40   60   ms

FIGURE A7.1 **Miniature postsynaptic potentials.**
Miniature endplate potentials (right mEPPs) recorded at the frog neuromuscular junction as shown **(a)**. Recordings are obtained in the presence of a low external Ca^{2+} concentration **(b)**. Adapted from Fatt P and Katz B (1952) Spontaneous subthreshold activity at motor nerve endings. *J. Physiol. (Lond.)* **117**, 109–128, with permission.

duration with the same doses and to approximately the same extent. Normally, miniature potentials are well below the firing level of the muscle cell and so remain localized and produce no contraction. Since they disappear after the motor nerve has been cut and in the presence of botulinum toxin, they are evoked by presynaptic release of Ach. The interpretation is that motor nerve terminals at rest are in a state of intermittent secretory activity: they liberate small quantities of ACh at random intervals at an average rate of about one per second.

Miniature endplate potentials, being the smallest recorded event with a relatively constant amplitude and an all-or-nothing characteristic, were named *quanta* (with reference to quantum physics) by Del Castillo and Katz in 1954. Considering the fast rise time of miniatures, they hypothesized that miniatures arise from the synchronous action of a packet of a large number of ACh molecules at a time: 'At this stage, the characteristic presynaptic vesicles were revealed by electron microscope and the suggestion arose that they could be the subcellular particles in which the transmitter is stored and from which it is released in an all-or-none fashion.' This was confirmed by the observation of exocytosis at active zones of the neuromuscular junction (Couteaux and Pécot-Dechavassine 1970).

However, direct visualization of exocytosis at a neuromuscular junction was achieved only 20 years later by using electron microscopy combined with new methods to rapidly freeze nerve terminals (Heuser *et al.* 1979). The tissue is freeze-fractured during synaptic activity at precise times following nerve stimulation. A metal replica of the presynaptic membrane after fracture is observed under the electron microscope. Images of exocytosis are observed only in the presence of 4-aminopyridine (a blocker of K^{+} channels which prolongs depolarization of the presynaptic membrane; see Section 7.3.5), so that important ACh release occurs. In the absence of drugs, rearrangement of presynaptic intramembranous particles is the most common ultrastructural observation. This low probability of observation of exocytosis in the absence of 4-aminopyridine is not surprising, considering the low probability of exocytosis at a synaptic complex at a given time (see Appendix 7.2).

Similar spontaneous miniature potentials are recorded from synapses of the central nervous system. They are called 'miniature postsynaptic potentials' (mEPSPs and mIPSPs) or 'miniature currents' (mEPSCs and mIPSCs) (**Figure A7.1b**).

The theory of vesicular release of neurotransmitter, or the quantal nature of chemical transmission, states that one quantum equals one vesicle. The size of a quantum is designated by q.

A7.1.2 The quantal composition of EPSPs and IPSPs

At the neuromuscular junction, a quantum produces a 0.5–1 mV miniature endplate potential (mEPP). In response to a presynaptic action potential and in the presence of control concentrations of external Ca^{2+}, an endplate potential (EPP) is recorded: it has a much larger amplitude (50–70 mV) than miniatures. The quantal theory assumes that EPPs are made up of n quanta released simultaneously (200–300 mEPPs). This evidence is obtained by lowering external Ca^{2+} concentration and thus reducing the amplitude of evoked EPPs. If one lowers the normal Ca^{2+} concentration and adds magnesium to the muscle bath, the amount of acetylcholine delivered by an impulse can be reduced to a very low level, and under these experimental conditions, the quantal composition of the endplate potential becomes immediately apparent (see Appendix 7.2). This result has been confirmed in various synapses of vertebrates and invertebrates.

APPENDIX 7.2 THE PROBABILISTIC NATURE OF NEUROTRANSMITTER RELEASE: THE NEUROMUSCULAR JUNCTION AS A MODEL

Neurotransmitter release is probabilistic. In response to a presynaptic action potential, each docked synaptic vesicle has a probability p to fuse with the presynaptic plasma membrane (i.e. to undergo exocytosis). This probability p varies from 0 to 1 ($0 < p < 1$). This means that even when all the proper conditions are fulfilled, the synaptic transmission can fail (see **Figure 7.3a**). What fails exactly is vesicle exocytosis. Of course, when chemical transmission at a single synapse is achieved by only one active zone (mammalian CNS synapses), failures are much more commonly observed than when it is achieved by a large number of active zones (neuromuscular junction).

The first studies on the probability of neurotransmitter release were performed on the neuromuscular junction, this preparation offering the possibility of simultaneously recording the activity of pre- and postsynaptic elements and of manipulating the parameters related to neurotransmitter release, here acetylcholine release. At the motor endplate level, the number of active zones is estimated at between 300 and 1000. For this reason, the recorded postsynaptic response is global, representing the summation of evoked responses at many active zones.

In the absence of any nerve stimulation, miniature endplate postsynaptic potentials of 0.5–1.0 mV average amplitude are recorded. They occur randomly. These miniature endplate potentials, being the smallest recorded event and having a relatively constant amplitude, were named 'quanta' by Del Castillo and Katz. They proposed that each quantum corresponds to the content of one synaptic vesicle. The size of a quantum is q.

In contrast, in response to nerve stimulation, the probability of synaptic vesicle exocytosis is very high and the size of the endplate potential is in the order of tens of millivolts. In order to test whether EPSPs are made up of quanta, the size of EPSPs is reduced by immersing the preparation in an extracellular medium containing a low Ca^{2+} concentration. In these conditions, following motor nerve stimulation, one can record postsynaptic depolarizations (motor endplate potentials) having a low and variable amplitude and also numerous failures (absence of postsynaptic response) (**Figure A7.2a**). These amplitude variations are in graduated steps, each one corresponding to a quantum of amplitude q (**Figure A7.2b**).

The postsynaptic response is constantly a multiple of 0, 1, 2, . . . , x quanta, x always being a whole number. If one admits that a quantum corresponds to the release of a synaptic vesicle, it appears that 0, 1, 2, . . . , x synaptic vesicles are released (x being a natural number as each synaptic vesicle is an entity). This produces fluctuations in the amplitude of the postsynaptic response. The distribution of these fluctuations on a graph shows the presence of regularly spaced peaks, the first three peaks clearly corresponding to amplitudes $1q$ (0.4 mV), $2q$ (0.8 mV) and $3q$ (1.2 mV), but the presence of other peaks ($4q$, . . . , xq) is less evident (**Figure A7.2c**). The first two amplitudes ($1q$ and $2q$) are more frequent than others. In other words, the probability of recording postsynaptic potentials of amplitude $1q$ or $2q$ is greater than that of recording potentials of amplitude $3q$ or above in the presence of a low external Ca^{2+} concentration.

The demonstration that postsynaptic potentials are composed of discrete units has necessitated the application of statistical tests to the experimental data. Poisson's Law was applied to test that each response is made up of discrete (0, 1, 2, 3, . . .) units. As stated earlier, the results shown in **Figure A7.2** were obtained in conditions where p is reduced (reduced extracellular Ca^{2+} concentration). This is a necessary condition for the use of the Poisson distribution. In other words, each time an action potential invades the axon terminals, it causes the release of a few vesicles out of a very large available population. In the Poisson distribution, the probability of observing a postsynaptic potential composed of x miniature potentials $p(x)$ is:

$$p(x) = \frac{e^{-m} m^x}{x!}$$

(a) Postsynaptic potentials

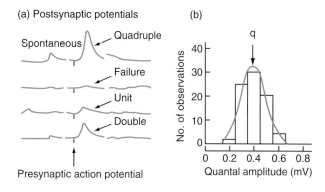

(b)

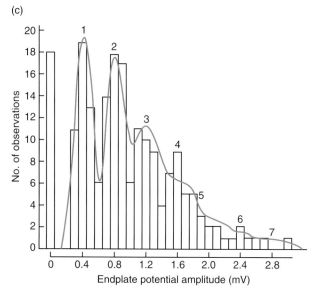

FIGURE A7.2 **Demonstration of the probabilistic nature of acetylcholine release at the neuromuscular junction.**
(a) Recordings of spontaneous miniatures and evoked endplate potentials (mEPP and EPP). The nerve-muscle preparation is bathed in a low-Ca^{2+} high-Mg^{2+} concentration medium. In such condition, spontaneous mEPPs have nearly the same unitary amplitude whereas evoked EPPs have an amplitude that is a multiple of the mEPP amplitude. **(b)** The distribution of mEPP amplitude is unimodal and their average amplitude is $q = 0.4$ mV. **(c)** Histogram of evoked EPPs recorded in response to a presynaptic action potential. The cases where no postsynaptic response is recorded (failures) are represented at 0 mV (18 cases). The theoretical distribution calculated from Poisson's Law (represented by the green line) fits the distribution of the amplitude of recorded EPPs (histogram). Part (a) adapted from Liley AW (1956) The quantal component of the mammalian endplate potential. *J. Physiol. (Lond.)* **133**, 571–587, with permission. Parts (b) and (c) from Boyd IA and Martin AR (1956) The endplate potential in mammalian muscle. *J. Physiol. (Lond.)* **132**, 74–91, with permission.

where m is the mean number of quanta (miniature potentials) that compose the postsynaptic response (i.e. the average number of released vesicles in response to a presynaptic spike). N is the total number of observations (the number of recordings of the postsynaptic potential

in response to a presynaptic spike) and $n(x)$ is the number of times that the recorded postsynaptic potentials is composed of x miniature potentials (amplitude of the postsynaptic response = xq). The probability that a postsynaptic potential is composed of x miniature potentials, $p(x)$, is equal to the number of times this event is observed, $n(x)$, over the total number of experiments, N:

$$p(x) = \frac{n(x)}{N} \qquad (1)$$

When N is large enough, $Np(x)$ is close to the observed number of responses which contain x quanta (which are made up of a summation of x miniatures).

The difficulty here is to determine the value of m, the mean number of quanta that compose the postsynaptic response (also called the average quantal content). To determine m, two methods can be used.

First, given that the amplitude of miniature potentials is a unit, one can calculate:

$$m = \frac{\text{average amplitude of evoked responses}}{\text{average amplitude of miniature potentials}}$$

This method is used in Chapter 8.

The second option is the so-called failure method. In conditions where p is artificially reduced (reduced external Ca^{2+} concentration), the number of times the synaptic transmission fails is high: numerous presynaptic action potentials are not followed by vesicle release. In these cases of failure, $x = 0$ (the postsynaptic responses of null amplitude composed of 0 miniature potentials). From equation (1), $p(0)$ is the number of failures over the total number of stimulations N; it is large and equal to:

$$p(0) = e^{-m} = \frac{\text{number of failures}}{\text{number of simulations}} = n(0)/N$$

Therefore $m = \ln(N/n(0))$. The $n(0)/N$ ratio, determined from experimental results (**Figure A7.2c**), leads to an easy deduction of m.

With m known, p can be calculated for each value of x, and a theoretical curve is drawn, showing the distribution of postsynaptic potentials. **Figure A7.2c** shows the correlation between experimental data and the Poisson distribution. Quantal acetylcholine release at the neuromuscular junction is a valid model. The following hypothesis has been proposed.

Acetylcholine release is a discontinuous quantal phenomenon, each quantum corresponding to the total content of one synaptic vesicle. The probability p that the postsynaptic response will be composed of $1, 2, \ldots, x$ quanta depends on the experimental conditions (composition of the extracellular medium) and on the frequency of presynaptic activity. Once again, the results in **Figure A7.2** have been obtained in a medium where p is reduced, a condition necessary to the use of the Poisson distribution.

FURTHER READING

Baumert M, Maycox PR, Navone F *et al.* (1989) Synaptobrevin: an integral membrane protein of 18,000 daltons in small synaptic vesicles of rat brain. *EMBO J.* **8**, 379–384.

Bennett MK, Calakos N, Scheller RH (1992) Syntaxin: a synaptic protein implicated in docking of synaptic vesicles at presynaptic active zones. *Science* **257**, 255–259.

Bollmann JH, Sakmann B, Borst JGG (2000) Calcium sensitivity of glutamate release in a calyx-type terminal. *Science* **289**, 953–957.

Burger PM, Mehl E, Cameron PL, Maycox PR *et al.* (1989) Synaptic vesicles immunoisolated from rat cerebral cortex contain high levels of glutamate. *Neuron* **3**, 715–720.

Calakos N and Scheller RH (1996) Synaptic vesicles, docking and fusion: a molecular description. *Physiological Reviews* **76**, 1–29.

Chapman ER (2002) Synaptotagmin: a Ca^{2+}-sensor that triggers exocytosis? *Nature Rev. Mol. Cell. Biol.* **3**, 1–11.

Jackson MB, Chapman ER (2006) Fusion pores and fusion machines in Ca^{2+}-triggered exocytosis. *Annu. Rev. Biophys. Biomol. Struct.* **35**, 135–160.

Leveque C, el Far O, Martin-Moutot N *et al.* (1994) Purification of the N-type calcium channel associated with syntaxin and synaptotagmin: a complex implicated in synaptic vesicle exocytosis. *J. Biol. Chem.* **269**, 6306–6312.

Littleton JT, Barnard JO, Titus SA *et al.* (2001) SNARE-complex disassembly by NSF follows synaptic-vesicle fusion. *PNAS* **98**, 12233–12238.

Matthew WD, Tsavaler L, Reichardt LF (1981) Identification of a synaptic vesicle-specific membrane protein with a wide distribution in neuronal and neurosecretory tissue *J. Cell. Biol.* **91**, 257–269.

Miledi R (1973) Transmitter release induced by injection of calcium into nerve terminals. *Proc. R. Soc. Lond. B. Biol. Sci.* **183**, 421–425.

Oyler GA, Higgins GA, Hart RA *et al.* (1989) The identification of a novel synaptosomal-associated protein, SNAP-25, differentially expressed by neuronal subpopulations. *J Cell Biol.* **109**, 3039–3052.

Prado VF, Martins-Silva C, de Castro BM *et al.* (2006) Mice deficient for the vesicular acetylcholine transporter are myasthenic and have deficits in object and social recognition. *Neuron* **51**, 601–612.

Rettig J, Heinemann C, Ashery U *et al.* (1997) Alteration of Ca^{2+} dependence of neurotransmitter release by disruption of Ca^{2+} channel/syntaxin interaction. *J. Neurosci.* **17**, 6647–6656.

Rothman JE (1994) Intracellular membrane fusion. *Adv. Second Messenger Phosphoprotein Res.* **29**, 81–96.

Schneggenburger R and Neher E (2000) Intracellular calcium dependence of transmitter release rates at a fast central synapse. *Nature* **406**, 889–893.

Sheng ZH, Westenbroek RE, Catterall WA (1998) Physical link and functional coupling of presynaptic calcium channels and the synaptic vesicle docking/fusion machinery. *J. Bioenerg. Biomembr.* **30**, 335–345.

Sheng ZH, Yokoyama CT, Catterall WA (1997) Interaction of the synprint site of N-type Ca^{2+}-channels with the C2B domain of synaptotagmin I. *Proc. Natl Acad. Sci. USA* **94**, 5405–5410.

Söllner T, Whiteheart SW, Brunner M *et al.* (1993) SNAP receptors implicated in vesicle targeting and fusion. *Nature* **362**, 318–324.

Song H, Ming G, Fon E *et al.* (1997) Expression of a putative vesicular acetylcholine transporter facilitates quantal transmitter packaging. *Neuron* **18**, 815–826.

Weber T, Zemelman BV, McNew JA *et al.* (1998) SNAREpins: minimal machinery for membrane fusion. *Cell* **92**, 759–772.

Whiteheart SW, Rossnagel K, Buhrow SA *et al.* (1994) N-ethymaleimide-sensitive fusion protein: a trimeric ATPase whose hydrolysis of ATP is required for membrane fusion. *J. Cell Biol.* **126**, 945–954.

Zhang X, Kim-Miller MJ, Fukuda M *et al.* (2002) Ca^{2+}-dependent synaptotagmin binding to SNAP-25 is essential for Ca-triggered exocytosis. *Neuron* **34**, 599–611.

The ionotropic nicotinic acetylcholine receptors

The nicotinic acetylcholine receptor (nAChR) is a glyco-protein present at nicotinic cholinergic synapses. The preparation that has been used most extensively to study the nicotinic receptor is the electric organ of the electric ray, *Torpedo* (Torpedo nAChR; **Figure 8.1a**), or of the electric eel. In part because this preparation is extremely rich in nicotinic receptors, and because snake venom α-toxins had been identified as highly selective

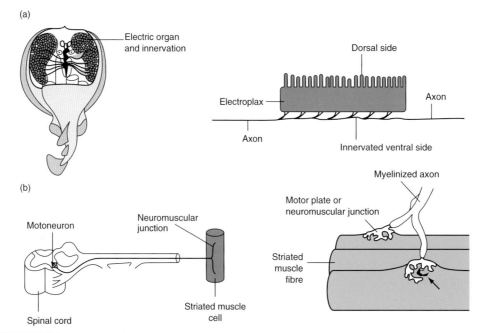

FIGURE 8.1 **Examples of preparations in which nicotinic receptors have been extensively studied.**
(**a**) The electric organ of the electric ray. On a dissected *Torpedo* (left) we can see the electric organs and their innervation. These organs constitute electroplax membranes (right) which are modified muscle cells that do not contract. Nicotinic receptors are present at the command neuron's synapse level, on the ventral side of the postsynaptic membrane of the electroplax. The electroplax are simultaneously activated and the summation of their electric discharges can be of the order of 500 V. (**b**) The neuromuscular junction. Striated muscle cells are innervated by motoneurons whose cell bodies are located in the ventral horn of the spinal cord (horizontal section, left). In mammals, each muscle cell is innervated by one nerve fibre. As the axon makes contact with the muscle cell, it loses its myelin sheath and divides into several branches that are covered by unmyelinated Schwann cells. The thick arrow (right) points to one terminal that has been lifted to show the postsynaptic folds where nicotinic receptors are located.

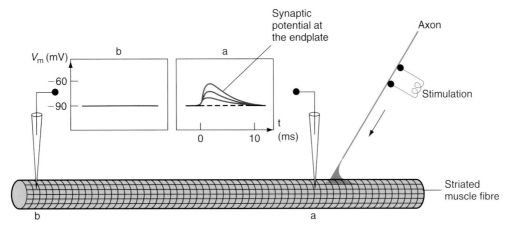

FIGURE 8.2   The endplate potential.
The axon of a motoneuron is stimulated and the postsynaptic response is recorded in the presence in the bath of a low Ca (0.5 mM) and a high Mg (6 mM) concentration. Two intracellular electrodes are implanted in the muscle fibre, one close to the neuromuscular junction (**a**) and the other more than 5 mm away (**b**). The evoked response is a transient depolarization called an endplate potential.

markers of nAChRs. In mammals, nAChRs have been mostly studied at the neuromuscular junction (muscle nAChR; **Figure 8.1b**) but also in the peripheral nervous system (synapses between pre- and postganglionic neurons of the autonomic nervous system), and more recently in the central nervous system where they are also present (neuronal nAChR; see Appendix 8.1).

8.1   OBSERVATIONS

The axon of a motoneuron is stimulated in the presence of a low Ca^{2+} concentration (0.5 mM) and a high Mg^{2+} concentration (6 mM) in the extracellular medium to reduce synaptic transmission. In this condition, a postsynaptic depolarizing potential is recorded with the use of an intracellular electrode implanted in the muscle fibre at the level of the neuromuscular junction (current clamp mode) (**Figure 8.2a**). It is an excitatory postsynaptic potential (EPSP), also called endplate potential (EPP). Its amplitude varies with the intensity of stimulation. In contrast, when the postsynaptic recording electrode is far from the endplate, no response is recorded (**Figure 8.2b**).

The low Ca^{2+} and high Mg^{2+} concentrations in the extracellular medium is a necessary condition to avoid muscle fibre contraction during recording. The number of active zones per neuromuscular junction is in fact so high (around 300 to 1000) that a presynaptic axonal spike always triggers a very large EPP that in turn depolarizes the muscle membrane to the threshold potential for voltage-gated Na^+ channels opening and thus evokes a postsynaptic action potential (not shown) and muscle fibre contraction. The elimination of the

muscle fibre action potential by lowering the release of acetylcholine from presynaptic terminals allows the recording of EPP in isolation.

Questions

- When released in the synaptic cleft, to which type(s) of postsynaptic receptor do the molecules of acetylcholine bind?
- How is the binding of acetylcholine transduced into a depolarization of the postsynaptic membrane?
- Why is the postsynaptic depolarization recorded only at the level of the neuromuscular junction?

8.2   THE TORPEDO OR MUSCLE NICOTINIC RECEPTOR OF ACETYLCHOLINE IS A HETEROLOGOUS PENTAMER $\alpha_2\beta\gamma\delta$

The only type of acetylcholine receptor present in the postsynaptic membrane of neuromuscular junctions or at electrical synapses of *Torpedo* is the nicotinic receptor (nAChR). It is named 'nicotinic' after its sensitivity to nicotine (nicotine is an agonist of nAChRs).

8.2.1   Nicotinic receptors have a rosette shape with an aqueous pore in the centre

Under the electron microscope, the nicotinic receptor of the neuromuscular junction, located in the postsynaptic muscular membrane, has a rosette shape with an 8–9 nm diameter and a central depression of diameter

(a)

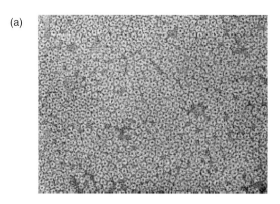

(b)

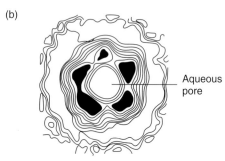

Aqueous pore

(c)

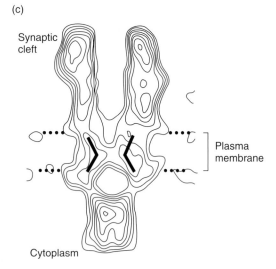

Synaptic cleft

Plasma membrane

Cytoplasm

FIGURE 8.3   The nicotinic receptor has a rosette shape.
(a) Membrane surface of *Torpedo* electric cells (electroplax). Each rosette constitutes one nicotinic receptor. (b) This computer reconstructed image of a single nicotinic receptor provides a more detailed view (superior view). (c) Electron microscopic analysis of tubular crystals of *Torpedo* nAChR viewed from the side. Part (a) from Cartaud J, Benedetti EL, Sobel A, Chargeux JP (1978) A morphological study of the cholinergic receptor protein from Torpedo narmorata in its membrane environment and in its detergent-extracted purified form. *J. Cell Sci.* 29, 313–337, with permission. Part (b) from Bon F *et al.* (1982) Orientation relative de deux oligornères constituant la forme lourde du récepteur de l'acétylcholine chez la torpille marbrée. *C. R. Acad. Sci.* **295**, 199, with permission. Part (c) from Unwin N (1993) The nicotinic acetylcholine receptor at 9 Å resolution. *J. Mol. Biol.* **229**, 1101–1124, with permission.

1.5–2.5 nm. This depression corresponds to the channel portion of the protein (**Figure 8.3**). Each rosette is made up of five regions of high electronic density arranged around an axis perpendicular to the plane of the plasma membrane. In transverse section the rosette appears as a cylinder 11 nm long, extending beyond each side of the membrane (6 nm towards the synaptic cleft and 1.5 nm towards the cytoplasm).

8.2.2   The four subunits of the nicotinicreceptor are assembled as a pentamer $\alpha_2\beta\gamma\delta$

The nicotinic receptor is normally purified from the electric organ of *Torpedo* or the electric eel. A 290–300 kD glycoprotein is obtained when this purification is performed on an affinity column using an agarose bound cholinergic ligand (**Figure 8.4**). When this glycoprotein is incorporated into a planar lipid bilayer or into lipid vesicles, it presents the same functional characteristics as the native receptor: when acetylcholine is present in the extracellular side at a concentration of 10^{-5}–10^{-4} mol 1^{-1}, it induces the passage of cations across the bilayer.

The nicotinic receptor is composed of four glycopolypeptide subunits $\alpha,\beta,\gamma,\delta$

In the presence of the detergent SDS (sodium dodecyl sulphate), the 290–300 kD protein dissociates into four different subunits, which migrate on a polyacrylamide gel as molecules with apparent molecular weights of 38 kD (α), 49 kD (β), 57 kD (γ) and 64 kD (δ) (**Figure 8.5a**). The same experiment carried out with nicotinic receptors obtained from the neuromuscular junction shows very similar results (**Figure 8.5b**).

Genes coding for each subunit of the nicotinic receptor of the electric ray and of the mammalian receptor have been cloned. When the corresponding mRNAs are injected into *Xenopus* oocytes, functional nicotinic receptors are synthesized and incorporated into the oocyte membrane. It has, therefore, been confirmed that the subunits α, β, γ and δ are sufficient to obtain a functional nicotinic receptor containing the acetylcholine receptor sites and the elements that form the ionic channel (**Figure 8.6**):

- an NH_2 terminal region which forms a large hydrophilic domain of 210–224 amino acids and carries the glycosylation sites;
- it is followed towards the COOH end by three hydrophobic sequences (M1, M2 and M3) of 20–30 residues each with short connecting hydrophilic loops;

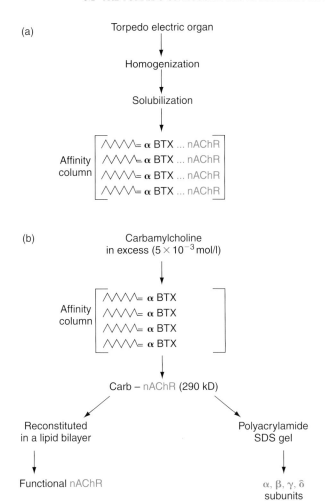

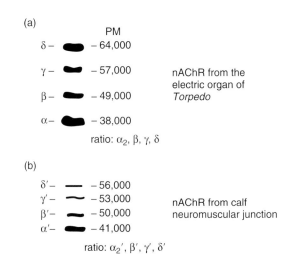

FIGURE 8.5 **Separation of the different nicotinic receptor subunits on a gel.**
The subunits of the purified nicotinic receptor have been dissociated with the detergent sodium dodecyl sulphate (SDS) and separated on a polyacrylamide gel: subunits of (**a**) the nicotinic receptor of electric organ, and of (**b**) the calf neuromuscular junction. The four subunits obtained, α, β, γ, δ and α', β', γ', δ', have similar molecular weights. From Anholt R, Lindstrom J, Montal M (1984) The molecular basis of neurotransmission: structure and function of the nAChR. In Martinosi A (ed.) *The Enzymes of Biological Membranes*, New York: Plenum Press, with permission.

cytoplasmic side and the four hydrophobic sequences are membrane-spanning segments. Each subunit therefore crosses the membrane four times and the carboxy terminal tail is oriented towards the synaptic cleft.

8.2.4 Each α-subunit contains one acetylcholine receptor site located in the hydrophilic NH$_2$ terminal domain

Before the structure of the nicotinic receptor was known, it had been demonstrated that two acetylcholine molecules had to be bound to the receptor in order to initiate an ionic flux (see Section 8.3). It seemed logical that these two sites had to be located on identical subunits; i.e. one on each α-subunit. Additionally, based on the organization of the hydrophilic sequences, it was proposed that this site is located in the large hydrophilic NH$_2$ terminal domain which is exposed to the synaptic cleft (**Figures 8.6b** and **8.7**).

This proposal has been confirmed by covalent binding studies of cholinergic agonists on α-subunits, isolated either from nicotinic receptor-rich membranes, or expressed in frog oocytes from the corresponding mRNA. Of the four subunit types α, β, γ and δ, the α-subunits have been shown to be the main contributors to cholinergic agonist binding.

FIGURE 8.4 **Stages of affinity column purification of the nicotinic receptor.**
(**a**) The electric organ of the electric ray is homogenized and membrane proteins solubilized. The resulting extract is run through an affinity column, on to whose sepharose (^^^=) a nicotinic cholinergic ligand α-bungarotoxin (α-BTX) has been covalently bound. Owing to their affinity to α-BTX, the nicotinic receptors bind to it. (**b**) In order to recover the nicotinic receptors, another nicotinic ligand, carbamylcholine (Carb), is run in excess through the column to displace the binding of α-BTX to the receptor. Carbamylcholine-bound nicotinic receptor is obtained at the outflow of the column. Carbamylcholine is eliminated by dialysis and the nicotinic receptor is thus obtained in an isolated form. The nicotinic receptor can then be reincorporated into a lipid bilayer to study its functional characteristics. It may also be treated with a detergent (SDS) to dissociate its subunits.

- a second large hydrophilic domain of about 150 residues containing functional phosphorylation sites;
- a fourth hydrophobic sequence and a short carboxy terminal tail.

A model of the transmembrane organization common to all subunits has been proposed (**Figure 8.6b**): the hydrophilic NH$_2$ terminal domain is located on the extracellular side of the membrane (in the synaptic cleft), the second hydrophilic domain is located on the

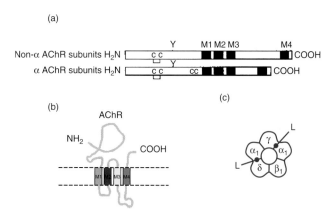

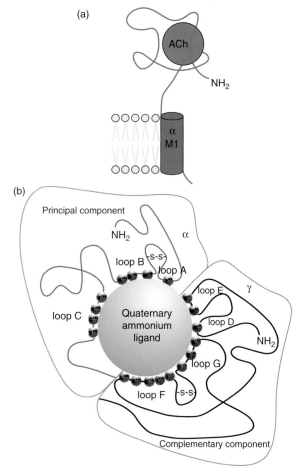

FIGURE 8.6 Muscle-type nicotinic receptor (nAChR) model. (a) Schematic representation of the primary sequence of s the α (α1–α9) and non-α (β1–β4, γ, ε, and δ) subunits of the AChR. M1–M4, transmembrane domains; CC, Cys-Cys bridge between C_{128} and C_{148} of the α1 nAChR subunit; CC, Cys-Cys pair found in the α subunits from both muscle- and neuronal-type nAChRs; Y, oligosaccharide groups. (b) Diagram of the tertiary organization of nAChRs. Each nAChR subunit includes: (1) a long NH_2-terminal hydrophilic extracellular region; (2) four highly hydrophobic domains named M1, M2, M3, and M4. M1–M2 and M2–M3 are connected by minor hydrophilic stretches; and (3) a major hydrophilic segment facing the cytoplasm. Additionally, the M4 domain orientates the COOH-terminus to the synaptic side of the membrane. (c) Schematic representation of the oligomeric organization of muscle-type nAChR. The hypothetical pentameric nAChR is formed by two α subunits and three non-α subunits. The two ligand binding sites (L) are located at the interfaces of one α subunit and one non-α subunit. For instance, the muscle-type AChR presents a high-affinity ACh binding site at the $\alpha\beta$ subunit interface and another low-affinity ACh locus at the $\alpha\gamma$ subunit interface. Parts (a,b) adapted from Arias HR (2000) Localization of agonist and competitive antagonist binding sites on nicotinic acetylcholine receptors. *Neurochemistry International* **36**, 595–645.

FIGURE 8.7 Model of the acetylcholine binding site. (a) Lateral and (b) Top view of a schematic model for the proposed folding of the extracellular domain involved in the binding site of several quaternary-ammonium compounds, including the native neurotransmitter ACh and the competitive antagonist curare. The low-affinity ACh binding site is located in the $\alpha\gamma$ interface. Its counterpart, the high-affinity ACh binding site, which is located in the $\alpha\delta$ interface is not included in this model. The large sphere represents a quaternary ammonium-containing molecule. The principal component of the ligand binding site is located in the α subunit which contributes with loops A, B, and C. The residues involved in the binding site are represented by small spheres in one letter code. Loop A is mainly formed by residue Y_{93}. Loop B is molded by amino acids W_{149}, Y_{152} and probably G_{153}. Loop C is shaped by residues Y_{190}, C_{192}, C_{193}, and Y_{198}. The disulfide bond indicated in the $\alpha\gamma$ subunit as S-S represents the link between C_{128} and C_{142}. The complementary component of the ligand binding site is located on either the γ or the δ subunit. Part (b) adapted from Arias HR (2000) Localization of agonist and competitive antagonist binding sites on nicotinic acetylcholine receptors. *Neurochemistry International* **36**, 595–645.

The next step was to determine which amino acids are part of the acetylcholine receptor site. To this end, labelled cholinergic ligands are used. These ligands are able to bind covalently to the acetylcholine receptor sites. One of the most used is MBTA (4-(N-maleimido) benzyl trimethylammonium iodide) which binds covalently to α-subunit receptor sites after reduction of disulphide bridges. Once labelled, the α chain is sequenced and the labelled regions identified. In this way a region containing cysteines 192 and 193 was identified and proposed as one of the potential sites of interaction with cholinergic ligands (**Figure 8.7b**).

Other data have provided additional evidence of the participation of cysteine residues 192 and 193 in the acetylcholine receptor site. In the first place, these cysteine residues are present only in the α-subunits. Furthermore, when frog oocytes are injected with

mRNA coding for α-subunits that have been mutated at the level of cysteines 192 and 193 (serines replaced for cysteines), the α-subunits obtained are unable to bind cholinergic ligands.

However, all these results have the shortcoming of having been obtained from preparations previously treated with disulphide bond-reducing agents (such as dithiothreitol). This treatment is necessary in order to allow the covalent binding of the cholinergic ligand MBTA to the receptor site but it alters the receptor site selectivity for cholinergic ligands.

In order to obtain a more detailed map of the native protein's acetylcholine receptor site, a labelled photoactivated cholinergic ligand has been used: ^{3}H-DDF (para-*N,N*-dimethylamino benzene diazonium fluoroborate). ^{3}H-DDF is a competitive antagonist of acetylcholine which, once photoactivated, binds covalently (irreversibly) to the acetylcholine receptor sites. This reaction is carried out on the whole nicotinic receptor channel and the α-subunits are then isolated, the segments labelled by ^{3}H-DDF are purified and their sequence analyzed. This led to the demonstration that the residues tyr (Y) 93, trp (W) 149, tyr (Y) 190, cys (C) 192 and cys (C) 193, all labelled by ^{3}H-DDF, are part of the acetylcholine receptor site. This labelling is in fact inhibited by other nicotinic agonists and competitive antagonists (**Figure 8.7b**). This result is valid for the nicotinic receptor of the electric organ as well as for that of the neuromuscular junction. Site-directed mutagenesis in combination with patch clamp techniques have extended the results obtained with the use of the photolabelling approach (**Figure 8.7b**).

8.2.5 The pore of the ion channel is lined by the M2 transmembrane segments of each of the five subunits

The ion channel can be considered as functionally equivalent to the active site of allosteric enzymes: its states (open, closed, blocked) are determined by the effectors of the receptor (binding of agonists, competitive antagonists and non-competitive antagonists; see Appendix 6.1). Concerning the channel structure, the question is which of the four hydrophobic membrane spanning segments M1 to M4 (**Figure 8.6**) are part of the walls of the ionic channel. On the basis of the hypothesis that non-competitive inhibitors bind to a high-affinity site located inside the open ion channel (channel blockers; see Section 8.6.3), photoactivable non-competitive inhibitors are used to label residues participating in the walls of the ion channel (this is a similar approach to that used for determination of the ACh binding site). Radioactive chlorpromazine activated with ultraviolet light labels serine, leucine and threonine residues from the M2 membrane-spanning segment from all subunits of the *Torpedo* acetylcholine receptor. These results point to a contribution of the

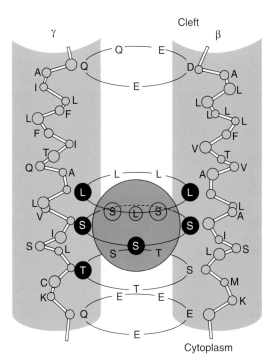

FIGURE 8.8 Characteristic amino acid residues along the M2 segment of the subunits of the nAChR.
The M2 membrane-spanning segments are symmetrically arranged around the central axis of the molecule (two of them are represented). The relative position of the α-carbons of the amino acids is shown as one-letter code. E, glutamic acid, S, serine, T, threonine, L, leucine, Q, glutamine. Adapted from Revah F, Galzi JL, Giraudat J *et al.* (1990) The noncompetitive blocker [^{3}H]-chlorpromazine labels three amino acids of the acetylcholine receptor γ subunit implications for the α-helical organization of the M2 segments and the structure of the ion channel. *Proc. Natl Acad. Sci. USA* **87**, 4675–4679, with permission.

M2 membrane-spanning segment to the walls of the ion channel.

In the M2 segments of nAChR subunits there are remarkable amino acids which, in the proposed model, form rings, assuming that the M2 segments of each of the subunits are symmetrically arranged around central axis of the molecule: a cytoplasmic ring of negatively charged amino acids that repel negative ions, a hydrophobic ring of leucines, a ring of serines, a ring of threonines and again a ring of negatively charged amino acids that repel negative ions (**Figure 8.8**).

Site-directed mutagenesis of some amino acids located in the M2 segment confirmed the contribution of M2 segments to the regulation of ion transport through the nicotinic channel. Chimeric cDNAs are constructed to add or substitute amino acids in the M2 segment. The results obtained will be explained in detail in the section describing the study of ionic selectivity (Section 8.3.2).

8.3 BINDING OF TWO ACETYLCHOLINE MOLECULES FAVOURS CONFORMATIONAL CHANGE OF THE PROTEIN TOWARDS THE OPEN STATE OF THE CATIONIC CHANNEL

8.3.1 Demonstration of the binding of two acetylcholine molecules

It has been demonstrated that two acetylcholine molecules must bind to the receptor to trigger the opening of the channel and allow cations to flow through. The proof of this has been obtained from dose–response curves. The response to acetylcholine (i.e. the flux of cations measured at very short intervals after the application of acetylcholine), or the opening probability of the channel (see below), is proportional to the square of the acetylcholine concentration:

$$\text{Response} = f(\text{ACh})^2$$

However, this demonstration is obscured by the consequences of receptor desensitization, which have to be eliminated from the recordings (see Section 8.4). For this reason this demonstration will not be presented in detail here. The conformational change of the protein towards the open state is clearly favoured when two acetylcholine molecules bind to the receptor. The following model accounts for these observations (however, as we shall see in Section 8.4, this model is in fact much more complex):

$$R \rightleftharpoons AR \rightleftharpoons A_2R \rightleftharpoons A_2R^* \rightarrow \textit{cationic current}$$

where R is the nicotinic receptor in its closed configuration; R* is the nicotinic receptor in its open configuration; and A is acetylcholine.

The rate of isomerization between R and R* lies in the microsecond to millisecond timescale. The passage of cations through the open channel is the result of the conformational change ($A_2R \rightleftharpoons A_2R^*$). Electrophysiological techniques (patch clamp recordings of the unitary cationic current flowing through a single channel) can be used to study this flow of cations. Based on the results obtained with electrophysiological techniques, we shall look at the properties of the nicotinic channel and of the protein conformational changes.

8.3.2 The nicotinic channel has a selective permeability to cations: its unitary conductance is constant

When the unitary current crossing a nicotinic channel in the presence of acetylcholine is recorded in patch clamp, all the preparations tested show an inward current at

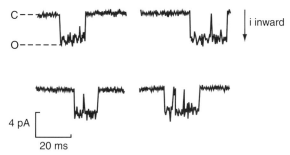

FIGURE 8.9 Patch clamp recording of nicotinic receptor activity in rat sympathetic neurons. Channel activity is recorded in the attached-cell configuration. While the membrane is kept at a negative potential an inward current is recorded in the presence of acetylcholine under physiological conditions. C, closed channel; O, open channel. Adapted from Colquhoun D, Ogden DC, Mathie A (1987) Nicotinic acetylcholine receptors of nerve and muscle: functional aspects. *TIPS* **8**, 465–472, with permission.

negative holding potentials (and under physiological ionic conditions) (**Figure 8.9**).

The nicotinic current reverses at 0 mV

If the imposed membrane potential (V_m) is varied between $-100\,\text{mV}$ and $+80\,\text{mV}$ while recording unitary currents with the patch clamp technique (i_{ACh}), one can trace an i_{ACh}/V curve (**Figure 8.9**). This curve is approximately linear between -80 and $+80\,\text{mV}$. The measured current is inward for negative voltages and outward for positive voltages.

The i_{ACh}/V curve crosses the voltage axis at a value where the current is zero. This value is called the *reversal potential of the nicotinic response*, or E_{ACh}. The value of this reversal potential is close to $0\,\text{mV}$ in the experimental conditions of **Figure 8.10** but may vary slightly towards negative voltages depending on the preparation.

The unitary conductance is constant

The linear i_{ACh}/V relationship observed in **Figure 8.10b** (between -80 and $+80\,\text{mV}$) is described by the equation: $i_{ACh} = \gamma_{ACh}(V_m - E_{ACh})$, where V_m is the membrane potential, E_{ACh} is the reversal potential of the nicotinic response, and γ_{ACh} is the conductance of a single nicotinic channel, or its unitary conductance. The value of γ_{ACh} is given by the slope of the linear i_{ACh}/V curve. It has a constant value at any given membrane potential. This value varies between 35 and 55 pS depending on the preparation and is a fundamental property of a nicotinic channel.

The nicotinic channel is a cationic channel

The reversal potential of the nicotinic response ($E_{ACh} = 0\,\text{mV}$) does not correspond to the equilibrium potentials of any of the ions in solution. It is not a Na^+ channel because, in the experimental conditions of **Figure 8.10**, $E_{Na} = 58 \log(160/3) = +100\,mV$. It is not a K^+ channel either, since $E_K = 58 \log(3/160) = -100\,mV$. And it is not a Cl^- channel because, if the chloride ions are replaced by large anions that cannot cross the channel, such as SO_4^{2-}, no reversal potential change of the nicotinic response is observed.

By performing extracellular ionic substitution experiments, it has been shown that the nicotinic channel is permeable to Na^+, K^+, Ca^{2+} and Mg^{2+} ions. However, Ca^{2+} and Mg^{2+} ions contribute only a small fraction to the nicotinic current, which is essentially due to the flux of Na^+ and K^+ ions through the open channel. If different cations cross the same channel and have similar permeabilities, we define:

$$E_{cations} = 58 \log([cations]_e/[cations]_i),$$

where $[cations] = [Na^+] + [K^+]$. In our case we obtain $E_{cations} = 0\,mV$. In other words, $E_{cations} = E_{ACh} = 0\,mV$.

In which direction do Na^+ and K^+ ions cross the open nicotinic channel at different membrane potentials?

When the membrane is at a voltage of $-80\,mV$, the Na^+ ion driving force is inward and equal to $-180\,mV$, while the K^+ ion driving force is outward and equal to

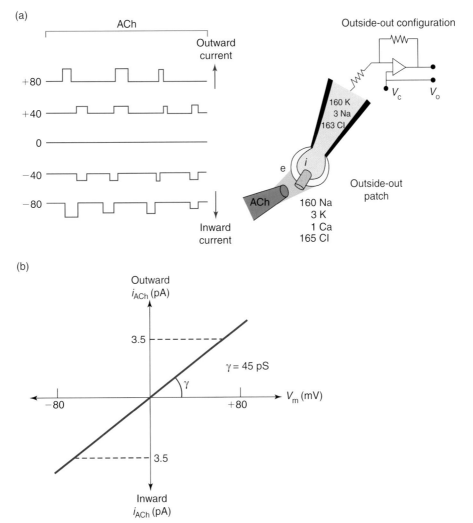

FIGURE 8.10   Nicotinic unitary current recorded in patch clamp (outside-out configuration) at various membrane potentials (from $-80\,mV$ to $+80\,mV$).
(a) Nicotinic unitary current recorded at different membrane potentials in response to the application of acetylcholine (ACh). A downward deflection indicates an inward current and an upward deflection indicates an outward current. (b) i_{ACh}/V curve obtained from the average values of i_{ACh} at each membrane potential V_m. The curve reverses at $0\,mV$ and the slope corresponds to the unitary conductance γ.

+20 mV (**Figure 8.11**). If channels open, more Na^+ will enter the cell than K^+ ions will leave it. The net flux of positively charged ions is, thus, inward: an inward current is recorded (**Figures 8.9** and **8.10a**).

If the same reasoning is followed for different membrane potential values, the same result is obtained as with the i_{ACh}/V curve: the unitary current is inward for negative membrane potentials (net flux of positive

charges is inward), and the unitary current is outward for positive membrane potentials (net flux of positive charges is outward) (**Figures 8.10a** and **8.12**).

Effect of a decrease in $[Na^+]_e$

When extracellular Na^+ ions are partially replaced with a non-permeant substance such as sucrose (without changing the osmotic pressure), the reversal potential of the nicotinic response shifts towards more negative potentials (**Figure 8.13**). This is explained by the fact that, at this point, extracellular Na^+ ions contribute less to the nicotinic current, and the nicotinic reversal potential E_{ACh} shifts towards the K^+ equilibrium potential (E_K).

Substitution of K^+ ions for extracellular Na^+ ions

When almost all extracellular Na^+ ions are replaced by extracellular K^+ ions, the I/V curve obtained superposes on the control I/V curve (**Figure 8.14**). Thus, extracellular K^+ ions can replace extracellular Na^+ ions; i.e. the nicotinic channel does not distinguish between Na^+ and K^+ ions. In other words, it presents similar permeabilities for both ions. For this reason the reversal potential of the nicotinic response is independent of the relative concentrations of extracellular Na^+ and K^+ ions. It depends solely on the sum of these concentrations.

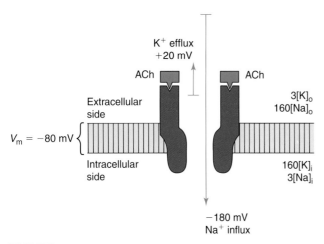

FIGURE 8.11 Determination and vectorial representation of the Na^+ ion driving force ($V_m - E_{Na}$) and K^+ ion driving force ($V_m - E_K$) for a membrane potential of $-80\,mV$.
Observe that 90% of the current is due to Na^+ ions. The net flux of positive charges is inward. This explains why ACh induces an inward current at $V_m = -80\,mV$.

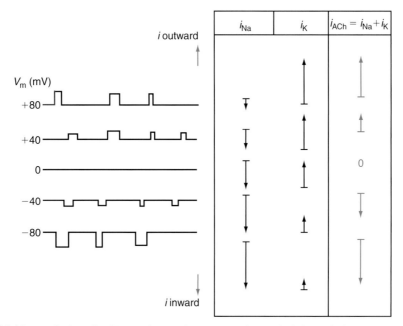

FIGURE 8.12 Evolution of sodium and potassium currents (i_{Na} and i_K) through the nicotinic channel as a function of membrane potential V_m.
The current induced by acetylcholine i_{ACh} corresponds to the sum of two currents: $i_{ACh} = i_{Na} + i_K$. When $i_{Na} = -i_K$ the current is zero. This occurs at the reversal potential of the nicotinic response ($V_m = E_{ACh}$). A deflection of the traces (left) or an arrow of the chart (right) in the downward direction represents an inward current, and in the upward direction an outward current.

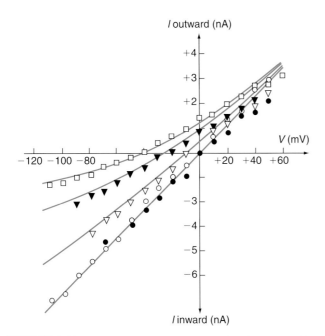

FIGURE 8.13   Effect of lowering the external Na$^+$ concentration on the I_{ACh}/V curve.

The composition of the external environment is: 5 mM of K$^+$; 0.1 mM of Ca^{2+}; and 21 ($\square$), 46 ($\blacktriangledown$), 96 ($\triangledown$) or 146 ($\bullet$ and $\bigcirc$) mM of Na$^+$. Each point represents the average of 25 measurements of I_{ACh}. When [Na$^+$]$_e$ decreases, the reversal potential of the nicotinic response shifts towards the K$^+$ equilibrium potential. Control E_{ACh}: ($\bullet$, $\bigcirc$) 58 log(146 + 5)/[cations]$_i$; ($\triangledown$) 58 log(96 + 5)/[cations]$_i$; ($\blacktriangledown$) 58 log(46 + 5)/[cations]$_i$; ($\square$) 58 log(21 + 5)/[cations]$_i$. Adapted from Linder TM, and Quastel DMJ (1978) A voltage clamp study of the permeability change induced by quanta of transmitter at the mouse endplate. *J. Physiol. (Lond.)* **281**, 535–556, with permission.

Mutations in the M2 membrane-spanning segment can convert ion selectivity from cationic to anionic

The question was: do substitutions and/or additions of amino acids within (or near) the M2 segment from a nicotinic α-subunit (here the neuronal α_7-subunit) with homologous amino acids of the glycine receptor suffice to convert α subunit ion-channel selectivity from cationic to anionic? (The glycine receptor is a receptor channel selectively permeable to anions, Cl$^-$ ions.) The M2 sequences of α-subunits of a cationic channel (nAchR) and anionic channels (GlyR and GABA$_A$R) show similarities at the level of the threonine (244) and leucine (247) rings and differences at the level of rings of negative amino acids, Glu 237 and Glu 258 (**Figure 8.15a**). A chimeric cDNA encoding the α_7-subunit of neuronal nAChR is constructed in which, in the M2 segment, a proline residue is added at position 236 bis, and amino acids at positions 237, 240, 251, 254, 255 and 258 are exchanged with those found in the M2 segment of the glycine receptor α-subunit (nAChRα_7^*; **Figure 8.15a**). Interestingly, glutamates (E) 237 and 258 which form negative rings in the nAChR (repelling negative ions)

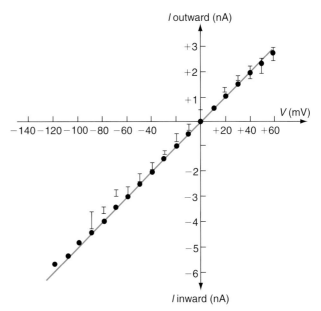

FIGURE 8.14   Replacing external K$^+$ for Na$^+$ ions has no effect.

Control I/V curve ($\bullet$) 146 mM of Na$^+$ and 5 mM of K$^+$, and (I) 2 mM of Na$^+$ and 149 mM of K$^+$. Observe that K$^+$ ions can replace Na$^+$ ions without affecting the I/V curve. The nicotinic channel does not distinguish between these two cations. Adapted from Linder TM and Quastel DMJ (1978) A voltage clamp study of the permeability change induced by quanta of transmitter at the mouse endplate. *J. Physiol. (Lond.)* **281**, 535–556, with permission.

are exchanged with alanine (A) and asparagine (N) residues, respectively. The chimeric cDNA is injected into oocytes that express homomeric mutated nAChR* (formed by five identical mutated α_7-subunits). The current recorded with double-electrode voltage clamp (see Appendix 4.2) is compared with that recorded in oocytes expressing wild-type (non-mutated) homomeric AChR.

Ionic currents recorded in response to ACh application (100 μM, 2 s duration) from oocytes expressing wild-type homomeric α_7nAChR (in the presence of a Ca^{2+} chelator inside the oocyte) reverses around +3 mV. Substitution of 90% of the external chloride ions did not change the value of the reversal potential. These data thus support the conclusion that wild-type homomeric αnAChR, like native nAChR, is selective for cations.

Ionic currents recorded in response to ACh application (100 μM, 2 s duration) from oocytes expressing mutated homomeric α^*nAChR reverses around −20 mV. Substitution of 90% external chloride ions by isethionate (an impermeant anion) shifts the reversal potential towards positive voltage (around +30 mV) (**Figure 8.15b**, left). This shift is well described by the Goldman–Hodgkin relationship for chloride specific channels (**Figure 8.15b**, right). This indicates that

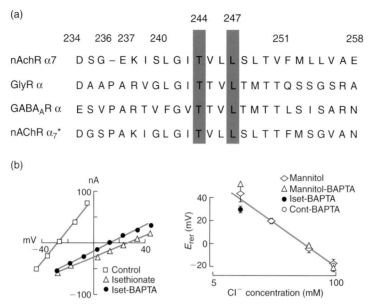

FIGURE 8.15 **Mutations in the M2 segment of an α-subunit of the nAChR convert ion selectivity of the homomeric nAChR from cationic to anionic.**
(a) Comparison of M2 sequences from subunits of the cation selective nicotinic α7 receptor subunit with those of the anion-selective glycine α1, GABA$_A$ α1 and β1 and mutated nicotinic α7* receptor subunits. (b, left) I/V relationship of the α7* mutant receptor is first determined in control conditions (control) with 2 s ACh (100 μM) applications (outside-out patch clamp recording of a patch of membrane containing a large number of nicotinic receptors called macropatch). Then, 90% of chloride ions of the extracellular medium were replaced by the non-permeant anion isethionate and the I/V curve determined (isethionate). This last experiment was also performed in the presence of a chelator of Ca^{2+} ions (BAPTA) injected inside the cell (iset-BAPTA) in order to reduce secondary currents that could be triggered by the entry of Ca^{2+} ions through the nicotinic receptors. (b, right) Reversal potential values as a function of the logarithm of external chloride concentration (92, 50.5, 19.75 and 9.5 μM external chloride after substitution of NaCl by mannitol or isethionate). The solid line corresponds to the theoretical Nernst relation. From Galzi JL, Devillers-Thiéry A, Hussy N (1992) Mutations in the channel domain of a neuronal nicotinic receptor convert ion selectivity from cationic to anionic. *Nature* **359**, 500–505, with permission.

ACh-activated currents are almost entirely carried by chloride ions in oocytes expressing mutated homomeric nAChR*. Then, introducing appropriate amino acid residues from the putative channel domain of a chloride-selective GlyR α-subunit into that of a cation selective nAChR α-subunit allows the design of an ACh-gated channel now selective for chloride. This confirms that the M2 segment forms the walls of the channel and strongly suggests that the exchanged residues face the lumen of the channel.

8.3.3 The time during which the channel stays open varies around an average value τ_o, the mean open time, and is a characteristic of each nicotinic receptor

When recording in patch clamp from myotubes (embryonic muscle cells) or from denervated muscle cells, in the presence of very small doses of acetylcholine, openings of the nicotinic channels separated by periods of silence are observed (**Figure 8.16**). The nicotinic receptor switches between states in which the channel is closed and the unitary current is zero (R, AR, A$_2$R), and a state in which the channel is open and shows a measurable unitary current (A$_2$R*) (**Figure 8.17**). These conformational changes can be modelled as:

$$R \rightleftharpoons AR \rightleftharpoons A_2R \rightleftharpoons A_2R^*$$

Closed channel states —— Open channel state

where A is acetylcholine or any nicotinic agonist; R is the receptor in the closed conformation; and R* is the receptor in the open conformation.

In **Figure 8.16a** one observes that the periods during which the channel is open, t_o, are variable. To obtain the mean open time of the channel, τ_o, one can build a frequency histogram of the different t_o. The exponential curve obtained provides the value of τ_o (see **Figure 8.16b** and Appendix 4.3). The functional significance of this value is as follows: during a time equal to τ_o the channel has a high probability of being open.

(a)

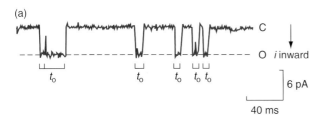

(b)

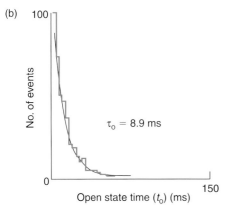

$\tau_o = 8.9$ ms

FIGURE 8.16 Patch clamp recording (attached-cell configuration) of myotube nicotinic receptor channel activity ($V_m = -170$ mV).
(a) Myotubes (embryonic muscle cells) are recorded in the presence of a low concentration of acetylcholine (200 nM). At this concentration, the channels open during periods t_o. This recording does not correspond to a single nAChR because one finds approximately 100,000 nAChR per patch. The repeated openings (downward deflections) correspond, therefore, to the opening of different nAChR. However, all the nAChR being identical, it seems as though the activity of the same nAChR was recorded. (b) The mean open time τ_o can thus be calculated. C, closed channel; unitary current is zero. O, open channel; inward unitary current (downward deflections).

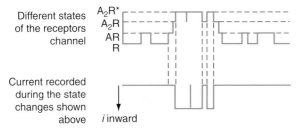

FIGURE 8.17 Correlation between the nicotinic current and the states of the channel.
The channel opens only (inward current, lower trace) when the protein is in the A_2R^* state. The rapid fluctuations between states A_2R^* and A_2R correspond to short-lived closures. When the receptor channel loses one or two of its acetylcholine molecules, the closures last longer. Adapted from Colquhoun D, Ogden DC, Mathie A (1987) Nicotinic acetylcholine receptors of nerve and muscle: functional aspects. *TIPS* **8**, 465–472, with permission.

τ_o is a characteristic of the nicotinic receptor channel type

Nicotinic receptors from the electric organ of *Torpedo* and from the calf neuromuscular junction can be studied in patch clamp after the expression of the corresponding mRNA injected into *Xenopus* oocytes (outside-out configuration). Recording of such channels shows that electric organ and neuromuscular junction channels present very similar conductances (40 and 42 pS) but very different mean open times ($\tau_o = 0.6$ and 7.6 ms, respectively).

Another example is given by the study of nicotinic receptors from fetal or adult bovine muscle. A study of the subunit structure of the bovine muscle nAChR showed the presence of the α, β, γ and δ-subunits as in the case for *Torpedo* electroplax nAChR. In addition, a novel subunit termed the ε-subunit has been discovered by cloning and sequencing the DNA complementary to the muscle mRNA encoding it. The ε-subunit shows higher sequence homology with the γ-subunit than with any other subunit. In order to study the properties of the γ and ε-subunits, various combinations of the subunit specific mRNAs are injected in *Xenopus* oocytes and their functional properties are studied in the presence of acetylcholine.

Figure 8.18a shows recordings of ACh-activated single channels from outside-out patches isolated from oocytes injected with the α, β, γ and δ-subunit-specific mRNAs (left) or with the α, β, ε and δ-subunit-specific mRNAs (right). The conductance and mean open time τ_o (**Figures 8.18c, d**) of the channels formed in a given oocyte differ in relation to the mRNA combin-ation with which it was injected. This suggests that a single subunit can change the conductance and gating properties of the nAChR channel.

To compare the two classes of nAChR channels produced in *Xenopus* oocytes with native bovine nAChR channels, the ACh-activated channels of fetal and adult bovine muscle are recorded. **Figure 8.18b** shows ACh-activated single currents from outside-out patches of native fetal (left) and adult (right) bovine muscle. nAChR single-channel current in fetal muscle is similar to that of nAChRγ whereas the nAChR single-channel current in adult muscle is similar to that of nAChRε (compare **Figures 8.18a, b**). This suggests that the nAChR channel in fetal muscle is assembled from α, β, γ and δ-subunits whereas the endplate channel in adult muscle is assembled from the α, β, ε and δ-subunits. To study this developmental change in the contents of the five nAChR-subunit mRNAs in bovine muscle, total RNA is extracted from the diaphragm muscle at various stages of fetal and postnatal development. It is then subjected to blot hybridization analysis using the respective cDNA probes. The results show that the contents of the γ- and ε-subunit mRNAs varies markedly during muscle development showing reciprocal changes: the γ-subunit mRNA is abundant at earlier fetal stages (3–5 months' gestation), but is hardly or not detectable

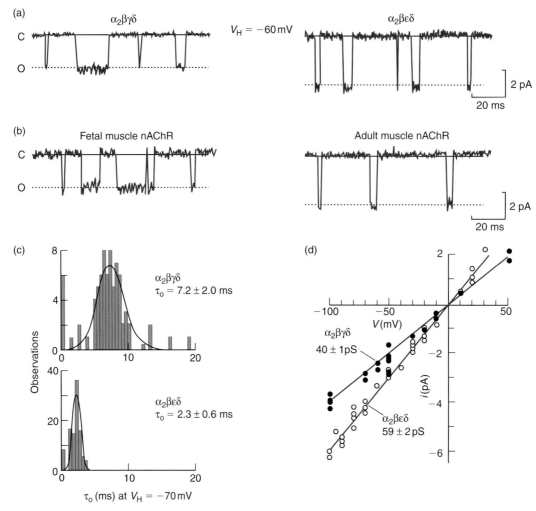

FIGURE 8.18 **The subunit structure participates in determining the nicotinic channel conductance and mean open time.**
See text for explanations. C, closed channel; O, open channel. Adapted from Mishina M, Takai T, lmoto K *et al.* (1986) Molecular distinction between fetal and adult forms of muscle acetylcholine receptor. *Nature* **321**, 406–411, with permission.

after birth; conversely, considerable amounts of ε-subunit mRNA appear only at postnatal stages and is not detectable at earlier fetal stages (3–4 months' gestation). Therefore, the replacement of the γ-subunit by the ε-subunit in the nAChR complex is responsible for the changes in the properties of the nAChR channel that occur during muscle development. This phenomenon of subunit replacement during development cannot be generalized to all the other mammalian nAChR.

8.4 THE NICOTINIC RECEPTOR DESENSITIZES

During the recording of a nicotinic receptor channel with the patch clamp technique (whole-cell configuration)

in the presence of a high and constant concentration of acetylcholine, there is a progressive diminution of the total current I_{ACh} (**Figure 8.19a**). This decrease in current corresponds to the progressive desensitization of the nicotinic receptor present in the membrane.

When recording unitary nicotinic currents (outside-out or cell-attached configuration) in the presence of a strong concentration of acetylcholine, there are repeated openings separated by long periods of silence (**Figure 8.19b**). These sequences of openings are known as *unitary current bursts*. Within a burst, the protein rapidly fluctuates between the closed and open states, symbolized as follows:

$$R \rightleftharpoons AR \rightleftharpoons A_2R \rightleftharpoons A_2R^*$$

$$\underbrace{\qquad\qquad}_{\text{Closed channel states}} \quad \underbrace{\qquad}_{\text{Open channel state}}$$

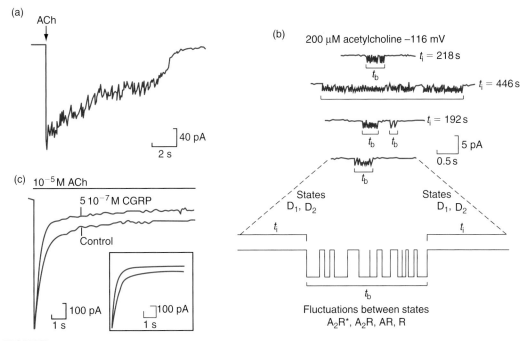

FIGURE 8.19   Nicotinic receptor desensitization.
(a) Patch clamp recording (whole-cell configuration) of the nicotinic current from adrenal chromaffin cells ($V_m = -70$ mV). In the presence of a high concentration of acetylcholine (20 μM) the inward current reaches a peak of 235 pA, and then decreases despite a constant acetylcholine concentration. This current corresponds to the sum of several unitary currents crossing all the activated nicotinic channels. The decrease in current is due to the desensitization of a large fraction of cellular nicotinic receptors. **(b)** Single-channel recordings (cell-attached or outside-out configurations) illustrating another consequence of desensitization. The membrane patch is exposed to a high concentration of acetylcholine (200 μM) for an extended period. Under these conditions, nicotinic receptors present in the patch desensitize. After a certain time, one of the channels reopens and fluctuates between the states A_2R^*, A_2R, AR and R during a time t_b (duration of the burst of openings) before desensitizing again for a duration t_i (interburst duration). The traces shown correspond to segments of a continuous recording. The duration of the desensitized periods t_i between two successive traces is indicated at the end of each trace (218, 466 and 192 s). **(c)** Cultured muscle cell recording (whole-cell configuration). Nicotinic current evoked by the application of 10 μM acetylcholine recorded in the presence of 500 nM CGRP (calcitonin gene related peptide) and in the absence of the peptide (control) ($V_m = -60$ mV). The nicotinic current reaches a peak with a 200 ms delay and then begins to decrease. The sum of two exponentials can describe this decrease in current. CGRP increases the speed of the fast component. Part (a) from Clapham DE and Neher E (1984) Trifluoperazine reduces inward ionic currents and secretion by separate mechanisms in bovine chromaffin cells. *J. Physiol. (Lond.)* **353**, 541–564, with permission. Part (b) from Colquhoun D, Ogden DC, Mathie A (1987) Nicotinic acetylcholine receptors of nerve and muscle: functional aspects. *TIPS* **8**, 465–472, with permission. Part (c) from Mulle C, Benoit P, Pinset C *et al.* (1988) Calcitonin gene-related peptide enhances the rate of desensitization of the nicotinic acetylcholine receptor in cultured mouse muscle cells. *Proc. Natl Acad. Sci. USA* **85**, 5728–5732, with permission.

The long silent periods correspond to desensitization of the receptor in the presence of acetylcholine. In the desensitized state, the nicotinic receptor is refractory to activation. Consequently, the channel does not open despite the fact that two molecules of ACh are bound to the receptor.

In summary, desensitization is a phenomenon that renders the nicotinic receptor incapable of being activated by its agonists. The desensitized nicotinic receptor presents two main characteristics: (i) a high affinity for ACh and (ii) a closed ionic channel: the unitary current is equal to 0 pA (long silent periods).

$$A_2R \;\rightleftharpoons\; A_2R^* \;\underset{k_{-3}}{\overset{k_3}{\rightleftharpoons}}\; A_2D_1$$

$$k_{-4} \big\updownarrow k_4$$

$$A_2D_2$$

These two states D1 and D2, which are closed channel states, are distinguished from one another by their

TABLE 8.1 Affinity constants for acetylcholine

State	Affinity constant for ACh K_D
R	10 μM to 1 mM
D_1	1 μM
D_2	3–10 nM

affinity constants for ACh (**Table 8.1**) and by the rate constants $k_3 = 0.01 \, \text{s}^{-1}$ and $k_4 = 1 \, \text{s}^{-1}$.

The concept of nicotinic receptor desensitization has been proposed by different authors, notably by Katz and Thesleff, from measurements of the time course of the global synaptic response during an iontophoretic application of acetylcholine.

The process of desensitization appears slowly and is slowly reversible. This is an intrinsic property of the protein. In order to study in patch clamp the states R and R* of the nicotinic receptor channel, it is necessary to choose conditions under which desensitization is negligible. To this end, researchers work with very low doses of acetylcholine (of the order of the nanomoles per litre) because high doses favour the conformational change of the protein into desensitized states (see **Figure 8.16**). An alternative approach is the use of high doses of acetylcholine (of the order of micromoles per litre; see **Figure 8.19b**). In this case, the receptors desensitize (silent periods known as interburst periods) and eventually one or several of the channels open and close repetitively (opening bursts) before re-desensitizing. Desensitization can be minimized by excluding the first and the last opening during the bursting periods, and thus the values calculated for τ_o are then related only to the R* state.

The rate of desensitization of the nicotinic receptor seems to be related to its state of phosphorylation. In fact, studies of the ionic flux through nicotinic receptors incorporated into liposomes have shown that an increase in the level of phosphorylation of the receptors by cyclic AMP augments the desensitization rate of these receptors. In the neuromuscular junctions a peptide present in the motoneurons is released at the same time as acetylcholine. This peptide, CGRP (calcitonin gene-related peptide), is capable of increasing the level of cyclic AMP in cultured embryonic muscle cells, consequently increasing the number of phosphorylated nicotinic receptors. In patch clamp recordings (whole-cell configuration) of embryonic muscle cells, the simultaneous application of this peptide and acetylcholine accelerates the rapid phase of desensitization of the nicotinic receptors (**Figure 8.19c**). In unitary recordings (cell-attached configuration), CGRP decreases the opening frequency of the nicotinic channels (while at the same time leaving unaffected their mean open time and unitary conductance). These effects are mimicked

by the application of substances that augment the intracellular cyclic AMP level (such as forskolin). The following hypothesis has been proposed. CGRP activates a specific membrane receptor. This leads to an increase in the intracellular cyclic AMP concentration and an activation of protein kinase A. Protein kinase A, directly or indirectly, phosphorylates certain subunits of the nicotinic receptor, leading to a rapid desensitization of the receptors.

Generalization

Nicotinic receptors are allosteric receptors as defined by Monod, Wyman and Changeux (MWC model, 1965). The MWC model hypothesizes the following:

- nAChRs are oligomers made up of a finite number of identical subunits that occupy equivalent positions and as a consequence possess at least one axis of rotational symmetry.
- nAChRs can exist spontaneously in the four freely interconvertible and discrete conformational states described above (R, R*, D_1, D_2), even in the absence of ligand. The closed states are R, AR and A_2R; the open states are R*, AR* and A_2R^*; the desensitized states are D_1, AD_1, A_2 and D_2, AD_2, A_2D_2.
- The affinity and activity of the stereospecific sites carried by the nAChR differ between these four states.

This gives a tetrahedral model for interactions between the four conformational states instead of the sequential model previously described for simplicity:

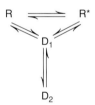

As a result, gating of the nAChR cannot be viewed solely as a ligand-triggered process but as reflecting an intrinsic structural transition of the receptor molecule, which may even occur in the absence of ligand. Moreover, at low agonist concentrations, desensitized states can be stabilized under conditions of negligible channel opening.

8.5 nACHR-MEDIATED SYNAPTIC TRANSMISSION AT THE NEUROMUSCULAR JUNCTION

A nicotinic synaptic current is evoked by a brief augmentation of the concentration of acetylcholine in the synaptic cleft. This increase, caused by the asynchronous release of synaptic vesicles, is brief because (**Figure 8.20a**): (i) the release is brief; and (ii) acetylcholine rapidly disappears from the synaptic cleft. In fact, when acetylcholine is released into the synaptic cleft, it may either bind to acetylcholine-gated channels, diffuse out of the synaptic cleft, or be rapidly degraded by acetylcholinesterase.

During the analysis of synaptic currents induced by the release of endogenous acetylcholine, the desensitized states of the receptor can be neglected because of the rapid elimination of acetylcholine from the synaptic cleft (in the order of microseconds). The model for this is:

$$\text{Acetylcholine release} \begin{cases} \text{Diffusion} \\ A + R \rightleftharpoons AR \rightleftharpoons A_2R \rightleftharpoons A_2R^* \\ \text{Degradation} \end{cases}$$

This is very different from what occurs during the recording of the activity of a nicotinic receptor in a patch of membrane in the inside-out configuration, when the acetylcholine is continuously present in the patch pipette, or the recording of the activity of a nicotinic receptor channel in a patch of membrane in the outside-out configuration in response to acetylcholine pressure applied from another pipette. Even the shortest applications in this case are of the order of tens of milliseconds.

8.5.1 Miniature and endplate synaptic currents are recorded at the neuromuscular junction

Two main types of postsynaptic currents can be recorded at any synapse: spontaneous and evoked synaptic currents. The former is recorded in the absence of presynaptic stimulation whereas the latter represents the response to a presynaptic stimulation. Spontaneous synaptic currents can be further divided into those evoked by spontaneous presynaptic spikes in a Ca^{2+}-dependent manner, and miniature currents evoked even in the absence of spontaneous spikes (in the presence of TTX). Miniature currents have a very small amplitude and correspond to the release of one or a few synaptic vesicles (see Appendix 7.1).

Miniature currents

Miniature currents are the currents recorded at the neuromuscular junction in the total absence of stimulation of the motor nerve and in the presence of TTX (**Figure 8.20b**). These currents are evoked by the spontaneous liberation of acetylcholine from the presynaptic terminal. Thus, if an innervated muscle fibre in the

absence of nerve stimulation is recorded from under voltage clamp (**Figure 8.20b**), from time to time a miniature current will be recorded (**Figure 8.20c**). The recorded current is due to the spontaneous release of a synaptic vesicle or quantum of acetylcholine (1 vesicle = 1 quantum; see Chapter 7). This current is inward at $-80\,mV$ and has a maximum amplitude of about $4\,nA$.

How many receptor channels are opened by acetylcholine at the peak of a miniature current? Knowing the amplitude of the unitary current and the amplitude of a miniature current at the same membrane potential, we can calculate the number of nicotinic receptor channels opened by acetylcholine at the peak of the miniature response.

At $V_m = -80\,mV$, we have:

$$i = 2.5\,pA, \text{ and } I = 4\,nA = 4 \times 10^3\,pA;$$

i.e. $4.10^3/2.5 = 1600$ nicotinic receptor channels opened by acetylcholine.

Since two molecules of acetylcholine are needed to open one channel, the average number of acetylcholine molecules released is $1600 \times 2 = 3200$ molecules of ACh per vesicle. In other words, 1 quantum equals about 3000 molecules of ACh.

What is the time course of a miniature current? The time it takes to reach the maximal amplitude of the miniature current is approximately $100\,\mu s$, while it takes longer to disappear. The decrease of the current has an exponential time course with a time constant of the order of milliseconds (**Figure 8.20e**). This current decrease depends solely on τ_o (**Figure 8.20d, e**).

Motor endplate current

The endplate current is recorded (in voltage clamp) at the neuromuscular junction while the motor nerve is being stimulated (**Figure 8.21**). At $V_m = -80\,mV$ the current is inward and has an amplitude of approximately $400\,nA$.

How many channels are opened at the peak of the motor endplate current? The motor endplate current is composed of $400\,nA/4\,nA = 100$ miniature currents produced by $1600 \times 100 = 16 \times 10^4$ nicotinic receptors opened by released acetylcholine.

What is the time course of the motor endplate current? Approximately 100 vesicles are released in an asynchronous manner by the stimulated presynaptic terminal. This is the reason why the time it takes to reach the maximal or peak amplitude of this current is relatively longer than the time it takes to reach the peak of a miniature current ($300\,\mu s$ instead of $100\,\mu s$).

In current clamp recordings, this endplate inward current depolarizes the postsynaptic membrane thus evoking the endplate potential recorded in **Figure 8.2**.

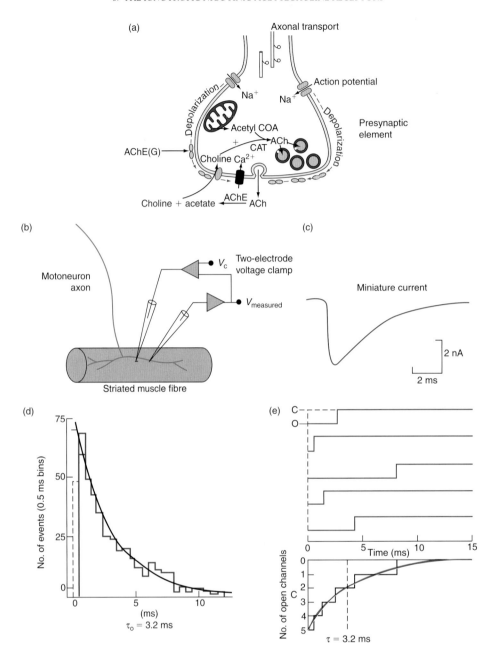

FIGURE 8.20 **(left) A cholinergic presynaptic terminal and miniature nicotinic current.**
(a) Functional scheme of the presynaptic component of the neuromuscular nicotinic cholinergic synapse. The enzymes choline acetyl transferase (CAT) and acetylcholinesterase (AChE) are synthesized in the cell body of the motoneuron and carried to axon terminals via anterograde axonal transport. Acetylcholine (Ach) is synthesized in axon terminals from choline and acetyl coenzyme A (acetylCOA). About 50% of the released choline is recaptured by the presynaptic terminals. Acetylcholine is actively transported from the cytoplasm into synaptic vesicles via a vesicular Ach carrier using the H^+ gradient as an energy source (antiport). (b) Miniature current recorded in two-electrode voltage clamp from a normally innervated muscle fibre in the absence of stimulation and in the presence of TTX. (c) The neuromuscular junction miniature current I_{ACh} is the current crossing N nicotinic channels activated by spontaneously released endogenous acetylcholine. The current rising phase is fast owing to the quasi-synchronous activation of the N nicotinic channels. The exponential falling phase of the current is slower ($V_m = -80\,mV$). (d) Since the opening of the channels is synchronous, their closure appears after a variable time t_0 whose distribution is exponential. The mean open time of the N channels is $\tau_o = 3.2\,ms$. (e) The same value of τ_o is obtained for the time constant of the falling phase (τ) of the total current I_{ACh}. In the example given in (d), $N = 5$, but N is in fact always larger (see text). Parts (c) and (d) adapted from Colquhoun D (1981) How fast do drugs work? *TIPS* **2**, 212–217, with permission.

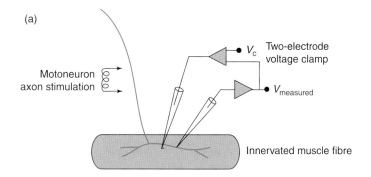

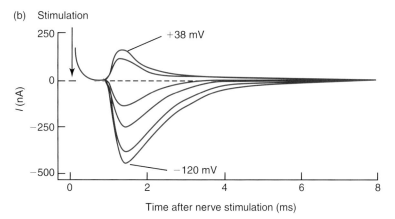

FIGURE 8.21   **Motor endplate current.**
(**a**) Current recorded in response to a stimulation of the nerve fibre under two-electrode voltage clamp. (**b**) This current is inward for negative membrane potentials and outward for positive voltages. As in the case of the unitary current, the motor endplate current reverses around 0 mV (frog's neuromuscular junction). Muscular action potentials are blocked by voltage clamping the corresponding muscle region. The muscle contraction induced by the inward current can be blocked by the destruction of T tubules (with a hyperosmotic shock). Adapted from Magleby KL and Stevens CF (1972) A quantitative description of end plate currents. *J. Physiol. (Lond.)* **223**, 173–197, with permission.

8.5.2 Synaptic currents are the sum of unitary currents appearing with variable delays and durations

There is a variable delay in the appearance of current flow through each one of the postsynaptic receptor channels. The reason for this is that the synaptic vesicles are released in an asynchronous manner. Furthermore, acetylcholine molecules must diffuse for a certain time before they reach a free receptor channel. We have seen that the concentration of acetylcholine in the synaptic cleft decreases so rapidly that a receptor channel has very few chances of being reopened a second time by binding again two acetylcholine molecules.

Determination of the value of the total synaptic current, I_{ACh}, at steady state is:

$$I_{ACh} = N p_o i_{ACh},$$

where N is the number of nicotinic channels in the membrane, i_{ACh} is the unitary current, and p_o is the open-state probability of the channel and depends on

the acetylcholine concentration and on the receptor channel opening (β) and closing (α) rate constants.

In the model below we have the following rate constants:

$$\mathrm{R} \underset{k_{-1}}{\overset{k_1}{\rightleftharpoons}} \mathrm{AR} \underset{k_{-2}}{\overset{k_2}{\rightleftharpoons}} \mathrm{A_2R} \underset{\alpha}{\overset{\beta}{\rightleftharpoons}} \mathrm{A_2R^*}$$

where α and β are the closing and opening rate constants of the channel. The channel's probability of being in the open state at steady state (see Appendix 4.3) is then:

$$p_o = \beta'/(\beta' + \alpha),$$

where β' is the apparent opening rate constant of the channel, which depends on β but also on the rate constants of the preceding stages k_1 and $k_2(\beta' = f[\mathrm{ACh}])$. Thus, if α is short (i.e. the mean open time τ_o is long (because $\tau_o = 1/\alpha$)), then p_o is high (it approaches its maximum value), and I_{ACh} is large. The falling phase of

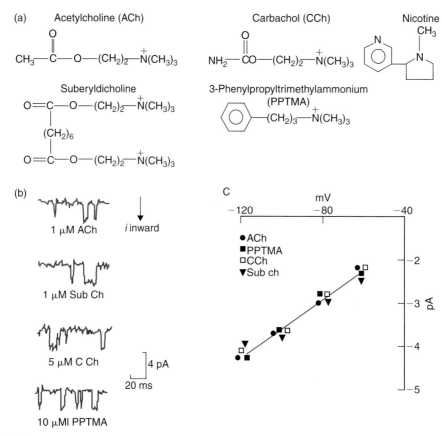

FIGURE 8.22   Unitary currents evoked by nicotinic agonists.
(a) Structure of the nicotinic agonists tested. (b) Inward unitary currents evoked by different nicotinic agonists
($V_m = -80$ mV) recorded in patch clamp (outside-out configuration) from isolated rat myotubes. Solutions (in
mM): intracellular 150 KCl, 5 Na_2EGTA, 0.5 $CaCl_2$; extracellular 135 NaCl, 5.4 KCl. (c) i/V curves built from
results similar to those shown in (b), but at different membrane potentials, are completely superimposable. The
slope of each curve gives a unitary conductance γ of approximately 34 pS. Adapted from Gardner P, Ogden DC,
Colquhoun D (1984) Conductances of single ion channels opened by nicotinic agonists are indistinguishable.
Nature **289**, 160–163, with permission.

the total current is exponential, with a time constant
equal to τ_o.

Let us assume that at a time t a certain number of ionic
channels are opened more or less synchronously (this is
the case of miniature currents). Because acetylcholine
disappears very rapidly from the synaptic cleft, a channel has very few chances of reopening. Each one of the
open channels has an opening duration t_o. We have seen
that the duration t_o during which each channel remains
open can be described by an exponential distribution
(see **Figure 8.16**). **Figure 8.20e** shows that the total current crossing N channels decreases exponentially with a
time constant equal to τ_o.

In conclusion, the falling phase of the total current is
not due to the progressive disappearance of acetylcholine from the synaptic cleft (because it disappears
with a time constant of the order of microseconds) but
depends only on τ_o, an intrinsic property of the channel.

8.6   NICOTINIC TRANSMISSION PHARMACOLOGY

8.6.1   Nicotinic agonists

Nicotinic receptor agonists (see Appendix 6.1) bind to
the same receptor site as acetylcholine and favour the
conformational changes of the protein towards the open
state. These agonists are, for example, suberyldicholine,
carbachol and PTMA (phenyl trimethyl ammonium)
(**Figure 8.22a**). The application of one of these on a
patch of muscle membrane (outside-out configuration)
leads to the onset of a current whose amplitude at each
membrane potential tested is equal to the current
evoked by acetylcholine. However, the duration of the
openings of the channel depends on the agonist used
(**Figures 8.22b, c**).

8.6.2 Competitive nicotinic antagonists

Competitive antagonists (see Appendix 6.1) bind to the same receptor site as acetylcholine but *do not favour* its conformational change towards the open state. By binding to the acetylcholine receptor sites, competitive antagonists prevent acetylcholine from binding to its receptor sites and activating the nAChR. They decrease the number of sites available to acetylcholine and, therefore, decrease or completely block (depending on the dose used) the nicotinic cholinergic response. A distinction is made between competitive antagonists whose effect is reversible ((+)tubocurarine) and those whose effect is irreversible (DDF).

The application of (+)tubocurarine on a patch of muscle membrane in the outside-out recording configuration and in the presence of a low dose of acetylcholine causes a drop in the opening frequency of the channels. This occurs because (+)tubocurarine reduces the number of receptor sites available for acetylcholine. It should be noted, however, that the amplitude of the unitary current i_{Ach} evoked in the presence or absence of (+)tubocurarine is identical. The application of (+)tubocurarine on an isolated nerve–muscle preparation induces a reduction in the amplitude of the spontaneous miniature currents (currents evoked by the endogenous and spontaneous liberation of acetylcholine). The effect of (+)tubocurarine can be reversed by elevating the concentration of acetylcholine applied or released. Since the binding of (+)tubocurarine to the nicotinic receptor is reversible, increasing the acetylcholine concentration will increase the probability that the receptor sites are occupied by acetylcholine.

The binding of α-bungarotoxin (venom from the snake *Bungarus multicinctus*) to the nicotinic receptor of the neuromuscular junction is very stable. For this reason, this toxin is used as a marker of acetylcholine receptor sites in this preparation. This labelling permits localizing and counting of these receptors. Labelled α-bungarotoxin also allows identification of the receptor during its purification process (see Figure 8.4).

8.6.3 Channel blockers

Channel blockers are substances that bind to the aqueous pore of the open receptor, preventing the passage of cations through it. Among these substances are procaine and its derivatives (QX 222, lidocaine, benzocaine), and also histrionicotoxin and chlorpromazine.

Application of acetylcholine in the presence of benzocaine to a muscle membrane patch in the outside-out recording configuration evokes the onset of opening bursts. These bursts of unitary currents are due to the numerous fluctuations of the receptor between its

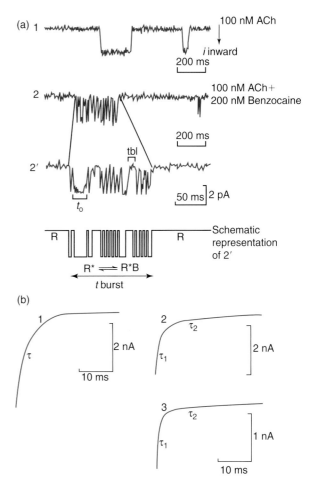

FIGURE 8.23 Benzocaine effect on the time course of ACh evoked unitary and miniature nicotinic currents.
(a) Patch clamp recording (cell-attached configuration) of nicotinic unitary currents evoked by the application of ACh (i_{ACh}). (1) In the presence of 100 nM of ACh, the channels open for a mean duration of $\tau_o = 19$ ms ($V_m = -110$ mV). (2) In the presence of 100 nM of ACh + 200 μM of benzocaine, bursts of openings ($V_m = -130$ mV) are recorded. (2') The same recording as (2) but with a different timescale and with an added diagram of the openings and closings of the channel. t_o, time during which the channel stays open; t_{bl}, time during which the channel is blocked by benzocaine; t_{burst} duration of a burst of openings. The histograms of t_o and of t_{bl} are described by a single exponential with the following average values: $\tau_o = 2.8$ ms, and $\tau_{bl} = 3.5$ ms (extrajunctional muscle membrane of 4- to 6-week-old muscle cells).
(b) Two-electrode voltage clamp recording of miniature nicotinic currents. Each curve corresponds to the average of 8 to 14 miniature currents ($V_m = -100$ mV). (1) In the absence of benzocaine, the miniature currents reach their maximum in approximately 1 ms. Their decrement is described by a single exponential with a time constant $\tau = 3.8$ ms. (2) In the presence of extracellular benzocaine (300 μM, 15 min), the peak amplitude of miniature currents decreases, and the falling phase is described by two exponentials with time constants $\tau_1 = 1.0$ ms and $\tau_2 = 7.6$ ms. (3) In the presence of a higher concentration of benzocaine (500 μM, 17 min), the amplitude of the miniature current peak is further diminished and the time constants of the falling phase become $\tau_1 = 0.7$ ms and $\tau_2 = 11.6$ ms (frog cutaneous pectoris muscle). Parts (a) and (b) adapted from Ogden DC, Siegelbaum SA, Colquhoun D (1981) Block of acetylcholine-activated ion channels by an uncharged local anesthetic. *Nature* **289**, 596–598, with permission.

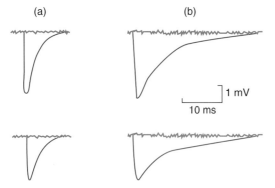

FIGURE 8.24 Effect of the acetylcholinesterase inhibitor prostigmine on the duration of miniature currents.
Recordings (**a**) in the absence of and (**b**) in the presence of prostigmine (10^{-6} g ml^{-1}), showing that prostigmine augments the duration of the miniature current. Adapted from Katz B and Miledi R (1973) The binding of acetylcholine to receptors and its removal from the synaptic cleft. *J. Physiol. (Lond.)* **231**, 549–574, with permission.

open and its closed state (**Figure 8.23a**). The following model describes this process:

$$R \rightleftharpoons R^* \rightleftharpoons R^*B,$$

where R represents the nicotinic receptor in its closed state, R* the nicotinic receptor in its open state, and R*B the nicotinic receptor in the open but blocked state. In the R* state, the cations cross the aqueous pore, while in the states R and R*B the ions cannot cross it.

In the presence of benzocaine, a change in the kinetics of the falling phase of the current (**Figure 8.23b**) is observed during the recording of spontaneous miniature currents of an isolated nerve–muscle preparation ($V_m = -100$ mV). In the absence of benzocaine, the falling phase is described by a single exponential with a time constant of $\tau = 3.8$ ms. In the presence of benzocaine, the falling phase is described by two exponentials with time constants $\tau_1 = 1.0$ ms and $\tau_2 = 7.6$ ms. The explanation of this effect is that the channels opened by acetylcholine are very rapidly blocked by benzocaine, which quickly blocks the unitary current. Thus, one observes a fast initial decrement of the miniature current (with a shorter time constant than in the absence of benzocaine). The channels then reopen and reblock repeatedly. This increases the duration of the miniature current and one observes a second slower decrementing phase (with time constant τ_2).

8.6.4 Acetylcholinesterase inhibitors

These inhibitors have a reversible effect, as in the case of prostigmine, or an irreversible effect, as in the case of DFP (difluorophosphate). The application of prostigmine to an isolated nerve–muscle preparation significantly

increases the miniature current duration (**Figure 8.24**). In the presence of prostigmine, acetylcholine molecules degrade much more slowly, and thus are able to bind repeatedly and trigger the reopening of nicotinic receptors. This repeated binding considerably increases the duration of the miniature current falling phase. As we have already seen (see Section 8.5), the miniature current time constant in the absence of acetylcholinesterase inhibitors reflects the nicotinic receptor average open time. However, the average open time of the nicotinic receptor is clearly not the same in the presence of prostigmine.

8.7 SUMMARY

Upon release of acetylcholine in the synaptic cleft of the neuromuscular junction, two molecules of acetylcholine bind to each postsynaptic nAChRs, at specific sites. Upon ACh binding, each nAChR undergoes fast activation leading to an open-channel state, and a slow desensitization reaction leading to a closed-channel state refractory to activation. nAChR opening occurs in the millisecond range, fast desensitization in the 0.1 s range and slow desensitization in the minute range.

To which type(s) of postsynaptic receptor do the molecules of acetylcholine bind?

At the neuromuscular junction, ACh binds to the muscle-type nAChR. Muscle nAChRs have a fixed composition $[\alpha 1]_2[\beta 1][\delta][\gamma$ or $\varepsilon]$ in vertebrates. Each subunit contains a large N-terminal hydrophilic domain exposed to the synaptic cleft, followed by three transmembrane segments (M1 to M3), a large intracellular loop and a C-terminal transmembrane segment (M4). Acetylcholine binding sites are located at the interface between α and non-α-subunits in the N-terminal regions. The ion channel is lined by the M2 segment from each of the five subunits.

How is the binding of acetylcholine transduced into a depolarization of the postsynaptic membrane?

Upon binding of two molecules of ACh, nAChRs undergo a fast conformational change to a state where the pore is open. The pore is selectively permeable to cations. At -90 to -80 mV, the resting potential of muscle fibres, the electrochemical gradient ($Vm - E_{ion}$) for Na$^+$ and Ca^{2+} ions is large whereas that for K$^+$ ions is small. As a result, there is a net inward flux of cations through open nAChRs measured as a transient inward current. This inward current transiently depolarizes the muscular membrane and thus evokes the endplate potential (EPP). EPP is transient (it lasts several milliseconds) owing to the rapid closing of nAChRs and the rapid elimination of ACh from the synaptic cleft by degradation by acetylcholinesterases. In physiological conditions, desensitization of nAChRs does not play a major role.

Why is the postsynaptic depolarization restricted to the postsynaptic membrane?

nAChRs are restricted to the membrane of the postsynaptic element where they are anchored by cytoskeletal proteins. Therefore, the ACh-evoked postsynaptic inward current is triggered at the level of the postsynaptic element only. It is then passively conducted along the postsynaptic muscular membrane. This conduction is decremental (see **Chapter 15**) such as that at several millimetres from the junction, the EPP amplitude is nearly close to zero.

The function of the nicotinic receptor is to ensure rapid synaptic transmission. This is achieved by converting the binding of two acetylcholine molecules into a rapid and transient increase in cationic permeability. This permeability increase is made possible by conformational changes of the receptor channel: it transiently switches from the state in which the channel is closed into a state in which the channel is open. The nicotinic acetylcholine receptor also presents allosteric binding sites, topographically distinct from the neurotransmitter binding site, to which a variety of pharmacological agents and physiological ligands can bind. In doing so they regulate the transitions between the different states of the nAChR.

APPENDIX 8.1 THE NEURONAL NICOTINIC RECEPTORS

To identify the presence of cholinergic nicotinic synapses in the central nervous system, researchers recorded responses of central neurons to the local application of nicotinic agonists. **Figure A8.1a** shows a current-clamp recording from a neuron in a rat brain slice (interneuron of the CA1 region of the hippocampus, see **Figure 19.5**), illustrating the membrane potential change in response to ACh application. ACh evoked a depolarization sufficient to bring the membrane potential of the neuron to threshold for action potential firing. When this interneurone was voltage clamped to $-70\,mV$, ACh evoked an inward current (**Figure A8.1b**). These data demonstrate that ACh can excite inhibitory interneurones of the hippocampal formation by activation of an inward current at near-resting membrane potentials. This effect of ACh is by direct action on the interneurone (postsynaptic effect) because it was not inhibited by blockers of synaptic transmission, including tetrodotoxin, glutamate receptor antagonists (CNQX and APV), and GABA$_A$ receptor antagonist (bicuculline). The inward current response to ACh and the resulting excitation suggested the involvement of cationic receptor channels such as nAChRs. To further investigate this, ACh was applied to interneurones voltage clamped to different membrane potentials to obtain a current-voltage relationship (**Figure 8.1c**). The membrane potential at which ACh produced zero net current flow was between -10 and $0\,mV$, a reversal potential consistent with the activation of a non-selective cation conductance by ACh. To investigate the pharmacology of the response to ACh, nicotinic receptor-selective ligands were utilized. Application of the nAChR agonist nicotine mimicked the effect of ACh in activating inward current responses (**Figure 8.1d**). Mecamylamine, a nAChR antagonist caused a marked inhibition of the inward current response to ACh (**Figure 8.1e**). The response to ACh was also inhibited by α-bungarotoxin (**Figure 8.1f**). In contrast the muscarinic receptor antagonist atropine ($10\,\mu M$) had no effect on the inward current response to ACh (not shown).

To identify whether central nAChRs are permeable Ca^{2+} ions, single-channel currents evoked by ACh in pure external Ca^{2+} medium are recorded in outside-out patches from freshly dissociated neurons of the rat central nervous system (habenula nucleus) (**Figure A8.2a**). The presence of a current shows that Ca^{2+} permeates these neuronal nAChRs channels. Similarly, in whole-cell recordings, ACh also evokes an inward current when all external cations are replaced with Ca^{2+} (**Figure A8.2b**). When habenula neurons are loaded with FURA-2 (see Appendix 5.1), an increase of [Ca^{2+}]$_i$ up to the micromolar range is observed upon acetylcholine application. This increase is reversibly abolished when Ca^{2+} is removed from the perfusion medium. To exclude a possible involvement of Ca^{2+} entry through voltage-gated Ca^{2+} channels opened by the depolarization subsequent to the activation of nAChRs, the membrane is clamped at $-60\,mV$. In this condition, application of high K$^+$ ($140\,mM$) external medium, which would depolarize a poorly clamped neuronal membrane, yields no detectable increase of [Ca^{2+}]$_i$. In contrast, application of nicotine evokes an inward whole-cell current and a concomitant rapid increase of [Ca^{2+}]$_i$ up to the micromolar range (**Figure A8.2c**).

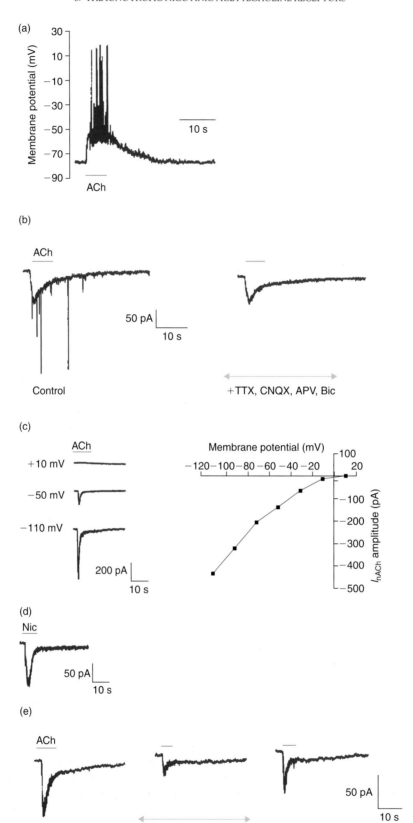

FIGURE A8.1

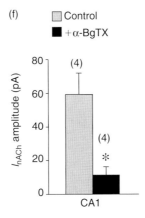

FIGURE A8.1 **Direct excitation of hippocampal interneurones by ACh-activated cation current and pharmacological characterization of the effect of ACh.**
(**a**) Current-clamp recording (using potassium gluconate intracellular solution) from a CA1 interneurone, showing the depolarization and action potential firing in response to ACh (100 μM, 5 s, bars). (**b**) Voltage-clamp recording (CA1 interneurone, cell held at -70 mV) demonstrating that the effect of ACh (100 μM, 5 s, bar) persists in the presence of inhibitors of synaptic transmission (TTX, CNQX, APV, Bic). (**c**) Membrane potential was changed from -110 to 10 mV in order to obtain a steady-state current voltage relationship for the effect of ACh (100 μM, 5 s, bar). Example traces at membrane potentials -110, -50 and 10 mV are shown (left). (**d**) Effect of the nAChR agonist nicotine (100 μM, 5 s) on a CA1 interneurone voltage clamped to -70 mV. (**e**) Effect of nAChR antagonists on the responses of interneurones to ACh. Mecamylamine (Mec, 10 μM) reversibly inhibited the response to ACh (100 μM, 5 s, bars) in this voltage-clamped CA1 interneurone by 75%. (**f**) Pre-treatment of slices with α-bungarotoxin (α-BgTx, 100 μM, 10 min minimum) significantly reduced ACh responses of CA1 interneurones; control slices were matched from the same rats. The response to ACh averaged -59 ± 13 pA (n = 4) in control slices and -11 ± 5 pA (n = 4) in α-BgTX-treated slices, corresponding to approximately 80% inhibition, *P < 0 05. From Jones S and Yakel JL (1997) Functional nicotinic ACh receptors on interneurons of the rat hippocampus. *J. Physiol.* **504**, 603–610, with permission.

nAChRs have now been identified in many presynaptic and postsynaptic membranes of CNS synapses. Neuronal nAChRs form a heterogeneous family of subtypes form by five subunits encoded by nine $\alpha(\alpha_2 - \alpha_{10})$ and three $\beta(\beta_2 - \beta_4)$ subunit genes. Two main subfamilies of neuronal nAChRs have been identified so far (**Figure A8.3**):

- Neuronal homomeric nAChRs which bind α-bungarotoxin and consist of α_7, α_8, α_9 or α_{10} subunits. They can be homopentameric (α_7, α_8 and α_9) or heteropentameric (α_7 and α_8 or α_9 and α_{10}). Their relative Ca/Na permeability (pCa/Na) is around 10, showing that these receptors are permeable to Ca^{2+} ions as determined by using current and fluorescence measurements (for memory pCa/Na of muscle nAChRs is around 0.2). For native neuronal nAChRs from α_7 gene product, another property is the very rapid desensitization of the whole-cell current. For example, in neurons of the chick peripheral nervous system (ciliary ganglion), fast perfusion of nicotine evokes a large, quickly desensitizing current, strongly depressed by α-bungarotoxin (**Figure A8.4**). The ligand-binding sites on the homopentameric receptors are present at the interface formed by opposite sides of the same subunit and it is thought that they have five identical ligand-binding sites per receptor molecule (**Figure A8.3**, right).

- Neuronal heteromeric nAChRs which do not bind α-bungarotoxin. They are only heteropentameric and consist of α and β subunits. Their relative Ca/Na permeability is around 2. The heteropentameric receptors have two binding sites per receptor molecule located at the interface between an α and β subunit. These receptors usually comprise: (i) two α subunits carrying the principal component of the ACh-binding site (α_2, α_3, α_4 or α_6); (ii) two non-α subunits carrying the complementary component of the ACh-binding site (β_2 or β_4); and (iii) a fifth subunit that does not participate in ACh binding (α_5, β_3 but also β_2 or β_4) (**Figure A8.3**, right).

The identification of the subtype composition of native neuronal nAChRs is currently based on a combination of technical approaches, such as *in situ* hybridization, single-cell PCR, pharmacology and electrophysiology.

Central nicotinic receptors are mostly localized in the membrane of axon terminals (presynaptic nAChRs) where, via the increase of the intracellular concentration of Ca^{2+} ions, they can enhance the release of the

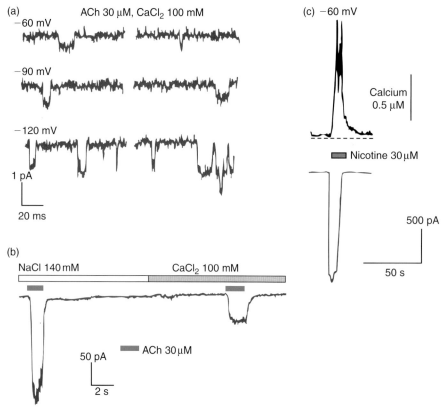

FIGURE A8.2   Ca²⁺ permeates neuronal nAChR channels.
(a) Single-channel currents evoked by ACh in outside-out patches of habenula neurons at three holding potentials (-60, -90, -120 mV) in the presence of a pure $CaCl_2$ external medium. (b) Whole-cell currents evoked by ACh in control (140 mM NaCl, 1 mM $CaCl_2$) and pure $CaCl_2$ medium. (c) Increase in $[Ca^{2+}]i$ (top trace) and whole-cell current (bottom trace) evoked by a 10 s application of nicotine (30 μM) at $VH = -60$ mV. Adapted from Mulle C, Choquet D, Korn H, Changeux JP (1992) Calcium influx through nicotinic receptor in rat central neurons: its relevance to cellular regulation. *Neuron* **8**, 135–143, with permission.

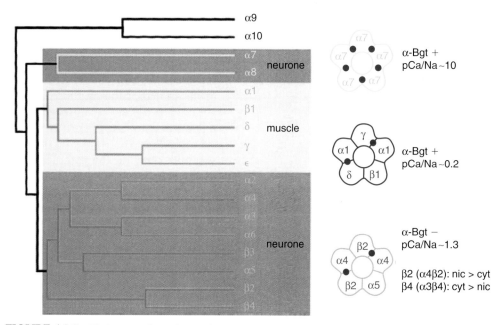

FIGURE A8.3   Phylogeny of nAChRs and properties of receptors.
Red dots indicate the agonist binding sites. pCa/Na, relative calcium/sodium permeability. From Le Novere N, Corringer PJ, Changeux JP (2002) The diversity of subunit composition in nAChRs: evolutionary origins, physiologic and pharmacologic consequences. *J. Neurobiol.* **53**, 447–456, with permission.

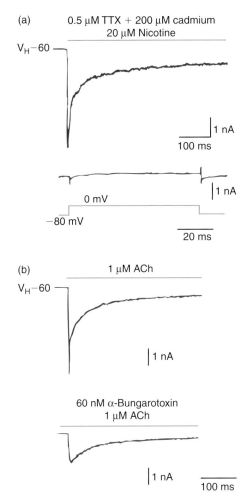

FIGURE A8.4   The neuronal nAChR-mediated whole-cell current sensitive to α-bungarotoxin rapidly desensitizes.
(**a**) Rapidly-decaying whole-cell current evoked by nicotine in the presence of blockers of Na^+ and Ca^{2+} voltage-gated channels (top trace). The bottom traces show the absence of voltage-activated currents in response to a depolarizing pulse to 0 mV in these conditions (TTX + Cd^{2+}). (**b**) The whole-cell current evoked by ACh is sensitive to ʃ-bungarotoxin (incubated for 2 h before recording). Adapted from Zhang ZW, Vijayaraghavan S, Berg DK (1994) Neuronal acetylcholine receptors that bind α-bungarotoxin with high-affinity function as ligand-gated ion channels. *Neuron* **12**, 167–177, with permission.

transmitter present in the vesicles of the axon terminal. Only rare cases of central synapses with a postsynaptic localization of nAChRs have been reported. In that case nAChRs participate in fast excitatory synaptic transmission (**Figure A8.1**). In contrast to the NMDA subtype of glutamate receptors that provide a current with an accompanying Ca^{2+} influx only when the membrane is depolarized enough to relieve the Mg^{2+} block (see Chapter 10), the nAChRs do the same but at negative membrane potentials, in a situation of strong driving forces for Na^+ and Ca^{2+} ions.

FURTHER READING

Abakas MH, Kaufmann C, Archdeacon P, Karlin A (1995) Identification of acetylcholine receptor channel-lining residues in the entire M2 segment of the α subunit. *Neuron* **13**, 919–927.

Changeux JP and Edelstein SJ (1998) Allosteric receptors after 30 years. *Neuron* **21**, 959–980.

Corringer PJ, Bertrand S, Bohler S *et al.* (1998) Critical elements determining diversity in agonist binding and desensitization of neuronal nicotinic acetylcholine receptors. *J. Neurosci.* **15**, 648–657.

Couturier S, Bertrand D, Matter JM *et al.* (1990) A neuronal nicotinic acetylcholine receptor subunit (α_7) is developmentally regulated and forms a homo-oligo-meric channel blocked by α-BTX. *Neuron* **5**, 847–856.

Czajikowski C and Karlin A (1995) Structure of the nicotinic acetylcholine-binding site: identification of acidic residues in the δ subunit with 0.9 nm of the α subunit-binding site disulfide. *J. Biol. Chem.* **270**, 3160–3164.

Dani JA and Bertrand D (2007) Nicotinic acetylcholine receptors and nicotinic cholinergic mechanisms of the central nervous system. *Annu. Rev. Pharmacol. Toxicol.* **47**, 20.1–20.31.

Jones S, Sudweeks S, Yakel JL (1999) Nicotinic receptors in the brain: correlating physiology with function. *Trends Neurosci.* **22**, 555–61.

Katz B and Miledi R (1973) The binding of acetylcholine to receptors and its removal from the synaptic cleft. *J. Physiol. (Lond.)* **231**, 549–574.

Leonard RJ, Labarca CG, Charnet P *et al.* (1988) Evidence that the M2 membrane-spanning region lines the ion channel pore of the nicotinic receptor. *Science* **242**, 1578–1581.

Monod J, Wyman J, Changeux JP (1965) On the nature of allosteric transitions: a plausible model. *J. Mol. Biol.* **12**, 88–118.

Murray N, Zheng YC, Mandel G *et al.* (1995) A single site on the ε-subunit is responsible for the change in ACh receptor channel conductance during skeletal muscle development. *Neuron* **14**, 865–870.

Roerig B, Nelson DA, Katz LC (1997) Fast synaptic signalling by nicotinic acetylcholine and serotonin 5-HT3 receptors in developing visual cortex. *J. Neurosci.* **17**, 8353–8362.

Unwin N (2003) Structure and action of the nicotinic acetylcholine receptor explored by electron microscopy. *FEBS Letters* **555**, 91–95.

9

The ionotropic GABA$_A$ receptor

K. Krnjevic and S. Schwartz observed in 1967 that γ-aminobutyric acid (GABA), applied by microiontophoresis on intracellularly recorded cortical neurons, evokes a hyperpolarization of the neuronal membrane which has properties similar to synaptically evoked hyperpolarization. This led to the hypothesis that GABA mediates inhibitory synaptic transmission in the *adult* vertebrate central nervous system.

GABA released by presynaptic terminals activates two main types of receptors: (i) ionotropic receptors called GABA$_A$ and GABA$_C$, which have different pharmacological properties (GABA$_C$ receptors mainly present in the retina, will not be explained here); and (ii) metabotropic receptors, i.e. receptors coupled to GTP-binding proteins called GABA$_B$ receptors.

The aim of the present chapter is to study how a GABA$_A$ receptor converts the binding of two GABA molecules to a rapid and transient change of the membrane potential in the *adult* vertebrate central nervous system. GABA$_A$ receptors of the developing brain will be explained in Chapter 20.

9.1 OBSERVATIONS AND QUESTIONS

In the rat hippocampus, in response to a single spike evoked in a presynaptic GABAergic interneuron, a transient hyperpolarization of the membrane mediated by GABA$_A$ receptors is recorded in the postsynaptic pyramidal cell (in the presence of blockers of GABA$_B$ receptors). This hyperpolarization of synaptic origin is called the 'single-spike inhibitory postsynaptic potential' (single-spike IPSP). This GABA$_A$-mediated IPSP has the following characteristics (**Figure 9.1**). It is:

- totally blocked in the presence of bicuculline thus showing it is mediated by GABA$_A$ receptors;
- potentiated by diazepam (a benzodiazepine, an anxiolytic and anticonvulsant drug);
- potentiated by pentobarbitone sodium (a barbiturate, a sedative and anticonvulsant drug).

These observations raise several questions:

- How does a GABA$_A$ receptor mediate the binding of GABA into a transient hyperpolarization of the membrane; i.e. what are the permeant ions, does GABA induce the entry of negatively charged ions or does it induce the exit of positively charged ions?
- Do benzodiazepines and barbiturates act directly on GABA$_A$ receptors? Are there selective and distinct binding sites on the receptor for each of these drugs? How do they potentiate the hyperpolarizing effect of GABA? Are there other modulators of GABA$_A$ receptors?

9.2 GABA$_A$ RECEPTORS ARE HETERO-OLIGOMERIC PROTEINS WITH A STRUCTURAL HETEROGENEITY

9.2.1 The diversity of GABA$_A$ receptor subunits

The first subunits to be identified were α_1 and β_1. A GABA$_A$ receptor purified on affinity columns from calf, pig, rat or chick brain, dissociates in the presence of a detergent into two major subunits which migrate

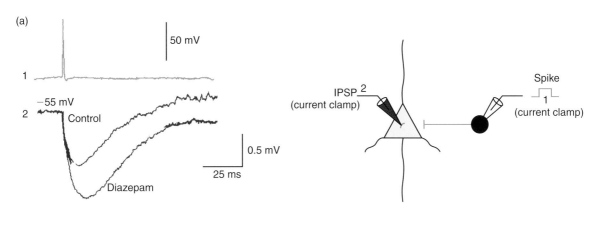

(a)

50 mV

−55 mV

Control

Diazepam

0.5 mV

25 ms

IPSP 2
(current clamp)

Spike

1

(current clamp)

(b)

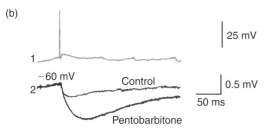

25 mV

−60 mV   Control

0.5 mV

50 ms

Pentobarbitone

FIGURE 9.1 GABA$_A$ receptor-mediated IPSPs recorded in pyramidal cells of the hippocampus and two examples of their modulation.
The activity of a pair of connected neurons, a presynaptic GABAergic interneuron and postsynaptic pyramidal cell, is recorded with two intracellular electrodes. **(a)** A single spike is triggered in the presynaptic GABAergic interneuron in response to a current pulse (1). It evokes an IPSP (averaged single-spike IPSP; 2 control) in the postsynaptic pyramidal neuron. This IPSP is increased in amplitude and duration in the presence of a benzodiazepine (2; diazepam 1 μM). **(b)** Similar experiment in a different cell pair with pentobarbitone sodium (250 μM), a barbiturate. Adapted from Pawelzik H, Bannister AP, Deuchars J *et al.* (1999) Modulation of bistratified cell IPSPs and basket cell IPSP by pentobarbitone sodium, diazepam and Zn^{2+}: dual recordings in slices of adult hippocampus. *Eur. J. Neurosci.* **11**, 3552–3564, with permission.

on polyacrylamide gels and have apparent molecular masses of 53 kD (α_1-subunit) and 56 kD (β_1-subunit). Electrophoretic studies based on receptors purified from different regions of the central nervous system show the presence of multiple bands corresponding to apparent molecular masses of 48–53 kD and of 55–57 kD. This strongly suggests the occurrence of several isoforms of α and β subunits.

Demonstration of the structural heterogeneity of the GABA$_A$ receptor came from cloning of cDNAs for different α and β subunits. The diversity of GABA$_A$ receptor subunits (α, β, γ, δ, ε, π, ρ) and the existence of different isoforms for the different subunits was then revealed. All subunits are similar in size, contain about 450–550 amino acids and are strongly conserved among species (**Figure 9.2a**). A high percentage of sequential identity (70–80%) is found between sub-unit isoforms (between α and between β isoforms for example). Sequential identity is also found, but to a lesser extent (30–40%), between subunit families. To date, nineteen isoforms of mammalian GABA$_A$ receptor subunits

have been cloned: α_{1-6}, β_{1-3}, γ_{1-3}, δ, ε, π, θ, ρ_{1-3}. This multiplicity of subunits provides a daunting number of potential subunit combinations. The common elements of the subunit structure include (**Figure 9.2b**):

- a large N-terminal hydrophilic domain exposed to the synaptic cleft (extracellular);
- then four hydrophobic segments named M1 to M4, each composed of approximately 20 amino acids which form four putative membrane-spanning segments.

The segment M2 of each of the subunits composing the GABA$_A$ receptor is thought to line the channel (as for the nAChR) and to contribute to ion selectivity and transport. Apparently, a small number of amino acids within the M2 sequence is responsible for anionic versus cationic permeability (see Section 8.3.2 and **Figure 8.15**).

There is also a large, poorly conserved hydrophilic domain separating the M3 and M4 segments, located in the cytoplasm, which contains putative phosphorylation sites.

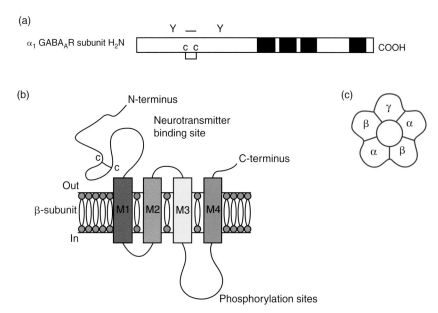

FIGURE 9.2   The GABA$_A$ receptor subunits.
(a) Schematic illustration of the primary sequence of the α1 subunit from the GABA$_A$R; C-C, Cys-Cys bridge; Y, oligosaccharide groups. **(b)** Model of transmembrane organization of GABA$_A$ receptor subunits. The large hydrophilic N-terminal domain is exposed to the synaptic cleft and carries the neurotransmitter site and glycosylation sites. The four M segments span the membrane. The hydrophilic domain separating M3 and M4 faces the cytoplasm and contains phosphorylation sites. The C-terminus is extracellular. **(c)** Schematic representation of the oligomeric organization of a GABA$_A$ receptor. The hypothetical pentameric GABA$_A$R is formed by two α subunits, two β subunits and one γ subunit. Part (a) adapted from Arias HR (2000) Localization of agonist and competitive antagonist binding sites on nicotinic acetylcholine receptors. *Neurochemistry International* **36**, 595–645. Part (b) from Lovinger DM (1997) Alcohols and neurotransmitter gated ion channels: past, present and future. *Naunyn-Schmiederberg's Arch. Pharmacol.* **356**, 267–282, with permission.

9.2.2 Subunit composition of native GABA$_A$ receptors and their binding characteristics

To form a GABA$_A$ receptor, with the large number of known subunits taken five at a time, thousands of combinations are possible. Two approaches are currently used to elucidate which subunit combinations exist:

- a comparative study of the functional properties of receptors expressed in oocytes or in transfected mammalian cells from known combinations of cloned subunits (with the restriction that *Xenopus* oocyte does not automatically assemble a channel composed of all injected subunits);
- a comparative study of the distribution of the various subunit mRNAs in the brain using the *in situ* hybridization technique (see Appendix 6.2).

Surprisingly, the transient expression in transfected cells of identical α or β subunits gives functional homomeric GABA$_A$ receptors; i.e. receptors which induce a current in the presence of GABA. This current is blocked by GABA$_A$ antagonists and potentiated by barbiturates but is unaffected by benzodiazepines. These properties can be attributed to the conserved structural features of all the subunits. However, these channels resulting from expression of single subunits are assembled inefficiently (are rare and slightly detectable) and it is unlikely that native receptors are formed from identical subunits. Expression of a γ-subunit together with an α- and a β-subunit in transfected cells gives rise to the expression of a GABA$_A$ receptor with all the features of the homomeric receptor with in addition the sensitivity to benzodiazepines. This does not necessarily imply that the receptor site for benzodiazepines is situated on the γ-subunit, but that expression of the latter is required for the action of benzodiazepines. The major receptor subtype of the GABA$_A$ receptor in adults consists of α_1, β_2 and γ_2 subunits and the most likely stoichiometry is two α subunits, two β subunits and one γ subunit ($\beta\,\alpha\,\beta\,\alpha\,\gamma$) (**Figure 9.2c**). In summary, functional receptors can consist of α and β subunits alone; however, native receptors include an additional 'modulatory' subunit as γ, δ, ε, π or δ.

In contrast, ρ-subunits seem not to combine with other classes of GABA$_A$ receptor subunits. They are capable of forming functional homo-oligomeric receptors. Such receptor assemblies derived from various isoforms of the ρ-subunit are present in the retina and classified as GABA$_C$ receptors, a specialized set of GABA$_A$ receptors, permeable to chloride ions, selectively activated by a GABA analogue, the cis-4-aminocrotonic acid (CACA) and insensitive to bicuculline, benzodiazepines and barbiturates.

9.3 BINDING OF TWO GABA MOLECULES LEADS TO A CONFORMATIONAL CHANGE OF THE GABA$_A$ RECEPTOR INTO AN OPEN STATE; THE GABA$_A$ RECEPTOR DESENSITIZES

9.3.1 GABA binding site

The amino acids identified by site-directed mutagenesis to affect channel activation by GABA are in the β-subunit: glutamate (E) 155 and nearby residues tyrosine (Y) 157 and threonine (T) 160. For example, cysteine (C) substitution of $\beta_2 Glu^{155}$ alters both channel-gating properties and impairs agonist binding as it results in spontaneously open GABA$_A$ channels. A model of the GABA$_A$R agonist-binding site predicts that β_2 Glu^{155} interacts with the positively charged moiety of GABA.

In the α_1-subunit, the mutation of phenylalanine (F) 64 to leucine (L) impairs activation of the GABA channel indicating a role for this α-subunit residue in GABA binding. Therefore, identified domains of the β-subunit and of the neighbouring α-subunit contribute to the GABA-binding site. It is thought that most $\alpha\beta\gamma$ receptors are pentameric with a stoichiometry 2α, 2β and 1γ, which is consistent with data indicating that GABA sites are located at the interface between α and β subunits and that there are two GABA sites per receptor (see below). GABA$_A$ receptors have a relatively low affinity for GABA, of the order of 10–20 μM.

9.3.2 Evidence for the binding of two GABA molecules

Analysis of dose–response curves suggests the binding of two GABA molecules prior to opening of the channel. The response studied, the peak amplitude of the total current I_{GABA} evoked by GABA in whole-cell patch clamp recording, is proportional to the square of the dose of GABA (but only at low doses of GABA):

$$I_{GABA} = f \, [GABA]^2.$$

At very low doses of GABA, when receptor desensitization is negligible, it seems that upon binding of two GABA molecules to the receptor, the conformational change of the receptor channel to an open state is favoured. These observations can be accounted for by the following model:

$$2G + R \underset{k_{-1}}{\overset{k_1}{\rightleftharpoons}} G + GR \underset{k_{-2}}{\overset{k_2}{\rightleftharpoons}} G_2R \underset{\alpha}{\overset{\beta}{\rightleftharpoons}} G_2R^*$$

where G is GABA; R is the GABA$_A$ receptor in the closed configuration; GR or G$_2$R is the mono- or doubly liganded GABA$_A$ receptor in the closed configuration; and G$_2$R* is the doubly liganded GABA$_A$ receptor in the open configuration.

9.3.3 The GABA$_A$ channel is selectively permeable to Cl$^-$ ions

The reversal potential of the GABA current varies with the Cl$^-$ equilibrium potential, E_{Cl}

The ionic selectivity of the channel is studied in outside-out patch clamp recordings from cultured spinal neurons. This patch clamp configuration allows control of the membrane potential as well as the composition of the intracellular fluid. When the intracellular and extracellular solutions contain the same Cl$^-$ concentration (145 mM), the unitary current evoked by GABA reverses at $E_{rev} = 0$ mV (**Figures 9.3a–c** and **9.5a**). In this case E_{Cl} is also equal to 0 mV:

$$E_{Cl} = -58 \log(145/145) = 0 \text{ mV}.$$

If part of the intracellular Cl$^-$ is replaced with non-permeant anions such as isethionate (HO-CH$_2$-CH$_2$-SO$_3$), for a 10-fold change in intracellular Cl$^-$ concentration a shift in the reversal potential of approximately 56 mV is observed (**Figure 9.3d**). This value approaches very closely that of 58 mV predicted by the Nernst equation for E_{Cl} at 20°C:

$$E_{Cl} = -58 \log(145/14.5) = -58 \text{ mV}.$$

Finally, changes in extracellular Na$^+$ or K$^+$ concentration have very little effect on the reversal potential of the GABA$_A$ response. Taken together, these results demonstrate that the GABA$_A$ channel is selectively permeable to Cl$^-$.

In physiological extracellular and intracellular solutions, the GABA$_A$ current recorded in isolated spinal neurons reverses at -60 mV

Using the technique of patch clamp recording one can record the unitary currents (i_{GABA}) across the GABA$_A$ channel (spinal neurons in culture, cell-attached

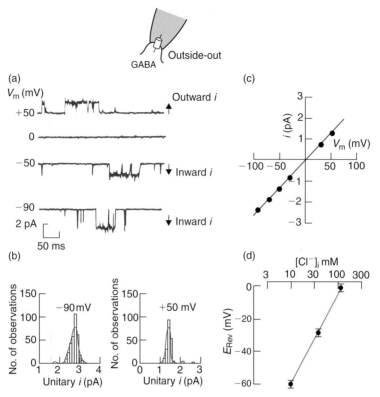

FIGURE 9.3   **Variations of the reversal potential of the GABA$_A$ response as a function of the Cl$^-$ equilibrium potential.**

The single-channel current i flowing across the GABA$_A$ channel is recorded in cultured mouse spinal neurons (outside-out patch-clamp recording; equal concentrations of Cl$^-$ on both sides of the patch: 145 mM). **(a)** In the presence of GABA (10 mM), the single-channel current i is outward at V_m = +50 mV (upward deflection), null at V_m = 0 mV and inward at V_m = −50 mV or −90 mV (downward deflections). **(b)** The distribution of single channel currents i in different patches of membrane held at V_m = −90 mV (left) and +50 mV (right) shows the existence of a single peak of current of −2.70 ± 0.17 pA and 1.48 ± 0.10 pA, respectively. These two values give a single channel conductance γ equal to 30 pS ($\gamma = i/V_m$ as E_{rev} = 0 mV). **(c)** i/V curve obtained by averaging the most frequently observed single-channel currents. It is a straight line according to the equation $i = \gamma(V_m - E_{rev})$. The relationship is linear between V_m = −90 mV and +50 mV and the slope is γ = 30 pS. **(d)** Reversal potential of the GABA$_A$ response (in mV) as a function of the intracellular Cl$^-$ concentration [Cl$^-$]$_i$ (in mM). Each point represents the mean value of E_{rev} from four different cells. Note that, at the three [Cl$^-$]$_i$ tested, E_{rev} (experimental value) is very close to E_{Cl} (calculated by the Nernst equation):

[Cl$^-$]$_i$ (mM)	E_{Cl} (mV)	E_{rev} (mV)
14.5	−58	−56
45	−29	−28
145	0	0

Parts (a)–(c) adapted from Borman J, Hamill OP, Sakmann B (1987) Mechanism of anion permeation through channels gated by glycine and γ-aminobutyric acid in mouse cultured spinal neurones, *J. Physiol. (Lond.)* **385**, 246–286, with permission. Part (d) from Sakmann B, Borman J, Hamill OP (1983) Ion transport by single receptor channels, *Cold Spring Harbor Symposia in Quantitative Biology* **XLVIII**, 247–257, with permission.

configuration). The GABA present in the solution inside the recording pipette (5 μM) evokes outward single-channel currents at −30, 0 and +20 mV (**Figure 9.4a**). The magnitude of the single-channel current increases with depolarization suggesting that the reversal potential for the GABA$_A$ response is negative to −30 mV. The i/V curve, obtained by plotting the unitary current i_{GABA} against the membrane potential V, shows in this experiment a reversal potential of the GABA-induced current around −60 mV (**Figure 9.4**). Thus, at a potential close to the resting membrane potential (−60 mV) the current evoked by GABA is not detectable.

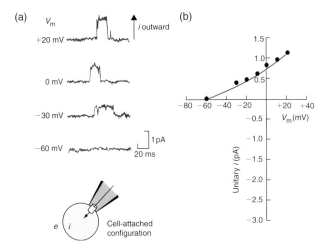

FIGURE 9.4 **Activity of a single GABA_A receptor channel in physiological solutions.**
The single-channel GABA_A current recorded in cell-attached configuration in rat spinal neurons. **(a)** At $V_m = -30, 0$ and $+20$ mV respectively, the GABA present in the patch pipette at a concentration of $10\,\mu M$ elicits an outward current (upward deflection). This current increases with depolarization of the patch. At $V_m = -60$ mV, no current is recorded. **(b)** i/V curve obtained by plotting the amplitude of the recorded unitary current i (pA) against the membrane potential V_m (mV). The intracellular medium is the physiological cytosol and the extracellular or intrapipette solution contains 144.6 mM Cl$^-$. The intracellular Cl$^-$ concentration is estimated at 13 mM, which gives a value of -60 mV for the Cl$^-$ reversal potential: $E_{Cl} = -58 \log (144.6/13) = -60$ mV. As all Na$^+$ ions are replaced by K$^+$ ions in the extracellular solution, the K$^+$ concentration is similar in both solutions, which gives a reversal potential for the K$^+$ current near 0 mV. The membrane potential values indicated in (a) and (b) are evaluated on the basis that the value 0 mV is the potential at which the K$^+$ currents across the K$^+$ channels are zero. From Sakmann B, Bormann J, Hamill OP (1983) Ion transport by single receptor channels, *Cold Spring Harbor Symposia in Quantitative Biology* XLVIII, 247–257, with permission.

At potentials more depolarized than rest (e.g. -30 mV) (**Figure 9.5b**), an outward current is recorded whose magnitude increases with depolarization of the postsynaptic membrane.

As E_{Cl} is close to the resting membrane potential (-60 mV) in physiological intracellular and extracellular solutions, the electrochemical gradient for the Cl$^-$ ions ($V - E_{Cl}$) for $V = E_{Cl} = -60$ mV is close to 0 mV. The net flux of Cl$^-$ ions at a potential close to rest is therefore null or very small: no current is recorded even though the GABA_A channels are open. On the other hand, as the membrane potential depolarizes, the net flux of Cl$^-$ ions becomes inward. An inward net flux of negative charges corresponds to an outward current. At potentials more depolarized than V_{rest}, an outward current is recorded (**Figure 9.4**). At potentials more hyperpolarized than -60 mV, i_{GABA} is inward (the net flux of Cl$^-$ ions is outward) but of very small amplitude.

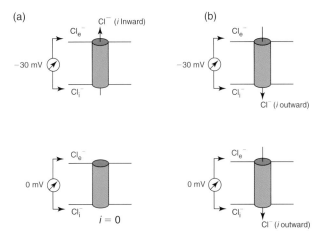

FIGURE 9.5 **Variations of the flux of Cl$^-$ ions as a function of membrane potential and intracellular and extracellular Cl$^-$ concentrations.**
(a) Symmetrical media: $[Cl^-]_e = [Cl^-]_i = 145$ mM; $E_{Cl} = 0$ mV. **(b)** Physiological media: $[Cl^-]_e = 145$ mM; $[Cl^-]_i = 14.5$ mM; $E_{Cl} = -58$ mV.

9.3.4 The single-channel conductance of GABA_A channels is constant in symmetrical Cl$^-$ solutions, but varies as a function of potential in asymmetrical solutions

Experiments have been performed on mouse spinal neurons in culture, in conditions of equal intra- and extracellular Cl$^-$ concentration (145 mM) to minimize rectification (variation of the conductance γ as a function of membrane potential). Histograms of single-channel currents evoked by GABA and recorded at $V = +50$ mV (outward i_{GABA}) and $V_m = -90$ mV (inward i_{GABA}) show at each potential a single peak of current equal to $+1.48$ and -2.7 pA respectively (**Figure 9.3b**). From these values of i_{GABA}, the mean single-channel conductance γ_{GABA} can be calculated, as $i_{GABA} = \gamma_{GABA} (V_m - E_{rev})$ and $E_{rev} = 0$ mV in these conditions. A value of 30 pS is obtained for both experimental conditions. This value of γ_{GABA} is also the slope of the i_{GABA}/V curve obtained by averaging the most frequent single-channel current i_{GABA} recorded at each membrane potential studied. This curve, based on the equation $i = \gamma (V - E_{rev})$ is linear between -90 mV and $+50$ mV and has a slope of 30 pS (**Figure 9.3c**).

However, in physiological conditions, when Cl$^-$ concentration is approximately 10-fold lower in the intracellular than in the extracellular fluid, the unitary conductance γ_{GABA} varies with the membrane potential. The conductance in fact decreases progressively as the outward Cl$^-$ current decreases (**Figure 9.4b**). This phenomenon is called *rectification*. This rectification (non-symmetrical inward and outward currents) results

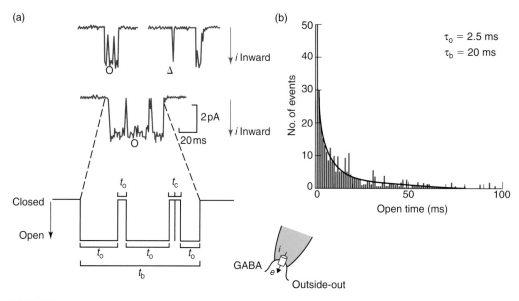

FIGURE 9.6 **Mean open time of GABA$_A$ channels.**
Patch-clamp recording of the activity of the GABA$_A$ receptor channels from chromaffin cells of the adrenal medulla (outside-out configuration). The intracellular and extracellular Cl$^-$ concentrations are similar and the membrane potential is maintained at -70 mV. **(a)** Inward unitary currents through a single GABA$_A$ channel evoked by GABA ($10\,\mu$M). Brief openings (triangle) and bursts of openings (O, long duration openings interrupted by short closures defined in this experiment as less than 5 ms). **(b)** Histogram of open times measured in a homogeneous population of channels (mean value of $i = -2.9$ pA). The open times plotted on the graph represent the duration of short openings (t_o) and the duration of bursts of openings (t_b). The histogram is described by the sum of two exponentials with decay time constants of $\tau_o = 2.5$ ms and $\tau_b = 20$ ms. τ_o corresponds to the mean open time of short openings and τ_b to the mean open time of bursts of openings. From Borman J, Clapham DE (1985) γ-aminobutyric acid receptor channels in adrenal chromaffin cells: a patch clamp study, *Proc. Natl Acad. Sci. USA* **82**, 2168–2172, with permission.

from the difference in Cl$^-$ concentration on either side of the membrane.

9.3.5 Mean open time of the GABA$_A$ channel

With patch clamp recording of GABA$_A$ channels (in chromaffin cells of the adrenal medulla or cultured hippocampal neurons, in outside-out configuration), two types of openings are observed in the presence of low concentrations of GABA (**Figure 9.6a**):

(i) brief openings, and (ii) longer duration openings interrupted by brief periods of closure: such a group of repeated openings and closures is called a burst of openings.

Brief openings (triangles)

Brief openings have a mean duration, τ_o, of 2.5 ms and contribute little to the total current.

Bursts of openings (open circles)

A burst is defined as a sequence of openings each one having a duration t_o, separated by brief closures of duration t_c. Brief durations are defined as less than 5 ms in the example illustrated in **Figure 9.6**. The

duration of each burst t_b is $\Sigma t_o + \Sigma t_c$ and its mean duration τ_b is equal to 20–50 ms depending on the preparation used (**Figure 9.6b**). The openings and the brief closures observed within each burst in the presence of GABA are thought to correspond to fluctuations of the receptor between the double-liganded open state and the double-liganded closed state (before the two molecules of GABA leave the receptor site). Thus, upon a single activation by two molecules of GABA, the double-liganded receptor would open and close several times:

$$\begin{array}{ccc} G_2R & \rightleftharpoons & G_2R^* \\ \text{Closed channel} & & \text{Open channel} \\ t_c & & t_o \end{array}$$

$$t_b$$

Silent periods

Silent periods separate single openings or bursts; they are periods during which the channel is closed and the

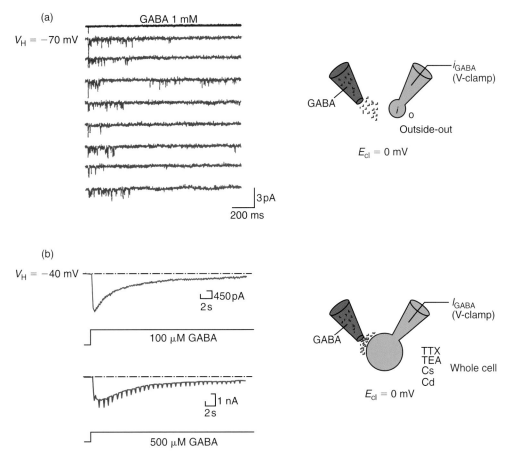

FIGURE 9.7   Desensitization of the GABA$_A$ receptor.
(a) Outside-out patch excised from a cell transfected with $\alpha_1\beta_2\gamma_2$ cDNAs. A 2 ms pulse of GABA (1 mM) evokes single GABA$_A$ channel activity. **(b, top)** Patch clamp recording (whole-cell configuration, $V_H = -40$ mV) from a chick cerebral neuron. A prolonged application of GABA at high concentration (100 μM) evokes a total current I_{GABA} which decreases with time to almost zero. The total current I_{GABA} corresponds to the sum of the unitary currents i_{GABA}, passing through the open GABA$_A$ channels, while the other currents have been blocked with TTX and TEA as well as with Cs$^+$ and Cd^{2+} ions. **(b, bottom)** The same experiment in the presence of 500 μM of GABA and with hyperpolarizing voltage steps applied at a constant rate. The decrease in amplitude of the step current during the GABA$_A$ response shows that the decrease of I_{GABA} is associated with a decrease in G_m (as $i_{step} = Gm \, V_{step}$, V_{step} being constant), a decrease of i_{step} implies a decrease of G_m. There are symmetrical Cl$^-$ concentrations in (a) and (b). Part (a) from Zhu WJ, Wang JF, Corsi L, Vicini S (1998) Lanthanum-mediated modification of GABA$_A$ receptor deactivation, desensitization and inhibitory synaptic currents in rat cerebellar neurons, *J. Physiol. (Lond.)* **511**, 647–661, with permission. Part (b) from Weiss DS, Barnes EM, Hablitz JJ (1988) Whole-cell and single-channel recordings of GABA-gated currents in cultured chick cerebral neurons, *J. Neurophysiol.* **59**, 495–513, with permission.

unitary current is zero. In the presence of very low concentrations of GABA (when the receptor has a low probability to desensitize) they correspond to the G$_2$R, GR and R states of the GABA$_A$ receptor. Recordings in outside-out configuration show a and R states of the GABA$_A$ receptor. Opening characteristics of the GABA$_A$ receptor are very similar to those of the nicotinic receptor (see Section 8.3.3). However, the short openings observed within bursts are approximately twice as abundant in the case of the GABA$_A$ receptor. This is explained by the fact that opening (β) and closing (α) rate constants have much closer values in the case of

the GABA$_A$ receptor (see Appendix 9.1) than in the case of the nicotinic receptor (see Appendix 4.3).

9.3.6 The GABA$_A$ receptor desensitizes

Recordings in outside-out configuration show a rundown of the frequency of opening of the GABA$_A$ channels upon prolonged application of GABA (0.5 μM), whereas neither the intensity of the unitary current nor the mean open time of the channels τ_o appears to be affected (**Figure 9.7a**). Considering that $I = Np_o i$, if p_o (open probability of the channel) decreases as a result

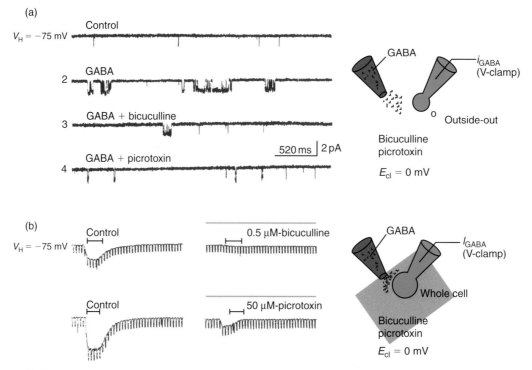

FIGURE 9.8 Antagonists of the GABA_A receptor.
(a) Activity of an outside-out patch from spinal cord neurons recorded in symmetrical Cl⁻ solutions. GABA (2 μM) is applied in the presence of bicuculline (0.2 μM) or picrotoxin (10 μM). **(b)** The same experiment in the whole-cell configuration. Steps are applied as in Figure 9.7 to evaluate membrane conductance G_m during the response. Adapted from MacDonald RL, Rogers CJ, Twyman RE (1989) Kinetic properties of the GABA_A receptor main conductance state of mouse spinal cord neurones in culture. *J. Physiol. (Lond.)* **410**, 479–499, with permission.

of a decrease in the frequency of opening events, the current I_{GABA} decreases even though i_{GABA} remains constant. Similarly, G_{GABA}, the total conductance, decreases (since $G = Np_o\gamma$). As the GABA_A receptors gradually desensitize, p_o becomes progressively smaller with time, I_{GABA} as well as G_{GABA} gradually decrease to practically zero. This is confirmed in recordings of the total current I_{GABA} evoked by a prolonged application of GABA: the amplitude of I_{GABA} decreases with time as well as G_{GABA} (**Figure 9.7b**). This rundown of the GABA_A response which increases with increasing concentrations of GABA is largely attributed to the desensitization of the GABA_A receptor. Therefore, upon long-lasting activation by GABA, the doubly liganded receptor goes into at least one desensitization state:

States of the doubly
liganded receptor: $G_2R \underset{\alpha}{\overset{\beta}{\rightleftharpoons}} G_2R^* \underset{k_{-3}}{\overset{k_3}{\rightleftharpoons}} G_2D$

State of the channel   closed   open   closed

9.4 PHARMACOLOGY OF THE GABA_A RECEPTOR

Benzodiazepines, barbiturates and neurosteroids enhance GABA_A receptor current, whereas bicuculline, picrotoxin and β-carbolines reduce GABA_A current, by binding to specific sites on the GABA_A receptor channels.

9.4.1 Bicuculline and picrotoxin reversibly decrease total GABA_A current; they are respectively competitive and non-competitive antagonists of the GABA_A receptor

Excised outside-out patches are obtained from spinal cord neurons and held at −75 mV to prevent spontaneous openings of voltage-gated channels (**Figure 9.8a**). Recordings are performed in symmetrical chloride solutions. Prior to GABA application, occasional brief spontaneous currents are recorded (1). Following GABA

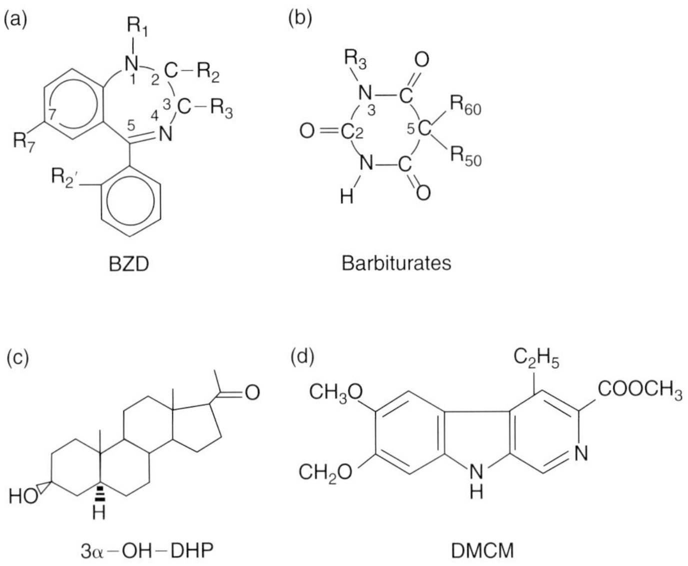

FIGURE 9.9 **Allosteric modulators of the GABA$_A$ receptor.**
(a) Structure of benzodiazepines (BZD). For diazepam, the radicals are: R$_1$=CH$_3$, R$_2$=O, R$_3$=H, R$_7$=Cl, R$_2'$=H. **(b)** Structure of barbituric acid derivatives. For pentobarbital, the radicals are: R$_{5a}$=ethyl, R$_{5b}$=H and R$_3$=phenyl. **(c)** Structure of a neurosteroid, allopregnanolone (3α-OH-DHP). **(d)** Structure of a β-carboline, methyl 6, 7-dimethoxyl-4-ethyl β-carboline 3 carboxylate (DMCM).

application, bursting inward chloride currents are evoked (2). These GABA-induced bursting currents are reversibly reduced in frequency by the concomitant application of bicuculline (3) or picrotoxin (4) (**Figure 9.8b**). In whole-cell recordings, this effect is recorded as a decrease of the amplitude of the total current I_{GABA}: if p_o decreases, $Np_o \, i_{GABA} = I_{GABA}$ decreases.

When the dose of GABA is increased and the dose of antagonist is kept constant, the inhibition by bicuculline is reduced whereas that by picrotoxin is unchanged (not shown). This shows that bicuculline is a competitive antagonist whereas picrotoxin is a non-competitive antagonist. Bicuculline binds to the same receptor sites as GABA. It is selective for the GABA$_A$ receptor and therefore serves as a good tool to identify GABA$_A$-mediated responses.

Picrotoxin in contrast binds to the ionic channel (it is a channel blocker). Its binding site involves the M2 segment, the region thought to line the chloride ion channel. The exact location of picrotoxin binding to ionophore is still unknown but its sensitivity to mutations in residues 2, 3 and 6 of M2 suggests that the site contains residues 2–6. Mutation of the highly conserved threonine residue at the 6' position in M2 of either α$_1$, β$_2$ or γ$_2$ subunits of α$_1$β$_2$γ$_3$ GABA receptors abolishes antagonism by picrotoxin at concentrations up to 100 μM. Importantly, while amino acid composition of residue 2 is variable in different ionotropic receptors, the composition of residue 6 is highly

conservative, implying that it is crucial for picrotoxin binding to ionophore, and most likely representing the epicentre of its binding pocket. Both bicuculline and picrotoxin are potent convulsants when administered intravenously or intraventricularly.

9.4.2 Benzodiazepines, barbiturates and neurosteroids reversibly potentiate total GABA$_A$ current; they are allosteric agonists at the GABA$_A$ receptor

Benzodiazepines and barbiturates are two classes of clinically active agents (**Figure 9.9a, b**). Barbiturates are hypnotic and anti-epileptic agents and the benzodiazepines are anxiolytic agents, muscle relaxants and anticonvulsants. Various progesterone metabolites that are synthesized in the brain and thus called neurosteroids act directly on the GABA$_A$ receptor. One example is allopregnanolone (**Figure 9.10c**).

Benzodiazepines, barbiturates and neurosteroids bind to the GABA$_A$ receptor at specific receptor sites

First, it was shown that co-expression of α or β subunits with a γ-subunit is required for the positive modulation of GABA-evoked Cl$^-$ currents by benzodiazepines and for photoaffinity labelling of the benzodiazepine receptor site (see Section 9.2.2). With

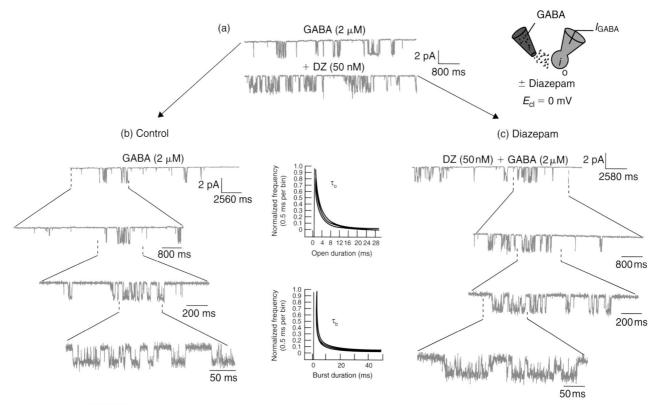

FIGURE 9.10 GABA_A single-channel current in the presence or absence of diazepam.
(a) Bursting inward currents in outside-out patches from spinal cord neurons evolved by GABA alone or GABA with diazepam (DZ). **(b, c)** The same experiment at increasing time resolution to demonstrate typical features of the unitary current. Open duration/frequency and bursts duration/frequency histograms for GABA are not significantly altered by addition of diazepam from 20 to 1000 nM (middle histograms). From Rogers CJ, Twyman RE, MacDonald RL (1994) Benzodiazepine and β-carboline regulation of single GABA_A receptor channels of mouse spinal neurones in culture. *J. Physiol. (Lond.)* **475**, 69–82, with permission.

site-directed mutagenesis, His 101 and Gly 200 amino acid residues of the α-subunit and Phe77, A79, Thr 81, Met 130 amino acid residues of the γ2-subunit are reported to be key determinants of the benzodiazepine site of the rat GABA_A receptors. However, GABA_A receptor assemblies derived from the α4- or α6-subunits fail to bind conventional benzodiazepines such as diazepam, flunitrazepam and clonazepam. This suggests that benzodiazepine agonists act via the benzodiazepine binding site of GABA_A receptors containing α_{1,2,3,5} in combination with a β-subunit and a γ2-subunit. Aside from these regions responsible for benzodiazepine binding, there are separated domains required for coupling benzodiazepine binding to potentiation of the GABA_A current that would be localized close to M1 and M2 segments. However, results based on site-directed mutagenesis need to be interpreted in the light of the possibility that amino acids identified may not be directly at the binding site but could be affecting the modulation of the compound at a distant site.

Benzodiazepines, barbiturates and neurosteroids potentiate the GABA_A response

To test whether *benzodiazepines* have a direct effect on the GABA_A receptor channel and to identify their effect, the unitary current i_{GABA} is recorded in voltage clamp ($V_H = -75$ mV) in outside-out patches (cultured spinal cord neurons). In the presence of GABA, the benzodiazepine diazepam (DZ, 50 nM) increases the opening frequency of the channel but does not change i_{GABA} amplitude nor the time spent by the channel in the open configuration at each opening (**Figure 9.11**).

Consistent with this finding, diazepam decreases the mean closed time τ_c; i.e. the time spent by the channel in the closed configuration (see **Figure 9.12c**). With decreasing or increasing doses, the effect of diazepam was less pronounced (U-shaped concentration dependency). Diazepam (50 nM) also increases the burst frequency without changing the mean burst duration (τ_b) nor the mean number of openings per burst. All the currents evoked by GABA alone or GABA with

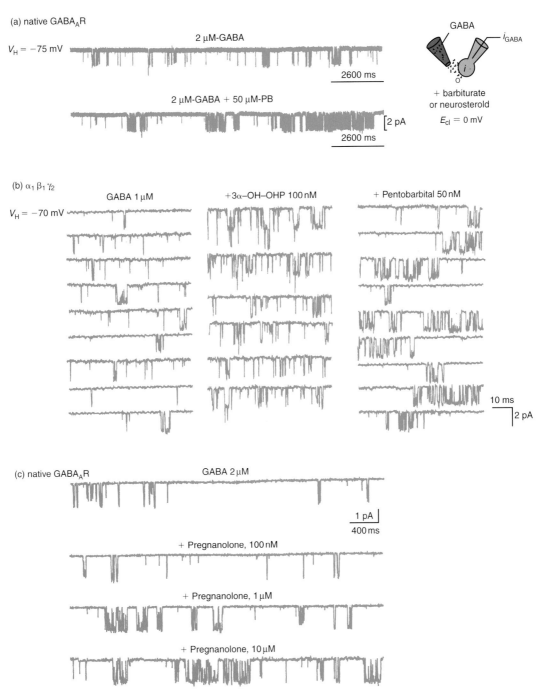

FIGURE 9.11 **(left) GABA_A single-channel current in the presence or absence of barbiturates or neurosteroids.**
(a) Unitary GABA_A currents in outside-out patches of spinal cord neurons evoked by GABA alone or GABA with phenobarbitone (PB). **(b)** Unitary GABA_A currents recorded in outside-out patches excised from $\alpha_1\beta_1\gamma_2$ transfected cells evoked by GABA alone (left) or in combination with the neurosteroid 3α-OH-DHP (centre) or pentobarbital (right). **(c)** Unitary GABA_A currents in outside-out patches of spinal cord neurons evoked by GABA alone or GABA with pregnanolone at increasing concentration. Part (a) from MacDonald RL, Rogers CJ, Twyman RE (1989) Barbiturate regulation of kinetic properties of the GABA_A receptor channel of mouse spinal neurones in culture, *J. Physiol. (Lond.)* **417**, 483–500, with permission. Part (b) from Twyman RE, MacDonald RL (1992) Neurosteroid regulation of GABA_A receptor single channel kinetic properties of mouse spinal cord neurons in culture, *J. Physiol. (Lond.)* **456**, 215–245, with permission. Part (c) from Puia G *et al.* (1990) Neurosteroids act on recombinant human GABA_A receptors, *Neuron* **4**, 759–765, with permission.

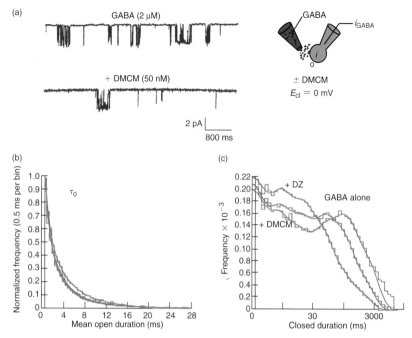

FIGURE 9.12   GABA$_A$ single-channel current in the presence or absence of a β-carboline.
(a) Unitary GABA$_A$ currents in outside-out patches of spinal cord neurons evoked by GABA alone or GABA with the β-carboline DMCM. **(b)** Histograms for open durations are plotted for GABA (2 μM, uppermost curve, τ_o = 4.10 ±0.03 ms) and for GABA and DMCM (20–100 nM, τ_o' = 4.40 ±0.07 ms). **(c)** Closed duration-frequency histograms for GABA (2 μM), for GABA with diazepam (DZ, 100 nM) and for GABA with DMCM (100 nM). DZ shifts long closed durations to shorter durations while DMCM shifts long closed durations to longer ones. From Rogers CJ, Twyman RE, MacDonald RL (1994) Benzodiazepine and β-carboline regulation of single GABA$_A$ receptor channels of mouse spinal neurones in culture, *J. Physiol. (Lond.)* **475**, 69–82, with permission.

diazepam are blocked by bicuculline, thus showing that they are mediated by the GABA$_A$ receptor. Since the i_{GABA}/V relationship shows that the unitary conductance is unchanged, it is hypothesized that diazepam alters the gating properties of the GABA$_A$ receptor channel: it increases the probability of the channel being in the open state, p_o.

The equation $Np_o i_{GABA} = I_{GABA}$ tells us that when p_o increases, I_{GABA} increases. Therefore, benzodiazepines should increase the amplitude of the total current I_{GABA} recorded in the whole-cell configuration (see **Figure 9.13**). The I/V curves show that the total currents I_{GABA} evoked in the presence or absence of benzodiazepines reverse at the same potential (not shown). This indicates that the potentiation of I_{GABA} by these drugs is not the result of a change in the ion selectivity of the channel. Benzodiazepine agonists bind to a site distinct from that of GABA and increase GABA-gated Cl$^-$ conductance by allosterically decreasing the GABA concentration needed to elicit half-maximal channel activity (EC$_{50}$).

Barbiturates also increase the bicuculline-sensitive current I_{GABA} but via a different mechanism: they do not increase the frequency of GABA$_A$ channel openings,

instead they increase the duration of single openings and bursts of native GABA$_A$ receptors (**Figure 9.11a**) or transfected α$_1$ β$_1$ γ$_2$ receptors (**Figure 9.11b**, right). An increase in the time spent in the open configuration at each opening results in an increase of the probability of the channel being in the open state and therefore in an increase of I_{GABA} (see **Figure 9.13**).

Neurosteroids at physiological concentrations increase the total bicuculline-sensitive current I_{GABA}. To study the mechanism of action, the effect of neurosteroids on single GABA$_A$ channel activity is recorded. Transformed human embryonic kidney cells 293 are transfected with the GABA$_A$ subunit combination α$_1$ β$_1$ γ$_2$. The activity of expressed single GABA$_A$ channels is recorded in the outside-out configuration with symmetrical Cl$^-$ concentrations on both sides of the patch of membrane. Allopregnanolone (3 α-OH-DHP) in combination with GABA increases the GABA$_A$ channel activity: it increases the number of active channels in the patch and the channel open probability (**Figure 9.11b**). On native GABA$_A$ receptors of spinal cord neurons, pregnanolone in combination with GABA also increases the duration of single and burst openings. Either one or both these effects on frequency of openings

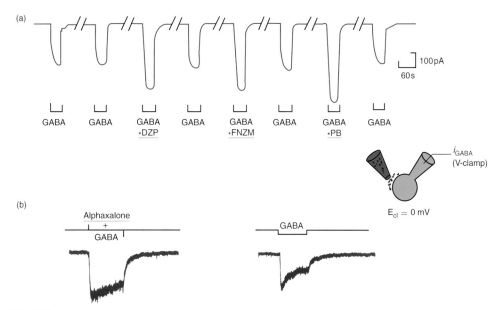

FIGURE 9.13 Potentiation of the total current I_{GABA} by benzodiazepines, barbiturates and neurosteroids.

(a) GABA$_A$ receptors are expressed in transfected cells with the α, β and γ cDNA subunits. This model is interesting because these cells do not normally express GABA receptors (neither GABA$_A$ nor GABA$_B$); the GABA applied in the bath activates therefore only the number (N) of GABA$_A$ receptors expressed. On the other hand, it is a model where the presynaptic release of GABA is excluded because of the absence of synapses. The total current I_{GABA} is recorded using the patch-clamp technique (whole-cell configuration). The total current recorded at $V_m = -60$ mV (GABA, 10 μM) is inward and is significantly potentiated by the simultaneous application of the two benzodiazepines, diazepam (DZP, 1 μM) and flunitrazepam (FNZM, 1 μM) and by pentobarbital (PB, 50 μM). **(b)** Whole-cell GABA$_A$ current evoked by 10 μM of GABA in spinal neurons in culture in the absence (right) or in the presence (left) of the neurosteroid alphaxalone (3α-hydroxy-5α-pregnane-11,20-dione, 1 μM). In (a) and (b), E_{Cl} is estimated to be near 0 mV. Part (a) adapted from Pritchett DB, Southeimer H, Shivers BD, *et al.* (1989) Importance of a novel GABA$_A$ receptor subunit for benzodiazepine pharmacology, *Nature* **338**, 582–585, with permission. Part (b) from Barker JL, Harrison NL, Lange GD, Owen DG (1987) Potentiation of γ-aminobutyric-acid-activated chloride conductance by a steroid anaesthetic in cultured rat spinal neurones, *J. Physiol. (Lond.)* **386**, 485–501, with permission.

or opening duration result in an increase of p_o and thus an increase I_{GABA} (see **Figure 9.13**).

In conclusion, benzodiazepines, barbiturates and neurosteroids can be considered as allosteric agonists of the GABA$_A$ receptor: they modulate the efficacy of activation of the receptor by GABA. They act via distinct receptor sites on the GABA$_A$ receptor and via different mechanisms.

9.4.3 β-carbolines reversibly decrease total GABA$_A$ current; they bind at the benzodiazepine site and are inverse agonists of the GABA$_A$ receptor

β-carbolines such as methyl-6,7-dimethoxyl-4-ethyl-β-carboline-3-carboxylate (DMCM) are convulsant and anxiogenic drugs. They bind to the benzodiazepine receptor site but have reverse effects: they are called 'benzodiazepine inverse agonists' (**Table 9.1**).

TABLE 9.1 Pharmacology of GABA$_A$ receptors

	GABA site	Benzodiazepine site
Selective agonists	Muscimol Isoguvacine	Flunitrazepam
Inverse agonists	–	β-carbolines
Competitive antagonists	Bicuculline Gabazine	–
Channel blocker	Picrotoxin	–

The activity of outside-out patches of spinal cord neurons in culture is recorded in voltage clamp. When DMCM is applied with GABA, it decreases the number of GABA$_A$ receptor openings compared with what is observed with GABA alone (**Figure 9.12a**). DMCM (20–100 nM) reduces single openings as well as burst frequency. However, the number of openings per burst is unchanged. The times spent by the GABA$_A$ channel in the open or bursting states are also unchanged but the time spent in the closed state is increased, consistent with a decrease in opening frequency (**Figure 9.12b, c**). Like diazepam, DMCM does not alter GABA$_A$ receptor single-channel conductance nor single-channel open or bursts properties. DMCM decreases p_o and thus decreases I_{GABA}. Since burst frequency, but not intraburst opening frequency, is altered, it is unlikely that receptor channel opening rates (α and β) are altered by diazepam or DMCM.

9.5 GABA$_A$-MEDIATED SYNAPTIC TRANSMISSION

9.5.1 The GABAergic synapse

To identify a synapse as GABAergic, several techniques are used such as immunocytochemistry for the GABA synthetic enzyme, glutamate decarboxylase (GAD). Now to identify a synaptic response as mediated by GABA$_A$ receptors, the simplest test is the block by bicuculline. Also benzodiazepines significantly potentiate and prolong GABA$_A$-mediated IPSPs (see **Figure 9.1a**). Because of the selective action of benzodiazepines on the GABA$_A$ channel, these substances can be used experimentally together with bicuculline to identify a GABA$_A$ response.

When GABA is released in the synaptic cleft, it can (see **Figure 9.14**):

- bind to postsynaptic GABA$_A$ receptors;
- bind to postsynaptic or presynaptic GABA$_B$ receptors;
- bind to glial and neuronal transporters and thus be taken up by presynaptic elements or glial cells;
- diffuse away from the synaptic cleft.

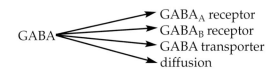

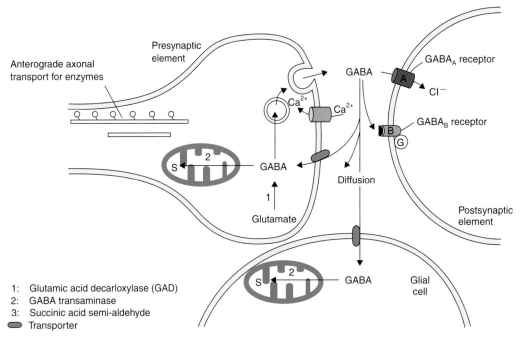

FIGURE 9.14 The GABAergic synapse.
Functional scheme of a GABAergic synapse where the ionotropic (receptor-channel) GABA$_A$ receptors and the metabotropic (G-protein linked) GABA$_A$ receptors are co-localized. Presynaptic receptors are omitted. In order to study in isolation the GABA$_A$ response, GABA$_A$ receptors are selectively blocked. GABA is formed by the irreversible decarboxylation of glutamate catalyzed by the enzyme glutamic acid decarboxylase (GAD) and is metabolized by the mitochondrial enzyme GABA transaminase (GABA-T) into succinic acid semi-aldehyde. Enzymes are synthesized in the soma and carried to axon terminals via fast anterograde axonal transport. GABA is synthesized in the cytoplasm and transported actively into synaptic vesicles by a vesicular carrier. A percentage of the GABA released in the synaptic cleft is uptaken into presynaptic terminals and glial cells by GABA transporters which co-transport Na$^+$Cl$^-$. These transports are inhibited by nipecotic acid or β-alanine.

9.5.2 The synaptic GABA$_A$-mediated current is the sum of unitary currents appearing with variable delays and durations

When GABA binds to postsynaptic GABA$_A$ receptors at resting potential, it generally evokes an inhibitory postsynaptic potential (IPSP).

From single GABA$_A$ current to IPSC

In a slice of hippocampus for example, GABAergic afferents are still spontaneously active and evoke spontaneous synaptic GABA$_A$-mediated postsynaptic currents (called IPSCs) that can be recorded in whole-cell configuration in voltage clamp. Since they are blocked by bicuculline, they are mediated by postsynaptic GABA$_A$ receptors. In symmetrical Cl$^-$ solutions, at a holding potential of -70 mV, GABA$_A$-mediated currents are inward (outward flow of Cl$^-$ ions) (**Figure 9.15a**). When IPSCs are observed at a higher magnification and a faster time base, unitary current steps can be identified. For example, during a very small IPSC (**Figure 9.15b**), the peak is an integer of 7 unitary currents. During a larger IPSC, steps cannot be identified at the level of the peak but are clear during the decay phase (**Figure 9.15c**). IPSCs result from the activation of N postsynaptic GABA$_A$ receptors.

Single-spike IPSC

When simultaneous intracellular recordings of a pair of connected neurons are performed, a presynaptic GABAergic interneuron and a postsynaptic pyramidal cell, each presynaptic action potential can be correlated with the postsynaptic current that it evokes, called *single-spike IPSC*. This postsynaptic current is outward at -30 mV in control intracellular Cl$^-$ concentration. Its amplitude varies at each trial due to the fact that a presynaptic action potential activates, at each trial, a variable number of active zones from the total number of active zones established by the presynaptic axon with the postsynaptic membrane (see Appendix 7.2).

Single-spike IPSP

The same experiment performed in current-clamp mode allows recordings of the postsynaptic variation

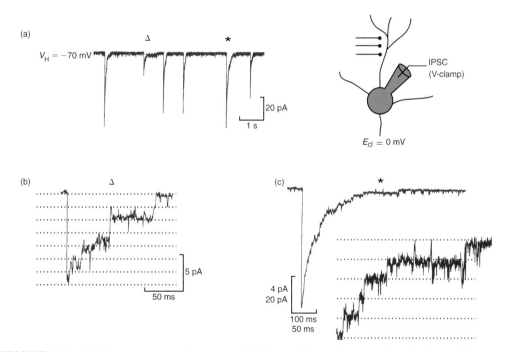

FIGURE 9.15 **Inhibitory postsynaptic currents (IPSCs) result from the summation of unitary GABA$_A$-mediated currents.**
(a) Spontaneous IPSCs are recorded from a cerebellar granule cell (whole-cell configuration, voltage-clamp mode). IPSCs are inward at -70 mV since E_{cl} in this experiment is at 0 mV. **(b, c)** Two IPSCs of different amplitude from trace (a) are enlarged to show that unitary step currents can be resolved in their decay phase. Ionotropic glutamatergic transmission is pharmacologically blocked. From Brickley SG, Cull-Candy SG, Farrant M (1999) Single-channel properties of synaptic and extrasynaptic GABA$_A$ receptors suggest differential targeting of receptor subtypes. *J. Neurosci.* **19**, 2960–2973, with permission.

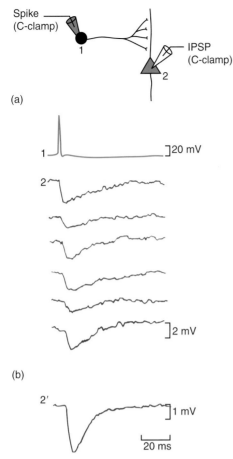

FIGURE 9.16 Characteristics of a single-spike IPSP mediated by GABA_A receptors.
(a) The activity of a pair of connected presynaptic GABAergic interneuron and postsynaptic pyramidal cell of the hippocampus *in vitro* is recorded with two intracellular electrodes. In response to each action potential evoked in neuron 1 (presynaptic GABAergic interneuron), a single-spike IPSP is recorded in neuron 2 (postsynaptic pyramidal neuron) with a mean latency of 0.7 ± 0.2 ms. Each single-spike IPSP results from the activation of several boutons which are not all activated at each trial as shown by their variable amplitude (mean amplitude $= -2.1 \pm 0.7$ mV). They are blocked by picrotoxin (10^{-4}M, not shown), a GABA_A channel blocker. **(b)** Average of 20 IPSPs obtained by triggering each trace from the peak of the presynaptic action potential. Adapted from Miles R, Wong RKS (1984) Unitary inhibitory synaptic potentials in the guinea-pig hippocampus *in vitro*, J. Physiol. **356**, 97–113, with permission.

of potential resulting from the postsynaptic GABA_A-mediated current (**Figure 9.16**). In response to a single presynaptic action potential (trace 1) an inhibitory postsynaptic potential called *single-spike IPSP* (traces 2; $V_m = -62$ to -66 mV) is recorded. It is a transient hyperpolarization of the membrane. The amplitude of this single-spike IPSP varies in amplitude at each trial, from 0.6 to 4.2 mV. This is due to the fact that a presynaptic action potential activates, at each trial, a variable number of active zones from the total number of

active zones established by the presynaptic axon. The conductance G_{IPSP} can be calculated knowing I_{IPSP}, V and E_{Cl}. In this experiment, $G_{IPSP} = 6.7$ nS ± 2.3 nS at the peak of the IPSP. Knowing that the unitary conductance of the GABA_A channel is estimated in these neurons to be $\gamma_{GABA} = 20$–30 pS, it can be deduced that approximately 300 GABA_A channels are open at the peak of the single-spike IPSP ($G_{IPSP} = Np_o \gamma_{GABA}$).

9.5.3 Generalization: the consequences of the synaptic activation of GABA_A receptors depend on the relative values of E_{Cl} and V_m (see Appendix 9.2)

It has been explained above that synaptically released GABA evokes a transient hyperpolarization of the postsynaptic membrane, called IPSP, via the activation of GABA_A receptors. However, this is not always the case.

When E_{Cl} is more hyperpolarized than V_{rest}, GABA_A receptor activation leads to a hyperpolarizing postsynaptic potential (IPSP) and inhibition of the postsynaptic activity

In recordings with electrodes filled with potassium acetate (instead of potassium chloride, in order not to change the intracellular concentration of Cl^-), depending on the neuron studied, E_{Cl} can be more hyperpolarized than the postsynaptic resting membrane potential (V_{rest}) and a hyperpolarizing postsynaptic potential (IPSP) is recorded in response to the stimulation of GABAergic afferent fibres (**Figures 9.16b** and **9.17a**).

When E_{Cl} is close to V_{rest}, activation of GABA_A receptor leads to a 'silent inhibition' of postsynaptic activity

When E_{Cl} is close to V_{rest}, the electrochemical gradient for Cl^- is very weak or null and no IPSP is observed even though GABA_A channels are open. However, GABA still has an inhibitory effect on postsynaptic activity, as any EPSP occurring during the effect of GABA is strongly inhibited (see Section 13.2.3). This inhibitory effect of GABA is called a *shunting effect*. It is due to an increase in the membrane conductance (G_{IPSP}) during the silent inhibition owing to the opening of GABA_A channels. If this effect is large, the membrane resistance decreases and any other synaptic current evoked at this time will produce only a small change in membrane potential (according to Ohm's Law, when I_{EPSP} is constant but R_m decreases, then V_{EPSP} decreases). The silent GABA_A inhibition reduces the amplitude of postsynaptic depolarizations and consequently prevents the generation of postsynaptic action

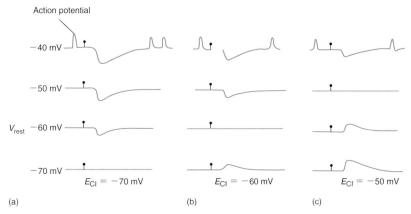

FIGURE 9.17 **Reversal potential of GABA$_A$-mediated IPSPs as a function of membrane potential.**
E_{Cl} is **(a)** more negative (-70 mV), **(b)** equal (-60 mV) or **(c)** more positive (-50 mV) than the resting membrane potential of the neuron. A resting membrane potential for the neuron of -60 mV and a threshold for action potential generation at -40 mV are assumed. Action potentials are truncated owing to their large amplitude. In the three cases represented, inhibition of activity is produced during the IPSP (upper trace).

potentials (**Figure 9.17b**). It must be noted that this shunting effect of GABA is always present, whatever the effect of GABA (hyperpolarization, silence or depolarization), and it attenuates other concomitant depolarizations or hyperpolarization.

When E_{Cl} is more depolarized than V_{rest} but below the threshold for action potential generation, GABA$_A$ receptor activation leads to a depolarizing current and an inhibition of postsynaptic activity

When E_{Cl} is more depolarized than V_{rest}, the activation of GABA$_A$ receptors causes an inward current (outward flow of Cl$^-$ ions) and a depolarization of the membrane. This depolarization does not excite the membrane; i.e. does not trigger action potentials when E_{Cl} is more negative than $V_{threshold}$. As long as E_{Cl} remains below the threshold for activation of voltage-dependent Na$^+$ channels, postsynaptic activity is inhibited (**Figure 9.17c**). In this case GABA$_A$ current adds to other depolarizing current flowing through the membrane but once V_m becomes more depolarized than E_{Cl}, the GABA$_A$ current becomes hyperpolarizing.

When E_{Cl} is more depolarized than V_{rest} and above the threshold for action potential generation, GABA$_A$ receptor activation leads to a depolarizing current and an excitation of postsynaptic activity

When E_{Cl} is more depolarized than the threshold for activation of voltage-dependent Na$^+$ channels, GABA$_A$ receptor activation causes a postsynaptic depolarization or EPSP and evokes spikes. Note that such a

depolarizing and *excitatory* effect of GABA is present in very young GABAergic synapses (see Chapter 20).

9.5.4 What shapes the decay phase of GABA$_A$-mediated currents?

GABA$_A$ receptor-mediated IPSCs peak rapidly (in 0.5–5 ms) and usually decay with two time constants ranging from a few milliseconds to tens or hundreds of milliseconds. To determine the factors responsible for the duration of IPSC is interesting since IPSC decay determines the time course of the resistive shunt or hyperpolarization that prevents neuronal firing in response to excitatory inputs.

The duration of the synaptic current is determined by the period during which each activated GABA$_A$ receptor remains in the G$_2$R* state. Various factors may determine the duration of the G$_2$R* state and therefore the time course of the postsynaptic GABA$_A$ response:

Case 1: The GABA$_A$ receptor deactivates due to unbinding of GABA and does not reopen because GABA has disappeared from the cleft:

$$G_2R^* \rightarrow G_2R \rightleftharpoons 2G + R$$

Case 2: The GABA$_A$ receptor deactivates due to unbinding of GABA but it reactivates several times before closing since removal of GABA from the synaptic cleft by neuronal/glial uptake or diffusion is very slow:

$$G_2R^* \rightleftharpoons G_2R \rightleftharpoons 2G + R$$

(a)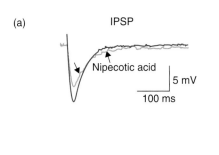

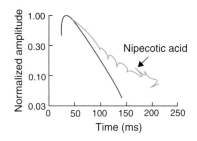

(b)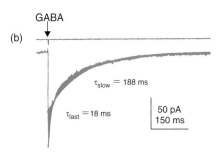

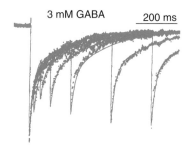

FIGURE 9.18   Modulation of the decay phase of GABA_A-mediated IPSPs and current.
(a) Current clamp recording from hippocampal neurons in rat brain slices. An orthodromic IPSP is recorded in response to stimulation of GABAergic afferents. This IPSP is sensitive to bicuculline (not shown). The IPSP is recorded before and after (arrows) application of nipecotic acid (1 mM for 40 min, left). Each trace represents the average of six responses. Nipecotic acid decreases the amplitude of the IPSP, which obscures the analysis of its effect on the time course of the IPSP. After normalizing the IPSPs to the same peak (right), the analysis of their time course reveals that nipecotic acid prolongs the later phase of the repolarization. **(b)** GABA_A receptor current is evolved in outside-out macropatches of cultured hippocampal neurons by a brief GABA pube (1 mM for 5 ms). The patch current decay is well fitted by the sum of two exponentials with time constants as indicated (left). Pairs of GABA pulses are given at 25, 50, 100, 200, 400 and 600 ms intervals (right). The second response in each pair is reduced at short interpulse intervals. Part (a) from Dingledine R, Korn SJ (1985) γ-aminobutyric acid uptake and termination of inhibitory synaptic potentials in the rat hippocampal slices. *J. Physiol. (Lond.)* **366**, 387–409, with permission. Part (b) from Jones MV, Westbrook GL (1995) Desensitized states prolong GABA_A channel responses to brief agonist pulses, *Neuron* **15**, 181–191, with permission.

Case 3: The GABA_A receptor desensitizes:

$$G_2R^* \rightleftharpoons G_2D$$

If the decay of the synaptic current is due to rapid closure of the GABA_A channels without any reopenings (case 1), this implies that the concentration of GABA decreases very rapidly in the synaptic cleft after its release (as a consequence, for example, of rapid diffusion or very efficient uptake mechanisms). In this case, the channels have very little chance of being reactivated by the repeated binding of two molecules of GABA: the time constant of the decay of the postsynaptic current is τ_o; i.e. the mean open time of the channels which depends only on the closing time constant of the channels (in the absence of desensitization). GABA_A receptors have a low affinity for GABA (10–20 μM) and display brief openings and bursts (0.2–25 ms).

Assuming a brief increase of GABA concentration in the synaptic cleft, these properties predict an IPSC decay of no more than 10–25 ms, much shorter than is frequently observed. This discrepancy could indicate

that GABA stays longer in the cleft (case 2) or synaptically activated GABA_A receptors visit desensitized states (case 3).

If the decay of the synaptic current is due to the slow disappearance of GABA from the vicinity of the receptors (removal due to GABA uptake and/or diffusion), this implies, in contrast, a prolonged presence of GABA in the synaptic cleft. The prolonged presence of GABA may result from a slow mechanism of uptake or from a restricted diffusion of GABA away from the synaptic cleft. Hence, GABA may reactivate GABA_A channels which have previously opened and the decay time constant will exceed the value of τ_o. However, uptake inhibitors have very little effect on GABA_A synaptic responses.

When applied by micro-iontophoresis in hippocampal slices, nipecotic acid, a neuronal and glial uptake inhibitor, has only a weak effect on the duration of the postsynaptic potential (IPSP) evoked by stimulation of GABAergic afferent fibres: it prolongs the later phase of the IPSP (**Figure 9.18a**). Thus, as mentioned earlier, the uptake process may be too slow to have much influence on the time course of the IPSP.

If the decay of the synaptic current is due to the transition of the channel to a desensitized state, the decay time constant would be related to the kinetics of desensitization. Moreover if there is a prolonged presence of the GABA in the synaptic cleft, and rapid kinetics of desensitization and recovery from desensitization, the channels may reopen once they have recovered from desensitization. To test this hypothesis, GABA is very briefly applied (for 1–10 ms) to outside-out patches of pyramidal neurons. The ensemble average of patch currents decay with bi-exponential kinetics similar to that of the IPSC (**Figure 9.18b**). When pairs of 1–3 ms GABA pulses are given at variable intervals, the second pulse evokes a smaller peak current than the first pulse, demonstrating that channels do not return to the unbound state immediately after closing but rather enter an agonist-insensitive (desensitized) state. This suggests that after a brief pulse, GABA occupies receptors long enough for many channels to accumulate in desensitized state(s). A rapidly-equilibrating desensitized state(s) A prolongs the $GABA_A$-mediated IPSC and provides a mechanism for low-affinity receptors to support long-lasting currents.

In the case of the nicotinic channel, the removal of acetylcholine is very rapid and desensitization of the nAChR is slow relative to deactivation and does not affect the shape of the endplate current. The kinetic properties of the deactivation of the channel mostly determine the time course of the endplate current (see Section 8.5.2). In the case of the $GABA_A$ receptor, the situation appears to be more complex. For very low-amplitude postsynaptic currents, the mean open time of the $GABA_A$ channel may determine their time course, the released GABA being rapidly removed by diffusion from the cleft. However, for larger postsynaptic currents, evoked by a greater presynaptic release of GABA, the slower removal of GABA from the synaptic cleft together with the desensitization of the channel that recovers over the course of deactivation can prolong synaptic currents.

9.6 SUMMARY

The $GABA_A$ receptor channel belongs to the cys-loop family of ligand-gated channels and is activated by γ-aminobutyric acid, the neurotransmitter at numerous synapses in the mammalian central nervous system. The $GABA_A$ receptor pentamer is formed by a combination of α, β, γ, δ, ε, π, θ subunits. Native receptors consist of α and β subunits and usually include an additional subunit, typically γ, δ, or ε. The $GABA_A$ receptor comprises the GABA receptor sites on its surface, the elements that make the ionic channel selectively permeable

to chloride ions, as well as all the elements necessary for interactions between different functional domains. Thus, the GABA receptor sites and the chloride channel are part of the same unique protein.

How does $GABA_A$ receptor mediate the binding of GABA into a transient hyperpolarization of the membrane?

The fixation of GABA to the $GABA_A$ receptor induces a conformational change of the receptor and opening of the $GABA_A$ channel. This channel opens in bursts (it rapidly fluctuates between closed and open bi-liganded states). It is permeable to Cl^- ions. Depending on the membrane potential and the concentration of Cl^- in the extracellular and intracellular media, there is an influx or an efflux of Cl^-. In adult neurons, there is generally an influx of Cl^-; i.e. an outward current which hyperpolarizes the membrane, in particular when the membrane is previously depolarized by an EPSP. This hyperpolarization is the inhibitory postsynaptic potential (IPSP) mediated by $GABA_A$ receptors. Even when the membrane is not hyperpolarized by GABA (when $E_{Cl} = V_m$), the opening of $GABA_A$ channels is inhibitory: opening of $GABA_A$ channels reduces membrane resistance and thus reduces the depolarizing effect of concomitant inward currents ($\Delta V_m = R_m \Delta I_{inward}$; when ΔI_{inward} is constant and R_m diminishes, ΔV_m is reduced).

The precise measurement of E_{Cl} and V_{rest} (Appendix 9.2) in many different central neurons is important to understand the behaviour of the membrane in response to GABA.

Do benzodiazepines and barbiturates act directly on $GABA_A$ receptors? Are there selective and distinct binding sites on the receptor for each of these drugs? How do they potentiate the hyperpolarizing effect of GABA? Are there other modulators of $GABA_A$ receptors?

Aside from the GABA receptor sites, the $GABA_A$ receptor contains a variety of topographically distinct receptor sites capable of recognizing clinically active substances, such as benzodiazepines (anxiolytics and anticonvulsants), barbiturates (sedatives and anticonvulsants), neurosteroids and ethanol (see **Table 9.1**). There are selective allosteric binding sites on $GABA_A$ receptors such as (i) the benzodiazepine site that recognizes the allosteric agonists benzodiazepines and the inverse agonists β-carbolines, (ii) the barbiturate site, and (iii) the ethanol site. Benzodiazepines increase the opening frequency of the $GABA_A$ receptor whereas barbiturates increase the duration of each opening; they both potentiate $GABA_A$ current. In contrast, inverse agonists depress GABA$_A$ current. GABA$_A$ receptors are probably a pentameric assembly derived from a combination of three different types of subunits. This leads not only to structural heterogeneity but also to pharmacological heterogeneity of the GABA$_A$ receptors, especially regarding the sensitivity to benzodiazepines.

FURTHER READING

Campo-Soria C, Chang Y, Weiss DS (2006) Mechanism of action of benzodiazepines on GABA$_A$ receptors. *British Journal of Pharmacology* **148**, 984–990.

Colquhoun D and Hawkes AG (1977) Relaxations and fluctuations of membrane currents that flow through drug-operated channels. *Proc. R. Soc. Lond. B Biol. Sci.* **199**, 231–262.

Ernst M, Bruckner S, Boresch S, Sieghart W (2005) Comparative models of GABA$_A$ receptor extracellular and transmembrane domains: important insights in pharmacology and function. *Mol. Pharmacol.* **68**, 1291–1300.

Gurley D, Amin J, Ross PC, Weiss DS, White G (1995) Point mutations in the M2 region of the α, β or γ subunit of the GABA$_A$ channel that abolish block by picrotoxin. *Receptors Channels* **3**, 13–20.

Kash TL, Jenkins A, Kelley JC et al. (2003) Coupling of agonist binding to channel gating in the GABA(A) receptor. *Nature* **421**, 272–275.

Krnjevic K and Schwartz S (1967) The action of γ-aminobutyric acid on cortical neurones. *Exp. Brain Res.* **3**, 320–336.

Mody I and Pearce RA (2004) Diversity of inhibitory neurotransmission through GABA(A) receptors. *Trends Neurosci.* **27**, 569–575.

Newell JG, McDevitt RA, Czajkowski C (2004) Mutation of glutamate 155 of the GABA$_A$ receptor β_2 subunit produces a spontaneously open channel: a trigger for channel activation. *J. Neurosci.* **24**, 11226–11235.

Polenzani L, Woodward RM, Miledi R (1991) Expression of mammalian γ-aminobutyric acid receptors with distinct pharmacology in *Xenopus* oocytes. *Proc. Natl Acad. Sci. USA* **88**, 4318–4322.

Pritchett DB, Sontheimer H, Shivers DB et al. (1989) Importance of a novel GABAA receptor subunit for benzodiazepine pharmacology. *Nature* **338**, 582–585.

Schofield PR, Darlison MG, Fujita N et al. (1987) Sequence identity and functional expression of the GABA-A receptor shows a ligand-gated receptor super-family. *Nature* **328**, 221–227.

Sedelnikova A, Erkkila B, Harris H, Zakharkin SO, Weiss DS (2006) Stoichiometry of a pore mutation that abolishes picrotoxin-mediated antagonism of the GABA$_A$ receptor. *J. Physiol.* **577**, 569–577.

APPENDIX 9.1 MEAN OPEN TIME AND MEAN BURST DURATION OF THE GABA$_A$ SINGLE-CHANNEL CURRENT

The conformational changes of the GABA$_A$ channel are modelled as follows:

$$2G + R \underset{k_{-1}}{\overset{k_1}{\rightleftharpoons}} GR \underset{k_{-2}}{\overset{k_2}{\rightleftharpoons}} G_2R \underset{\alpha}{\overset{\beta}{\rightleftharpoons}} G_2R^*$$

Upon application of GABA, the conformational change of the GABA$_A$ receptor towards the G$_2$R* state is favoured (opening of the channel). However, it has been found that the GABA$_A$ receptor closes and opens rapidly several times upon opening: these are bursts of openings (see **Figure 9.7**). The short-duration closures represent the fluctuation of the receptor between the G$_2$R and G$_2$R* states. During these bursts, the receptor returns much less frequently to the GR state, and even less frequently to the R state, before reopening.

The mean open time is $\tau_o = 1/\alpha$. When the receptor is in the G$_2$R* state, it can only transfer to the G$_2$R state with a rate constant at α (this is true when desensitization is negligible). τ_o is calculated experimentally from the different open times t_o within each burst.

The mean closed time within bursts is $\tau_c = 1/(\beta + 2k_{-2})$. When the receptor is in the G$_2$R state, it can either reopen with the rate constant β or transfer to the GR state with a rate constant $2k_{-2}$. The average number of short closures per burst is $nf = \beta/2k_{-2}$ (see Colquhoun and Hawkes 1977).

Once the values of τ_o, τ_c and nf are experimentally defined, the values of α and β can be deduced: $\alpha = 50\,s^{-1}$; $\beta = 330\,s^{-1}$.

Note that α and β differ by a factor of about 6 (whereas for the acetylcholine nicotinic receptor nAChR, they differ by a factor of 40). This illustrates numerically the fact that fluctuations between the double-liganded open and closed states are more frequent for the GABA$_A$ receptor, the probability that the channel opens in a given time being only six times greater than the probability that the channel closes.

APPENDIX 9.2 NON-INVASIVE MEASUREMENTS OF MEMBRANE POTENTIAL AND OF THE REVERSAL POTENTIAL OF THE GABA$_A$ CURRENT USING CELL-ATTACHED RECORDINGS OF SINGLE CHANNELS

Whether GABA depolarizes or hyperpolarizes neurons, depends on the value of the resting membrane

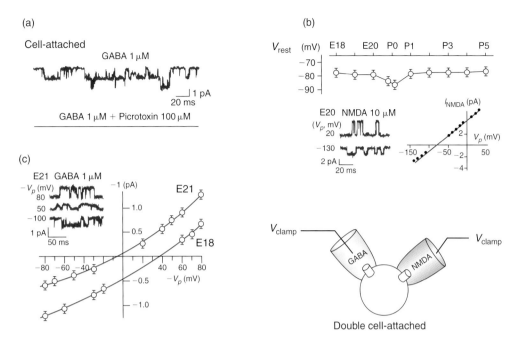

FIGURE A9.1 (a) Cell-attached recordings of single GABA$_A$ channels with 1 μM of GABA in the patch pipette (upper trace); the unitary currents were not observed in the presence of GABA$_A$ antagonist picrotoxin (100 μM) together with GABA in the patch pipette (lower trace). (b) Age-dependence of the resting membrane potential (V_{rest}) of CA3 pyramidal cells inferred from the reversal potential of single-channel NMDA currents recorded in a cell-attached mode (n = 84 cells; 4 to 12 patches for each point). (c) I–V relationships of the currents through GABA$_A$ channels in two cells, at E21 and E18; their reversal potential corresponds to the GABA$_A$ driving force (DF_{GABA}).

potential (V_{rest}) and the reversal potential of the GABA$_A$ receptor-mediated responses ($E_{GABA(A)}$). The difference between $E_{GABA(A)}$ and V_{rest} is called GABA driving force ($DF_{GABA(A)}$). GABA causes neuronal depolarization or hyperpolarization if $DF_{GABA(A)}$ is positive or negative, respectively. When $E_{GABA(A)}$ equals V_{rest}, $DF_{GABA(A)}$ is null and the GABA$_A$ receptor-mediated response is isoelectric and associated with a drop in the membrane resistance without any evident change in the membrane potential. Therefore, to determine the action of GABA in a given neuron, one has to measure $E_{GABA(A)}$ and V_{rest}. However, conventional intracellular and whole-cell recording techniques introduce numbers of errors in the measurements of $E_{GABA(A)}$ and V_{rest}. These include alterations in the intracellular ionic composition, liquid junction potentials, space-clamp problems and neuronal depolarization due to the short-circuit effect of the conductance between the electrode and cell membrane. These errors are most important in small cells but they are not negligible also in large neurons. To overcome the problems of conventional techniques, cell-attached recordings of ionic channels can be used to measure $E_{GABA(A)}$ and V_{rest} from intact cells in a non-invasive manner. Such

measurements were used in the investigation of the action of oxytocin on GABA signalling at birth (**Figure A9.1**).

In cell-attached configuration of patch clamp recordings, the patch of membrane under the recording electrode contains a variety of ionic channels. Addition of GABA to the pipette solution causes activation of currents through GABA$_A$ channels (**Figure A9.1a**). The amplitude and direction of these currents is determined by Ohm's Law and depends on the driving force acting on ions across GABA$_A$ channels and their conductance. In cell-attached recordings, the driving force on ions across GABA$_A$ channels is: $DF_{GABA(A)} = -V_p$, where V_p is the pipette potential. Therefore, $DF_{GABA(A)} = V_p$ at the reversal potential of the currents through GABA$_A$ channels. $DF_{GABA(A)}$ thus can be directly obtained from the current-voltage relationships of the currents through GABA$_A$ channels in cell-attached recordings (**Figure A9.1c**). In addition, the slope of the current-voltage functions provides an estimate of the conductance of GABA$_A$ channels.

V_{rest} can be measured using similar methodology by studying the reversal potential of the currents with a known reversal potential. For example, it is known that the reversal potential of currents through NMDA channels is close to 0 mV (see Chapter 10). Therefore, in cell-attached recordings, $DF_{NMDA} = -V_{rest}$ (Fig. 1E). Simultaneous recordings of GABA$_A$ and NMDA channels from the same neuron (with double cell-attached recordings) enable to determine $DF_{GABA(A)}$ and V_{rest} in the same neuron. Knowing these two parameters, we can calculate $E_{GABA(A)}$ since $E_{GABA(A)} = DF_{GABA(A)} + V_{rest}$ (**Figure A9.1b**). Other channels with a known reversal potential (such as potassium channels) can also be used to measure V_{rest}.

CHAPTER

10

The ionotropic glutamate receptors

From the original observations of Curtis and collaborators (1961) it is known that glutamate has a depolarizing effect on neurons. Glutamate is with GABA the major neurotransmitter in the vertebrate central nervous system. It activates two main types of postsynaptic receptors: (i) ionotropic glutamate receptors (iGluRs) that are ligand-gated channels, and (ii) metabotropic glutamate receptors (mGluRs) that are receptors coupled to GTP- binding proteins. The latter type is covered in Chapter 12.

The tightly regulated glutamate release from one neuron is detected by the glutamate receptors localized in the postsynaptic membrane of the adjacent neuron. Specificity of synaptic signalling by glutamate in space and time is conferred by the precise positioning of synapses and by the neuron-specific expression of a subset of genes encoding glutamate receptors.

10.1 THERE ARE THREE DIFFERENT TYPES OF IONOTROPIC GLUTAMATE RECEPTORS. THEY HAVE A COMMON STRUCTURE AND ALL PARTICIPATE IN FAST GLUTAMATERGIC SYNAPTIC TRANSMISSION

10.1.1 Ionotropic glutamate receptors are named after their selective or preferential agonist. They share a common structure

Based on agonist preference, iGluRs can be pharmacologically categorized into those that form receptors that are activated by the synthetic agonist N-methyl-D-aspartate (NMDA) and those that are not (**Figure 10.1a, b**). Non-NMDA receptors are further categorized by their affinity

for the synthetic agonist α-amino-3-hydroxy-5-methyl-4-isoxazole propionate (AMPA) versus the naturally occurring neurotoxin kainate. Agonists and antagonists that selectively act on one or the other have been developed in recent years.

Despite varying degrees of sequence identity, the major structural features are conserved in all known iGluR subunits. NMDA, AMPA and kainate receptors are glycoproteins composed of several subunits. To date, molecular cloning has identified 18 cDNAs, four for AMPA receptor subunits (termed GluR1-4), five for kainate receptor subunits (termed GluR5-7, plus KA1 and KA2) and seven for NMDA receptor subunits (termed NR1, NR2A-D, NR3A-B) (**Figure 10.1c**). iGluR subunits have in common a large extracellular N-terminus domain and four hydrophobic segments. Immunocytochemical and biochemical studies have indicated that the C-terminus is intracellularly located. When N-glycosylation consensus sequences were introduced at different sites along the entire length of a GluR1 subunit, to test which part of the protein was extracellularly located (glycosylation at a particular site is taken as a proof of its external location), the hypothesis was put forward that the receptor has only three transmembrane domains, corresponding to M1, M3 and M4. In this model, M2 does not span the membrane but is considered to lie in close proximity to the intracellular surface of the plasma membrane and to have a hairpin structure (P). Furthermore, the entire region between M3 and M4 is extracellular. In summary, a typical iGluR subunit consists of a bilobed amino-terminal domain (ATD), a two-domain ligand-binding core (S1, S2), three membrane-spanning segments (M1, M3, M4) and a re-entrant pore loop (P or M2), and a cytoplasmic carboxy-terminal domain (CTD) of variable length. This model is applicable to all iGluR subunits (**Figure 10.2**).

(a)

iGluRs	Non-NMDA receptors		NMDA receptors	
	AMPA R	**Kainate R**	Glu site	Gly site
Selective agonists		ATPA (GluR5)	NMDA	Glycine D-serine
Non-selective agonists	Glutamate AMPA Kainate		Glutamate	–
Selective antagonists	GYKI 53655 NBQX (1 µM)	LY 293558 (GluR5)	D-APV (or AP5)	5,7-dichloro-kynurenate
Non selective Antagonists	CNQX DNQX NBQX (>1µM)		–	–
Channel blockers	–	–	MK 801 ketamine	–

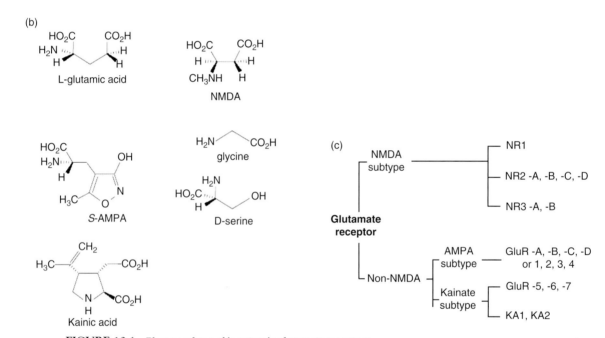

(b)

L-glutamic acid

NMDA

S-AMPA

glycine

D-serine

Kainic acid

(c)

FIGURE 10.1   **Pharmacology of ionotropic glutamate receptors.**
(a) *Key:* AMPA, α-amino-3-hydroxy-5-methyl-4-isoxalonepropionate; ATPA, (*RS*)-2-amino-3-(3-hydroxy-5-tert-butylisoxazol-4-yl) propionate; CNQX, 6-cyano-7-nitroquinoxaline-2,3-dione; D-APV, D-2-amino-5-phosphonopentanoate; DNQX, 6,7-dinitroquinoxaline-2,3-dione; GYKI 53655, atypical 2,3-benzodiazepines; MK 801, 5-methyl-10,11-dihydro-5H-dibenzocyclohepten-5,10-imine maleate; NBQX, 2,3-dihydroxy-6-nitro-7-sulphamoyl-benzo(F) quinoxaline; NMDA, *N*-methyl-D-aspartate. **(b)** Agonists of iGluRs. **(c)** Family of ionotropic glutamate receptor subunits. Part (c) adapted from Wollmuth LP, Sobolevsky AI (2004) Structure and gating of the glutamate receptor ion channel. *Trends in Neurosciences* 27, 321–328.

10.1.2 The binding site

Firstly, researchers identified two extracellular domains called S1 and S2 as responsible for agonist specificity. Across iGluR subunits, these regions show a high degree of sequence homology. This ligand-binding domain consists of two regions termed S1 (N-terminal of M1 transmembrane spanning domain) and S2 (between regions M3 and M4). Both these domains participate to S1 and S2 (**Figure 10.2**). Then researchers designed a water-soluble mini-receptor, able to bind ligands with affinities and selectivities similar to those

(a)

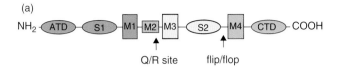

(b)

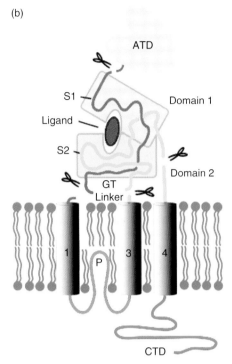

FIGURE 10.2   Schematic diagrams of iGluR subunit topology. (a) See text. (b) The amino terminus, which begins the amino-terminal domain (ATD), is located extracellularly. The ligand-binding core, also located in the extracellular space is composed of discontinuous polypeptide segments S1 and S2. The ion channel is formed by the membrane-embedded domains 1, P (M2), 3 and 4 while the carboxy-terminal domain (CTD) is located on the inside of the cell. The iGluR S1S2 constructs are generated by deleting the ATD, coupling the end of S1 to the beginning of S2 via a Gly-Thr (GT) linker and deleting the final transmembrane segment by ending the polypeptide near the end of S2. For non-NMDA and NMDA receptor NR2 subunits, the S1S2 complex forms the glutamate-binding site, whereas for NMDA receptor NR1, it forms the glycine-binding site. The large ATD and CTD domains are not shown to scale. Part (a) from Wollmuth LP and Sobolevsky AI (2004) Structure and gating of the glutamate receptor ion channel. *Trends in neurosciences* **27**, 321–328. Part (b) from Gouaux E (2004) Structure and function of AMPA receptors. *J. Physiol* **554**, 249–253, with permission.

observed with intact AMPA receptors. It consisted of the agonist binding core, the S1S2 domains linked together by a short, flexible, peptide linker. They made this S1S2 construct from one AMPA receptor secreted as a soluble protein from either insect or E. coli cells. Once this type of construct was produced in sufficiently large quantities in a functionally active state, they could obtain well-ordered crystals and perform structural analysis by X-ray diffraction.

The crystal structure of the GluR2 S1S2-kainate complex showed that the receptor fragment has a 'clamshell'-like structure and that the ligand resides in a cavity formed between both domains; i.e. in the cleft between each shell. Furthermore, it allowed defining the location of the amino acid residues that interact with the agonists glutamate, AMPA, kainate and the competitive antagonist DNQX. But the important discovery was that the 'clamshell' in the presence of AMPA and glutamate was around 20 degrees more closed in comparison to the state in the presence of the antagonist DNQX, whereas in the presence of the partial agonist kainate the degree of closure was intermediate. This suggested that the fundamental conformational change involved in the activation of the ion channel was the closure of the clamshell. Similar studies of the binding core of NMDA subunits revealed that it too retained a general clamshell-like architecture reminiscent of the other AMPA and kainate subunit-binding pockets.

In summary, the overall fold of the iGluR ligand-binding domains for AMPA, kainate and NMDA receptors is nearly identical. Key amino-acid side-chains that interact with the ligand α-amino and α-carboxy groups are the same in all four iGluR families. What differs are the amino acids that interact with the glutamate γ-carboxy group or, in the case of NR1 subunits, prevent the binding of glutamate. In the agonist-bound complex of all iGluRs, the ligand is buried in the interior of the protein but the volume of the ligand-binding cavity varies substantially. Finally, the ligand-binding cores of iGluRs are believed to assemble as dimers of dimers.

10.1.3 The three ionotropic receptors participate in fast glutamatergic synaptic transmission

In the cat neocortex, in response to a single spike in a presynaptic pyramidal neuron, a transient depolarization of the membrane mediated by ionotropic glutamate receptors is recorded in the postsynaptic interneuron. This depolarization of synaptic origin is called a *single-spike excitatory postsynaptic potential* (single-spike EPSP) (**Figure 10.3a**). Note the small amplitude of this EPSP. It is insensitive to APV (the antagonist of NMDA receptors) and totally abolished by CNQX (an antagonist of both AMPA and kainate receptors) (not shown), thus showing that it results from the activation of postsynaptic non-NMDA receptors.

In the rat hippocampus, a large-amplitude EPSP is recorded from an interneuron in response to the stimulation of the presynaptic glutamatergic pyramidal neuron. This EPSP presents two components: an early one abolished by CNQX and a late one abolished by APV.

(a) Non-NMDA-mediated EPSP

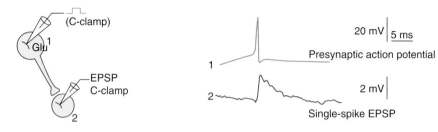

(b) NMDA and non-NMDA-mediated EPSP

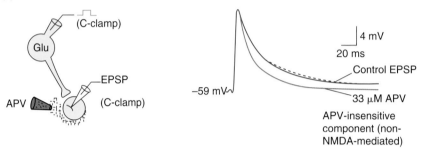

(c) Kainate-mediated EPSP

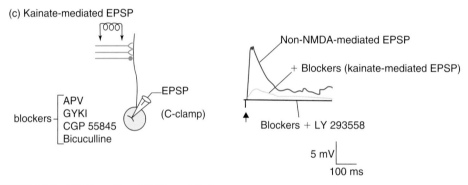

FIGURE 10.3   iGluR-mediated EPSPs and their components.

(a) Current clamp recordings of a presynaptic glutamatergic pyramidal neuron and a postsynaptic interneuron in the cat neocortex *in vitro*. A spike triggered by a current pulse in the presynaptic neuron evokes a single-spike EPSP in the postsynaptic interneuron (intracellular recordings). **(b)** The same experiment in the hippocampus *in vitro*. In response to a stronger stimulation of the presynaptic pyramidal neuron and in the absence of external Mg^{2+}, a postsynaptic EPSP of larger amplitude (control) with a fast rising phase and a long duration (sometimes up to 500 ms) is recorded. The same experiment in the presence of APV (APV, 33 μM). Picrotoxin is added to the extracellular solution in order to block $GABA_A$ synaptic receptors. **(c)** In response to the stimulation of afferent fibres, a control EPSP is recorded from a hippocampal interneuron. In the presence of blockers of NMDA (APV), AMPA (GYKI 53655), $GABA_B$ (CGP 55845) and $GABA_A$ (bicuculline) receptors, a low-amplitude component is still present. It is mediated by kainate receptors since it is totally blocked by LY 293558. Part (a) from Buhl EH, Tamas G, Szilagyi T *et al.* (1997) Effect, number and location of synapses made by single pyramidal cells onto aspiny interneurons of cat visual cortex. *J. Physiol.* (*Lond.*) **500**, 689–713, with permission. Part (b) from Forsythe ID, Westbrook GL (1988) Slow excitatory postsynaptic currents mediated by N-methyl-D-aspartate receptors on cultured mouse central neurons. *J. Physiol.* (*Lond.*) **396**, 515–533, with permission. Part (c) from Cossart R, Esclapez M, Hirsch J *et al.* (1998) GluR5 kainate receptor activation in interneurons increases tonic inhibition of pyramidal cells. *Nature Neurosci.* **1**, 470–478, with permission.

In the presence of APV there is a clear reduction in the EPSP duration (**Figure 10.3b**). The early component of the EPSP which remains is APV-insensitive; i.e. mediated by non-NMDA receptors as it is abolished by CNQX, a non-selective antagonist of AMPA and kainate receptors (not shown).

In the same preparation, the non-NMDA component recorded in the presence of APV can be further separated in AMPA-mediated and kainate-mediated thanks to GYKI 53655, a selective antagonist of AMPA receptors. The kainate component of the EPSP is thus revealed (**Figure 10.3c**). It is a very small-amplitude component that is antagonized by LY 293558, a selective antagonist GluR5-containing kainate receptor.

These observations raise several questions:

- Do all ionotropic glutamate receptors have the same properties, for example the same sensitivity to glutamate, the same ionic permeability?
- Do they have the same conditions of activation, deactivation, and desensitization?
- Do postsynaptic elements contain more than one iGluR type?

10.2 AMPA RECEPTORS ARE AN ENSEMBLE OF CATIONIC RECEPTOR-CHANNELS WITH DIFFERENT PERMEABILITIES TO Ca^{2+} IONS

10.2.1 The diversity of AMPA receptors results from subunit combination, alternative splicing and post-transcriptional nuclear editing

Cloning studies have demonstrated that AMPA selective ionotropic glutamate receptors are built from the four closely related subunits GluR1, GluR2, GluR3 and GluR4. The four predicted polypeptide sequences, each approximately 900 amino acids in length, revealed similarities between 70% (GluR1 and 2) and 73% (GluR2 and 3). These subunits when expressed *in vitro* constitute a high-affinity ³H AMPA and low-affinity kainate receptor type of glutamate-gated ion channel. These different subunits are abundantly and differentially expressed in the brain, as revealed by *in situ* hybridization studies (see Appendix 6.2). Although these different iGluR subunits exhibit some ability to form homomeric channels when expressed by themselves in *Xenopus* oocytes or cultured mammalian cells, it is considered likely that channels are formed *in vivo* by different combinations of subunits. Thus, with four receptor subunits there are already a very large number of potential combinations.

The diversity of iGluRs results not only from subunit combinations but also from two genetic processes: *alternative splicing* and *editing* of the pre-messenger RNA (or primary transcript). Alternative splicing concerns the 38-amino-acid sequence preceding the most C-terminal putative transmembrane domain M4 of each of the four receptor subunits (**Figure 10.2a**).

This small segment has been shown to exist in two versions (with different amino-acid sequences), designated 'flip' and 'flop', and encoded by adjacent exons of the receptor genes. As a consequence, each of the four subunits exists in two molecular forms (GluR1 flip and GluR1 flop, GluR2 flip and GluR2 flop…). When these splicing derivatives are expressed in oocytes, the proteins exhibit different properties (see Section 10.2.4). Native iGluRs may be composed of heteromeric assemblies of different subunits which contain either flip or flop sequences.

Editing is a post-transcriptional change of one or more bases in the pre-mRNA such that the codon(s) encoded by the gene and the codon(s) present in the mRNA differ. It has been established that the sequences necessary for editing lie in the introns. Thus, only primary transcripts can be edited. Therefore, editing is not a regulatory mechanism for mature mRNA, but results from post-transcriptional nuclear editing. In AMPA receptors, editing concerns only the GluR2 subunit. GluR2 subunit possesses an arginine (R) in the M2 putative membrane spanning segment at position 586; whereas in GluR1, GluR3 and GluR4 subunits, glutamine (Q) lies in the homologous position. This functional critical position is referred to as the Q/R site (see **Figure 10.2a**). The arginine codon (CGG) is not found in the GluR2 gene. It is introduced into the GluR2 mRNA by an adenosine to inosine conversion in the respective glutamine codon (CAG) of the GluR2 transcript by a double-stranded RNA adenosine deaminase. The inosine is subsequently read as a guanosine, resulting in the change of codon identity (CGG). Editing is developmentally regulated such that 99% of the GluR2 subunit in postnatal stages is in the edited (R) form. The consequences of the GluR2 editing at the Q/R site are analyzed in Section 10.2.3.

10.2.2 The native AMPA receptor is permeable to cations and has a unitary conductance of 8 pS

When quisqualate (1–10 μM) is applied in the extracellular milieu of cultured central neurons recorded in the outside-out patchclamp configuration, an inward unitary current, i_q, is observed at $V_m = -60$ mV (**Figure 10.4a**). If the application of quisqualate is repeated on different membrane patches recorded at the same membrane potential, one observes that the recorded inward unitary currents are not a homogeneous population. The unitary conductance values corresponding to the four peaks of **Figure 10.4b** are calculated from the different iq, recorded with the equation $i_q = \gamma_q (V - E_{rev})$ given that $E_{rev} = 0$ mV. One set of similar unitary currents appears with a higher frequency than the others,

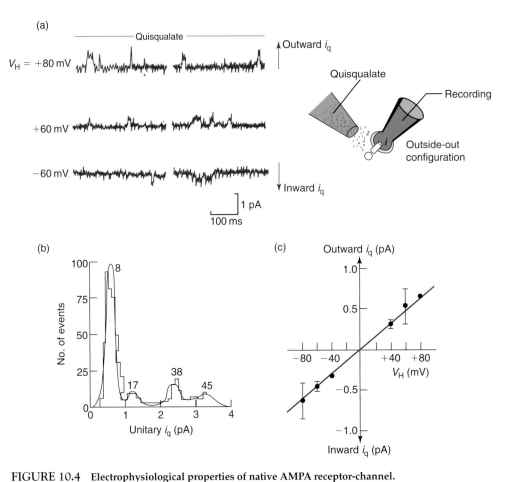

FIGURE 10.4 Electrophysiological properties of native AMPA receptor-channel.
(a) Patch clamp recording (outside-out configuration) of the activity of a quisqualate-activated channel. When the membrane is held at $-60\,mV$, the unitary current i_q is inward (downward deflection). At $+60\,mV$ or $+80\,mV$, i_q is outward (upward deflection). **(b)** Unitary current amplitude histogram (in pA). Currents recorded at the same voltage but from different membrane patches. **(c)** i_q/V curve obtained from the averages of unitary currents recorded from a homogeneous population of channels (8 pS population). Intrapipette solution (in mM): 140 CsCI, 5 K-EGTA, 0.5 CaCl$_2$; extracelluar solution: 140 Nacl, 2.8 KCl, 1 CaCl$_2$. Parts (a) and (c) from Ascher P, Nowak L (1988) Quisqualate and kainate-activated channels in mouse central neurons in culture. *J. Physiol. (Lond.)* **399**, 227–245, with permission. Part (b) from Cull-Candy SG, Usowicz MM (1987) Patch clamp recording from single glutamate-receptor channels. *TIPS* **8**, 218–224, with permission.

the one of $\gamma_q = 8\,pS$. Plotting the mean unitary current amplitude of this population as a function of the membrane potential, we obtain an i_q/V relation (**Figure 10.4c**) that follows the equation $i_q = \gamma_q\,(V - E_{rev})$. Between $-80\,mV$ and $+80\,mV$ this curve is linear (i.e. voltage-independent), has a slope of $8\,pS$, and shows a quisqualate current reversal potential of approximately $0\,mV$.

These results show that the majority of the channels activated by quisqualate have a unitary conductance of $8\,pS$. This conductance is only slightly or not at all sensitive to membrane potential variations. Additionally, the unitary current i_q reverses at a value close to $0\,mV$ if the extracellular and intracellular environments

contain similar concentrations of monovalent cations (**Figure 10.4c**). This suggests that the quisqualate-activated channel is permeable to cations.

To test this hypothesis, the reversal potential of the current is recorded at different intracellular and extracellular concentrations of Na$^+$ and K$^+$ ions. When the extracellular Na$^+$ concentration is lowered from 140 to $50\,mM$ by replacing Na$^+$ ions with choline ions, the reversal potential becomes negative (it shifts from $0\,mV$ to about $-20\,mV$). When Cs$^+$ ions substitute for intracellular K$^+$ ions, the reversal potential is not affected. These results suggest that the quisqualate-activated channel is permeable to Na$^+$, K$^+$ and Cs$^+$ ions, and impermeable to choline ions.

The quisqualate-activated channel shows a low permeability to divalent cations, especially to Ca^{2+}. Thus, variations by a factor of 20 of the extracellular Ca^{2+} concentration have no effect on the reversal potential of the quisqualate current recorded from neurons in the whole-cell configuration. Likewise, only small changes of the photometrically recorded intracellular Ca^{2+} concentration (see Appendix 5.1) can be measured during a quisqualate-evoked response at a constant voltage of –60 mV. It should be noted that it is essential to carry out these recordings in cells maintained at membrane potentials lower than the activation threshold of the voltage-sensitive Ca^{2+} channel in order to prevent Ca^{2+} inflow through these channels.

In conclusion, the native quisqualate-activated channel, recorded in spinal neurons in culture, is permeable to monovalent cations: the application of quisqualate at a membrane potential of $V_m = -60$ mV evokes a unitary inward current that results from the inflow of Na^+ ions and an outflow of K^+ ions through the same channel (the Na^+ inflow is stronger than the K^+ outflow). This AMPA receptor is a classic cationic channel receptor: it has a negligible permeability to Ca^{2+} ions ($P_{Ca}/P_{Na} = 0.1$) and its conductance is only weakly voltage-dependent.

However, studies performed in other preparations show that some AMPA receptors are permeable to Ca^{2+} ions. The above pioneering electrophysiological experiments which characterized the properties of native AMPA receptor-channels were carried out before molecular cloning of glutamate receptor subunits had been achieved, which showed that the Ca^{2+} permeability of AMPA receptor-channels varies with their subunit composition. In the example of **Figure 10.4**, the native AMPA receptors studied probably contained in their structure the edited form of GluR2 (GluR2(R)) as explained below.

10.2.3 AMPA receptors are permeable to Na^+, K^+ and Ca^{2+} ions unless the edited form of GluR2 is present; in the latter case, AMPA receptors are impermeable to Ca^{2+} ions

The mRNA editing at the Q/R site for AMPA receptors concerns only the GluR2 subunit. This change from glutamine (Q) to arginine (R) (**Figure 10.5a**) has a major consequence on the channel permeability: those channels with a GluR2(R) edited subunit are Ca^{2+}-impermeable while those with a GluR2(Q) non-edited subunit or with no GluR2 subunit are Ca^{2+}-permeable.

To show this, the current response to glutamate of homomeric iGluRs expressed in transfected cells is studied in the presence of extracellular solutions containing Na^+ or Ca^{2+} as the only cations. In cells expressing the GluR2(R) subunit only, the glutamate-evoked current is present in high Na^+ solution and nearly absent in high Ca^{2+} solution, indicating that this homomeric channel has a low divalent/monovalent permeability ratio. Moreover, heteromeric GluR2(R) + GluR1 subunit association forms Ca^{2+}-impermeable oligomeric channels in oocytes (**Figure 10.5b**). The situation is different in the absence of the GluR2(R) subunit since homomeric GluR1, GluR3 or heteromeric GluR1 + GluR3 channels allow the influx of Ca^{2+}. Therefore, the presence of a positively charged side-chain of one amino acid (R) determines the divalent/monovalent permeability ratio and thus GluR2(R) would dominate the properties of ion flow through the heteromeric GluR channel. The positive charge of arginine at the Q/R site hinders the permeation of divalent cations in AMPA channels.

10.2.4 The presence of flip or flop isoforms plays a role on the amplitude of the total AMPA current

Each of the GluR1–GluR4 subunits exists in two different forms, flip and flop, created by alternative splicing of a 115-base-pair region immediately preceding M4 (**Figures 10.2** and **10.5a**). Nine amino acids in this region are different between flip and flop versions. When these splicing derivatives are expressed in oocytes, the AMPA receptors exhibit different properties: AMPA receptors incorporating the flip sequence allow more current entry into the cell in response to glutamate than receptors containing solely flop modules (**Figure 10.5c**). Native AMPA receptors may be composed of heteromeric assemblies of different subunits which contain either flip or flop sequences. For example, native AMPA receptors expressed in rat cerebellar granule cells are known to be composed of both flip and flop forms of GluR2 and GluR4, with the flop isoform increasing with age.

10.3 KAINATE RECEPTORS ARE AN ENSEMBLE OF CATIONIC RECEPTOR-CHANNELS WITH DIFFERENT PERMEABILITIES TO Ca²⁺ IONS

Kainic acid is a powerful neurotoxin, which kills neurons by means of overexcitation. It is isolated from a seaweed known for its potency at killing intestinal worms. The word 'kainic' is derived from the Japanese *kaininso* which means the 'ghost of the sea'. At nanomolar concentrations, kainate is a preferential agonist of

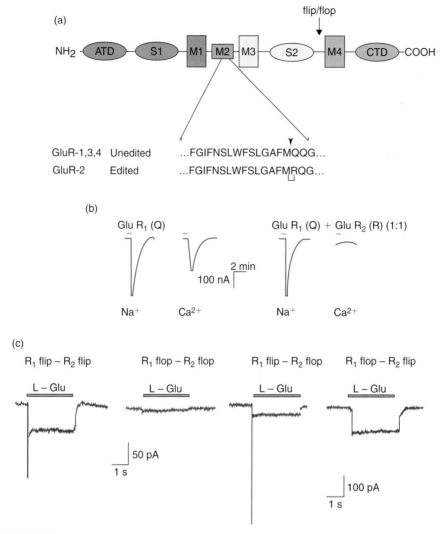

FIGURE 10.5   Functional properties of AMPA receptors.
(a) Linear representation of the predicted structure of AMPA receptors. The regions where alternative splicing and Q/R occur are indicated. **(b)** Comparison of whole-cell currents evoked by pulse application (25 s, bars) of a glutamate agonist to homomeric GluR1 (Q) (left) or heteromeric GluR1(Q) + GluR2(R) (right) channels expressed in oocytes and recorded in normal Ringer (Na$^+$) and Ca^{2+}-Ringer (Ca^{2+}) solutions. Oocytes were injected with a single GluR subunit cRNA (2 ng) or a combination of two types of GluR subunit cRNA (2 ng + 2 ng for 1:1 combination). Intrapipette solution (in mM): 250 CsCl, 250 CsF, 100 EGTA. Na-external solution (in mM): 115 NaCl, 2.5 KCl, 1.8 CaCl$_2$, 10 Hepes; Ca^{2+}-external solution (in mM): 10 CaCl$_2$, 10 Hepes. **(c)** Whole-cell recordings of the inward current evoked by rapid application of an L-glutamate agonist (300 μM) at a holding potential of −60 mV in cultured mammalian cells engineered for the transient expression of the flip forms or the flop forms of GluR1 and GluR2. Part (a) from Sommer B, Keinänen K, Verdoorn TA *et al.* (1990) Flip and flop: a cell-specific functional switch is glutamate-operated channels of the CNS. *Science* **249**, 1580–1585, with permission. Part (c) from Hollmann M, Hartley M, Heinemann S (1991) Ca^{2+} permeability of KA-AMPA-gated glutamate receptor channels depends on subunit composition. *Science* **252**, 851–853, with permission.

kainate receptors but at higher concentrations it also activates AMPA receptors (see **Figure 10.1a**).

10.3.1   The diversity of kainate receptors

A family of kainate receptors has been cloned and five subunits termed GluR5, GluR6, GluR7, KA1 and KA2 have been identified (see **Figure 10.1c**). GluR5 to GluR7 may represent subunits with a low-affinity kainate-binding site and a dissociation constant in the range of 50–100 nM, whereas KA1 and KA2 correspond to subunits with a high-affinity kainate-binding site (K_D of 5–15 nM). GluR5 to GluR7 are of similar size and share 75–80% amino acid sequence identity with

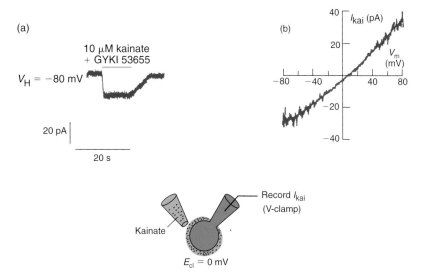

FIGURE 10.6   Whole-cell kainate current in cerebellar granule cells.
(a) Whole-cell current (I_{kai}) evoked by the application of 10 μm of kainate in the presence of GYKI 53655 (100 μM) in a concanavalin A-treated granule cell. **(b)** I_{kai}/V relationship. I_{kai} is measured during 500 ms voltage ramps from −80 to +80 mV. From Pemberton KE, Belcher SM, Ripellino JA, Howe JR (1998) High affinity kainate-type ion channels in rat cerebellar granule cells. *J. Physiol. (Lond.)* **510**, 401–420, with permission.

each other and around 40% with AMPA receptors. KA1 and KA2 share 70% amino-acid sequence identity with each other and around 40% with either GluR1–GluR4 or GluR5–GluR7.

Kainate receptors are tetrameric combinations of five subunits. Of these, GluR5-7 can form functional homomeric or heteromeric receptors. In contrast, KA1 or KA2 expression does not generate agonist-sensitive channels. They participate in heteromeric receptors, partnering any of the GkuR5-7 subunits.

Like those for AMPA receptors, the mRNAs encoding for kainate receptors are subject to editing and/or alternative splicing. GluR5 and GluR6 mRNA can be edited at a Q/R site in the M2 segment (see **Figure 10.2a**), but not GluR7, KA1 and KA2. In analogy with the GluR2 subunit of AMPA receptors, the presence of an arginine residue results in receptors that have low permeability to Ca²⁺. However, in contrast to GluR2, the Q/R site editing is incomplete during development and significant amounts of both edited and non-edited versions of GluR5 and GluR6 coexist in adult brain. Alternative splicing of GluR5, GluR6 and GluR7 further adds to receptor diversity whereas the KA1 and KA2 subunits do not undergo any known process of alternative splicing or RNA editing.

10.3.2 Native kainate receptors are permeable to cations

One way to study kainate channels in isolation is to apply kainate (or the agonist SYM 2081) in the presence

of a selective antagonist of AMPA receptors, particularly the 2,3 benzodiazepine GYKI 53655. Granule cells of the rat cerebellar cortex are dissociated and plated in culture dishes. Whole-cell recordings are performed in voltage clamp mode in order to record the total kainate current (I_{kai}). Concanavalin A is added in the extracellular solution to reduce kainate receptor desensitization. Kainate 10 μM in the presence of GYKI 53655 evokes an inward current when the membrane is held at –80 mV (**Figure 10.6a**). The current/voltage curve obtained by varying the holding potential is shown in **Figure 10.6b**. I_{kai} reverses around 0 mV, suggesting that it is carried by cations (in the presence of control extra- and intracellular solutions). Experiments in high Na⁺ or Cs⁺ solutions have confirmed the cationic permeability of kainate channels. Editing of either M1 or M2 sites changes the Ca²⁺ permeability of GluR6 subunits (**Figure 10.7a**).

Editing at the Q/R site affects the I_{kai}/V relationship in hippocampal neurons. In cells expressing GluR6(Q), the non-edited form of GluR6 at the M2 site, as shown by single-cell RT-PCR, the current/voltage relationship shows a strong inward rectification (**Figure 10.7b**); whereas in a cell expressing the edited form, GluR6(R), the curve is almost linear (**Figure 10.7c**). This shows a clear relationship between the rectification properties of native kainate receptors and editing of Q/R site in the GluR6 subunit mRNA. One of the possible physiological consequences for the presence of rectification is that in a membrane depolarized to around –20 mV, the GluR6(Q)-mediated current is extremely small.

(a)

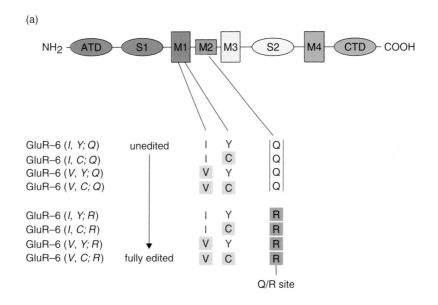

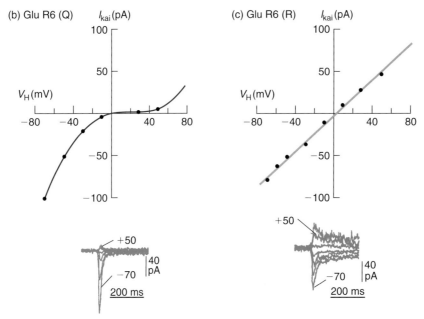

FIGURE 10.7 Correlation of functional properties of native kainate receptors and RNA editing of the Q/R site of the GluR6.

(a) Eight GluR6 variants generated by editing codons in the M1 and M2 segments. An approximate relative frequency for each variant is indicated on the right. **(b, c)** Current/voltage relationships of the kainate-induced current in two different hippocampal neurons, **(b)** expressing homomeric GluR6(Q) and **(c)** expressing homomeric GluR6(R). Kainate (300 μM) is rapidly applied while holding the membrane potential at different voltages, from −70 to +50 mV. Insets show the current traces at these different voltages. Part (a) from Köhler M, Burnashev N, Sakmann B, Seeburg PH (1993) Determinants of Ca²⁺ permeability in both TM1 and TM2 of high affinity kainate receptor channels: diversity by RNA editing. *Neuron* 10, 491–500, with permission. Part (b) from Ruano D, Lambolez B, Rossier J *et al.* (1995) Kainate receptor subunits expressed in single cultured hippocampal neurons: molecular and functional variants by RNA editing. *Neuron* 14, 1009–1017, with permission.

10.4 NMDA RECEPTORS ARE CATIONIC RECEPTOR-CHANNELS HIGHLY PERMEABLE TO Ca^{2+} IONS; THEY ARE BLOCKED BY Mg^{2+} IONS AT VOLTAGES CLOSE TO THE RESTING POTENTIAL, WHICH CONFERS STRONG VOLTAGE DEPENDENCE

10.4.1 Molecular biology of NMDA receptors

Molecular cloning has identified to date seven cDNAs encoding NR1, NR2A-D and NR3A-B subunits of the NMDA receptor (see **Figure 10.1c**), the deduced amino acid sequences of which are 18% (NR1 and NR2), 55% (NR2A and NR2C), 70% (NR2A and NR2B) or around 25% (NR3 and NR1 or NR3 and NR2) identical. When the *Xenopus* oocyte system and transfected mammalian cells are employed to study the functional properties of these subunits, large currents are measured only in oocytes co-expressing NR1 and NR2 subunits. Actually, native NMDA receptors are obligate hetero-oligomers and functional NMDA receptors are composed of the constitutive NR1 subunit and one or more of four different NR2 subunits. The presence and role of NR3 in native NMDA receptors are poorly understood.

Recombinant heteromeric NMDA receptors display different properties depending on which of the four NR2 subunits are assembled with NR1. Site-directed mutagenesis has revealed that the NR2 subunit carries the binding site for glutamate whereas the homologous domains of the NR1 subunit carries the binding site for the co-agonist glycine (see **Figure 10.2b**).

The crystal structures of the ligand-binding core of NR2A bound to glutamate and NR1-NR2A heterodimer bound to glycine and glutamate gave important data. The first one defined the determinants of glutamate and NMDA recognition whereas the second one suggested a mechanism for ligand-induced ion channel opening. Also, the crystal structure of the NR1 subunit glycine-binding domain (NR1S1S2) revealed that it too retains a general clamshell-like architecture reminiscent of the other AMPA and kainate subunit-binding pockets. Why does the binding core of NR1 recognize glycine and not glutamate? One amino acid in the NR1 subunit glycine-binding pocket, a Trp side-chain at position 731, has a key role in selectivity by occupying space required for the glutamate γ-carboxy group; as a result, glutamate cannot bind to the NR1 subunit.

NR1 and NR2 subunits carry in the M2 segment (which forms a re-entrant P loop) an asparagine residue in a position homologous to the Q/R site of AMPA receptors and NR2 subunits carry an asparagine at the N + 1 site (see **Figure 10.11a**). Expression of modified subunits in *Xenopus* oocytes showed that these asparagines are crucial for the particular properties of divalent ion permeation of NMDA channels.

10.4.2 Native NMDA receptors have a high unitary conductance of 50 pS or a lower one of 20–35 pS depending on subunit composition

The experiments related in this section allowed the characterization of the electrophysiological properties of native NMDA channels. They were carried out before molecular cloning of NMDA receptor subunits had been achieved. As has already been pointed out, the NMDA channel is blocked by extracellular Mg^{2+} ions at voltages close to the resting potential of the cell. Mg^{2+} ions block the channel in the open state thus preventing the passage of other ions. The concentrations of Mg^{2+} that produce a significant block are similar to the concentrations of Mg^{2+} normally present in the extracellular milieu. For the sake of clarity, we shall first look at the conductance and permeability properties of the NMDA channel in the absence of extracellular Mg^{2+}. Subsequently, we shall consider the nature of the changes that occur in a medium containing Mg^{2+} ions.

Let us look at patch clamp recordings (outside-out configuration) of cultured central neurons in the absence of Mg^{2+} ions. The application of NMDA (10 µM) in the extracellular milieu at $V = -60$ mV induces an inward unitary current, i_N (**Figure 10.8a**). The i_N/V relation obtained under these conditions is linear (between -80 mV and $+60$ mV) and is described by the equation $i_N = \gamma_N (V - E_{rev})$ (**Figure 10.8b**). The slope of this curve corresponds to the unitary conductance of the NMDA channel, γ_N. The average value of γ_N is in the range 40–50 pS. The unitary conductance of NMDA channels is only slightly voltage-dependent in the absence of Mg^{2+} ions.

We now know that the frequently described 50 pS openings are associated with NR2A- or NR2B-containing receptors whereas NR2C- or NR2D-containing receptors display low-conductance (20–35 pS) openings.

10.4.3 The NMDA channel is highly permeable to monovalent cations and to Ca^{2+}

The i_N/V relation shows that i_N reverses at a membrane potential value close to 0 mV when the extracellular and intracellular cationic concentrations are similar. This value suggests that the NMDA channel is permeable to cations. Replacing any of the monovalent cations (Na$^+$, K$^+$ or Cs$^+$) by another, induces only minor changes in the reversal potential; i.e. the channel

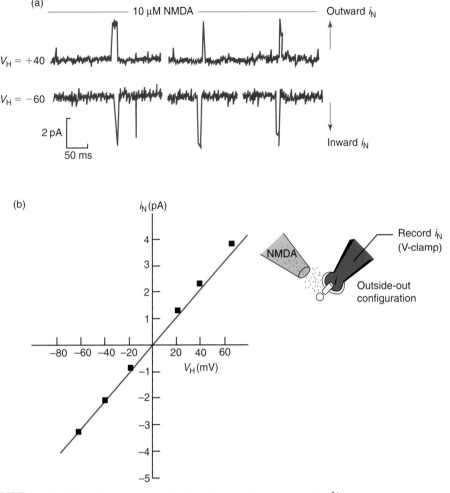

FIGURE 10.8   Unitary NMDA current in the absence of extracellular Mg^{2+}.
(a) Outside-out patch clamp recordings of the activity of a NMDA ($10\,\mu M$) activated channel at two holding potentials, -60 and $+40\,mV$. (b) i_N/V relation obtained from the averages of unitary currents i_N recorded from a homogeneous population of channels (a population that shows a $40-50\,pS$ unitary conductance). Intrapipette solution (in mM): 140 CsCl, 5K-EGTA, 0.5 $CaCl_2$; extracellular solution: 140 NaCl, 2.8 KCl, 1 $CaCl_2$. Part (a) from Ascher P, Bergestovski P, Nowak L (1988) *N*-methyl-D-aspartate-activated channels of mouse central neurons in magnesium free solutions. *J. Physiol. (Lond.)* **399**, 207–226, with permission. Part (b) from Cull-Candy SG, Usowicz MM (1987) Patch clamp recording from single glutamate-receptor channels. *TIPS* **8**, 218–224, with permission.

discriminates only slightly between the different monovalent cations.

In order to establish whether the NMDA channel is permeable to Ca^{2+} or not, two types of experiments have been performed. The first type of experiment consisted of photometric measurements (see Appendix 5.1) of the variations of intracellular Ca^{2+} in response to NMDA. To carry out these experiments and to prevent the activation of voltage-dependent Ca^{2+} channels, the activity of a spinal neuron is recorded in patch clamp (whole-cell configuration) at a holding potential of $-60\,mV$ (**Figure 10.9a**). Under these conditions, the intracellular Ca^{2+} concentration strongly increases

during an NMDA-evoked response. This augmentation is selectively blocked by the NMDAR antagonist APV or the NMDA channel blocker MK 801. Since it disappears in Ca^{2+}-free external solution, it clearly results from an influx of Ca^{2+} through NMDA channels and is not due to a release of these ions from intracellular storage pools of Ca^{2+}. Also, changes in the extracellular Ca^{2+} concentration are accompanied by changes in the reversal potential of the macroscopic NMDA current (**Figure 10.9b**). Furthermore, an inward single-channel current is recorded at $-60\,mV$ under conditions where Ca^{2+} is the only cation present in the extracellular milieu; i.e. when Ca^{2+} ions are the only

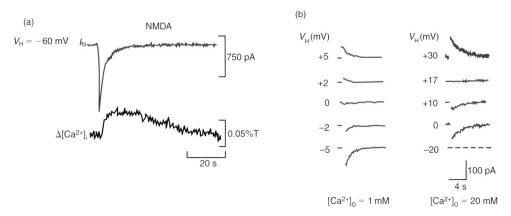

FIGURE 10.9   Optical measurements of intracellular Ca^{2+} concentration changes during a NMDA-evoked response.
The Ca^{2+}-sensitive dye Arsenazo III is used. The absorption coefficient of this dye varies at certain wavelengths when it complexes Ca^{2+} ions. The activity of cultured spinal neurons is recorded in the whole-cell patch clamp configuration in the absence of external Mg^{2+}. **(a)** Pressure application of $1\,\mu M$ of NMDA (20 ms) in the presence of 2.5 mM of Ca^{2+} in the extracellular milieu evokes an inward current I_N (top trace). During this response there is an increase of $[Ca^{2+}]_i$ (bottom trace). **(b)** Reversal potential of the whole-cell NMDA current (I_N) as a function of extracellular $[Ca^{2+}]_o$. Currents activated by the application of 1 mM of NMDA are recorded at different membrane potentials in the presence of 1 mM (left) or 20 mM (right) of $[Ca^{2+}]_o$. Part (a) from Mayer ML, MacDermott AB, Westbrook GL *et al.* (1987) Agonist- and voltage-gated calcium entry in cultured mouse spinal chord neurons under voltage clamp using Arsenazo III. *J. Neurosci.* **7**, 3230–3244, with permission. Part (b) from MacDermott AB, Mayer ML, Westbrook GL. *et al.* (1986) NMDA-receptor activation increases cytoplasmic calcium concentration in cultured spinal neurons. *Nature* **321**, 519–522, with permission.

ions that can carry this current. All these results indicate that Ca^{2+} ions actually carry part of the NMDA current.

What does molecular biology tell us about Ca^{2+} permeability?

The molecular substrate for Ca^{2+} permeability of NMDA channels is analyzed by exchanging (by site-directed mutagenesis) either glutamine (Q) or arginine (R) for asparagine (N) in the M2 domain of NR1 or NR2A subunits (see **Figure 10.11a**). Wild-type and mutant NR subunits are co-expressed by cells transfected with cDNAs. Whole-cell currents are activated by application of L-glutamate to transfected cells expressing heteromeric wild type or mutant NMDA receptors and differences in Ca^{2+} permeability are analyzed. Replacing the asparagine (N) by arginine (R) in the NR1 subunit generates 'mutant NR1-wildtype NR2A' channels that do not exhibit a measurable Ca^{2+} permeability (see **Figure 10.11b**). In high Ca^{2+} solution, glutamate evokes a small outward current at $V_H = -60\,mV$, indicating a low Ca^{2+} permeability of the mutant channel. Thus, when the positively charged arginine (R) occupies the critical position in M2 of the NR1 subunit, Ca^{2+} ions appear to be prevented from entering the channel, suggesting that the size and the charge of the amino acid present at this critical position in the M2 segment are important for Ca^{2+} permeability.

10.4.4 NMDA channels are blocked by physiological concentrations of extracellular Mg^{2+} ions; this block is voltage-dependent

Single NMDA channel current evoked by NMDA ($10\,\mu M$) is recorded in the presence of increasing concentrations of extracellular Mg^{2+} ions (in μM: 0, 10, 50, 100) (**Figure 10.10a**). At $V_H = -60\,mV$, NMDA evokes an inward unitary current whose amplitude remains constant at all the Mg^{2+} concentrations tested ($i_N = -2.7\,pA$). However, whereas in the absence of Mg^{2+} the NMDA channel opens for periods of several milliseconds (0), in the presence of Mg^{2+} (10, 50, 100 μM) the recordings show bursts of short openings during which the channel fluctuates between the open state (mean duration t_o) and blocked state (mean duration t_{bl}). At a Mg^{2+} concentration of 100 μM, the unitary current appears to decrease. This is due to the high Mg^{2+} blocking frequency which makes channel openings too brief to be resolved by the recording system. The repeated closures of the channel (which correspond to fluctuations between the open and the blocked states) strongly diminish the average time during which the channel is open. Note that when the recorded unitary current is outward at $V_H = +40\,mV$, the presence of Mg^{2+}, even at a concentration of 100 μM, has no effect on the channel open time. The most interesting aspect

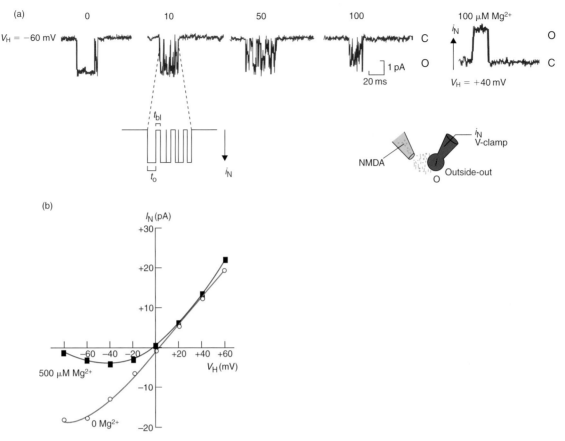

FIGURE 10.10   NMDA channel block by extracellular Mg²⁺ ions.

(a) Outside-out patch clamp recording of the activity of cultured central neurons. Application of NMDA (10 μM) in the absence of external Mg²⁺ ions (O) and in the presence of Mg²⁺ (10, 50, 100 μM) at $V_H = -60$ mV. At $V_H = +40$ mV, i_N is outward. **(b)** Voltage sensitivity of the NMDA response in the presence of extracellular Mg²⁺ ions. The total current I_N is recorded in the whole-cell patch clamp configuration in the absence (o), and in the presence (■) of 500 μM of Mg²⁺. Part (a) from Ascher P, Nowak L (1988) The role of divalent cations in the *N*-methyl-D-aspartate responses of the mouse central neurons in culture. *J. Physiol. (Lond.)* **399**, 247–266, with permission. Part (b) from Nowak L, Bregestovski P, Ascher P *et al.* (1984) Magnesium free glutamate-activated channels in mouse central neurones. *Nature* **307**, 463–465, with permission. C – Closed state; O – Open state.

of this block is its voltage-dependence. Mg²⁺ block becomes progressively stronger as voltage is made more and more hyperpolarized while at positive voltages (where the current i_N is outward) block by physiological Mg²⁺ concentrations is hardly visible.

The properties of Mg²⁺ block of unitary currents can be used to predict macroscopic currents (whole-cell configuration). Recall that $I_N = Np_o i_N$. We know that i_N is approximately constant at a given membrane potential irrespective of the Mg²⁺ concentration, and that the mean open time of each channel (and therefore the open state probability p_o of the channel as well) decreases as a function of the Mg²⁺ concentration. From the recordings in **Figure 10.10a** we can predict that at negative potentials, in the presence of Mg²⁺, the macroscopic current I_N will be small.

As a matter of fact, the I_N/V curve described by the equation $I_N = G_N (V - E_{rev})$ in the presence of extracellular Mg²⁺ ions (500 μM) is not linear. This nonlinearity appears at negative voltages; i.e. when the current is inward (**Figure 10.10b**). Furthermore, we observe that for V_H between −35 and −80 mV, the current amplitude diminishes (region of negative conductance) instead of increasing, as would be expected from the increasing electrochemical gradients for Na⁺ and Ca²⁺ and the decreasing electrochemical gradient for K⁺. This peculiar property of the I_N/V curve in this region of voltage is due to the block of the channel by Mg²⁺. Since Mg²⁺ ions are normally found in the extracellular milieu at concentrations of approximately 1 mM, at membrane potentials close to the resting potential a majority of NMDA channels are blocked.

Mechanism of action of Mg^{2+} ion block: a hypothesis

Mg^{2+} ions block open NMDA channels, thus preventing the passage of Na^+, Ca^{2+} and K^+ ions. The probability that a Mg^{2+} ion will enter the NMDA channel increases with the level of membrane hyperpolarization: the greater the electrical gradient, the stronger are the Mg^{2+} ions attracted into the channel. For this reason, the block of NMDA channels by Mg^{2+} is voltage-sensitive. This block can be symbolized as follows:

$$R + NMDA \rightleftarrows NMDA - R^* \overset{+Mg^{2+}}{\rightleftarrows} NMDA - R^*\text{-}Mg$$

where R is the NMDA receptor in the closed state, R* is the NMDA receptor in the open state, and R*-Mg is the open NMDA receptor blocked by Mg^{2+} ions. The reaction $R^* + Mg^{2+} \rightleftarrows R^*\text{-}Mg$ is strongly favoured to the right when $[Mg^{2+}]$ is increased and when V is hyperpolarized.

Why is the NMDA channel permeable to Ca^{2+} ions and blocked by Mg^{2+} ions?

One can separate the effects of cations into two groups:

- those, like Ca^{2+}, which pass through the NMDA channel (e.g. Ba^{2+}, Cd^{2+});
- those which mimic the Mg^{2+} effect, i.e. block the NMDA channel (e.g. Co^{2+}, Ni^{2+}, Mn^{2+}).

The difference between the ions that pass through the channel and those that block it coincides with the difference in the speed with which the water molecules surrounding these ions can exchange with other water molecules of the aqueous solution. In fact, this exchange is a thousand times faster for the group of permeable (Ca^{2+}-like) ions than for the group of blocking (Mg^{2+}-like) ions.

These differences have led to the suggestion that both ions can cross the channel but only in their dehydrated forms. The following model of the channel has been proposed. The channel has a large extracellular entrance and presents a narrow constriction towards the intracellular side through which the ions can cross only in their dehydrated form. The narrow constriction is formed by non-homologous asparagine residues: the N site of the NR1 subunit and an adjacent one, to the N site in NR2, the N+1 asparagine (**Figure 10.11a**). The size of hydrated Mg^{2+}, at least 0.7 nm, is larger than the estimated 0.55 nm pore size of NMDA receptor channels suggesting that Mg^{2+} block at the narrow constriction could arise by steric occlusion. Because of the slow rate of dehydration of the cations from the Mg^{2+} group these ions are trapped in the interior of the channel, thus blocking it. Another hypothesis for the blockade by Mg^{2+} ions is that a high-affinity binding site for Mg^{2+} ions exists inside the channel so that Mg^{2+} ions would cross the channel slowly and thus block it for the time of passage.

What does molecular biology tell us about Mg^{2+} block?

The molecular substrate for the hypothesis of a high-affinity site in the NMDA channel for Mg^{2+}-ion binding is analyzed by exchanging (by site-directed mutagenesis) either glutamine (Q) or arginine (R) for asparagine (N) in the M2 domain of NR1 or NR2A. Wild-type and mutant NR subunits are co-expressed by cells transfected with cDNAs. Whole-cell currents are activated by application of L-glutamate to transfected cells expressing heteromeric wild-type or mutant NMDA receptors, and differences in Ca^{2+} or Mg^{2+} permeability and channel block by extracellular Mg^{2+} are analyzed. Replacing the asparagine (N) by arginine (R) in the NR1 subunit generates 'mutant NR1-wild-type NR2A' channels that do not exhibit a measurable Ca^{2+} permeability (**Figure 10.11b**), as already pointed out. Whole-cell I_N/V curves in (1) divalent ion-free external solution and (2) after adding 0.5 mM of Mg^{2+} to the external solution superimpose almost completely, showing that the mutant channel is not blocked by external Mg^{2+} ions (**Figure 10.11c**). Conversely, the presence of glutamine (Q) instead of asparagine (N) in the NR2A subunit generated 'wild-type NR1-mutant NR2A' channels with increased Mg^{2+} permeability and thus reduced sensitivity to block by extracellular Mg^{2+}. The whole-cell current evoked by glutamate is recorded in (1) divalent ion-free Ringer and (2) after addition of 0.1 mM of Mg^{2+} as a function of membrane potential. The I_N/V relations show that the wild-type channel consisting of NR1 and NR2A subunits (**Figure 10.11d**) is blocked by external Mg^{2+}, while the mutant channel comprising wild-type NR1 and mutant NR2A (N595Q) subunits is permeable to Mg^{2+} since it is only slightly blocked by external Mg^{2+} (**Figure 10.11e**). Moreover, substitutions of the two adjacent asparagines in the NR2A subunit strongly reduce the block. These effects show little dependence on pore size thus suggesting that the block does not arise by steric occlusion.

10.4.5 Glycine is a co-agonist of NMDA receptors

Glycine is an amino acid that acts as an inhibitory neurotransmitter at certain central nervous system

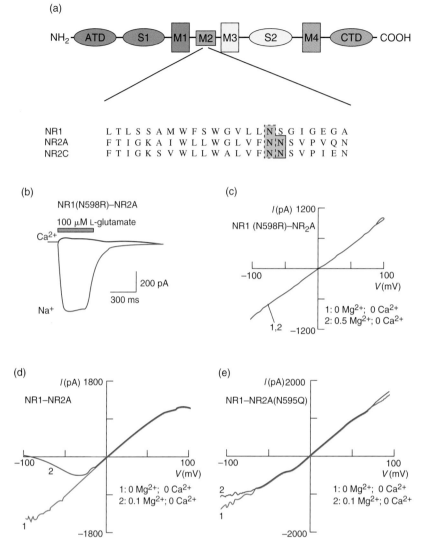

FIGURE 10.11 Permeability of NMDA channels to divalent cations.
(a) Linear representation of a NMDA receptor subunit. Hydrophobic regions (M1 to M4) are boxed. The M2 segment is expanded to list sequences of subunits belonging to NMDA receptors: NR1, NR2A, NR2C. They carry an asparagine residue (N) in a position homologous to the Q/R site of AMPA and kainate receptor-subunits. (b,c) Reduction of Ca^{2+} permeability and channel block by extracellular Mg^{2+} in a mutant channel where asparagine (N) in the M2 segment of the NR1 subunit is replaced by arginine (R). (b) Whole-cell current elicited by 100 μM of glutamate (bar) at $V_H = -60$ mV in high Na^+ (inward current) or high Ca^{2+} (small outward current) extracellular solution. (c) Whole-cell I_N/V relations in (1) divalent ion-free external solution and (2) after adding 0.5 mM of Mg^{2+} to the external solution. (d,e) Difference in channel block by extracellular Mg^{2+} between wild-type and mutant NMDA receptor-channels. In the NR2A (N595Q) subunit, one asparagine (N) in the M2 segment is replaced by glutamine (Q) by site-directed mutagenesis. The whole-cell current evoked by glutamate is recorded in (1) divalent ion-free Ringer and (2) after addition of 0.1 mM of Mg^{2+} as a function of membrane potential from (d) wild-type channels and (e) a mutant channel. Extracellular high Na^+ solution (in mM): 140 Nacl, 5 HEPES: high Ca^{2+} solution: 110 $CaCl_2$, 5 HEPES. Divalent ion-free Ringer's solution: 135 NaCl, 5.4 KCl, 5 HEPES. Part (a) from Wisden W and Seeburg PH (1993) Mammalian ionotropic glutamate receptors. *Curr. Opin. Neurobiol.* **3**, 291–298, with permission. Parts (b)–(e) from Burnashev N, Schoepfer R, Monyer H *et al.* (1992) Control of calcium permeability and magnesium blockade in the NMDA receptor. *Science* **257**, 1415–1419, with permission.

synapses in vertebrates. However, this amino acid also plays a role in modulating the NMDA response. The effect of glycine on the NMDA receptor channel has been discovered in cultured central neurons recorded in the whole-cell patch clamp configuration. When NMDA was applied by slow perfusion the response was much larger than when NMDA was rapidly perfused into the bath. The following hypothesis was proposed.

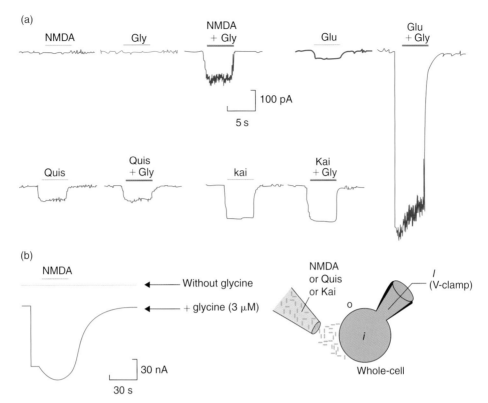

FIGURE 10.12 Potentiation of the NMDA response by glycine.
(**a**, top) Whole-cell currents evoked in cultured central neurons in response to $10\,\mu M$ of NMDA or $10\,\mu M$ of glutamate at $V_H = -50\,mV$ in the absence or presence of $1\,\mu M$ of glycine (Gly). (**a**, bottom) The same experiment with quisqualate (Quis) or kainate (Kai) applications. Glycine by itself does not trigger an inward current at any concentration, through either NMDA or non-NMDA channels. (**b**) Whole-cell inward current (66 ± 13 nA) evoked in *Xenopus* oocytes which express NMDA receptors, in response to $300\,\mu M$ of NMDA at $V_H = -60\,mV$ in the absence of presence of $3\,\mu M$ of glycine. In (**a**) and (**b**) the extracellular solution is devoid of Mg^{2+} ions. Part (a) from Johnson JW and Ascher P (1987) Glycine potentiates the NMDA response in cultured mouse brain neurons. *Nature* **325**, 529–531, with permission. Part (b) from Kleckner N and Dingledine R (1988) Requirements for glycine in activation of NMDA receptors expressed in *Xenopus* oocytes. *Science* **241**, 835–837, with permission.

The cultured cells (neurons and glia) tonically release a substance that accumulates in the bath due to the slow perfusion and potentiates the NMDA response. In order to characterize the active substance present in the medium, a variety of treatments were applied. It was established that its activity was still present after heating the medium to 90°C, and that its molecular weight is below 700 D. After testing the most common amino acids, glycine proved to be the most effective in reproducing the effects of the conditioning medium on the NMDA response (**Figure 10.12a**).

Patch clamp outside-out recordings showed that glycine potentiates the NMDA response by augmenting the NMDA receptor-channel opening frequency (thus increasing the open-state probability of the channel, p_o). The molecular mechanisms of this potentiating effect remained to be determined. Nevertheless, the fact that glycine has an effect on NMDA receptor-channels recorded from excised outside-out patches rules out

the mediation of its effect by a diffusible, intracellular second messenger. It was then suggested that glycine would in fact be indispensable for the activation of the NMDA receptor-channel by its agonists. When the activity of NMDA receptor-channels expressed in oocytes is recorded in the whole-cell patch clamp configuration, a current in response to NMDA application is observed only when glycine is also present in the bath (**Figure 10.12b**).

How can these results be interpreted from a physiological perspective? Glycine is in fact present at relatively high concentrations in the cerebrospinal fluid (several μM). This level is close to the concentration required to produce its maximum effect. However, a high-affinity glycine pump may lower extracellular glycine concentration at the level of glutamatergic synapses. These transporters, via the modulation of extracellular glycine concentration, could play a significant role in determining the NMDA response.

10.4.6 Conclusions on NMDA receptors

The NMDA receptor-channel is unique among the glutamate receptors in that it requires at least two conditions for its activation: the binding of both glutamate and the co-agonist glycine to the receptor and the depolarization of the membrane. It is a doubly gated channel: glutamate binds to NR2A-D subunits while glycine binds to NR1. It should be noted that the voltage sensitivity of the NMDA channels (a sensitivity due to an extrinsic ion, namely Mg^{2+}) differs radically from that of voltage-dependent Na^+ and Ca^{2+} channels. In the two latter cases, the voltage sensitivity is an intrinsic property of the protein, which does not require extracellular or intracellular ligands.

The voltage sensitivity of the NMDA channel has important physiological implications. Since these channels are blocked by Mg^{2+} ions at voltages close to the resting potential of the cell, does the presence of the neurotransmitter in the synaptic cleft suffice to evoke a postsynaptic NMDA response? Knowing that the non-NMDA and the NMDA receptors coexist in the postsynaptic membrane, what is the fraction of the synaptic current that is due to the activation of NMDA receptors? These questions are analyzed in the following section.

10.5 SYNAPTIC RESPONSES TO GLUTAMATE ARE MEDIATED BY NMDA AND NON-NMDA RECEPTORS

10.5.1 Glutamate receptors are co-localized in the postsynaptic membrane of glutamatergic synapses

The glutamatergic synapses typically exhibit an electron-dense postsynaptic density where ionotropic glutamate receptors are concentrated as shown by immunogold labelling. To date, among iGluRs, only NMDA and AMPA receptors can be labelled for electron microscopy observation. In contrast to iGluRs, metabotropic glutamate receptors (mGluRs) appear to occur at highest concentrations in the perisynaptic annulus; i.e. the narrow zone surrounding the postsynaptic specialization (**Figure 10.13**). The receptors localized in the postsynaptic specialization are directly apposed to the presynaptic active zone. The enrichment of iGluRs at the site of the postsynaptic specialization reflects their roles in mediating fast glutamatergic transmission. Such enrichment depends on anchoring synaptic proteins (see Chapter 7).

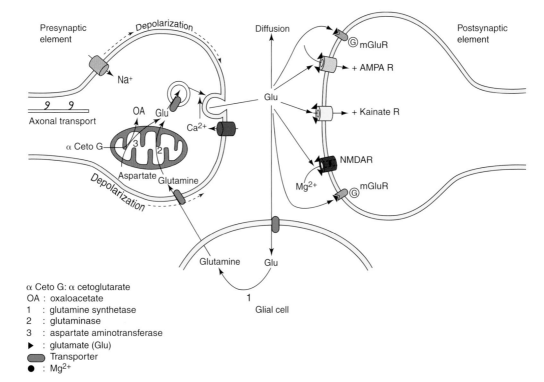

α Ceto G: α cetoglutarate
OA : oxaloacetate
1 : glutamine synthetase
2 : glutaminase
3 : aspartate aminotransferase
▶ : glutamate (Glu)
⬭ Transporter
● : Mg^{2+}

FIGURE 10.13 **The glutamatergic synapse.**
Functional scheme of a glutamatergic synapse where ionotropic and metabotropic glutamate receptors are co-localized. Presynaptic receptors are omitted. The enzymes (1 to 3) and mitochondria are carried to axon terminals via anterograde axonal transports. Glutamate synthesized in mitochondria of the presynaptic element is transported actively into synaptic vesicles by a vesicular carrier. A percentage of the glutamate released in the synaptic cleft is uptaken into presynaptic terminals and glial cells by transporters. Inset shows iGluRs antagonists. Inset from Mody I (1998) Interneurons and the ghost of the sea. *Nature Neurosci.* **1**, 434–436, with permission.

Glutamate released into the synaptic cleft diffuses to the postsynaptic membrane and binds to postsynaptic NMDA as well as non-NMDA receptors (**Figure 10.13**) and evokes a synaptic response which is a postsynaptic excitatory current (EPSC). In turn this EPSC depolarizes the membrane; i.e. it evokes an excitatory postsynaptic potential (EPSP; see **Figure 10.3**). *In vivo*, when the membrane potential is near the resting potential of the cell, a large fraction of the NMDA receptors are blocked by Mg^{2+} ions present in the synaptic cleft. Therefore, glutamate first activates non-NMDA receptors.

- Is the synaptic response to glutamate mediated by non-NMDA receptors only?
- If not, under which conditions are NMDA receptors activated by synaptically released glutamate and how do they contribute to the synaptic response?

To answer these questions it is necessary to differentiate, in the postsynaptic current, between the component due to the activation of non-NMDA and that due to NMDA receptors (**Figure 10.13**, inset).

10.5.2 The glutamatergic postsynaptic current is inward and can have at least two components in the absence of extracellular Mg^{2+} ions

A global postsynaptic inward current (excitatory postsynaptic current; EPSC), evoked by the stimulation of a presynaptic glutamatergic neuron, is recorded in the whole-cell configuration (voltage clamp mode). When the presynaptic neuron is stimulated in the absence of extracellular Mg^{2+} ions and the postsynaptic membrane is held at −46 mV, an inward current showing two components is recorded (**Figure 10.14a**):

- An early component: an initial peak of current of great amplitude and rapid inactivation.
- A late component: a current of smaller amplitude that inactivates slowly.

To identify the NMDA and non-NMDA components of the postsynaptic current we can make use of the different properties of the non-NMDA and the

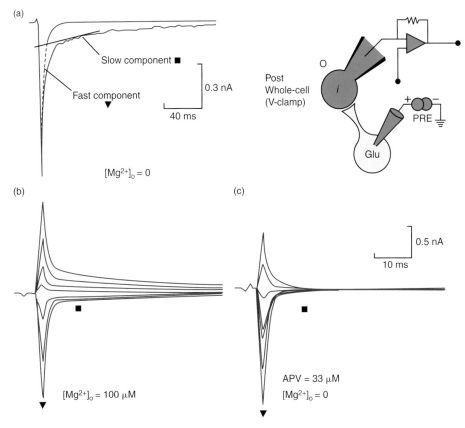

FIGURE 10.14 Postsynaptic inward current evoked by the stimulation of a glutamatergic presynaptic neuron. **(a)** Whole-cell postsynaptic inward current (EPSC) recorded $V_H = −46$ mV in the absence of Mg^{2+}, in response to the activation of a presynaptic glutamatergic neuron. The peak current decays with a time constant $\tau_1 = 4.2$ ms and the slow components decays with a time constant $\tau_2 = 81.8$ ms. **(b)** The EPSC is recorded at different V_H in the presence of Mg^{2+} (100 μM). **(c)** The EPSC is recorded at different V_H in the presence of 33 μM of D-APV and in the absence of extracellular Mg^{2+}. Picrotoxin (10–100 μM) is added to the extracellular solution to block GABAergic inhibitory synaptic activity. From Forsythe ID and Westbrook GL (1988) Slow excitatory postsynaptic currents mediated by *N*-methyl-D-aspartate receptors on cultured mouse central neurons. *J.Physiol. (Lond.)* **396**, 515–533, with permission.

NMDA channels summarized in **Figure 10.1a**. The non-NMDA component is not affected by different concentrations of extracellular Mg^{2+} ions, nor by the presence of APV, but disappears in the presence of CNQX, an antagonist of non-NMDA receptors. On the other hand, the NMDA component is present in a medium devoid of Mg^{2+} ions but disappears in the presence of APV, the competitive antagonist of NMDA receptors.

In the presence of Mg^{2+} ions the late component is largely attenuated at negative potentials but is present at all positive voltages (**Figure 10.14b**, black square). In the absence of Mg^{2+} ions and in the presence of APV in the extracellular environment (**Figure 10.14c**), the late component disappears at all voltages tested. These results strongly suggest that the late component of the synaptic current results from the activation of NMDA receptors (APV-sensitive and blocked by Mg^{2+} at negative voltages). The reverse experiment in the presence of CNQX is not shown.

Which receptor channels contribute to the non-NMDA component of the synaptic current?

The relative participation of AMPA and kainate receptors in the non-NMDA component of the synaptic glutamatergic current required the discovery of selective antagonists as CNQX is not selective of either one of these receptors. Only after the fortuitous discovery that the 2,3-benzodiazepine muscle relaxant GYKI 53655 is a specific AMPA antagonist (see **Figure 10.1a**) could physiologists begin to distinguish between the separate activation of AMPA and kainate receptors at synapses. In many glutamatergic synapses, the non-NMDA postsynaptic current results solely from AMPA receptors. But there are few places where synaptically activated kainate receptors were identified. One of them is the CA3 region of the hippocampus where pyramidal neurons and interneurons receive excitatory input from glutamatergic mossy fibres (see **Figure 6.3**).

EPSCs evoked by the stimulation of afferents are recorded from interneurons in the whole-cell configuration (voltage clamp mode) in the continuous presence of D-APV to block NMDA receptors. A large, rapidly decaying control EPSC with a small, long-lasting tail is recorded (**Figure 10.15a**). Application of GYKI, the AMPA antagonist, blocks the rapid component but the slow component is mostly unaffected. The subsequent addition of the AMPA/kainate antagonist CNQX blocks the GYKI-resistant slow component. When the synapse between afferent glutamatergic fibres and pyramidal CA1 neurons is now studied, no GYKI-resistant (kainate-mediated) component can be shown, although the EPSC in pyramidal cells is more than

twice as large as the EPSC in interneurons (**Figure 10.15b**). This indicates that, at this synapse in CA1, kainate receptors are absent and that the non-NMDA component of the EPSC is mediated only by AMPA receptors.

10.5.3 The glutamatergic postsynaptic depolarization (EPSP) has at least two components in the absence of extracellular Mg^{2+} ions

The postsynaptic response to the stimulation of the presynaptic neurons (same preparation as above) is recorded in whole-cell configuration (current clamp mode to leave the voltage free to vary) in the absence of Mg^{2+} ions. Such stimulation evokes a postsynaptic depolarization (EPSP), which results from the evoked synaptic inward current through postsynaptic ionotropic glutamate receptors. As in the case of the EPSC, the EPSP shows two identifiable components (see **Figure 10.3b**). In the presence of APV the early component is only slightly affected or not at all. This APV-insensitive early component is a result of the early synaptic inward current; i.e. the inward current through non-NMDA receptors. In these conditions (absence of Mg^{2+} ions, presence of APV), the duration of the EPSP is reduced. The difference between the APV-insensitive component of the EPSP and the total EPSP corresponds to the APV-sensitive component; i.e. the component resulting from the inward current through NMDA receptors.

In summary, the component resulting from the activation of the NMDA receptors has a slower rising phase and lasts longer than the component mediated by the non-NMDA receptors. Thus, when the NMDA receptors are activated, the peak of the EPSP is not always affected but the duration of the EPSP is much longer. As for EPSCs, the non-NMDA component of glutamatergic EPSPs is mediated either by AMPA receptors alone or by both AMPA and kainate receptors (see **Figure 10.3c**).

10.5.4 Synaptic depolarization recorded in physiological conditions: factors controlling NMDA receptor activation

The recordings of **Figure 10.14** have shown the presence of two components in the synaptic current and depolarization, when the extracellular medium is Mg^{2+}-free. What is the situation in physiological conditions, when the extracellular physiological milieu has a Mg^{2+} concentration of approximately 1 mM? Since at this Mg^{2+} concentration and at membrane potentials close to the resting potential of the cell, most of the NMDA

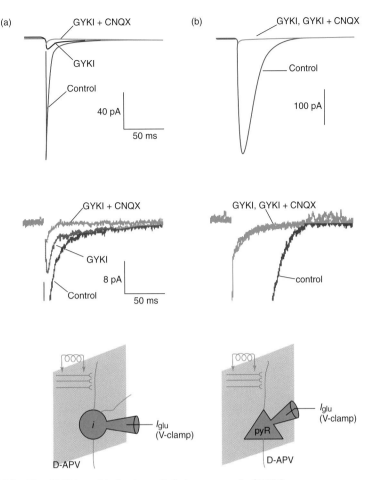

FIGURE 10.15   **The AMPA- and kainate-mediated component of EPSCs.**
Experiments performed in slices of the rat hippocampus (CA1 region). **(a)** Averaged EPSCs recorded from an interneuron (I) in control conditions (continuous presence of $100\,\mu M$ of D-APV), after bath application of $70\,\mu M$ of GYKI 53655 and after addition of $100\,\mu M$ of CNQX. Middle traces are the same EPSCs at high gain. **(b)** The same experiment performed in pyramidal cells (pyR). V_H in (a) and (b) is $-80\,mV$. From Frerking M, Malenka RC, Nicoll RA (1998) Synaptic activation of kainate receptors on hippocampal interneurons. *Nature Neurosci.* **1**, 479–486, with permission.

channels are closed, under which conditions will NMDA receptors participate to synaptic transmission?

It seems unlikely that the extracellular Mg^{2+} concentration *in vivo* will vary sufficiently to allow the 'unblocking'of NMDA receptors. However, depolarizations reduce the level of Mg^{2+} block of the NMDA channel. Thus, one can imagine that a depolarization of the membrane is precisely what allows the NMDA channels to become 'unblocked'. A depolarization can be the consequence of the activation of other receptors present in the postsynaptic membrane, such as non-NMDA receptors. It can also result from the activation of a subpopulation of NMDA receptors that are not blocked at the resting potential. This hypothesis can be summarized as follows.

When NMDA and non-NMDA receptors coexist in the postsynaptic membrane

When the glutamate concentration is sufficiently high to activate non-NMDA receptor channels, a current is generated through these channels and an APV-insensitive depolarization is recorded:

If this non-NMDA mediated depolarization is not strong enough to allow 'unblocking' of the NMDA receptors, only the early component of the depolarization (non-NMDA component) is recorded (**Figure 10.3a**).

If this non-NMDA mediated depolarization is sufficiently strong to 'unblock' NMDA receptors, it triggers the activation of an inward NMDA current through these channels and an additional depolarization of the membrane. This depolarization allows the 'unblocking' of additional NMDA receptors which, activated by glutamate, evoke an enhanced depolarization. The more depolarized the membrane, the higher the number of NMDA receptors activated by glutamate. Note that this regenerative phenomenon, due to the voltage sensitivity of the NMDA receptors (associated with the negative-slope region of the I_N/V curve), reminds us of a similar phenomenon observed with the action potential generating Na^+ channels. In the present case, an important postsynaptic depolarization made up of the non-NMDA early component and the NMDA late component is recorded. However, the NMDA component not only prolongs the EPSP but also allows a significant influx of Ca^{2+} ions. These ions have numerous roles: one of them is the activation of channels sensitive to intracellular Ca^{2+} ions. Another role of intracellular Ca^{2+} is as a second messenger. Consequently, it participates in the regulation of a number of intracellular Ca^{2+}-sensitive processes.

When NMDA receptors are the only receptors present in the postsynaptic membrane

In certain preparations the postsynaptic depolarization recorded in response to the endogenous release of glutamate shows only one component, the NMDA component. This has led to the assumption that not all the NMDA receptors are blocked by Mg^{2+} ions at resting membrane potential. For example, channels containing NR2A or NR2B are more sensitive to Mg^{2+} block compared with NR2C- or NR2D-containing channels. The mechanism of NMDA receptor activation in this case would be the following. When the glutamate concentration in the synaptic cleft is high enough to activate the few NMDA receptors that are not blocked by Mg^{2+} at the resting potential, a small inward current is activated. This current produces a small depolarization of the membrane which allows the 'unblocking' of additional NMDA receptors and, as in the previous example, this triggers a regenerative phenomenon. The resulting Ca^{2+} influx through the open NMDA channels triggers Ca^{2+}-dependent processes.

10.6 SUMMARY

Glutamate receptors comprise the glutamate receptor sites on their surface, the elements that make the ionic channel selectively permeable to cations, as well as all the elements necessary for interactions between different functional domains. Thus, the agonist receptor sites and the cationic channel are part of the same unique protein. The function of postsynaptic iGluRs is to mediate fast excitatory synaptic transmission by converting the binding of glutamate to a rapid and transient increase in cationic permeability and a subsequent membrane depolarization. NMDA, AMPA and kainate receptors are co-expressed in many neurons. Therefore, to study them separately, patch clamp techniques and the use of selective agonists for each receptor type have proven to be particularly useful.

Do all ionotropic glutamate receptors have the same properties; for example, the same sensitivity to glutamate, the same ionic permeability?

NMDA and kainate receptors have a higher affinity (around 1 μM) for glutamate than do AMPA receptors (250–1500 μM). All iGluRs are cationic channels permeable to Na^+ and K^+. Some AMPA receptors and all NMDA receptors are also permeable to Ca^{2+}.

Do all these receptors have co-agonists acting at modulatory sites?

No – to date, only NMDA receptors have been shown to contain a co-agonist binding site (for glycine).

What are the exact conditions of the activation of the different iGluRs?

AMPA receptors show fast-gating kinetics, desensitize strongly and are typically poorly permeable to Ca^{2+} ions. Therefore, once glutamate is released in the cleft, AMPA receptors open rapidly and briefly, allowing a transient K^+ efflux + Na^+ influx (recorded as an inward current at physiological membrane potentials) and sometimes a Ca^{2+} influx, through the postsynaptic membrane. This AMPA receptor-mediated current or EPSC generates a fast-rising EPSP (or a fast-rising EPSP component) with a time to peak in the order of 1 ms.

When kainate receptors are present in the postsynaptic element, they allow a transient K^+ efflux + Na^+ influx (recorded as an inward current at physiological membrane potentials) and sometimes a Ca^{2+} influx, through the postsynaptic membrane. The kainate receptor-mediated EPSC is smaller and slower (time to peak of the order of 5–10 ms) than the AMPA-mediated one, thus giving a slow-rising, low-amplitude EPSP component.

NMDA receptors are unusual ligand-gated channels because their activation not only requires the binding of two agonists, glutamate and glycine, but also demands the relief of Mg^{2+} block by depolarization. NMDA receptors have a complex role based on three properties: they open slowly (they require more than 2 ms to open) and remain open longer than AMPA receptors (they slowly deactivate and weakly desensitize); this slow time course allows the summation of responses to events tens of milliseconds apart. NMDA receptors function as a coincidence detector: NMDA current occurs only with coincident presynaptic release of glutamate and postsynaptic depolarization; i.e. only when agonist binding and cell depolarization take place simultaneously. Once open, NMDA channels allow a transient K^+ efflux + Na^+ influx + Ca^{2+} influx (recorded as an inward current at physiological membrane potentials) through the postsynaptic membrane. The resulting EPSC triggers a slow-rising, long-duration EPSP (or EPSP component) and the resulting increase of intracellular Ca^{2+} concentration triggers a cascade of molecular events in the postsynaptic cell.

Note that the risetime of EPSPs depends on the agonist binding rate and on the opening rate of postsynaptic receptors. The amplitude of EPSPs depends on the number of open channels in the postsynaptic element; i.e. on the number N of receptors present in the membrane, on the open probability (p_o) of the channel, and on the concentration of neurotransmitter in the synaptic cleft.

FURTHER READING

Armstrong N, Sun Y, Chen GQ, Gouaux E (1998) Structure of a glutamate receptor ligand-binding core in complex with kainate. *Nature* **395**, 913–917.

Burnashev N, Monyer H, Seeburg PH, Sakmann B (1992) Divalent ion permeability of AMPA receptor channels is dominated by the edited form of a single subunit. *Neuron* **8**, 189–198.

Chen PE and Wyllie DJA (2006) Pharmacological insights obtained from structure–function studies of ionotropic glutamate receptors. *British Journal of Pharmacology* **147**, 839–853.

Furukawa H, Singh SK, Mancusso R, Gouaux E (2005) Subunit arrangement and function in NMDA receptors. *Nature* **438**, 185–192.

Gregor P, Mano I, Maoz I et al. (1989) Molecular structure of the chick cerebellar kainate-binding subunit of a putative glutamate receptor. *Nature* **342**, 689–692.

Hirai H, Kirsch J, Laube B et al. (1996) The glycine binding site of the N-methyl-D-aspartate receptor subunit NR1: identification of novel determinants of co-agonist potentiation in the extracellular M3–M4 loop region. *Proc. Natl Acad. Sci.* **93**, 6031–6036.

Hume RI, Dingledine R, Heinemann SF (1991) Identification of a site in glutamate receptor subunits that controls calcium permeability. *Science* **253**, 1028–1031.

Kuner T and Schoepfer R (1996) Multiple structural elements determine subunit specificity of Mg^{2+} block in NMDA receptor channels. *J. Neurosci.* **16**, 3549–3558.

Kuusinen A, Arvola M, Keinänen K (1995) Molecular dissection of the agonist binding site of an AMPA receptor. *EMBO J* **14**, 6327–6332.

Laube B, Hirai H, Sturgess M et al. (1997) Molecular determinants of agonist discrimination by NMDA receptor subunits: analysis of the glutamate binding site on the NR2B subunit. *Neuron* **18**, 493–503.

Mayer ML (2005) Glutamate receptor ion channels. *Curr. Opin. Neurobiol.* **15**, 282–288.

Sheng M and Pak DT (1999) Glutamate receptor anchoring proteins and the molecular organization of excitatory synapses. *Ann. NY Acad. Sci. USA* **868**, 483–493.

Sommer B, Kohler M, Sprengel R, Seeburg PH (1991) RNA editing in brain controls a determinant of ion flow in glutamate-gated channels. *Cell* **67**, 11–19.

Stern-Bach Y, Bettler B, Hartley M, Sheppard PO, O'Hara PJ, Heinemann SF (1994) Agonist selectivity of glutamate receptors is specified by two domains structurally related to bacterial amino-acid binding proteins. *Neuron* **13**, 1345–1357.

Takumi Y, Matsubara A, Rinvik E, Ottersen OP (1999) The arrangement of glutamate receptors in excitatory synapses. *Ann. NY Acad. Sci. USA* **868**, 474–482.

Wada K, Dechesne CJ, Shimasaki S et al. (1989) Sequence and expression of a frog complementary DNA encoding a kainate binding protein. *Nature* **342**, 684–689.

Wollmuth LP and Sobolevsky AI (2004) Structure and gating of the glutamate receptor ion channel. *Trend. Neurosci.* **27**, 321–328.

11

The metabotropic GABA$_B$ receptors

11.1 GABA$_B$ RECEPTORS WERE ORIGINALLY DISCOVERED BECAUSE OF THEIR INSENSITIVITY TO BICUCULLINE AND THEIR SENSITIVITY TO BACLOFEN

Gamma-aminobutyric acid (GABA) is the primary inhibitory neurotransmitter in the mammalian central nervous system. It is found in virtually every area of the brain. It exerts fast and powerful synaptic inhibition by acting on GABA$_A$ receptors. These receptors are directly coupled to an integral chloride channel and produce inhibition by increasing the membrane chloride conductance. This form of synaptic inhibition is critical for maintaining and shaping neuronal communication.

However, like other neurotransmitters that activate fast, ionotropic responses lasting for milliseconds, GABA can also activate a second class of receptors which produce slow synaptic responses capable of lasting for seconds. The receptors producing these slow, metabotropic responses are designated GABA$_B$ receptors. GABA$_B$ receptors play a major role in regulating neurotransmission, which makes them potentially important therapeutic targets in the treatment of a variety of neurological conditions, including epilepsy, spasticity, pain and psychiatric illness. GABA$_B$ receptors are G protein coupled to a number of cellular effector mechanisms, including adenylyl cyclase, voltage-dependent calcium channels and inwardly rectifying potassium channels. These different effectors enable GABA$_B$ receptors to produce, not only inhibition, but a diversity of other effects on neuronal function. Thus, GABA$_B$ receptors enable GABA to modulate neuronal activity in a fashion that is not possible through GABA$_A$ receptors alone.

This chapter will focus on GABA$_B$ receptors and the different effects that these receptors can have on cellular function.

The discovery of GABA$_B$ receptors was made possible by the development in the early 1970s of the compound β-parachlorophenyl GABA (baclofen). Baclofen is a GABA analogue which can be orally administered and will penetrate the blood–brain barrier. It was hoped that after gaining access to the brain this compound would act on GABA receptors and be an effective anticonvulsant. Indeed, baclofen did mimic many of the actions of GABA and was found to reduce skeletal muscle tone and inhibit spinal reflex activity, making it a successful agent in treating spinal cord spasticity. Yet, despite these similarities with GABA, several important differences between the actions of GABA and baclofen were reported, the most notable of which was the insensitivity of the actions of baclofen to the classical GABA antagonist, bicuculline.

It was at this time that Norman Bowery and his colleagues found that application of GABA decreased the release of norepinephrine from a preparation of the rat isolated atrium. Interestingly, this effect of GABA was insensitive to bicuculline and was not mimicked by classical GABA agonists, such as isoguvacine and THIP. Bowery and his colleagues found similar results when they measured the effect of GABA on the release of norepinephrine in another peripheral preparation, the rat isolated anococcygeus muscle. In both of these preparations the GABA analogue, baclofen, mimicked the action of GABA by depressing the release of norepinephrine in

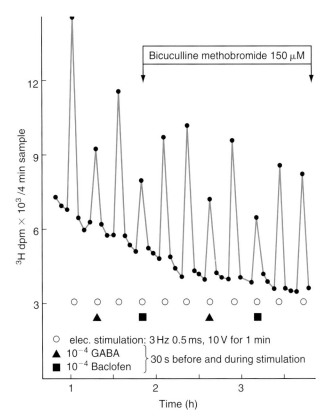

FIGURE 11.1 GABA and baclofen suppress ³H-norepinephrine release from the rat atrium.
The release of ³H-norepinephrine was assessed by taking samples of the superfusate every 4 minutes and measuring the tritium content (in dpm) by liquid scintillation spectrometry. Electrical stimuli were delivered to the tissue at times indicated by the open circles. These stimuli caused the release of ³H-norepinephrine and so increased the tritium content of the sample. GABA (filled triangles) and baclofen (filled squares) reduced the release of ³H-norepinephrine by the stimulus. The effect of these drugs was insensitive to co-application of bicuculline methobromide. From Bowery, N.G., Doble, A., Hill, D.R., Hudson, A.L., Turnbull, M.J., Warrington, R. (1981) Structure/activity studies at a baclofen-sensitive, bicuculline-insensitive GABA receptor. In DeFeudis, F.V. and Mandel, P. (eds), *Amino Acid Neurotransmitters*. New York: Raven Press, with permission.

a dose-dependent manner (**Figure 11.1**). Furthermore, neither the effect of GABA nor that of baclofen appeared to be mediated by an increase in chloride conductance, suggesting that a receptor other than the classical GABA receptor was responsible for the presynaptic inhibition of norepinephrine release.

To determine whether this bicuculline-insensitive action of GABA was confined to the periphery, Bowery and co-workers tested the effect of GABA and baclofen on potassium-evoked norepinephrine release from brain slices. They found that, as in the periphery, GABA suppressed norepinephrine release by acting on a bicuculline-insensitive receptor that was separate from the classical bicuculline-sensitive GABA receptor. This action of GABA was mimicked by the GABA analogue

baclofen, but not by other known GABA agonists. Radio-ligand receptor binding in brain using ³H-GABA demonstrated two distinct binding sites for GABA with different distributions. These results led Bowery and his co-workers in 1981 to propose the existence of a new class of GABA receptor, which they termed the GABA_B receptor, while designating the classical GABA receptor as the GABA_A receptor.

GABA_B *receptor pharmacology*

GABA is the endogenous agonist at both GABA_A and GABA_B receptors. GABA_B receptors are pharmacologically distinguished from GABA_A receptors by their insensitivity to the GABA_A antagonist bicuculline and their selective activation by the prototypic agonist baclofen (**Figure 11.2**). Baclofen activates GABA_B receptors in a stereospecific manner with the (−) isomer being about 100 times more potent than the (+) isomer. In contrast, GABA_B receptors are not sensitive to classical agonists at the GABA_A receptor, such as muscimol and isoguvacine, or to modulators of GABA_A receptors such as benzodiazepines, barbiturates and neurosteroids.

The discovery of selective GABA_B receptor antagonists with increased receptor affinity and improved pharmacokinetic profile has been an important element in establishing the significance and structure of GABA_B receptors. The first GABA_B receptor antagonists, phaclofen and 2-hydroxysaclofen (**Figure 11.2**), represented a major breakthrough in the study of GABA_B receptors even though they possessed relatively low potencies. Subsequently, Froestl and co-workers introduced CGP35348, the first GABA_B receptor antagonist capable of crossing the blood–brain barrier. This was soon followed by CGP36742, the first orally active GABA_B receptor antagonist. Although these compounds displayed rather low potency, Froestl and co-workers found that the substitution of a dichlorobenzene moiety into these antagonist molecules increased their affinities by about 10,000-fold. This breakthrough resulted in the production of a host of compounds, such as CGP52432, CGP55845, CGP64213 and CGP71872 (**Figure 11.2**), which had affinities in the nanomolar and even subnanomolar range. This series of compounds eventually led to the development of the radioiodinated, high-affinity antagonist [¹²⁵I]-CGP64213, which was used to clone GABA_B1.

11.2 STRUCTURE OF THE GABA_B RECEPTOR

Using a high-affinity antagonist, the structural properties of the GABA_B receptor were characterized

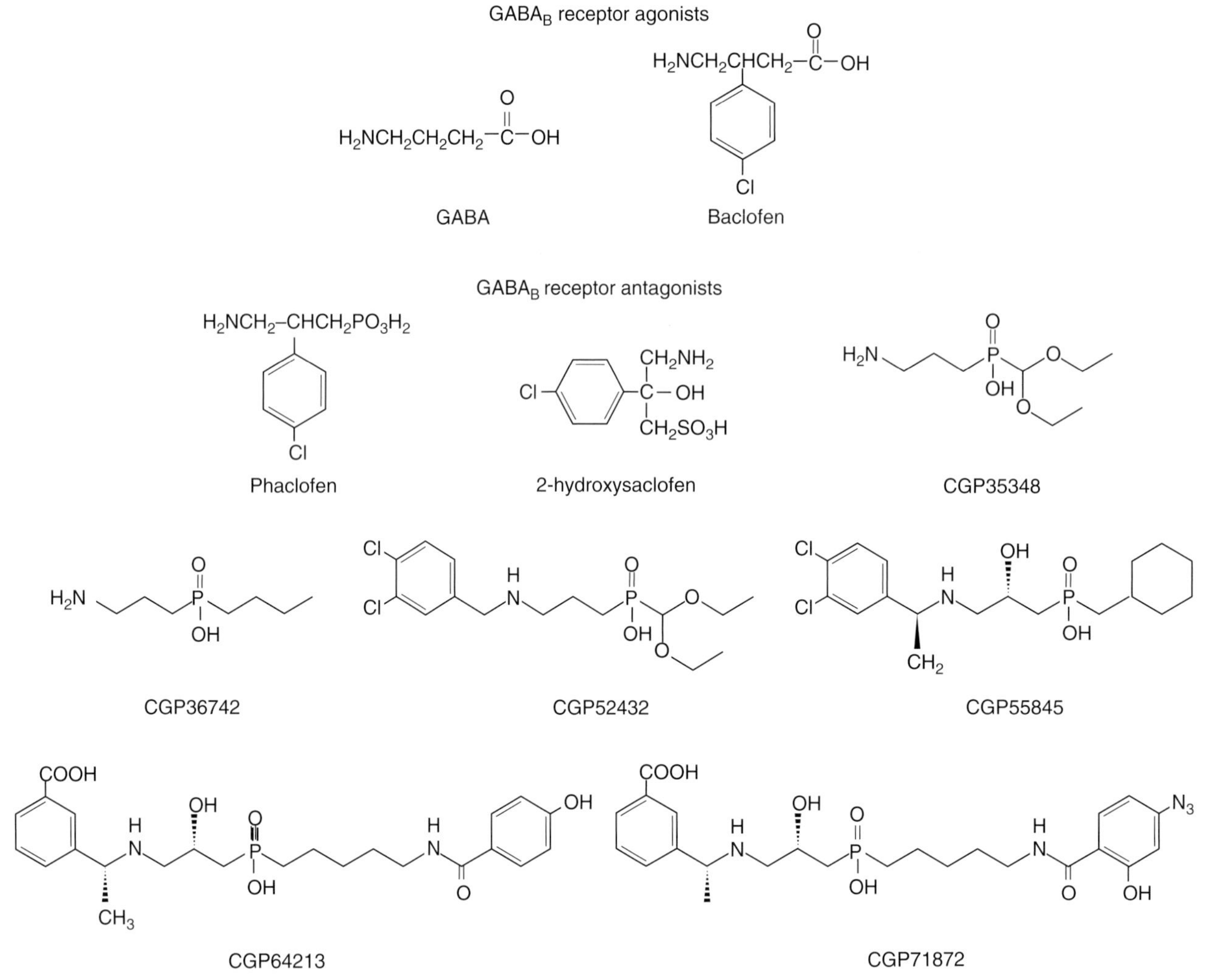

GABA_B receptor agonists

GABA

Baclofen

GABA_B receptor antagonists

Phaclofen

2-hydroxysaclofen

CGP35348

CGP36742

CGP52432

CGP55845

CGP64213

CGP71872

FIGURE 11.2   Structures of selected GABA_B receptor agonists and antagonists.
Adapted from Mott, D.D. and Lewis, D.V. (1994) The pharmacology and function of central GABA_B receptors. *Int. Rev. Neurobiol.* **39**, 97–223, with permission.

by expression cloning. Expression of a fully functional GABA_B receptor was found to require coupling between two separate and distinct gene products, GABA_{B1} and GABA_{B2}. GABA_B receptors are thus the first example of a functional heterodimeric metabotropic receptor.

11.2.1 GABA_B receptors belong to Family 3 G protein-coupled receptors

Cloning of the GABA_B receptor

In 1997 Bettler and colleagues successfully cloned the first GABA_B subunit, which they named GABA_{B1}. The derived sequence of GABA_{B1} indicated that it shares no significant sequence similarity to GABA_A or GABA_C receptors, but is distantly related to Family 3 G protein-coupled receptors (GPCRs). This family of receptors includes metabotropic glutamate receptors (mGluRs), the Ca^{2+}-sensing receptor, a family of pheromone receptors and certain mammalian taste receptors. Like other Family 3 GPCRs, GABA_{B1} subunits have several characteristic features, including a large extracellular amino terminus which plays a critical role in ligand binding, followed by seven closely spaced transmembrane domains, indicative of GPCRs. When compared to mGluRs, GABA_{B1} shares only 18–23% sequence homology, however hydrophobicity profiles indicate clear conservation of structural architecture between these receptors. The N terminal extracellular domain of both mGluRs and GABA_{B1} shares limited, but significant similarity with bacterial periplasmic amino-acid-binding proteins (PBP) such as the leucine-binding protein (LBP). However, the intracellular loops of the GABA_B receptor are not as well conserved as in other Family 3 receptors. In particular, most cysteine residues, which are highly conserved in other Family 3 receptors, are not conserved in $GABA_{B1}$.

11.2.2 GABA_B receptors are heterodimers

GABA_B1 receptors are non-functional

Whereas $GABA_{B1}$ displays binding and biochemical characteristics similar to those of native $GABA_B$ receptors, several important discrepancies were noted between these cloned receptors and native $GABA_B$ receptors. For example, the affinity of agonists, but not antagonists, was 100–150-fold lower for $GABA_{B1}$ than for native receptors. Most importantly, when expressed in cell lines $GABA_{B1}$ coupled only weakly to adenylyl cyclase and did not couple to other effector systems, such as calcium or potassium channels. The reason for the failure of $GABA_{B1}$ to produce functional receptors was examined using epitope-tagged versions of $GABA_{B1}$ to study the cellular distribution of the receptor protein. It was found that $GABA_{B1}$ was retained in the endoplasmic reticulum and therefore failed to reach the cell surface. Thus, it appeared as though $GABA_{B1}$ required additional information for functional targeting to the plasma membrane.

Fully functional GABA_B receptors require coupling between GABA_B1 and GABA_B2

The failure of $GABA_{B1}$ to produce functional $GABA_B$ receptors inspired an intensive search for other related genes, ultimately resulting in the discovery of a second $GABA_B$ receptor gene, termed $GABA_{B2}$. This receptor subtype was 35% homologous with $GABA_{B1}$ and exhibited many of the structural features of $GABA_{B1}$, including a large molecular weight, an extended extracellular N-terminus and seven transmembrane spanning domains.

Importantly, it was found that this receptor must be co-expressed with $GABA_{B1}$ to form a fully functional $GABA_B$ receptor. Co-expression of $GABA_{B1}$ and $GABA_{B2}$ resulted in efficient surface expression of the receptor and the agonist affinity of these heterodimeric receptors was similar to that of native $GABA_B$ receptors. This finding represented the first evidence for heterodimerization among GPCRs. Recombinant heteromeric $GABA_B$ receptors are fully functional and display robust coupling to all prominent effector systems of native $GABA_B$ receptors (**Figure 11.3**).

The existence of $GABA_B$ heterodimers in neurons was confirmed in immunoprecipitation experiments. In these experiments antibodies raised against $GABA_{B2}$ efficiently co-precipitated the $GABA_{B1}$ proteins from cortical membranes. Conversely, antibodies which recognize $GABA_{B1}$ co-precipitated the $GABA_{B2}$ receptor. Thus, native $GABA_B$ receptors appear to be heterodimers composed of $GABA_{B1}$ and $GABA_{B2}$ which interact in a stoichiometry of 1:1.

Genetic studies have found that mice lacking the $GABA_{B1}$ or $GABA_{B2}$ gene show a loss of all typical $GABA_B$ responses. These findings indicate that $GABA_{B1}$ and $GABA_{B2}$ alone can account for all of the classical $GABA_B$ functions and strongly suggest that the existence of additional obligatory receptor subunits is unlikely. In addition, $GABA_{B1}$ and $GABA_{B2}$ protein are substantially down-regulated in $GABA_{B2}$ and $GABA_{B1}$ knock-out mice, respectively. The degradation of the partner subunit in each of these knock-out mouse lines indicates the importance of heteromeric assembly for stable expression of each subunit. It also suggests that in wild-type mice virtually all $GABA_{B1}$ protein is associated with $GABA_{B2}$. These studies support the conclusion that native $GABA_B$ responses are predominantly mediated by heteromeric receptors derived from $GABA_{B1}$ and $GABA_{B2}$ genes.

11.2.3 Surface expression of GABA_B receptors requires coupling between GABA_B1 and GABA_B2

For both recombinant and native $GABA_B$ receptors, the interaction of $GABA_{B1}$ and $GABA_{B2}$ within the cell is critical for the correct assembly of the heterodimer on the cell surface. It is now known that $GABA_{B1}$ is prevented from travelling to the cell surface by an endoplasmic reticulum (ER)-retention signal in its cytoplasmic tail. The ER-retention signal in $GABA_{B1}$ is located in an α-helical coiled-coil domain in the carboxy-terminus of the peptide. A coiled-coil domain is also present in the carboxy-terminus of $GABA_{B2}$. $GABA_{B1}$ and $GABA_{B2}$ form a tightly coupled heterodimer via an interaction of these coiled-coil domains. Formation of this heterodimer masks the ER-retention signal in $GABA_{B1}$ from its ER-anchoring mechanism, allowing the heteromeric receptor to travel to the cell surface. The ER-retention motif therefore ensures that only correctly assembled receptor complexes traffic to the cell surface. ER-retention signals are also observed in other multisubunit proteins, where they serve as a quality-control mechanism.

$GABA_B$ receptors are the first functional heterodimers to be identified within the metabotropic class. Only among the ionotropic receptors have heterodimers been previously recognized (i.e. $GABA_A$ receptors). Other members of the Family 3 GPCRs, including metabotropic glutamate receptors and the Ca^{2+}-sensing receptor, have previously been reported

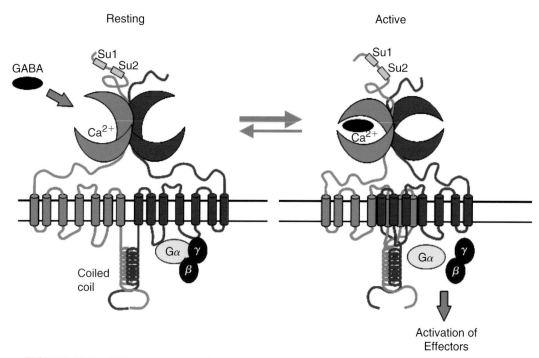

FIGURE 11.3   GABA_B receptors are functional heterodimers.

Agonist binding activates the GABA_B receptor heterodimer causing a conformational change that results in receptor coupling to effector systems. The GABA_{B1} (Green) and GABA_{B2} (Red) subunits interact with each other through their carboxy-terminal coiled-coil domains. GABA binds only to the GABA_{B1} subunit whereas G proteins bind to the GABA_{B2} subunit. In the inactive resting state the ligand-binding pocket in the GABA_{B1} extracellular domain is open and the transmembrane domains of GABA_{B1} and GABA_{B2} are apart. Agonist binding to GABA_{B1} causes the ligand-binding pocket to close and the receptor to become activated. The binding of GABA within the binding pocket and the closed conformation of the receptor are stabilized by calcium. The activated receptor undergoes a conformational change in which the transmembrane domains of GABA_{B1} and GABA_{B2} move closer together. This conformational change is necessary for the initiation of downstream effector signalling. The sushi domains in the amino-terminus of the GABA_{B1a} subunit are indicated (Su1 and Su2). These domains are important for receptor targeting in the cell and are not present in GABA_{B1b}. Adapted from Calver, AR, Davies, CH and Pangalos, M (2002) GABA_B receptors: from monogamy to promiscuity. *Neurosignals* **11**, 299–314, copyright S Karger AG, Basel, with permission.

to form homodimers. As opposed to the GABA_B receptor, dimer formation for these receptors has been shown to be caused by the disulfide interaction of cysteine residues in the extracellular N-terminal domain. These cysteine residues are absent in the GABA_B receptor, which dimerizes predominantly through an interaction of coiled-coil domains in the carboxy-terminus (**Figure 11.3**). That such closely related receptors have evolved different mechanisms of dimerization suggests that dimerization is important for this class of receptors.

11.2.4 The GABA-binding site is in the extracellular amino-terminal domain of GABA_{B1}

The ligand-binding domain for Family 3 GPCRs is located in the extracellular amino-terminus of the receptor in a region with significant homology to bacterial periplasmic-binding proteins, such as the

leucine-binding protein (LBP). Activation of GABA_B receptors occurs exclusively by GABA binding to the GABA_{B1} subunit. The GABA_{B2} subunit does not bind agonist or antagonist. The function of the extracellular domain of GABA_{B2} is not well understood, however it has been suggested to interact with and enhance the affinity of GABA_{B1} agonist-binding site for GABA.

To better understand the agonist-binding domain of GABA_{B1}, Bettler and colleagues constructed chimeric receptors which contain the amino-terminus of GABA_{B1} on the body of the mGlu_1 receptor. They found that these chimeric receptors and wild-type GABA_B receptors possessed similar binding affinities for GABA_B receptor ligands. Furthermore, radioiodinated antagonist binding affinities were also unaltered in GABA_{B1} truncation mutants in which the entire carboxy-terminus after the first transmembrane domain was deleted. Finally, when the amino-terminus of the GABA_{B1} subunit was produced as a soluble miniprotein it bound radiolabelled GABA$_B$ receptor antagonist with a similar affinity to control wild-type receptor. These studies indicate that, like the amino-terminal domain of other family 3 GPCRs, the amino-terminal domain of GABA$_{B1}$ is both necessary and sufficient for ligand binding.

Mutagenesis studies support the LBP-like domain in the N-terminus of GABA$_{B1}$ as a critical region for ligand binding. Mutation of several key residues in this area markedly alters the affinity of the GABA$_B$ receptor for antagonists, suggesting that the architecture of this region bears structural homology to that of LBP. Three-dimensional modelling of the GABA$_B$-binding domain based on the known crystal structure of LBP supports a Venus flytrap model for receptor activation. According to this model, the ligand-binding site is formed in a groove between two large globular domains in the N-terminus of GABA$_{B1}$. Activation of the receptor results from the closure of these two lobes upon agonist binding, similar to a Venus flytrap. Closure of these lobes produces a conformational change in the protein complex that allows activation of the G protein-coupled signalling system (**Figure 11.3**). This model is similar to that proposed for other members of Family 3 GPCRs.

Ligand binding to GABA$_B$ receptors requires the presence of divalent cations

Like other members of Family 3 GPCRs, such as the Ca^{2+}-sensing receptor and metabotropic glutamate receptor, GABA$_B$ receptors are sensitive to calcium. This differs from GABA$_A$ receptors which have no such requirement. Calcium binding to the GABA$_{B1}$ subunit of the receptor allosterically potentiates the action of GABA by stabilizing the closed conformational state of the agonist-binding domain (**Figure 11.3**). Interestingly, other divalent cations, including Hg^{2+}, Pb^{2+}, Cd^{2+} and Zn^{2+}, inhibit GABA receptor binding. The effects of calcium on the GABA$_B$ receptor are agonist-dependent in that calcium more strongly potentiates the effect of GABA than baclofen. Mutational analysis has revealed that a specific highly conserved serine residue (S269) in the GABA$_{B1}$ ligand-binding site is critical for the effect of calcium. Possibly, calcium binding to this residue helps to optimally position GABA in the agonist-binding pocket. Because GABA$_B$ receptors have a high sensitivity to calcium, the calcium binding site on GABA$_{B1}$ is saturated with calcium under normal physiological conditions. Therefore, calcium modulation of GABA binding to the GABA$_B$ receptor would potentially play a role only during times when extracellular calcium concentrations are low, such as during ischemia or epileptic seizures.

11.2.5 GABA$_{B2}$ subunits couple to inhibitory G proteins

Guanyl nucleotide binding proteins (G proteins) carry signals from activated membrane receptors to effector enzymes and channels. These molecules enable a single receptor to be functionally connected to a variety of different effector mechanisms in a single cell or to different effectors in different cells. Coupling of GABA$_B$ receptors and G proteins was originally deduced from binding studies of ^{3}H-GABA and ^{3}H-baclofen to crude synaptic membranes prepared using whole rat brain. In these experiments the addition of guanyl nucleotides, such as GTP, did not affect the binding of ^{3}H-GABA to GABA$_A$ receptors, but potently inhibited GABA$_B$ receptor binding (**Figure 11.4**). This effect was concentration-dependent and was not mimicked by adenosine 5′-triphosphate (ATP), indicating that it was specific for guanyl nucleotides. The inhibition of ligand binding produced by GTP was caused by a decrease in GABA$_B$ receptor affinity and not a decrease in the number of available GABA$_B$ receptors. It was concluded that the addition of GTP promoted the dissociation of the G protein from the receptor, causing the receptor to revert to its low-affinity conformation. Thus, GABA$_B$ receptors appeared to couple to G proteins.

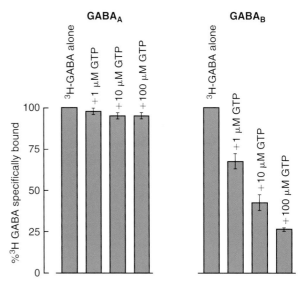

FIGURE 11.4 The effect of GTP on ^{3}H-GABA binding to GABA$_A$ and GABA$_B$ receptors.

^{3}H-GABA binding to crude synaptic membranes from whole rat brain was measured in the presence of either isoguvacine or baclofen to saturate GABA$_A$ and GABA$_B$ receptors, respectively. The addition of increasing concentration of GTP had no effect on GABA$_A$ receptor binding but produced a concentration-dependent inhibition of GABA$_B$ receptor binding. From Hill, D.R., Bowery, N.G., Hudson, A.L. (1984) Inhibition of GABA$_B$ receptor binding by guanyl nucleotides. *J. Neurochem.* **42**, 652–657, with permission.

The identity of the G proteins coupled to GABA_B receptors was established through two different experiments. First, it was observed that inhibition of GABA_B receptor binding by GTP was blocked by pertussis toxin. This demonstrated that GABA_B receptors are functionally coupled to the inhibitory G proteins, G_i and/or G_o. This finding was further confirmed using cloned heteromeric GABA_B receptors (GABA_B1/GABA_B2) expressed with chimeric G_q proteins in human embryonic kidney cells (HEK 293). Wild-type G_q protein activates phospholipase C (PLC). Ordinarily GABA_B receptors do not stimulate PLC activity, indicating that they do not couple to G_q protein. PLC activity produced by GABA_B receptor activation was then measured following the addition of chimeric G_q proteins in which the five carboxy-terminal residues of the $G_q\alpha$ subunit had been exchanged for those of either $G_i\alpha$, $G_o\alpha$, or $G_z\alpha$ protein. The five carboxy-terminal residues of the $G\alpha$ subunit are critical for coupling of G proteins to receptors. Only those chimeric G_q proteins containing the coupling sites of $G_i\alpha$ or $G_o\alpha$ protein were able to activate PLC, indicating that only G_i and G_o proteins interact with the GABA_B receptor.

A large number of studies have examined the molecular determinants of receptor-G protein coupling selectivity in GABA_B receptor subunits. It is now established that G proteins bind to the heptahelical region of the GABA_B2 subunit. In contrast, the heptahelical region of GABA_B1 does not bind G protein but enhances the efficiency of the interaction of G protein with the GABA_B2 subunit. Thus, the GABA_B receptor is an obligate heterodimer in which the GABA_B1 subunit is necessary for agonist binding while the GABA_B2 subunit is required for G protein signalling (**Figure 11.3**).

In Family 3 GPCRs the second intracellular loop (i2) plays a critical role in the interaction of the receptor with G proteins. In the GABA_B receptor the i2 loop in GABA_B2 is critical for G protein coupling as well. Sequence comparison between the i2 loops of GABA_B1 and GABA_B2 revealed several important differences. Thus, exchanging the i2 loop between GABA_B1 and GABA_B2 prevented receptor function. This finding indicated that the i2 loops on the GABA_B2 subunit needs to be correctly positioned relative to other intracellular domains for proper G protein coupling. Mutational analysis confirmed the importance of the i2 loop in GABA_B2. Furthermore, these studies found that mutation of a critical lysine residue (K686) in the third intracellular loop of GABA_B2 also suppressed coupling of G proteins to the GABA_B receptor. This lysine reside is important for functional coupling of G proteins to GABA_B receptors and appears to serve a similar function in other Family 3 GPCRs as well.

11.2.6 Molecular diversity of GABA_B receptors arises from GABA_B1 isoforms

When it became apparent that only two genes encoded all GABA_B receptor subunits, a search began for subunit isoforms. To date, no isoforms of GABA_B2 have been identified. In contrast, two predominant isoforms of GABA_B1, termed GABA_B1a and GABA_B1b have emerged. Numerous studies in recombinant systems have concluded that GABA_B1a and GABA_B1b isoforms exhibit no unique functional or pharmacological properties. GABA_B1a and GABA_B1b isoforms are generated by differential promoter usage within the GABA_B1 gene. GABA_B1a differs from GABA_B1b in having a longer amino-terminus that contains a pair of short consensus repeats, also known as sushi repeats (**Figure 11.3**). Sushi repeats were originally identified in complement proteins and are involved in protein-protein interactions. The function of these sequences in the GABA_B receptor is unknown; however they have been proposed to play a role in targeting GABA_B receptors to specific sites within the cell.

11.2.7 GABA_B receptors are located throughout the brain at both presynaptic and postsynaptic sites

GABA_B receptors can be found in most regions of the brain (**Figure 11.5**). In the majority of these areas the number of GABA_B receptors is either less than or equal to the number of GABA_A receptors. However, there are a few brain regions, such as the brainstem and certain thalamic nuclei, where GABA_B receptors can account for up to 90% of the total GABA binding sites. Autoradiography or antibody labelling of GABA_B1 and GABA_B2 suggests that these subunits are similarly distributed; however there are some brain regions, such as

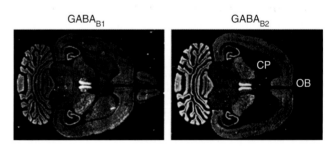

FIGURE 11.5 Distribution of GABA_B1 and GABA_B2 subunit mRNA in the rat brain.
GABA_B1 and GABA_B2 transcripts are present at similar levels throughout the brain, except in caudate putamen (CP) and olfactory bulb (OB) where GABA_B2 is less abundant than GABA_B1 mRNA. From Bettler, B., Kaupmann, K., Mosbacher, J. and Gassmann, M. (2004) Molecular structure and physiological functions of GABA_B receptors. *Physiol. Rev.* **84**, 835–867, used with permission.

the caudate putamen or olfactory bulb, where GABA$_{B1}$ is in much greater abundance than GABA$_{B2}$. The brain regions possessing the highest density of GABA$_B$ receptors are the thalamic nuclei, the molecular layer of the cerebellum, the cerebral cortex and the interpeduncular nucleus. GABA$_B$ receptors are also found in high density in laminae II and III of the spinal cord.

Both electrophysiological recordings and immunogold electron microscopic techniques have been used to investigate the subcellular localization of GABA$_B$ receptors. These receptors are present on presynaptic terminals, where they modulate the release of a variety of different neurotransmitters, and on postsynaptic membranes, where they inhibit excitatory neurotransmission. Presynaptically, GABA$_B$ subunits are located in the extrasynaptic membrane and near the active zones in presynaptic glutamatergic and GABAergic terminals, supporting a close link with the transmitter release machinery. Postsynaptically, GABA$_B$ receptors are located on both dendritic shafts as well as the extrasynaptic membrane of dendritic spines. Dendritic spines form the majority of excitatory synapses and the presence of GABA$_B$ receptors on these structures as well as on glutamatergic terminals suggests a close coupling of excitatory and inhibitory systems. The subcellular localization of GABA$_B$ receptors also appears to depend upon the brain region examined. For example, immunogold electron microscopic studies have revealed that in the cerebellum GABA$_B$ receptors are enriched in synapses, whereas in thalamic nuclei GABA$_B$ receptors are found in extrasynaptic membrane, having no enrichment in synapses. Presumably, GABA$_B$ receptors in extrasynaptic membrane or on excitatory terminals would be activated by GABA spilling over from neighbouring inhibitory terminals.

In situ hybridization techniques have suggested a differential localization of GABA$_{B1a}$ and GABA$_{B1b}$ to pre- and postsynaptic sites. These studies have suggested that GABA$_{B1a}$ is more closely associated with presynaptic receptors, whereas GABA$_{B1b}$ may participate in the formation of postsynaptic GABA$_B$ receptors. The role of presynaptic and postsynaptic GABA$_B$ receptors will be discussed in greater detail later in this chapter (see Section 11.6).

11.3 SUMMARY

Much has been learned about the GABA$_B$ receptor since its initial description more than 25 years ago. GABA$_B$ receptors represent the first example of a heterodimeric GPCR. These receptors require both GABA$_{B1}$ and GABA$_{B2}$ subunits for efficient surface expression and function. The GABA$_{B1}$ subunit contains the GABA binding domain, but this subunit is trapped in the endoplasmic reticulum by an ER-retention signal in its carboxy-terminal. The GABA$_{B2}$ subunit interacts with GABA$_{B1}$, masking this ER-retention signal and allowing the heterodimeric receptor to be trafficked to the surface. Once at the surface the GABA$_{B2}$ subunit links to the G protein, allowing for a fully functional receptor. Although GABA$_{B2}$ does not bind GABA, it interacts with the GABA binding domain on the GABA$_{B1}$ subunit, enhancing its affinity for GABA. Similarly, GABA$_{B1}$ interacts with the GABA$_{B2}$ subunit promoting its association with the G protein. Thus, the two subunits act in concert to link GABA binding to activation of downstream effectors. We will discuss effectors linked to GABA$_B$ receptors in the next section.

11.4 GABA$_B$ RECEPTORS ARE G-PROTEIN-COUPLED TO A VARIETY OF DIFFERENT EFFECTOR MECHANISMS

GABA$_B$ receptors have the potential to produce a variety of different neuronal responses because they are coupled through inhibitory G proteins to several intracellular effectors (**Figure 11.6**). These different effectors enable GABA, acting through GABA$_B$ receptors, to have a broader range of effects than it could by acting on GABA$_A$ receptors alone. The primary actions of GABA$_B$ receptor activation include modulation of adenylyl cyclase activity, inhibition of voltage-dependent calcium

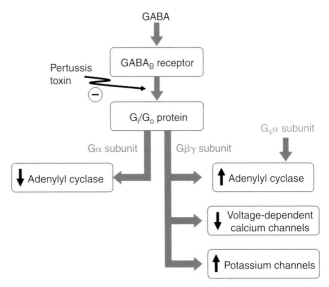

FIGURE 11.6 A schematic diagram depicting the major effector systems to which GABA$_B$ receptors are coupled.

channels and activation of inwardly rectifying potassium channels.

11.4.1 GABA$_B$ receptors regulate the activity of adenylyl cyclase

Adenylyl cyclase converts ATP to cyclic AMP. Cyclic AMP, in turn, activates several different target molecules, such as cyclic AMP-dependent protein kinase (protein kinase A or PKA) to regulate cellular functions, including gene transcription, cellular metabolism and synaptic plasticity. Nine isoforms of adenylyl cyclase (types I to IX) have been identified and all are expressed in neurons. The α subunit of G$_i$ and G$_o$ proteins inhibits several adenylyl cyclase isoforms, including types I, V, and VI.

GABA$_B$ receptors are negatively coupled to adenylyl cyclase through inhibitory G proteins

The ability of GABA$_B$ receptors to couple to inhibitory G proteins suggested that GABA$_B$ receptor activation would inhibit adenylyl cyclase activity through activation of G$_i$α and/or G$_o$α proteins. To test this hypothesis the effect of GABA$_B$ receptor activation on adenylyl cyclase activity was measured by the enzymatic conversion of [α-^{32}P]ATP to cyclic [^{32}P]AMP in crude synaptosomal preparations from a variety of regions of the rat brain. Application of baclofen or GABA caused a decrease in cAMP levels, reflecting a reduction in basal adenylyl cyclase activity (**Figure 11.7a**). This effect was blocked by the GABA$_B$ receptor antagonist, CGP 35348, indicating that it was mediated by GABA$_B$ receptors (**Figure 11.7b**). Application of pertussis toxin dramatically reduced the effect of baclofen on adenylyl cyclase. Since pertussis toxin selectively inactivates G$_i$/G$_o$ proteins, these results demonstrate that GABA$_B$ receptors are negatively coupled to adenylyl cyclase through one or both of these inhibitory G proteins.

Reconstitution experiments have also been used to demonstrate that GABA$_B$ receptors are negatively coupled to adenylyl cyclase through inhibitory G proteins. Purified phospholipids were combined with purified GABA$_B$ receptor, partially purified G$_i$/G$_o$ protein, partially purified adenylyl cyclase and GTP to form a reconstituted membrane preparation. This preparation was then incubated with forskolin, to activate the adenylyl cyclase, and either baclofen or GABA, to activate the GABA$_B$ receptors. In theory, during this incubation, the baclofen or GABA should bind to the GABA$_B$ receptor, causing a decrease in the formation of cAMP by adenylyl cyclase as compared to the level of cAMP formation in the absence of baclofen or GABA. This was exactly what happened (**Figure 11.7c**).

Furthermore, the inhibitory effect of baclofen and GABA on adenylyl cyclase was antagonized by the addition of the GABA$_B$ receptor antagonist, 2-hydroxysaclofen, demonstrating that the inhibition was mediated by GABA$_B$ receptors.

To demonstrate the necessity of each element in the preparation, partially reconstituted membrane preparations were prepared. As predicted, inhibition of cAMP formation by baclofen or GABA was not observed if either the GABA$_B$ receptor or the G$_i$/G$_o$ protein was omitted from the preparation. Furthermore, the omission of adenylyl cyclase resulted in the almost complete absence of cAMP formation. The inability of GABA$_B$ receptors to inhibit cAMP formation in the absence of G$_i$/G$_o$ protein further confirms that GABA$_B$ receptors can negatively couple to adenylyl cyclase through either or both of these G proteins.

GABA$_B$ receptors facilitate neurotransmitter-mediated activation of adenylyl cyclase

In contrast to its direct suppression of cAMP levels through the α subunit of G$_i$/G$_o$ proteins, GABA$_B$ receptor activation can also have another seemingly opposite effect on cAMP accumulation. When adenylyl cyclase is stimulated to produce cAMP by a G$_s$ protein-coupled receptor, GABA$_B$ receptor activation will enhance this increase in cAMP accumulation. For example, addition of baclofen enhances by two- to three-fold the increase in cAMP accumulation produced by norepinephrine (β receptors), adenosine (A2 receptors) or vasoactive intestinal peptide (VIP) receptors (**Figure 11.8**). This effect is contrary to the inhibition of adenylyl cyclase discussed above.

The mechanism of this effect lies in the ability of the βγ subunit from the G$_i$/G$_o$ protein, liberated by the activation of GABA$_B$ receptors, to synergize the interaction of G$_s$α with certain isoforms of adenylyl cyclase, specifically types II, IV and VII. The stimulatory action of GABA$_B$ receptors on cAMP levels represents a form of G protein cross-talk. It depends upon the simultaneous activation of GABA$_B$ receptors and a G$_s$-coupled GPCR and the expression of appropriate adenylyl cyclase isoforms. Under these conditions, adenylyl cyclase types II, IV and VII can act as molecular 'coincidence detectors'. The adenylyl cyclase responds only minimally to activation by a single signal but synergistically to the coincident arrival of dual signals through separate pathways.

Accordingly, G protein α and βγ subunits liberated by the activation of GABA$_B$ receptors could produce opposing effects on cAMP levels. The α$_i$/α$_o$ subunits could directly inhibit one isoform of adenylyl cyclase while the βγ subunits could synergize the G$_s$-mediated

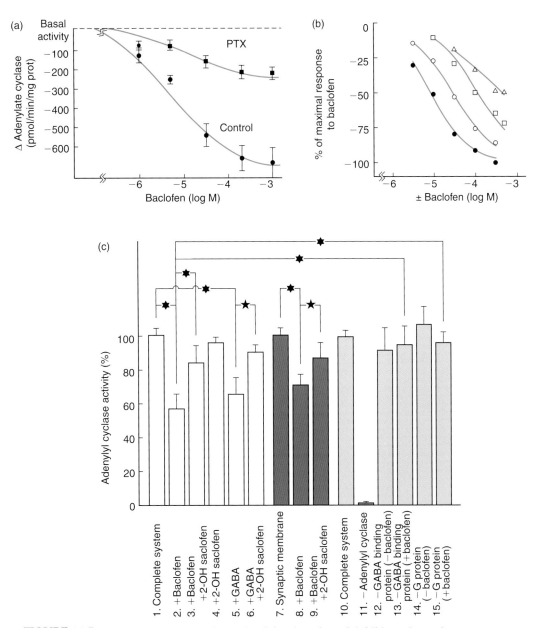

FIGURE 11.7 **GABA$_B$ receptors couple to adenylyl cyclase through inhibitory G proteins.**
(a) Adenylyl cyclase activity in membranes of cerebellar granule cells was measured by the conversion of [α-^{32}P]ATP to cyclic [^{32}P]AMP. In control preparations baclofen decreased the activity of adenylyl cyclase in a concentration-dependent manner (filled circles). Treatment of the membranes with pertussis toxin (PTX) antagonized the effect of baclofen on adenylyl cyclase activity (filled squares). **(b)** The inhibition of adenylyl cyclase activity by baclofen was antagonized in a concentration-specific manner by the GABA$_B$ receptor antagonist, CGP 35348. The inhibition produced by increasing concentrations of baclofen alone (filled circles) was compared to that observed in the presence of baclofen plus either 0.6 mM (open circles), 1.5 mM (open squares) or 5 mM (open triangles) CGP 35348. **(c)** The effect of baclofen and GABA on adenylyl cyclase activity measured in reconstituted membranes (open bars), synaptic membranes (dark bars) and partially reconstituted membranes (light bars). Note that the removal of the G protein from the reconstituted system (15) blocked the inhibitory effect of baclofen (2). See text for details. Significant differences are indicated by an asterisk (*) signifying $p < 0.05$ or a star (★) signifying $p < 0.01$. Part (a) from Xu, J. and Wojcik, W.J. (1986) Gamma aminobutyric acid B receptor-mediated inhibition of adenylate cyclase in cultured cerebellar granule cells: blockade by islet-activating protein. *J. Pharmacol. Exp. Ther.* **239**, 568–573, with permission. Part (b) adapted from Holopainen, I., Rau, C., Wojcik, W.J. (1992) Proposed antagonists at GABA$_B$ receptors that inhibit adenylyl cyclase in cerebellar granule cell cultures of rat. *Eur. J. Pharmacol. Mol. Pharmacol. Section* **227**, 225–228, with permission. Part (c) from Nakayasu, H., Nishikawa, M., Mizutani, H., Kimura, H., and Kuriyama, K. (1993) Immunoaffinity purification and characterization of γ-aminobutyric acid (GABA)$_B$ receptor from bovine cerebral cortex. *J. Biol. Chem.* **268**, 8658–8664, with permission.

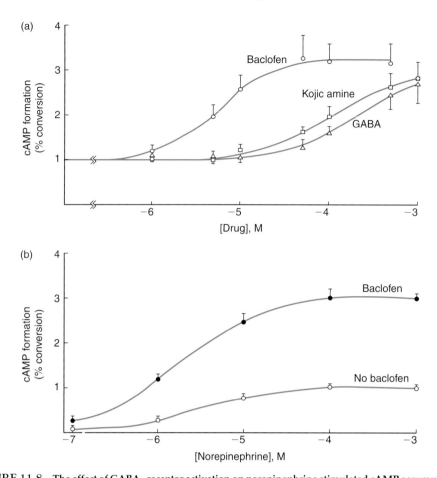

FIGURE 11.8 **The effect of GABA$_B$ receptor activation on norepinephrine stimulated cAMP accumulation.**

(a) The effect of different GABA$_B$ receptor agonists on the cAMP accumulation produced by 100 μM norepinephrine in rat brain cerebellar slices is shown. The GABA$_B$ agonists, baclofen (open circles), kojic amine (open squares) and GABA (open triangles) were applied at increasing concentrations in the presence of norepinephrine. All of these GABA$_B$ agonists enhanced cAMP formation produced by the norepinephrine. **(b)** Baclofen (100 μM) potentiates the cAMP formation induced by increasing concentrations of norepinephrine. The effect of norepinephrine alone (open circles) and norepinephrine plus baclofen (filled circles) is shown. From Karbon, E.W., Duman, R.S., Enna, S.J. (1984) GABA$_B$ receptors and norepinephrine-stimulated cAMP production in rat brain cortex. *Brain Res.* **306**, 327–332, with permission.

stimulation of a different adenylyl cyclase. Depending upon the overall balance between inhibitory and stimulatory effects, this could result in a net increase or decrease in cAMP accumulation. Through these mechanisms, GABA$_B$ receptors have the potential to regulate a variety of cAMP dependent mechanisms in neurons.

11.4.2 GABA$_B$ receptor activation inhibits voltage-dependent calcium channels

GABA$_B$ receptors are negatively coupled to voltage-dependent calcium channels. By inhibiting calcium entry through these channels GABA$_B$ receptors have the potential to modulate a variety of neuronal functions,

perhaps the most significant of which is the ability to regulate neurotransmitter release.

Heterodimeric GABA$_B$ receptors directly inhibit calcium currents

Inhibition of calcium currents by GABA$_B$ receptor activation was first observed in electrophysiological recordings made from neurons in the dorsal root ganglion (DRG). In this preparation baclofen was found to decrease the calcium-dependent plateau phase of the action potential (**Figure 11.9a**). The effect of baclofen was blocked by a GABA$_B$ antagonist, indicating that it was mediated by GABA$_B$ receptors.

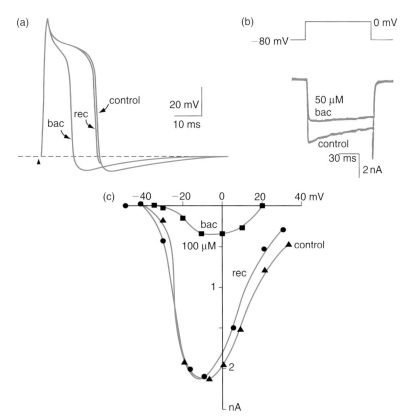

FIGURE 11.9   **Baclofen suppresses voltage-dependent calcium currents in DRG neurons.**

(a) The effect of baclofen on the action potential in a DRG neuron. Baclofen (100 μM; bac) reversibly depressed the calcium-dependent plateau phase of the action potential compared to control (con) or wash (rec). **(b)** In the same preparation 50 μM baclofen depressed the pharmacologically isolated calcium current (bottom). This current was evoked by a depolarizing voltage step from −80 mV to 0 mV (top). See text for details. **(c)** The current–voltage relationship for the voltage-dependent calcium current in DRG neurons is shown in control (con), in 100 μM baclofen (bac) and after 5 minutes of wash (rec). The current–voltage curve represents the amplitude of the calcium current evoked by a voltage step from the holding potential of −80 mV to a variety of test potentials. Baclofen markedly inhibited the calcium current. From Dolphin, AC, Huston, E, Scott, RH (1990). GABA$_B$-mediated inhibition of calcium currents: a possible role in presynaptic inhibition. In Bowery, NG, Bittiger, H and Olpe, H-R (eds), *GABA$_B$ Receptors in Mammalian Function*. Chichester: Wiley, with permission.

To confirm that GABA$_B$ receptor activation directly inhibited calcium channels the effect of baclofen on calcium currents in voltage-clamped DRG neurons was examined. The calcium current was pharmacologically isolated by application of blockers of sodium and potassium currents. Under these conditions a depolarizing voltage step from a holding potential of –80 mV evoked a sustained inward calcium current. Baclofen reversibly reduced the amplitude of this current, indicating that GABA$_B$ receptor activation depresses voltage-dependent calcium currents in DRG neurons (**Figure 11.9b,c**). Similar studies have subsequently confirmed that GABA$_B$ receptor activation can inhibit voltage-dependent calcium currents in many different types of both peripheral and central neurons.

A subsequent study addressed the question of whether heterodimerization of GABA$_B$ receptors is required for the coupling of these receptors to calcium channels in neurons. In this study GABA$_B$ expression constructs were injected into the nuclei of superior cervical ganglion (SCG) neurons, resulting in the expression of GABA$_B$ receptor protein. Baclofen had no effect on calcium currents in uninjected SCG neurons. However, the expression of heterodimeric GABA$_B$ receptors composed of GABA$_{B1}$ plus GABA$_{B2}$ resulted in a marked baclofen-mediated inhibition of calcium channel currents in these cells. The actions of baclofen were blocked by the selective GABA$_B$ receptor antagonist CGP62349, indicating that the effect was mediated by GABA$_B$ receptors. Injection of an antisense construct to block GABA$_{B1}$ expression markedly decreased GABA$_{B1}$ protein levels as well as the inhibitory effects of baclofen on calcium currents. These results suggest that heterodimeric assemblies of GABA$_{B1}$ and GABA$_{B2}$ are necessary for GABA$_B$ receptor-mediated inhibition of calcium channel currents.

GABA$_B$ receptors inhibit a variety of voltage-gated calcium channels

The voltage-dependent calcium current evoked in a given cell is typically produced by the activation of several different calcium channel types. Thus, partial suppression of this current by GABA$_B$ receptor activation could be produced by a partial inhibition of several different channel types or the complete inhibition of only a single type. Because of the different physiological functions of the various voltage-dependent calcium channels, it is important to determine the type(s) of calcium channel inhibited by GABA$_B$ receptors. This can be accomplished through the use of calcium channel antagonists that are specific for different calcium channels. The ability of a selective antagonist to occlude inhibition of a calcium current by baclofen indicates that the antagonist and baclofen are acting on the same subset of channels. Alternately, specific calcium channel antagonists can be used to pharmacologically isolate a single type of calcium current and the effect of GABA$_B$ receptor activation assessed. Finally, kinetic analysis of the calcium current inhibited by GABA$_B$ receptors can be used to determine the electrophysiological characteristics of the inhibited current, which can then be compared to the known properties of identified calcium channels.

Using these techniques, GABA$_B$ receptors have been shown to inhibit all types of calcium channels (**Figure 11.10**). Inhibition of N-type and P/Q-type calcium channels by GABA$_B$ receptors is observed in most neurons. In comparison, GABA$_B$ receptor-mediated inhibition of L-type channels is dependent upon the cell type. For example, it is observed in cerebellar granule neurons and hippocampal pyramidal neurons, but not in cerebellar Purkinje neurons, spinal cord neurons or thalamocortical neurons. Similarly, GABA$_B$ receptor-mediated inhibition of T-type calcium channels is also neuron-dependent. Baclofen suppresses current through these channels in DRG neurons and interneurons in the *stratum lacunosum moleculare* of the hippocampus, but not in thalamocortical neurons or pyramidal neurons of the hippocampus. Thus, GABA$_B$ receptor activation has the potential to inhibit a variety of different voltage-dependent calcium channels. The mechanisms that enable cell-type-dependent regulation of these calcium channels by GABA$_B$ receptors are not well understood.

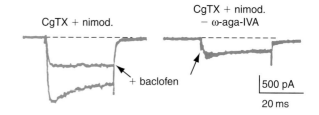

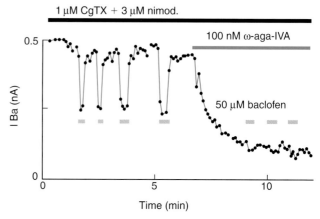

FIGURE 11.10 Baclofen suppresses the P-type calcium current in cerebellar Purkinje neurons. In the presence of 1 μM ω-conotoxin (CgTX) and 3 μM nimodipine (nimod.) to block N-type and L-type calcium channels a voltage step from −80 mV to +10 mV elicits an inward calcium current (top left). This current is partially inhibited by 50 μM baclofen. Application of the P-type calcium channel antagonist ω-agatoxin-IVA (ω-aga-IVA; 100 nM) partially blocks the current and occludes any further inhibition by baclofen (top right). The time course of the peak calcium channel current amplitude throughout the experiment is shown below. CgTX and nimod. are applied throughout the experiment (black bar), ω-aga-IVA is applied for the period of time indicated by the green bar. Note that ω-aga-IVA suppressed the calcium current and completely occluded any further inhibition by baclofen, demonstrating that baclofen was acting on the P-type calcium current. In this experiment barium was exchanged for calcium so the currents that were measured represent barium flux through calcium channels. From Mintz, IM and Bean, BP (1993). GABA$_B$ receptor inhibition of P-type Ca^{2+} channels in central neurons. *Neuron* **10**, 889–898, with permission.

Inhibition of calcium channels is dependent upon G$_i$/G$_o$ proteins

Several lines of evidence were used to demonstrate the involvement of G proteins in the inhibition of calcium channels by GABA$_B$ receptors (**Figure 11.11**). First, simply omitting GTP from the internal pipette solution during whole-cell recording gradually blocked the effect of baclofen on the calcium current. This occurred because, in the absence of a replacement supply, GTP slowly washed out of the cell during the experiment, thereby inactivating G proteins. Alternately, loading the cell with guanosine 5'-O-(2-thiodiphosphate) (GDP-β-S), a GDP analogue that inhibits the binding of GTP to G proteins, antagonized the effect of baclofen on

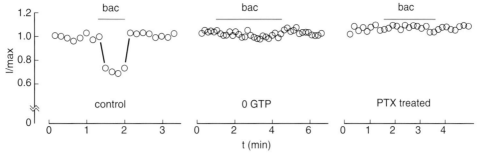

FIGURE 11.11   GABA_B receptors are coupled to calcium channels through inhibitory G proteins.
Calcium currents in cerebellar granule cells were evoked by stepping from a holding voltage of −80 mV to a test voltage of +10 mV. Current was expressed as a percentage of the maximal current in the cell at the beginning of the experiment. Bath application of baclofen (100 μM; bac) for the time indicated by the bar reduced the size of the calcium current (left). This inhibition was antagonized by removal of GTP from the internal pipette solution (centre) and by pretreatment of the neurons 12 to 16 hours earlier with pertussis toxin (PTX; right). These observations indicate that GABA_B receptors mediate inhibition of calcium channels through inhibitory G_i/G_o proteins. From Amico, C, Marchetti, C, Nobile, M, Usai, C (1995) Pharmacological types of calcium channels and their modulation by baclofen in cerebellar granules. *J. Neurosci.* **15**, 2839–2848, with permission.

calcium currents. Conversely, the effect of baclofen was enhanced when cells were loaded with guanosine 5'-O-(3-thiotriphosphate)(GTP-γ-S), a non-hydrolyzable GTP analogue that irreversibly activates G proteins. These findings support a role for G proteins in the coupling of GABA_B receptors to calcium channels. Furthermore, the observation that pertussis toxin blocks the inhibitory effect of baclofen on calcium channels indicated that GABA_B receptors coupled to calcium channels through inhibitory G_i/G_o proteins.

In theory, G_i/G_o proteins could inhibit calcium channels by physically interacting with the channel itself or by resulting in the production of a second messenger molecule which would diffuse to and inhibit the channel. Experimental evidence indicates that G proteins interact directly with N-type and P/Q-type channels. Evidence for this conclusion came from experiments using cell-attached patches in DRG neurons. It was found that baclofen, applied outside the patch pipette, did not affect the amplitude of calcium currents in cell-attached patches. However, in the same cell baclofen applied inside the patch pipette produced clear inhibition of the calcium current, demonstrating that baclofen was able to inhibit calcium currents in these cells. The inability of baclofen, applied outside the patch pipette, to inhibit calcium channels under the patch indicates that a diffusible second messenger was not involved in the inhibition.

G proteins inhibit calcium currents through a direct interaction of the βγ subunit with the calcium channel

To identify which G proteins are involved in receptor-mediated inhibition of calcium channels in native systems, a number of studies were performed with blocking antibodies and antisense oligonucleotides complementary to G protein subunits. These studies suggested that G_oα proteins were primarily responsible for the inhibition. However, other studies suggested a role for both G_oα and G_iα. This led to the hypothesis that the species involved was the Gβγ subunit, which is common to both of these G proteins, rather than any particular Gα subunit. This idea received experimental support when it was demonstrated that transfection of primary neurons or cell lines with Gβγ, but not Gα subunits, mimicked the effect of baclofen and produced tonic inhibition of the calcium current. Binding studies using purified Gβγ dimers and recombinant calcium channel subunits have further demonstrated a direct interaction between Gβγ subunits and calcium channels. It is now well established that G proteins inhibit N and P/Q-type calcium currents through a direct interaction of the Gβγ dimer with the calcium channel.

GABA_B receptors inhibit calcium channels by altering their voltage-dependence

It was originally proposed that GPCRs, such as GABA_B receptors, which inhibit calcium channels, do so by blocking and thereby reducing the number of functional channels. However, subsequent experiments demonstrated that strong depolarization of the neuronal membrane could overcome the transmitter-mediated inhibition of the calcium current. This result can not be explained by a mechanism which involves a reduction in the number of functional channels. Instead, it appears that receptor activation induces a large shift in the voltage-dependence of channel activation. Thus, following exposure to the transmitter, almost all of the channels are still fully functional and can be opened by a strong depolarization. However, a percentage of the channels undergo a shift in voltage-dependence so that they are no longer opened by small to moderate depolarizations. In practice this means that a transmitter, such as GABA, is able to inhibit the calcium current during activation by low to moderate depolarization, but the inhibitory effect of the transmitter is lost during strong depolarizations (**Figure 11.12**). In light of this observation, calcium channels have been proposed to exist in two states termed 'willing' and 'reluctant' to describe their ease of activation. These two states exist in equilibrium and according to the following model where $C_{willing}$ is the closed channel in the absence of transmitter, $C_{reluctant}$ is the closed channel in the presence of transmitter, and O is the open channel:

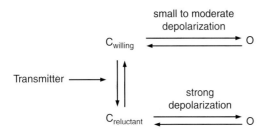

Calcium channels predominantly exist in the willing mode in the absence of transmitter and can be activated by small to moderate depolarizations. In contrast, activation of G_i/G_o coupled receptors, like GABA_B receptors, shifts the balance of equilibrium to favour the 'reluctant' state in which large depolarizations are required to open the channels.

11.4.3 GABA_B receptors activate potassium channels

GABA_B receptors activate potassium currents mediated by G-protein-activated inwardly rectifying potassium (Kir) channels (previously termed GIRK channels). Through these potassium channels, GABA_B receptors play a critical role in regulating neuronal excitability.

GABA_B receptors couple to inwardly rectifying (Kir3) potassium channels

The interaction of GABA_B receptors with potassium channels was initially suggested in experiments showing that baclofen produced a strong outward current

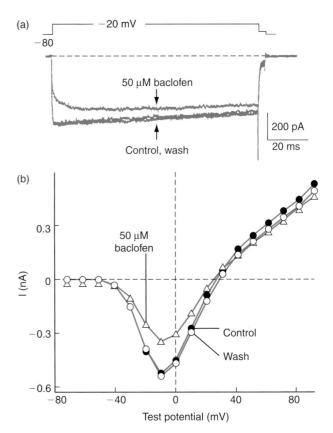

FIGURE 11.12   Inhibition of the calcium current by baclofen is voltage-dependent.
(a) In cerebellar Purkinje neurons a voltage step from −80 mV to −20 mV (top) in the presence of 1 μM ω-conotoxin and 3 μM nimodipine evokes a P-type calcium current (bottom). This current is suppressed by the subsequent application of baclofen (50 μM) and recovers following washout of the baclofen. (b) Inhibition of P-type calcium currents by baclofen is voltage-dependent. This graph shows the amplitude of the calcium current evoked by depolarizing steps to a variety of different test potentials in control (filled circles), 50 μM baclofen (open triangles) and after washout of the baclofen (open circles). The holding potential was −80 mV. Baclofen inhibited the current most effectively when it was evoked with voltage steps to potentials below +20 mV. Strong depolarizations were able to overcome the inhibition produced by baclofen. Adapted from Mintz, IM and Bean, BP (1993) GABA_B receptor inhibition of P-type Ca²⁺ channels in central neurons. *Neuron* **10**, 889–898, with permission.

and an increase in membrane conductance in voltage-clamped hippocampal pyramidal neurons (**Figure 11.13a,b**). A GABA_B receptor antagonist blocked both the outward current and conductance increase produced by baclofen, confirming that these effects were mediated by GABA_B receptors. The current–voltage curve for the baclofen-mediated current in these pyramidal neurons displayed inward rectification and reversed at a membrane potential of about −80 mV. This reversal potential corresponded well with the calculated equilibrium potential for potassium ions, suggesting that

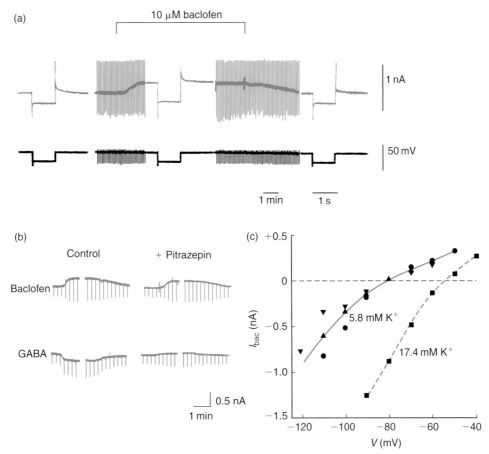

FIGURE 11.13 **GABA_B receptor activation produces an outward current that is mediated by potassium ions.** **(a)** In a hippocampal pyramidal neuron held in voltage clamp at a membrane potential of $-61\,\text{mV}$ baclofen ($10\,\mu\text{M}$) produces an outward current, seen as a gradual upward deflection in the current recording as the baclofen washed in to the preparation. Both the current (green) and voltage (black) recordings are shown. Voltage steps lasting 1 sec were delivered repetitively during the experiment to assess the membrane conductance. At three points in the experiment the recording was expanded to better show the response to the voltage step. Note that in the presence of baclofen the current response to the voltage step is larger, indicating an increase in membrane conductance. **(b)** Another voltage clamped pyramidal cell shows a similar outward current in response to baclofen (top left). Baclofen increased the membrane conductance of this neuron as indicated by an increase in the amplitude of the current deflection in response to repetitive voltage steps (downward deflections). In this cell GABA evoked an inward current and conductance increase (bottom left), suggesting that it was primarily acting on GABA_A receptors. Blockade of the GABA_A-linked chloride conductance with pitrazepin ($10\,\mu\text{M}$) had no effect on the baclofen-evoked current (top right), but caused the GABA response to become an outward current (bottom right). This occurred because blockade of the GABA_A receptor-mediated inward chloride current enabled the underlying GABA_B receptor-mediated outward current to become visible. **(c)** The current–voltage relationship for the baclofen-evoked current in a pyramidal cell when the extracellular concentration of potassium ions was $5.8\,\text{mM}$ (circles, triangles) and $17.4\,\text{mM}$ (squares). Altering the extracellular potassium concentration depolarized the reversal potential of the baclofen-evoked current by an amount predicted by the Nernst equation, indicating that this current was mediated by potassium ions. Adapted from Gähwiler, B.H. and Brown, D.A. (1985) GABA_B-receptor-activated K^+ current in voltage-clamped CA3 pyramidal cells in hippocampal cultures. *Proc. Natl Acad. Sci. USA* **82**, 1558–1562, with permission.

the current was caused by an increase in the potassium conductance of the membrane.

This was further confirmed by measuring the shift in the reversal potential of the GABA_B receptor-mediated current produced by an increase in the extracellular potassium concentration from $5.8\,\text{mM}$ to $17.4\,\text{mM}$ (**Figure 11.13c**). The reversal potential of the current depolarized $26\,\text{mV}$, an amount close to that predicted by the Nernst equation ($29\,\text{mV}$). This close agreement indicated that the GABA_B current was carried by potassium ions. This conclusion was further confirmed when it was observed that compounds that are known to block potassium channels, such as extracellular barium or intracellular cesium, also blocked the response to baclofen and GABA.

Subsequent studies have supported these early findings and have further demonstrated that GABA$_B$ receptors couple to the Kir3 family of inwardly rectifying potassium channels. Specifically, heteromeric potassium channels composed of Kir3.1/Kir3.2 (GIRK1/GIRK2) or Kir3.1/Kir3.4 (GIRK1/GIRK4) couple with high efficiency to GABA$_B$ receptors in a variety of heterologous systems. Another study, using mice whose Kir3.2 genes were genetically deleted, reported that GABA$_B$ receptor-mediated potassium currents were absent in CA3 hippocampal neurons. Analysis of Kir3 protein levels in these mice revealed a lack of Kir3.2 protein and a substantial reduction in Kir3.1 protein, indicating a critical role for Kir3 proteins in mediating the effect of GABA$_B$ receptor activation. Alternately, examination of the electrophysiological and pharmacological properties of the GABA$_B$ receptor-mediated potassium current in hippocampal CA3 pyramidal neurons revealed that this current shared similar properties to that mediated by Kir3.1/Kir3.2 or Kir3.1/Kir3.4 potassium channels. These studies point to the important role of Kir3 channels in mediating the effects of GABA$_B$ receptors. However, it has also been reported that in some neurons baclofen can induce linear or outwardly rectifying potassium conductances, suggesting that channels other than Kir3 can also contribute to the GABA$_B$ receptor-mediated potassium current.

GABA$_B$ receptors are coupled to potassium channels via inhibitory G proteins

Just as they are linked to their other effector systems, GABA$_B$ receptors are coupled to potassium channels through G proteins. This conclusion is based on the observation that GDP-β-S reduced the potassium current produced by baclofen. In contrast, GTP-γ-S mimicked the effect of baclofen. Exposure to pertussis toxin blocked the activation of potassium channels by both baclofen and GABA, indicating that the effect of GABA$_B$ receptors on potassium channels is achieved through either one or both of the inhibitory G proteins, G$_i$ and/or G$_o$.

The identity of the G protein through which GABA$_B$ receptors couple to potassium channels was further examined using heteromeric GABA$_B$ receptors (GABA$_{B1a}$/GABA$_{B2}$ or GABA$_{B1b}$/GABA$_{B2}$) expressed in HEK293 cells which stably expressed Kir3.1/Kir3.2 potassium channels. In these cells all endogenous G$_i$/G$_o$ protein activity was eliminated by pertussis toxin

treatment, preventing the GABA$_B$ receptors from activating any potassium current. The introduction of mutant pertussis toxin resistant G$_i$/G$_o$ proteins in these cells then allowed the determination of those G proteins which would rescue coupling between GABA$_B$ receptors and potassium channels. Interestingly, G protein coupling by GABA$_{B1a}$- and GABA$_{B1b}$-containing receptors was different. For receptors containing GABA$_{B1a}$, coupling to Kir3 channels was rescued only by the addition of G$_{oA}$ proteins. However, for receptors containing GABA$_{B1b}$, both G$_{oA}$ and G$_{i2}$ proteins rescued Kir3 coupling. Thus, both GABA$_{B1}$ subunits appear able to signal through G$_{oA}$ protein. However, the ability of GABA$_{B1b}$ to also signal through G$_{i2}$ protein suggests differences in receptor–effector coupling.

GABA$_B$ receptors are directly coupled to potassium channels by $\beta\gamma$ subunits of G proteins

Several lines of evidence indicate that G$_i$/G$_o$ proteins couple to potassium channels via their $\beta\gamma$ subunits. For example, application of purified G$\beta\gamma$ subunits, but not Gα subunits to the intracellular surface of excised patches of chick embryonic atrial cells activated G-protein-gated potassium channels, suggesting that $\beta\gamma$ subunits carried the functional signal. This suggestion was confirmed by subsequent binding studies which demonstrated a direct interaction between G$\beta\gamma$ subunits and Kir3.1, Kir3.2 and Kir3.4. Similarly, it was found, using the yeast two hybrid system, that the G protein β subunit bound directly with the amino-terminus of Kir3.1. Mutational analysis was then used to determine the binding site for G$\beta\gamma$ subunits on the Kir3 proteins and revealed that $\beta\gamma$ subunits bound to sites on both the amino- and carboxy-terminus of the Kir3 protein.

The ability of G$\beta\gamma$ dimers to activate Kir3 channels brought into question the molecular determinants of the interaction specificity. For example, in native tissues only G$_i$/G$_o$, and not G$_s$ proteins activate Kir3 channels and yet all of these G proteins release free $\beta\gamma$ subunits upon receptor stimulation. It is known that receptor specificity does not lie at the level of the $\beta\gamma$ subunit since a variety of different $\beta\gamma$ subunits have been shown to be equally effective at stimulating Kir3 channels. In a mammalian expression system (HEK293 cells) G$_i$/G$_o$-coupled receptors, but not G$_s$-coupled receptors activate Kir3.1/Kir3.2 channels, suggesting that the receptor specificity lies at the level of the G protein α subunit. This possibility was confirmed by the observation that G$_s$-coupled receptors could be made to stimulate potassium channels by swapping critical residues on the carboxy-terminus of the G$_s\alpha$ subunit with those of the G$_i\alpha$ subunit. Thus, G$\beta\gamma$ directly controls Kir3 channels; however Gα determines the specificity of receptor action.

The Gα subunit determines receptor specificity by facilitating the association between G$\beta\gamma$ and the Kir3 channel. Recent studies using fluorescence resonance energy transfer (FRET) and total internal reflected fluorescence (TIRF) microscopy reported that the G$\beta\gamma$ complex is closely associated with the Kir3 channel's cytosolic domains at rest and that upon GPCR activation the G$\beta\gamma$ dimer undergoes a change in its relative position on the channel to promote activation. The G$\beta\gamma$ association with the channel at rest depends on its interaction with Gα. Thus, the specificity of the interaction between G$\beta\gamma$ and the Kir3 channel results from the close association between these proteins and this interaction is facilitated by Gα.

GABA$_B$ receptors and Kir3 channels form a macromolecular signalling complex in lipid rafts

Recent studies indicate that GABA$_B$ receptors and their effector Kir3 channels form tight associations within lipid rafts. Lipid rafts are specialized plasma membrane microdomains enriched in certain lipids that can serve as platforms for signalling molecules. For example, G$_i$/G$_o$ proteins are enriched in the lipid raft fraction from cerebellar membranes. GABA$_B$ receptors and Kir3 channels are also enriched in lipid rafts. The presence of each of these proteins suggests that lipid rafts serve to cluster GABA$_B$ receptors with their effector and signalling systems. This suggestion received further support from studies using fluorescence resonance energy transfer (FRET) and fluorescently labelled proteins that report a tight association between GABA$_B$ receptors, Kir3 channels and G proteins in macromolecular signalling complexes. It has been proposed that certain GABA$_{B1}$ isoforms and Kir3 channel compositions preferentially coexist in these signalling complexes, providing a mechanism for functional heterogeneity within the GABA$_B$ system.

GABA$_B$ receptor-activated potassium channels display flickering behaviour

Single-channel potassium currents, activated by baclofen or GABA, can be recorded from cell-attached patches of cultured hippocampal neurons. They are blocked by GABA$_B$, but not GABA$_A$ receptor antagonists, indicating that they are GABA$_B$ receptor-dependent. These currents are potassium selective. Thus, alterations in the concentration of potassium ions in the pipette cause a corresponding shift in the reversal potential of the single-channel current. The single-channel current amplitude that occurs with highest probability is about

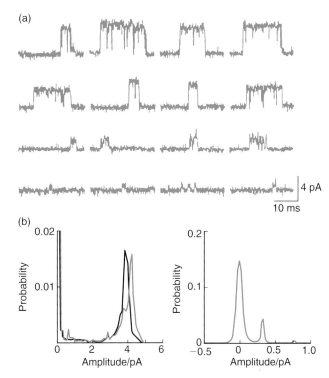

FIGURE 11.14 GABA$_B$ receptors activate single-channel currents with a mean amplitude of 4 pA.

(a) Examples of single-channel currents evoked by GABA$_B$ receptor activation in cultured hippocampal neurons. Currents were recorded in the cell-attached patch configuration and GABA was applied through the bath to the membrane outside of the patch. Currents in the lower two rows were selected for this figure because of their small amplitude. **(b)** Current amplitude probability histograms of the GABA$_B$ receptor-mediated single-channel current. These histograms were constructed from data collected from the same patch as in A. The graph on the left shows two histograms (black and green lines) representing the current amplitudes taken from two unbroken segments of data in this patch. Note that in both cases a channel with an amplitude of about 4 pA occurred with the greatest probability. The histogram on the right was taken from sections of the data which had the smallest currents. The smaller peak corresponds to elementary channel current with an amplitude of 0.36 pA. From Premkumar, LS, Chung, S-H, Gage, PW (1990) GABA-induced potassium channels in cultured neurons. *Proc. R. Soc. Lond. B* **241**, 153–158, with permission.

4 pA (**Figure 11.14**). This corresponds to a conductance of 67 pS.

Kir3 channels that are coupled to GABA$_B$ receptors exhibit complex behaviour with a number of different gating modes. Activation of GABA$_B$ receptors alters the gating of these channels such that the channel spends more time in modes with higher open probabilities. When activated, a prominent characteristic of these single-channel currents is a rapid flickering between open and closed states. This flickering appears to show a variety of different subconductance levels

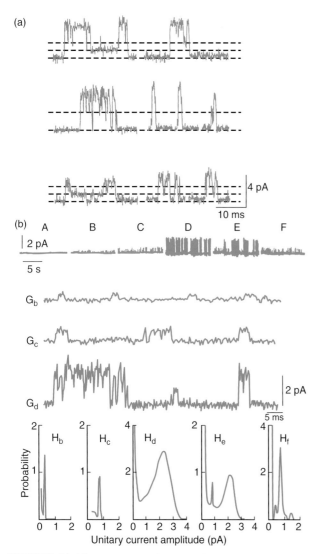

(Figure 11.15a). These different conductance levels are particularly prominent during wash on and wash out of the baclofen (Figure 11.15b). Histograms of the current amplitudes reveal many peaks which appear to occur at multiples of the smallest peak. This smallest peak represents an elementary current amplitude of 0.36 pA (Figure 11.14), corresponding to a conductance of 5–6 pS. Whereas this measurement reflects the conductance coupled to somatic GABA_B receptors, a subsequent study used non-stationary variance analysis to estimate the conductance linked to synaptically activated GABA_B receptors. This study reported a small unitary conductance in the range of 5–12 pS, in agreement with the elementary conductance observed at the single-channel level. This small subconductance state predominates during synaptic GABA_B currents.

In single-channel studies the kinetics of the current activation and deactivation, even during large events (>4 pA), are extremely rapid. Since the elementary current amplitude is only 0.36 pA, many of these elementary channels would need to open or close synchronously to cause these rapid transitions. However, it is extremely unlikely that these channels would behave independently in such a synchronized fashion. Therefore, it has been suggested that these elementary channels function cooperatively. According to this hypothesis, these elementary co-channels would form oligomers of varying size which would function as a single unit. Activation of the oligomer would cause many of the channels to open simultaneously. The flickering behaviour of the current would then represent the transient opening and closing of the elementary channels within the oligomer.

FIGURE 11.15 GABA_B single-channel currents have multiple subconductance states.
(a) Examples of single-channel currents evoked by GABA_B receptor activation in cell-attached patches of cultured hippocampal neurons. These currents were selected to emphasize different subconductance states of the channels. Dotted lines indicated different conductance levels. (b) In the presence of bicuculline to block GABA_A currents, exposure of a cell to GABA causes the slow development of GABA_B receptor-mediated single-channel currents in a cell-attached patch. These currents appear to go through several different conductance states until finally reaching their maximal amplitude. The panel in A represents the baseline response of the patch before the addition of agonist. The panels in B–D show activity in the patch at 25 sec (B), 1.5 min (C) and 4 min (D) after the addition of agonist. Panels E and F show patch activity after 5 and 10 minutes of wash, respectively. The records in the middle three rows represent an expansion of a portion of the panels shown in B (G_b), C (G_c), and D (G_d). Finally, the current amplitude probability histograms (H_b-H_f) were produced from data collected at the same times as panels B–F. Note the progressive increase in the GABA_B receptor-mediated current amplitude following the application of GABA. From Premkumar, LS, Chung, S-H, Gage, PW (1990) GABA-induced potassium channels in cultured neurons. *Proc. R. Soc. Lond. B* **241**, 153–158, with permission.

11.5 SUMMARY

GABA_B receptors are coupled through inhibitory G_i/G_o proteins to multiple effector systems. The primary effects of GABA_B receptor activation include inhibition of adenylyl cyclase, inhibition of voltage-dependent calcium channels and activation of inwardly rectifying potassium channels. Future studies may reveal other effector systems to which GABA_B receptors are also coupled. By coupling to these different effector systems, GABA_B receptors enable GABA to have a broader range of effects on neurons than it could by acting only on GABA_A receptors. The discussion so far has focused on the intrinsic properties of the GABA_B receptor and the effector systems to which they are coupled. We will now turn our attention to the role that these receptors play in synaptic activity.

11.6 THE FUNCTIONAL ROLE OF GABA$_B$ RECEPTORS IN SYNAPTIC ACTIVITY

GABAergic synapses in the central nervous system contain both GABA$_A$ and GABA$_B$ receptors capable of responding to the synaptic release of GABA. Once released, the lifetime of GABA in the synaptic cleft is very brief (milliseconds) both because the duration of the release is very short and the GABA that is released quickly diffuses away. In addition, there exists an avid uptake system to actively remove GABA from the synaptic cleft. These systems combine to tightly regulate GABA concentration in the synaptic cleft.

Synaptic activation of GABA$_A$ receptors produces a rapid, synchronous opening of chloride channels, resulting in a fast inhibitory postsynaptic current. In contrast, synaptic activation of GABA$_B$ receptors initiates a second messenger-mediated process which is considerably slower. Because of the delay inherent in the second messenger system, GABA has disappeared from the synaptic cleft before the GABA$_B$ receptor-mediated response even begins. Thus, the kinetics of this response are determined not by the binding/unbinding of GABA from the GABA$_B$ receptor but rather by the kinetics of the second messenger system involved. The effects of GABA$_B$ receptors are exerted by both post-synaptic and presynaptic receptors, which play very different roles in neuronal function. Postsynaptic GABA$_B$ receptors inhibit excitatory transmission primarily by hyperpolarization. In contrast, the primary functional effect of presynaptic receptors is to inhibit the release of neurotransmitter.

11.6.1 Postsynaptic GABA$_B$ receptors produce an inhibitory postsynaptic current (IPSC)

When stimulated by synaptically released GABA, postsynaptic GABA$_B$ receptors increase the potassium conductance of the neuronal membrane. For a neuron near its resting potential, this increase in potassium conductance produces a large hyperpolarization of the membrane which is seen in a whole-cell voltage clamp recording as an outward current. This outward current, termed an inhibitory postsynaptic current (IPSC) is produced by the summation of the elementary current flowing through each of the GABA$_B$ receptor-activated potassium channels (**Figure 11.16a**). The small elementary conductance of the GABA$_B$-coupled potassium channel suggests that a large number of these channels open during an average-sized GABA$_B$ IPSC. In the example shown in **Figure 11.16a**, the conductance of the GABA$_B$ IPSC is 1.25 nS. Therefore, based on an elementary conductance for GABA$_B$-coupled potassium channels of 5–12 pS, it can be calculated that approximately 150 channels opened at the peak of the GABA$_B$ IPSC.

The kinetics of the GABA$_B$ receptor-mediated response are slow

Because it is coupled through a second messenger system, the GABA$_B$ receptor-mediated hyperpolarization has a time course that is very different from that produced by an ionotropic receptor-channel, such as GABA$_A$ (**Figure 11.16b**). Measurements of the time required from stimulation of the presynaptic terminals to the initiation of the postsynaptic hyperpolarization have ranged from 20 to 50 ms. This onset latency is considerably longer than that of the GABA$_A$ receptor-mediated response (<3 ms). The risetime of the GABA$_B$-mediated current is also slow and it does not reach a peak for 130 to 300 ms. This risetime is much slower than the GABA$_A$ response which typically reaches a peak in 1–15 ms. The slower risetime of the GABA$_B$ response is thought to occur because of the asynchronous activation of potassium channels. Finally, the GABA$_B$ response decays back to baseline over the next 400 to 1300 ms. This slow rate of decay may reflect the rate of GTP hydrolysis, suggesting that it is the decline of activated G protein that ultimately terminates the response. In contrast, the GABA$_A$ response decays to baseline much more rapidly (80–220 ms). The prolonged duration of the GABA$_B$ response enables GABA to produce inhibition over a much longer period of time than it could by acting on GABA$_A$ receptors alone.

GABA$_B$ receptors are more sensitive than GABA$_A$ receptors to GABA

Dose-response curves reveal that GABA is much more potent in activating GABA$_B$ than GABA$_A$ receptors. Despite the higher sensitivity of GABA$_B$ receptors to GABA, activation of these receptors requires high intensity or repetitive stimulation of the neuronal network. This situation arises because GABA$_B$ receptors are mostly located extrasynaptically and are not activated until sufficient GABA is released to overcome local uptake systems and spill over onto the receptors. However, the higher sensitivity of GABA$_B$ receptors to GABA enables these receptors to respond to the low concentrations of GABA that are able to reach these extrasynaptic spaces.

The GABA$_B$ IPSC produces inhibition by hyperpolarizing the neuronal membrane

Whole-cell voltage clamp recordings reveal that the maximal peak conductance increase produced by

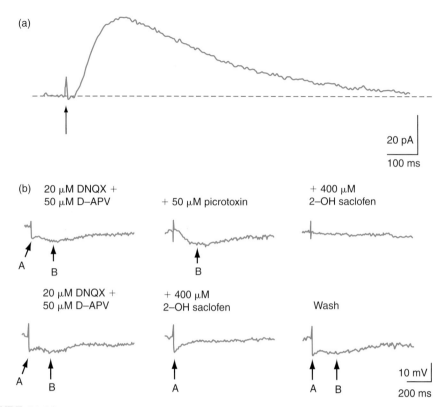

FIGURE 11.16 **Synaptically released GABA activates postsynaptic GABA$_B$ receptors to produce a slow IPSC.**
(a) Stimulation of inhibitory fibres evokes a stimulus artifact (arrow) followed by a GABA$_B$ receptor-mediated IPSC in a hippocampal neuron held at a potential of −60 mV in a whole-cell voltage clamp. The GABA$_B$ IPSC was pharmacologically isolated from the excitatory synaptic current using DNQX, which blocks AMPA/kainate receptors, and APV, which blocks NMDA receptors. It was also isolated from the GABA$_A$ inhibitory current using bicuculline which blocks GABA$_A$ receptors. Note the slow onset of the IPSC and its long latency. **(b)** GABA$_A$ and GABA$_B$ inhibitory postsynaptic potentials (IPSPs) were recorded in current clamp from a dentate gyrus granule cell. These hyperpolarizing potentials were evoked by stimulating inhibitory fibres. They were isolated from glutamatergic excitatory potentials by application of DNQX and APV. GABA$_A$ and GABA$_B$ IPSPs are indicated by arrows labelled 'A' and 'B', respectively. In control (top left) a stimulus evoked a stimulus artifact (upward deflection) followed by both a GABA$_A$ and a GABA$_B$ IPSP which can be seen as the fast and slow components of the hyperpolarizing response, respectively. Application of the GABA$_A$ antagonist, picrotoxin blocks the GABA$_A$ IPSP leaving only the slow GABA$_B$ IPSP (top centre). The GABA$_B$ antagonist 2-hydroxsaclofen blocks this GABA$_B$ IPSP. Similarly, in another cell application of 2-hydroxysaclofen to the control response (bottom left) blocks the GABA$_B$ IPSP, leaving an isolated GABA$_A$ IPSP (bottom centre). The effect of this antagonist is reversible (bottom right). Note the difference in the time course of the isolated GABA$_B$ IPSP (top centre) and the isolated GABA$_A$ IPSP (bottom centre). Part (a) from Mott, DD and Lewis, DV Unpublished observations. Part (b) from Mott, DD and Lewis, DV (1992) GABA$_B$ receptors mediate disinhibition and facilitate long-term potentiation in the dentate gyrus. *Epilepsy Res.* **Suppl. 7**, 119–134, with permission.

activation of GABA$_A$ receptors is much greater (5- to 10-fold) than that produced by activation of GABA$_B$ receptors. For example, in hippocampal pyramidal neurons the maximal conductance of the GABA$_A$ IPSC ranges from 90 to 140 nS. This compares to a range of 13 to 19 nS for the maximal conductance of the GABA$_B$ IPSC in these same cells. Similar differences between the maximal conductance values of GABA$_A$ and GABA$_B$ receptor-mediated currents have been reported in other brain regions.

Despite its relatively small conductance, the GABA$_B$ current produces a large hyperpolarization from rest in most neurons. This strong hyperpolarization occurs because activation of GABA$_B$ receptors drives the membrane potential towards the reversal potential for potassium ions. In physiological conditions the equilibrium potential for potassium ions (−80 to −98 mV) is quite negative relative to the resting membrane potential (−50 to −75 mV) of most cells. Therefore, even though the conductance of the GABA$_B$ IPSC is small, the driving force for potassium can be quite large. In fact, because of this large driving force and the long duration of the GABA$_B$ response, the GABA$_B$ IPSC can move an amount of charge that is close to that carried by the GABA$_A$ IPSC.

For example, in granule cells of the dentate gyrus about 8 pC of charge leave the cell during the GABA$_B$ IPSC. This compares favourably to the 9 to 35 pC that are carried by the GABA$_A$ response in these same cells.

The GABA$_A$ IPSC powerfully inhibits neuronal excitability both by hyperpolarizing the postsynaptic membrane and increasing its conductance. Hyperpolarization moves the postsynaptic membrane away from action potential threshold, whereas the conductance increase produced by the GABA$_A$ IPSC shunts the postsynaptic membrane thereby short-circuiting excitatory responses. This inhibition powerfully suppresses both voltage-dependent and voltage-independent excitatory currents and can not be overcome by depolarization. In contrast, the GABA$_B$ IPSC produces a large hyperpolarization with a fairly small conductance increase. Thus, it inhibits neurons primarily through hyperpolarization. This hyperpolarizing inhibition is effective in suppressing voltage-dependent currents, such as NMDA receptor-mediated responses. However, since it can be overcome by neuronal depolarization, it does not effectively inhibit voltage-independent currents. Inhibition produced by GABA$_B$ receptors has been suggested to be a more modulatory form of inhibition than that produced by GABA$_A$ receptors, enabling a fine-tuning of neuronal function. Thus, GABA$_B$ receptor-mediated inhibition differs in both kinetics and function from inhibition produced by GABA$_A$ receptors.

11.6.2 Presynaptic GABA$_B$ receptors inhibit the release of many different transmitters

GABA$_B$ receptors are located on presynaptic terminals where they inhibit the release of a variety of neurotransmitters, including GABA, glutamate, dopamine, serotonin and norepinephrine. Inhibition of transmitter release by synaptically released GABA is dramatically enhanced following pharmacological blockade of GABA uptake, indicating that released GABA has to overcome uptake in order to reach these presynaptic GABA$_B$ receptors. By activating presynaptic GABA$_B$ receptors, synaptically released GABA can inhibit transmitter release at the inhibitory terminal from which the GABA was originally released (homosynaptic depression) as well as at neighbouring inhibitory and/or excitatory terminals (heterosynaptic depression).

Presynaptic GABA$_B$ receptors inhibit the release of GABA

Inhibition of GABA release by presynaptic GABA$_B$ receptors has been especially well examined. It has been conclusively demonstrated that synaptically released GABA can feedback onto presynaptic GABA$_B$

receptors located on the activated GABAergic terminal as well as on other neighbouring GABAergic terminals. These presynaptic GABA$_B$ receptors can then suppress the subsequent release of GABA, causing both GABA$_A$ IPSCs and GABA$_B$ IPSCs to be smaller.

This effect can be clearly observed in a cortical neuron using whole-cell voltage clamp to record GABA$_A$ IPSCs (**Figure 11.17**). During the delivery of paired electrical stimuli to GABAergic axons, the first stimulus of the pair evokes the release of GABA, resulting in the production of a GABA$_A$ IPSC. However, this released GABA also activates presynaptic GABA$_B$ receptors on the inhibitory terminals suppressing the release of further GABA. Thus, a second identical stimulus delivered 300 ms later evokes a GABA$_A$ IPSC that is greatly

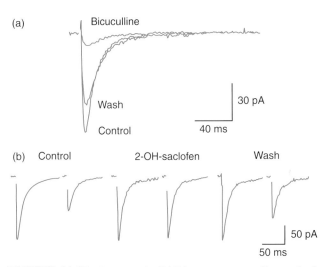

FIGURE 11.17 Presynaptic GABA$_B$ receptors mediate paired pulse depression of IPSCs.

(a) The pharmacologically isolated GABA$_A$ IPSC in a neuron in the somatosensory cortex is reversibly blocked by bicuculline. This GABA$_A$ IPSC was evoked by electrical stimulation of inhibitory fibres and recorded in whole-cell voltage clamp. It was isolated from the excitatory synaptic current by application of CNQX and APV, antagonists of AMPA/kainate and NMDA receptors, respectively. The neuron was held at a membrane potential of -70 mV, causing the GABA$_A$ IPSC to be an inward current. **(b)** Under these same conditions if paired stimuli are delivered 300 ms apart, the GABA$_A$ IPSC evoked by the second stimulus of the pair is reduced (left). This reduction of the second IPSC is blocked by the GABA$_B$ antagonist, 2-hydroxysaclofen (centre). This effect is reversible after washout of the antagonist (right). The cell recorded in this experiment was from a very young (postnatal day 10) rat. In animals of this young age the postsynaptic GABA$_B$ response is developmentally immature, whereas the presynaptic GABA$_B$ response is fully developed. This difference in development explains why no GABA$_B$ IPSC is evident in these recordings and gives further confirmation that postsynaptic GABA$_B$ receptors are not necessary for paired pulse depression of IPSCs. In older animals postsynaptic GABA$_B$ IPSCs are fully developed and both GABA$_A$ and GABA$_B$ IPSCs are depressed by presynaptic GABA$_B$ receptors (see **Figure 11.18**). Adapted from Fukuda, A., Mody, I., Prince, D.A. (1993) Differential ontogenesis of presynaptic and postsynaptic GABA$_B$ inhibition in rat somatosensory cortex. *J. Neurophysiol.* **70**, 448–452, with permission.

reduced. Application of the GABA_B antagonist, 2-hydroxysaclofen, blocks the presynaptic GABA_B receptors, preventing the reduction in the second IPSC. The ability of released GABA to act on presynaptic GABA_B receptors to suppress subsequent GABA_A IPSCs endows the GABAergic system with a powerful feedback mechanism capable of suppressing GABAergic inhibition in an activity-dependent manner.

The time course of the depression of GABA release is similar to the time course of the postsynaptic GABA_B IPSC.

Just like the postsynaptic effect of GABA_B receptors, the time course of the inhibition of GABA release by GABA_B receptors reflects a second messenger-coupled mechanism. Following the stimulation of an inhibitory pathway, the onset of the presynaptic inhibition is slow, reaching a peak about 200 ms after the initial stimulus (**Figure 11.18**). The duration of the effect is also quite prolonged and can extend for up to several seconds. Thus, although GABA has a brief lifetime in the synaptic cleft, the activation of presynaptic GABA_B receptors by this GABA enables it to modulate the subsequent release of transmitter for a more prolonged time.

GABA_B receptors suppress transmitter release by inhibiting voltage-dependent calcium channels

In most brain regions GABA_B receptors suppress neurotransmitter release through a voltage-dependent inhibition of N- and P/Q-type calcium channels. These calcium channels are expressed presynaptically and have been implicated in the control of neurotransmitter release. It has been reported that activation of GABA_B receptor inhibits up to 50% of the calcium current. However, because of the non-linear relationship between calcium concentration in the presynaptic terminal and transmitter release, this 50% reduction in calcium current is often sufficient to inhibit neurotransmitter release by more than 90%.

In theory, GABA_B receptors could suppress calcium currents by directly inhibiting calcium channels or by activating Kir3 channels that would hyperpolarize the presynaptic terminal and oppose the depolarization necessary for calcium channel activation. Two lines of evidence argue in support of a direct effect of GABA_B receptors on presynaptic calcium channels. First, at many terminals GABA_B receptor activation blocks some, but not all types of calcium channel. This differential inhibition suggests a direct effect of GABA_B receptors on distinct calcium channel types within the terminal. In contrast, the activation of a potassium conductance by GABA_B receptors would cause a general

decrease in all components of the calcium influx. Second, electrophysiological recordings from the giant nerve terminals (calyces of Held) in the medial nucleus of the trapezoid body have demonstrated that baclofen-mediated

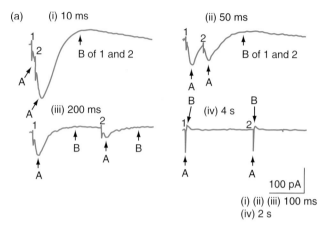

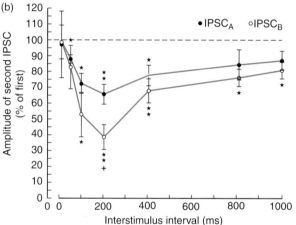

FIGURE 11.18 Paired pulse depression of IPSCs is maximal when stimuli are delivered 200 ms apart.

(a) Isolated GABA_A and GABA_B IPSCs were recorded in whole-cell voltage clamp from a dentate gyrus granule cell. Since the cell was held at a membrane potential of −80 mV, the GABA_A IPSC is an inward current whereas the GABA_B IPSC is an outward current. Inhibitory fibres were electrically stimulated to evoke IPSCs. Paired stimuli were delivered at increasing intervals to determine the time course of the inhibition of GABA release produced by presynaptic GABA_B receptors. Responses to paired stimuli at four different intervals are shown. In this cell suppression of the second IPSC was greatest when the stimuli were delivered 200 ms apart. GABA_A and GABA_B IPSCs are indicated by arrows labelled 'A' and 'B', respectively. **(b)** Graph of the averaged data obtained from six cells showing the time course of the suppression of the second IPSC. For both the GABA_A and GABA_B IPSC the second response of the pair was maximally depressed when the stimuli were delivered about 200 ms apart. Stars indicate a significant depression of the IPSC (★$P < 0.05$, ★★$P < 0.01$). The cross (+) indicates that the GABA_B IPSC was significantly more depressed than the GABA_A IPSC. From Mott, DD, Xie, CW, Wilson, WA, Swartzwelder, HS, Lewis, DV (1993) GABA_B autoreceptors mediate activity-dependent disinhibition and enhance signal transmission in the dentate gyrus. *J. Neurophysiol.* **69**, 674–691, with permission.

suppression of synaptic transmission was associated with a reduction in presynaptic calcium current but not with activation of a potassium current in the presynaptic terminal. These observations strongly argue that, at least at these synapses presynaptic GABA$_B$ receptors suppress transmitter release by directly inhibiting calcium channels.

Several observations suggested that presynaptic GABA$_B$ receptors in these giant terminals were coupled to calcium channels via G$_i$/G$_o$ proteins. First, loading of the presynaptic terminal with GDP-β-S blocked the effect of baclofen on calcium currents. In contrast, GTP-γ-S suppressed presynaptic calcium currents and occluded the effect of baclofen. Second, inhibition of calcium channels by baclofen was blocked by N-ethylmaleimide, a sulfhydryl alkylating agent which uncouples G$_i$/G$_o$ proteins from their receptors. These results directly indicate that at this giant synapse GABA$_B$ receptors suppress synaptic transmission by inhibiting presynaptic calcium channels and that GABA$_B$ receptors couple to these calcium channels via G$_i$/G$_o$ proteins.

As discussed previously (Section 11.4.3), GABA$_B$ receptor-mediated inhibition of calcium channels is voltage dependent. Strong depolarization of presynaptic terminals relieves the inhibition. The extent of depolarization of the presynaptic terminal and the level of GABA$_B$-mediated inhibition of transmitter release are therefore regulated by action potential frequency. During high-frequency activity, depolarization of the presynaptic terminal would relieve GABA$_B$ inhibition of calcium channels and restore neurotransmitter release. At inhibitory terminals the relief of GABA$_B$-mediated inhibition of GABA release during high-frequency activity may serve as a feedback mechanism to prevent overexcitation.

11.7 SUMMARY

GABA$_B$ receptors enable GABA to produce a variety of effects on neuronal function. These receptors are located both pre- and postsynaptically where they can be activated by synaptically released GABA. Postsynaptic GABA$_B$ receptors generate a slow inhibitory current which is carried by potassium ions. This current produces a hyperpolarizing inhibition which effectively inhibits voltage-dependent conductances, such as the NMDA receptor-mediated current. Presynaptic GABA$_B$ receptors inhibit the release of a variety of different neurotransmitters, including glutamate and GABA. The ability of GABA$_B$ receptors to regulate GABA release provides an important mechanism for the feedback control of both GABA$_A$ and GABA$_B$ inhibition. Thus,

by acting at both pre- and postsynaptic sites, GABA$_B$ receptors have the potential to produce profound changes in neuronal function.

FURTHER READING

Bettler B, Kaupmann K, Mosbacher J, Gassmann M (2004) Molecular structure and physiological functions of GABA$_B$ receptors. *Physiol. Rev.* **84**, 835–867.

Bettler B and Tiao JY (2006) Molecular diversity, trafficking and subcellular localization of GABA$_B$ receptors. *Pharmacol. Ther.* **110**, 533–543.

Bowery NG, Bettler B, Froestl W, Gallagher JP, Marshall F, Raiteri M, Bonner TI, Enna SJ (2002) International Union of Pharmacology. XXXIII. Mammalian gamma-aminobutyric acid$_B$ receptors: structure and function. *Pharmacol. Rev.* **54**, 247–264.

Couve A, Filippov AK, Connolly CN, Bettler B, Brown DA, Moss SJ (1998) Intracellular retention of recombinant GABA$_B$ receptors. *J. Biol. Chem.* **273**, 26361–26367.

De Koninck Y and Mody I (1997) Endogenous GABA activates small-conductance K+channels underlying slow IPSCs in rat hippocampal neurons. *J. Neurophysiol.* **77**, 2202–2208.

Filippov AK, Couve A, Pangalos MN, Walsh FS, Brown DA, Moss SJ (2000) Heteromeric assembly of GABA$_{B1}$ and GABA$_{B2}$ receptor subunits inhibits Ca^{2+} current in sympathetic neurons. *J. Neurosci.* **20**, 2867–2874.

Ikeda SR (1996) Voltage-dependent modulation of N-type calcium channels by G-protein beta gamma subunits. *Nature* **380**, 255–258.

Kaupmann K, Huggel K, Heid J, Flor PJ, Bischoff S, Mickel SJ, McMaster G, Angst C, Bittiger H, Froestl W, Bettler B (1997) Expression cloning of GABA$_B$ receptors uncovers similarity to metabotropic glutamate receptors. *Nature* **386**, 239–246.

Kaupmann K, Malitschek B, Schuler V, Heid J, Froestl W, Beck P, Mosbacher J, Bischoff S, Kulik A, Shigemoto R, Karschin A, Bettler B (1998) GABA$_B$-receptor subtypes assemble into functional heteromeric complexes. *Nature* **396**, 683–687.

Malitschek B, Schweizer C, Keir M, Heid J, Froestl W, Mosbacher J, Kuhn R, Henley J, Joly C, Pin J-P, Kaupmann K, Bettler B (1999) The N-terminal domain of γ-aminobutyic acid$_B$ receptors is sufficient to specify agonist and antagonist binding. *Mol. Pharmacol.* **56**, 448–454.

Parmentier M-L, Prézeau L, Bockaert J, Pin J-P (2002) A model for the functioning of Family 3 GPCRs. *Trends Pharmacol. Sci.* **23**, 268–274.

Schuler V, Lüscher C, Blanchet C, Klix N, Sansig G, Klebs K, Schmutz M, Heid J, Gentry C, Urban L, Fox A, Spooren W, Jaton AL, Vigouret JM, Pozza M, Kelly PH, Mosbacher J, Froestl W, Käslin E, Korn R, Bischoff S, Kaupmann K, van der Putten H, Bettler B (2001) Epilepsy, hyperalgesia, impaired memory, and loss of pre- and postsynaptic GABA$_B$ responses in mice lacking GABA$_{B1}$. *Neuron* **31**, 47–58.

Takahashi T, Kajikawa Y, Tsujimoto T (1998) G-protein-coupled modulation of presynaptic calcium currents and transmitter release by a GABA$_B$ receptor. *J. Neurosci.* **18**, 3138–3146.

White JH, Wise A, Main MJ, Green A, Fraser NJ, Disney GH, Barnes AA, Emson P, Foord SM, Marshall FH (1998) Heterodimerization is required for the formation of a functional GABA$_B$ receptor. *Nature* **396**, 679–682.

Yamada M, Inanobe A, Kurachi Y (1998) G protein regulation of potassium ion channels. *Pharmacol. Rev.* **50**, 723–757.

12

The metabotropic glutamate receptors

Earlier chapters have described how glutamate is the neurotransmitter of most excitatory synapses in the CNS. Initially, the actions of glutamate in the nervous system were thought to be solely mediated by ligand-gated ion channels, also called ionotropic glutamate receptors (iGluRs). However, in the mid-1980s several groups were looking for other types of glutamate receptors since glutamate was able to bind on brain membranes at specific sites different from the known iGluRs. At the same time, a new signalling cascade for G-protein coupled receptors was discovered. It involves a phospholipase C that degrades phospholipids from the plasma membrane to produce two second messengers: inositol triphosphate and diacylglycerol (**Figure 12.1a**).

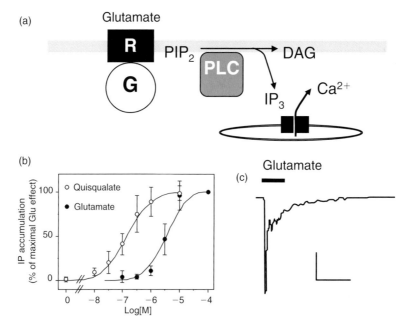

FIGURE 12.1   **Glutamate activates G-protein coupled receptors producing the second messenger inositol tri-phosphates (IP₃).**

(a) Metabotropic glutamate receptors (mGluRs) activate the enzyme PLC by stimulating specific G-proteins. This enzyme degrades the membrane phospholipid phosphatidyl-inositol-biphosphate (PIP$_2$) to produce IP$_3$ and di-acylglycerol (DAG). IP$_3$ then acts on intracellular receptor-channels and allows the release of Ca^{2+} from intracellular stores. **(b)** The first evidence for the existence of mGlu receptors came when glutamate was found to stimulate the production of IP$_3$ in cultured neurons and brain tissues. Adapted from Sladeczeck *et al.*, *Nature* 1985. **(c)** Further evidence for the existence of PLC-coupled mGluRs was the observation that glutamate activates Chloride currents through the release of Ca^{2+} from intracellular stores in *Xenopus* oocytes injected with rat brain mRNA (scale bars: vertical: 200 nA; horizontal: 20 s).

Then two groups found that glutamate was able to activate this transduction cascade by activating a receptor different from the iGluRs (**Figure 12.1b**), both in cultured neurons and in brain tissues. Indeed, this response was neither mimicked by iGluR selective agonists nor inhibited by iGluRs selective antagonists. Two years later, this receptor was successfully expressed in *Xenopus* oocytes after injection of rat brain mRNA, as illustrated by activation of an oscillatory chloride current in these cells (**Figure 12.1c**), a response typical for PLC-coupled receptors expressed in these cells. These data firmly demonstrated the existence of G-protein coupled (or 'metabotropic') glutamate receptors (mGluRs). Today it is well established that these receptors are involved in important brain physiological functions such as learning and memory, and play pivotal roles in neuropathologies such as pain, drug addiction, schizophrenia, anxiety, epilepsy, Alzheimer's and Parkinson's diseases. This chapter discusses the structural features and functions of mGluRs.

12.1 THE IDENTIFICATION OF THE EIGHT METABOTROPIC GLUTAMATE RECEPTOR SUBTYPES

A common way to clone a given gene is to utilize the function of the encoded protein to screen a library of cDNAs expressed in a heterologous system, such as *Xenopus laevis* oocytes or cell lines. The first mGluR was cloned by injecting small pools of cDNA into oocytes and was identified using electrophysiological responses resulting from inositol phosphate production. It was named mGluR1. Hydrophobicity analysis of the deduced amino acid sequence suggested seven transmembrane domains similar to all G-protein coupled receptors. However, mGluR1 shared no sequence homology to these receptors, indicating that it belonged to a distinct G-protein coupled receptor family. Cloning of other members of the mGluR family was then accomplished using low-stringency hybridization of DNA probes derived from mGluR1, and looking for mGluRs presumably containing similar sequences. This approach allowed the identification of seven additional genes encoding mGluR2 to mGluR8. Alternative splicing of many of these mGluR genes results in further diversity among mGluR proteins with different C-terminal intracellular tails. The sequencing of both human and mouse genomes did not reveal additional homologous proteins aside from the more distant GABA$_B$, calcium-sensing, basic amino acid and sweet and monosodium glutamate taste receptors.

The cloning of eight mGluR subtypes immensely expanded the study of these receptors as clones were expressed in heterologous systems to determine their coupling to second-messenger systems and establish their pharmacological profiles. Based on sequence homology and nature of the coupled G-protein, mGluR subtypes could be classified into three different groups (**Figure 12.2a**). The group-I mGluRs include mGluR1 and mGluR5. These receptors share approximately 60% sequence identity. They are most potently activated by quisqualate and are selectively activated by 3,5-dihydroxyphenylglycine (DHPG). The group-II mGluRs include mGluR2 and mGluR3 subtypes. They display about 70% sequence identity with each other, but less than 50% homology with the six other mGluR clones. They are potently and selectively activated by (2S,2'R,3'R)-2-(2',3'-dicarboxycyclopropyl)glycine (DCG-IV) and LY354740. The group-III mGluRs (mGluR4, mGluR6, mGluR7 and mGluR8) share approximately 70% sequence identity within the group, and less than 50% identity with the other four mGluRs. L-2-amino-4-phosphonobutyric acid (L-AP4) is the most potent and selective agonist of group-III mGluRs.

Several variants generated by alternative splicing have been identified. Among those reported in the literature, 3, 2 and 2 variants for mGluR1, mGluR5 and mGluR7, respectively, have been well characterized and demonstrated *in vivo* (**Figure 12.2b**). All these variants differ at the level of their intracellular tail that is the site of interaction of several intracellular proteins, which regulate the trafficking, function, and desensitization properties of these proteins.

12.2 HOW DO METABOTROPIC GLUTAMATE RECEPTORS CARRY OUT THEIR FUNCTION? STRUCTURE-FUNCTION STUDIES OF METABOTROPIC GLUTAMATE RECEPTORS

The main function of mGluRs is to activate G-proteins upon glutamate binding. Hydrophobicity analysis of various mGluRs suggests a large extracellular N-terminus with a signal peptide, followed by seven transmembrane domains, and a cytoplasmic C-terminal tail. While the seven transmembrane domains is a general feature of G-protein coupled receptors, the large N-terminal domain of mGluRs is structurally divergent. A simple early hypothesis postulated that the N-terminal domain is responsible for agonist binding, while the transmembrane domains and cytoplasmic loops are responsible for G-protein coupling and signalling. This

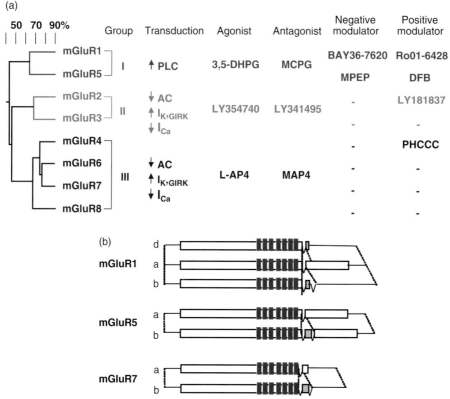

FIGURE 12.2   Classification, coupling properties and pharmacology of mGluRs.
(a) mGluRs can be divided into three groups based on amino-acid sequence similarity (left), signalling properties (centre) and pharmacology (right). Agonists and antagonists selective for each group of mGluRs are indicated. The recently identified negative and positive modulators are indicated. Note these modulators are
selective for a single mGluR subtype. **(b)** Schematic representation of the well characterized mGluR variants
resulting from alternative splicing of the pre-mRNA. The 7TM coding region is indicated with the 7 red vertical
bars. Identical mRNA sequences are linked by dotted lines.

was confirmed by constructing various chimeric receptors made of domains taken from mGluRs with different
properties. For example, a mGluR chimera made of the
extracellular domain of mGluR2 (group-II) followed by
the remainder of mGluR1 (group-I) amino acid
sequence has the agonist profile of group-II mGluRs
(activation by DCG-IV), but the coupling properties of
group-I mGluRs (stimulation of PI hydrolysis; see
below) (**Figure 12.3a**).

Further support for the idea that the N-terminal
domain mediates agonist binding is the finding that
this domain shares weak sequence homology with bacterial periplasmic amino acid binding proteins. Based
on the known crystal structure of these bacterial
proteins, the N-terminal domain of mGluRs was modelled as a 'Venus flytrap'-like structure, which clamps
together upon ligand binding. Consistent with this
model, point mutations of residues that are critical for
ligand binding, dramatically reduced agonist binding.
Further evidence for the N-terminal domain being
responsible for agonist binding was the demonstration
that this domain produced alone as a soluble protein

binds agonist with an affinity similar to that measured
on the full-length mGluR. Eventually, crystals were
obtained from this purified soluble protein and
allowed the resolution of the atomic structure of this
part of mGluR1 (**Figure 12.4**). This structure was solved
both in its empty form and with bound glutamate.
These studies confirmed the Venus flytrap bilobate
structure with the glutamate binding site located in the
cavity between the two lobes, and bring important
information on the understanding of the activation
process of this protein (see below). As proposed, based
on the homology with bacterial amino acid binding
proteins, this protein adopts a closed conformation in
the presence of glutamate.

Metabotropic glutamate receptors must somehow
convert the binding of glutamate into G-protein activation. Regions of G-protein coupled receptors that are
responsible for G-protein activation have been extensively studied, namely in β-adrenergic receptors and
rhodopsin. Specific regions of G-protein coupled
receptors determine the efficacy of transducing ligand
binding to G-protein activation, while other regions

(a)

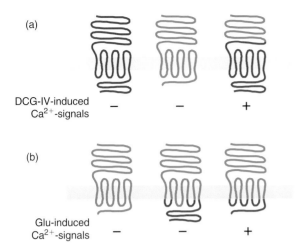

DCG-IV-induced
Ca^{2+}-signals   −   −   +

(b)

Glu-induced
Ca^{2+}-signals   −   −   +

FIGURE 12.3 Determination of the major functional domains of mGluRs using the chimeric approach.
(a) Swapping the 2 third N-terminal portion of the extracellular domain of the group-I mGluR1 (red) by those of the group-II mGluR2 (green) generated a chimeric receptor that couples to PLC, like the group-I mGluR1, but is activated by the group-II agonist DCG-IV. This indicates that the N-terminal domain of mGluRs is responsible for agonist recognition. Adapted from Takahashi K, Tsuchida K, Taneba Y *et al.* (1993) Role of the large extracellular domain of metabotropic glutamate receptors in agonist selectivity determination. *J. Biol. Chem.* **266**, 19341–19345, with permission. **(b)** Swapping most intracellular parts, except the second intracellular loop, of the adenylyl-cyclase-coupled mGluR3 (green) by those of the PLC-coupled mGluR1 (red) was not sufficient to allow the chimeric receptor to activate PLC and generate intracellular Ca^{2+} signals (central). However, swapping of the second intracellular loop and other intracellular parts resulted in chimeric receptors that activate PLClike mGluR1 (right). This indicates that the second intracellular loop plays a critical role in specifying PLC activation by mGluR1, whereas the other intracellular parts are required for an efficient coupling to PLC. Adapted from Gomeza J, Joly C, Kuhn R *et al.* (1996) The second intracellular loop of metabotropic glutamate receptor 1 cooperates with other intracellular domain to control coupling to G proteins. *J. Biol. Chem.* **271**, 2199–2205, with permission.

(a)

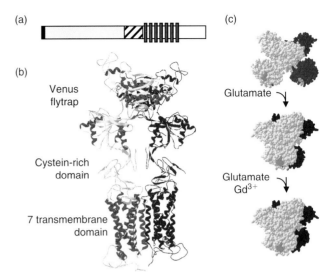

(b)

Venus
flytrap

Cystein-rich
domain

7 transmembrane
domain

(c)

Glutamate

Glutamate
Gd^{3+}

FIGURE 12.4 Structure and activation mechanism of mGluRs.
(a) Schematic view of the structural domains of mGluRs as determined by sequence homology searches: from the N-terminal end (left) to the C-terminus (right): signal peptide (in black), the Venus flytrap domain (grey), the cystein-rich domain (hatched), the seven transmembrane domain (with 7 vertical bars) and the intracellular C-terminal segment (white). **(b)** Structural view of a mGluR based on the solved structure of the Venus flytrap domain of mGluR1, on the 3D model of the cystein-rich region, and the solved structure of the prototypical G-protein coupled receptor rhodopsin. Note the receptor is a constitutive dimer: the subunit in the front is coloured according to its secondary structure (helices in red, strands in yellow and loops in grey), while the other subunit in the back is in blue. The dimer of Venus flytrap domain is that observed in the absence of bound agonist (inactive state), while the dimer of 7 Tms is based on the proposed dimerization mode of rhodopsin in its inactive state. **(c)** The three identified conformations of the dimer of Venus flytrap of mGluR1: top, the inactive unliganded form; upon binding of at least one molecule of glutamate, closure of the yellow subunit results in a major change in the relative orientation of the domains, leading to a partial activation of PLC; bottom, upon binding of glutamate in the blue subunit and cation (such as Gd^{3+}) binding at the interface between the two subunits, a third state is obtained in which both domains are closed, and this corresponds to the fully active state of the receptor.

specify which G-protein is activated. In most G-protein coupled receptors, the second and third intracellular loops play critical roles in coupling efficacy and specificity. Conversely, mGluRs chimera studies have identified the second intracellular loop as the major determinant of coupling specificity, while all of the other intracellular loops are involved in coupling efficacy. For example, a chimera of mGluR3 with the second intracellular loop and cytoplasmic tail of mGluR1 couples to phospholipase C (PLC), thus exhibiting the G-protein specificity of mGluR1 (**Figure 12.3b**).

At first glance, the regions responsible for G-protein coupling specificity lack of sequence homology between mGluRs and the other G-protein coupled receptors, suggesting a specific structure for the mGluRs. However, the second intracellular loop of mGluRs is predicted to form amphipathic α-helices similar to the third intracellular loop of the other

G-protein coupled receptors. Furthermore, mGluRs as all other G-protein coupled receptors recognize the same amino acid residues on G-protein α-subunits. Therefore the general structural strategy utilized to activate G-proteins is probably shared among all G-protein coupled receptors, although the specific regions involved in this process may differ.

Accordingly, the general structure of mGluRs consists of a Venus flytrap domain connected, through a cystein-rich region, to a prototypical seven transmembrane spanning domain (**Figure 12.4**). But how can the closure of the venus flytrap domain by glutamate activate the transmembrane domain leading to G-protein stimulation? Much information to answer that question came from the solved structure of the extracellular domain of mGluR1. Indeed, the structure revealed a dimer of Venus flytrap domains, their N-terminal lobes contacting each other through a hydrophobic interface. This was consistent with previous biochemical studies indicating that mGluRs are constitutive dimers linked by a disulfide bridge both in transfected cells and in neurons. The structure of the active dimer with bound glutamate revealed a major conformational change resulting from agonist binding. Not only were the Venus flytraps closed, but also an important rotation of one flytrap compared to the other was found (**Figure 12.4**). Such a movement was then proposed to force the seven transmembrane domains to also move one compared to the other. This was confirmed using energy transfer technology that allows to estimate the relative distance between proteins fused to the cyan fluorescent protein (CFP) and yellow fluorescent protein (YFP). More recent studies revealed that the change in conformation is different whether one or two glutamate molecules are bound to the dimer, and whether cations also interact with the complex. Indeed, a single agonist per mGluR1 dimer allows a partial activation of the PLC pathway, but full activation of the adenylyl cyclase. Two agonists per dimer are required for the full activation of PLC. This represents one of the first examples of clear evidence that different conformations of a G-protein coupled receptor exist, which leads to the activation of different signalling pathways (**Figure 12.4**).

12.3 HOW TO IDENTIFY SELECTIVE COMPOUNDS ACTING AT METABOTROPIC GLUTAMATE RECEPTOR? – TOWARDS THE DEVELOPMENT OF NEW THERAPEUTIC DRUGS

After the cloning of mGluRs, a search for selective ligands started with the aim of identifying their physiological roles, and to validate these receptors as new targets for the treatment of psychiatric and neurological diseases. At first, most laboratories synthesized glutamate derivatives and examined their effects on the eight mGluRs. These studies led to the discovery of both agonists and antagonists, some of which displayed a nanomolar affinity such as the group-II agonist, LY354740. However, most of these molecules, and especially the antagonists, display a low affinity and a poor selectivity. Indeed, due to the high conservation of the glutamate binding site, most of the identified compounds were only selective for mGluRs from one group, but none could be considered as really selective for a single receptor. These studies however bring further

information on the mechanism of activation of these receptors since most antagonists were found to possess extrafunctional groups that prevent the closure of the Venus flytrap domain.

Due to the difficulty in identifying potent and selective compounds derived from glutamate, pharmaceutical companies started high throughput screening campaigns based on functional assays using cell lines expressing these receptors. To do so, most companies used specific fluorescent probes that respond to intracellular Ca^{2+}, as an indication of PLC activation. Such signals could not only be recorded with the PLC-coupled group-I mGluRs, but also with the group-II and group-III receptors co-expressed with modified G-protein α- subunits allowing their coupling to PLC. Such assays were conducted in microplates containing 384 or even 1536 wells, such that millions of molecules could be tested for activity in a few weeks. This approach led to the discovery of very selective compounds with chemical structures totally different from that of glutamate. Not surprisingly, these molecules were found to bind at a site distinct from that of glutamate, not located in the Venus flytrap domain, but in the seven transmembrane domain (**Figure 12.5**). As such these molecules were called allosteric modulators.

Two types of allosteric modulators were identified. The first ones, called negative modulators, act as non-competitive antagonists, decreasing the maximal response generated by glutamate without affecting its potency (**Figure 12.5**). The first compounds identified were MPEP, a highly selective mGluR5 inhibitor, and BAY36-7620, a selective mGluR1 inhibitor (**Figure 12.2**). The second type of modulators facilitate the action of

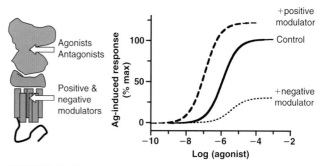

FIGURE 12.5 Modulation of mGluR function by positive and negative modulators.
On the left, a scheme of the general structure of a mGluR subunit shows that whereas agonists and antagonists bind to the Venus flytrap domain, allosteric modulators (both positive and negative) bind to the 7TM domain. On the right, the effects of the positive and negative modulators on the dose-effect of an agonist are illustrated. The positive modulators increase both the potency (the dose-response curve is shifted to the left) and efficacy (increase in the maximal effect of agonist) of agonists. The negative modulators act as non-competitive antagonists, decreasing the maximal effect of agonists.

glutamate by increasing both its potency and its efficacy, and have no agonist activity on their own (**Figure 12.5**). Like the negative modulators, these molecules are highly selective for a given mGluR (**Figure 12.2**). Such compounds, called positive modulators, offer a number of advantages over agonists for therapeutic applications. Whereas an agonist activates the receptor in any cells that express it, and in a sustained manner, the positive modulators only facilitate the action of glutamate when and where released in the brain. As such, these positive modulators maintain the normal biological rhythm of the targeted receptor, and do not permanently activate it. Accordingly, and in contrast to what is expected when using agonists, the signalling system (the receptor itself or the signalling network involved) targeted by these positive modulators will less likely desensitize. The advantage of using positive modulators for therapeutic intervention is well illustrated by the potent anxiolytic activity of benzodiazepines, which are positive modulators of the $GABA_A$ receptors, whereas the pure agonists are devoid of such effects.

12.4 WHAT BIOCHEMICAL MEANS DO METABOTROPIC GLUTAMATE RECEPTORS UTILIZE TO ELICIT PHYSIOLOGICAL CHANGES IN THE NERVOUS SYSTEM? – SIGNAL TRANSDUCTION STUDIES OF METABOTROPIC GLUTAMATE RECEPTORS

Once activated by glutamate, mGluRs initiate a host of intracellular biochemical cascades, which eventually change the physiological behaviour of the cell. As any G-protein coupled receptor, mGluRs activate heterotrimeric G-proteins. These consist of a GTP hydrolyzing α-subunit and a membrane-bound complex of β and γ subunits. Once the G-protein coupled receptors are activated, the G-protein α-subunit exchanges GDP with GTP and dissociates from the βγ complex. Both the activated GTP-bound α-subunit and the free βγ complex now initiate various downstream processes. The α-subunit activates membrane-bound enzymes, whereas βγ-subunits can directly modulate the activity of ion channels.

As mentioned earlier, the discovery of mGluRs was initiated by the finding that glutamate can stimulate PI hydrolysis. Phospholipids are major constituents of cell membranes that play important roles in intracellular signalling. Phospholipases (PLCs) hydrolyze phospholipids, including the PI phosphatidyl-inositol diphosphate (PIP2), into inositol triphosphate (IP3) and

membrane-bound diacylglycerol (DAG). Thus, a simple way to follow phospholipase C activity is to load cells with radioactively labelled inositol, which is then incorporated into membrane phospholipids, and determine the release of inositol phosphate from the membrane or organic fraction to the cytoplasmic or aqueous fraction. In neurons loaded with ^3H-inositol, glutamate increases the aqueous fraction of radiolabelled inositol phosphate, indicating PLC hydrolysis of PIP2.

Based on pharmacological studies of native and recombinant mGluRs, it is now known that group-I mGluRs couple to Gq protein, whose α-subunit activates PLC. This results in the production of IP3, which releases Ca^{2+} from internal stores. The released Ca^{2+}, along with the concomitantly synthesized DAG, activates protein kinase C (PKC). DAG may also be further processed to yield various lipid messengers, including arachidonic acid. Thus, group-I mGluRs can exert their effects through intracellular Ca^{2+}, lipid messengers and PKC activation. The rapid rise in intracellular Ca^{2+} concentration is now widely used to record the activity of PLC-coupled receptors using fluorescent Ca^{2+} binding molecules such as Fluo4.

Activation of PLC is not however an exclusive group-I mGluR pathway in native systems. For instance, in cultured glial cells, these receptors activate mitogen-activated protein kinase (MAPK), possibly through phosphoinositide 3-kinase (PI3K). In cerebellar neurons, mGluR1 can trigger a direct functional coupling between intracellular ryanodine-sensitive receptors and plasma membrane L-type Ca^{2+} channels

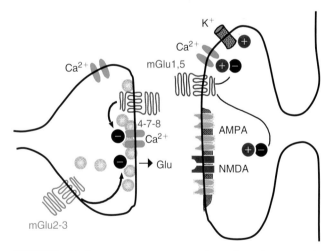

FIGURE 12.6 Schematic representation of a glutamatergic synapse with the modulatory roles of mGluRs.
Note group-I mGluRs (1 and 5) are mostly located in the postsynaptic spines, on the side of the postsynaptic density, from where they regulate postsynaptic ionotropic glutamate receptors, as well as Ca^{2+} and K^+ channels. Group-II (mGluR2) and group-III (mGluR4, 7 and 8) are mostly presynaptic, regulating neurotransmitter release by, at least, inhibiting Ca^{2+} channels.

through a G-protein-dependent, but unidentified pathway (**Figures 12.6–12.7**). In CA3 hippocampal neurons, functional pharmacological studies have shown that mGluR1 can mobilize a Src-family tyrosine kinase pathway, independently of G-protein activation.

The α-subunits of G_s and G_o/G_i proteins can stimulate and inhibit the adenylyl cyclase (AC) activity respectively. Group-II and group-III mGluRs couple to Gi/Go proteins. Similarly to PLC, AC activity was measured by loading living cells with radiolabelled adenine, which is then incorporated into intracellular ATP pools and finally cAMP. Nowadays several non-radioactive and high-throughput assays are being used to measure

cAMP produced in the cells. Sensors have also been developed that allow the visualization of cAMP concentration in living cells in real time. These sensors are made with cAMP binding proteins fused to fluorescent proteins. Activity of both group-II and group-III mGluRs inhibits forskolin-stimulated cAMP accumulation in heterologous expression systems, neuronal cultures and brain slices. A more relevant issue is the effect of group-II and group-III mGluRs on neurotransmitter-induced increases in cAMP accumulation in native systems. Group-II mGluR agonists potentiate cAMP accumulation induced by stimulation of β-adrenergic receptors and other G_s-coupled receptors. In this case, released

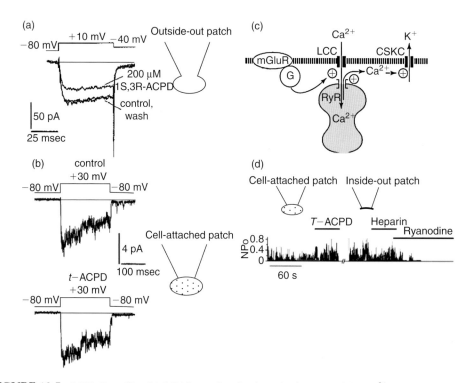

FIGURE 12.7   MGluR-mediated inhibition and activation of voltage-sensitive Ca^{2+} currents in neurons.
(a) The upper panel shows a recording of macroscopic Ca^{2+} currents from a rat CA3 pyramidal cell outside-out patch. The membrane was held at –80 mV and stepped to +10 mV, resulting in activation of voltage-sensitive Ca^{2+} channels. Application of the group-I mGluR agonist, t-ACPD, to the surface of the patch induced in a reversible reduction of the Ca^{2+} current. **(b)** However, when macroscopic Ca^{2+} currents were recorded in the cell-attached mode and the agonist was applied outside the patch, t-ACPD had no effect. These findings suggest that group-I mGluRs acted through a membrane delimited mechanism rather than readily diffusible second messenger. Adapted from Swartz KJ and Bean BP (1992) Inhibition of Ca^{2+} channels in rat CA3 pyramidal neurons by a metabotropic glutamate receptor. *J. Neurosci.* **12**, 4358–4371, with permission. **(c)** In cerebellar granule cells, mGluR1 triggers a tight coupling between ryanodine-sensitive receptors (RyR) and membrane L-type Ca^{2+} channels (LCC), in a G-protein-dependent manner. The release of Ca^{2+} from intracellular ryanodine-sensitive stores activates a Ca^{2+}-dependent K^+ conductance (CSKC) located in close proximity to LCC and RYRs. **(d)** The opening probability of L-type Ca^{2+} channels (NPo) was monitored in a cell-attached patch and after excision of the patch into the inside-out configuration. The agonist, t-ACPD (100 μM), induced opening of LCC recorded in the cell-attached configuration. LCC remained active in the excised patch and the activity was blocked by ryanodine, but not the IP3 receptor antagonist, heparin, when applied to the intracellular side of the recorded patch. This experiment confirmed the model described in C. Modified from Fagni L, Chavis P, Ango F (2000) Complex interactions mGluRs, intracellular Ca^{2+} stores and ion channels in neurons. *Trends in Neurosci.* **23**, 80–88.

G-protein $\beta\gamma$-subunits may actually potentiate type-II AC activation by G_s-protein α-subunits. Conversely, group-III mGluRs inhibit neurotransmitter-induced increases in cAMP, through G_o/G_i-protein α-subunit. Therefore, group-II and group-III mGluRs may exert their effects by altering intracellular cAMP levels. However, repeated attempts to find physiological roles for group-II and group-III mGluRs involving changes in intracellular cAMP in native systems have been unsuccessful, with one exception. In hippocampal glial cells, a form of glial-neuron signalling occurs, in which group-II mGluRs potentiate cAMP formation induced by β-adrenergic receptors. Cyclic AMP metabolites are then released by glia and activate adenosine receptors on nearby neurons, thus modulating synaptic transmission.

Alternative avenues are becoming increasingly evident. For instance, the group-III presynaptic receptor, mGluR7, mobilizes two separate cAMP-independent pathways and inhibits P/Q-type Ca^{2+} channels in hippocampal and cerebellar neurons. First, mGluR7 can

activate PLC, probably through Gi/Go protein $\beta\gamma$-subunits, which results in PKC activity. The activated PKC directly or indirectly inhibits the Ca^{2+} channels (**Figure 12.8b**). Second, the Ca^{2+}-calmodulin complex and G-protein $\beta\gamma$-subunits bind to the C-terminus of mGluR7 in a mutually exclusive manner. Thus, when Ca^{2+}-calmodulin interacts with the receptor, the released G-protein $\beta\gamma$-subunits can directly interact with and inhibit the Ca^{2+} channels (**Figure 12.8c**). Group-II and group-III mGluRs can activate GIRK (G-protein coupled inwardly rectifying K^+) channels expressed in *Xenopus* oocytes through direct interaction of endogenously released G-protein $\beta\gamma$-subunits with the channels. The relative dearth of evidence for group-II and group-III mGluRs modulation of cAMP levels leading to physiological effects may indicate that direct inhibition of P/Q-type Ca^{2+} channels or activation of GIRK channels by G-protein $\beta\gamma$-subunits plays an equal or even greater role than modulation of AC activity. Indeed group-II and group-III mGluRs seem to couple to these

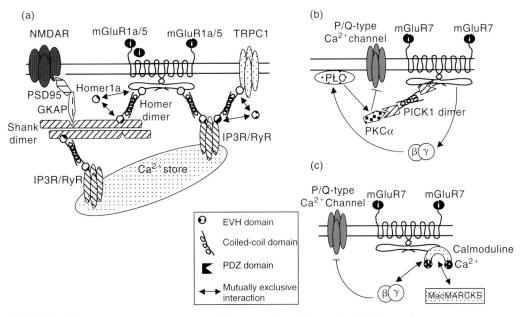

FIGURE 12.8 Current models for post- and presynaptic multiprotein mGluR complexes.
(a) Homer proteins bind to Ca^{2+}-permeable store-operated TRPC1 channel, postsynaptic mGluR1a and mGluR5, as well as two sites of IP3 and ryanodine receptors (IP3R/RyR). Homer can also interact with the shank protein, which associates with GKAP-PSD95-NMDA receptor complex through a PDZ interaction. The confinement of these receptors and channels by homer-based protein-protein interactions increases the functionality of the system and optimizes mGluR1a mGluR5 intracellular Ca^{2+} signalling. **(b)** The adaptor protein, PICK1, interacts with the C-terminus of mGluR7 and PKCa via its PDZ domain and dimerizes through its coiled-coil domain. Thus it physically links mGluR7 to PKCa. Stimulation of mGluR7 releases G-protein $\beta\gamma$-subunits, which results in phospholipase C (PLC) activation in neurons. The PLC pathway triggers PKC activity, which then directly or indirectly inhibits P/Q-type Ca^{2+} channels. Proper functioning of this cascade requires the integrity of the mGluR7-PICK1-PKCa complex, presumably because of the necessity for mGluR7 to be in close proximity to its effector, PKCa (see Perrog *et al.* (2002). *EmBOJ* **21**, 2990–2999) **(c)** Ca^{2+}-calmodulin complex and G-protein $\beta\gamma$-subunits undergo competitive binding on the C-terminus of presynaptic mGluR7. The Ca^{2+} influx from voltage-gated channels activates calmodulin, which then binds to mGluR7 and releases pre-bound G-protein $\beta\gamma$-subunits from mGluR7 C-terminus. Free G-protein $\beta\gamma$-subunits are thus available for direct inhibition of P/Q-type Ca^{2+} channels (see Bertzso *et al.* (2006) *J. Neurochem* **99**, 288–298). These examples illustrate the importance of mGluR multiprotein complexes in the proper function of these receptors.

channels more efficiently than to inhibition of cAMP formation in both expression systems and native neurons.

12.5 HOW IS THE ACTIVITY OF METABOTROPIC GLUTAMATE RECEPTORS MODULATED? – STUDIES OF mGLuR DESENSITIZATION

Soon after the discovery that group-l mGluRs underlie glutamate-stimulated PI hydrolysis, it was discovered that pre-incubation of neuronal cultures or brain slices with group-I agonists decreases the PI hydrolysis response to subsequent exposures of agonist. This phenomenon commonly occurs with many G-protein coupled receptors and is referred to as desensitization. This phenomena has been well characterized for many G-protein coupled receptors and involved interaction of the activated receptor with various intracellular proteins. First, the activated receptor is recognized by G-protein coupled receptor specific kinases (GRK) that phosphorylate the receptor at various places, especially within the C-terminal tail. The phosphorylated receptor gains affinity for a protein called β-arrestin (β-arrestin 1 and β-arrestin 2) that prevents G-protein activation. This allows the recruitment of additional proteins to the receptor and activation of new signalling pathways. Most importantly, the recruitment of β-arrestin permits incorporation of the receptor into specific membrane microdomains from where it will be internalized into endosomes. The internalized receptor will then either be recycled and re-targeted to the cell surface, or sent to lysosomes to be degraded.

Such a desensitization cascade has indeed been reported for group-I mGluRs, although differences were observed. Indeed, activated mGluR1 or mGluR5 were shown to recruit GRK and β-arrestin both in transfected cells and in neurons, resulting in the internalization of the receptor. However, GRK2 (one subtype of the GRKs) can inhibit receptor signalling without phosphorylating the receptor. This effect is independent of β-arrestin recruitment and likely results from a direct competition of GRK2 and the G-protein on the receptor. In contrast, GRK4 kinase activity is needed for this GRK to desensitize mGluR1 in Purkinje neurons. Other proteins such as optineurin, a protein that interacts with the protein huntingtin involved in Huntington disease, have also been shown to uncouple group-I mGluRs from its G-protein.

In addition to these direct desensitization and internalization mechanisms resulting from the activation of mGluRs, heterologous desensitization can also occur independently of the receptor activation. The most common mechanism for reduced coupling of the receptor to G-protein is phosphorylation of the receptor itself, as well demonstrated for many other G-protein coupled receptors. For example, β-adrenergic receptors desensitize when phosphorylated by cAMP-dependent protein kinase (PKA) activated either by the β-adrenergic receptor itself or by other receptors. Similarly, group-I mGluRs desensitization can be mediated by the direct phosphorylation of the receptor by protein kinase C (PKC) whether activated by the receptor itself or by other receptors. As such, this represents not only a mechanism for negative feedback, but also a mechanism for crosstalk between different neurotransmitter pathways that also utilize PKC-dependent signalling.

Much less is known about desensitization and internalization properties of group-II and group-III mGluRs. A few studies indicated that these receptors may not desensitize in neurons. However, mGluR4 and mGluR7 desensitization and internalization was observed after PKC-activation by mGuR7 itself or other receptors. GRK2 was also shown to uncouple mGluR4 from the MAPK pathway without affecting its ability to inhibit adenylyl cyclase.

12.6 METABOTROPIC GLUTAMATE RECEPTORS MODULATE NEURONAL EXCITABILITY

MGluRs modulate two major neuronal functions: excitability and synaptic transmission. Altered excitability is mainly exhibited by changes in the threshold for action potential firing or firing pattern. A powerful mechanism for altering neuronal excitability is to modulate K^+ channels. Potentiation of K^+ channel function leads to reduced excitability, while inhibition of these channels results in enhanced excitability. Spike accommodation is defined as a reduction in spike frequency during a suprathreshold depolarizing stimulus. This phenomenon strongly depends on K^+ channel activity and it is a mechanism by which neurons control their excitability. Fast and slower activating K^+ currents are involved in spike frequency maintenance and accommodation. One of these is $I_{K,AHP}$ (K^+ afterhyperpolarization current), which is activated by rises in intracellular Ca^{2+} occurring during repetitive firing. Inhibition of $I_{K,AHP}$ leads to reduced K^+ efflux and firing fails to accommodate. An early experiment showed that application of glutamate and related group-I mGluR agonists onto hippocampal neurons inhibits $I_{K,AHP}$ and spike accommodation (**Figure 12.9**). As the agonist did not affect the $I_{K,AHP}$ concomitant Ca^{2+} increase, it was

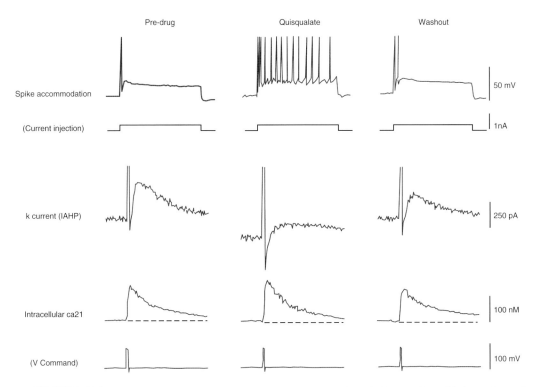

FIGURE 12.9 **Modulation of action potential accommodation by inhibition of a Ca^{2+}-dependent K$^+$ current, rather than Ca^{2+} influx.**
The top row shows the action potential firing pattern elicited by prolonged depolarizing current injection in hippocampal neurons. Only one to two action potentials normally fire even in the presence of a suprathreshold stimulus. Such accommodation is due to depolarization-induced Ca^{2+} influx and activation of $I_{K,AHP}$. The middle row shows the late outward K$^+$ current, $I_{K,AHP}$, under voltage clamp conditions, while the bottom row shows the Ca^{2+} influx as measured by fluorescence-based techniques. The spike accommodation is reversibly blocked by the group-I mGluR agonist, quisqualate, as is the $I_{K,AHP}$ current. However, the magnitude of the intracellular Ca^{2+} response is not altered by quisqualate, indicating that group-I mGluRs directly modulate the $I_{K,AHP}$ channel, rather than the Ca^{2+} influx that activates the channel. Adapted from Charpak S, Gahwiler BH, Do KQ, Knopfel T (1990) Potassium conductances in hippocampal neurons blocked by excitatory amino acid transmitters. *Nature* **347**, 765–767, with permission.

concluded that the inhibition of $I_{K,AHP}$ must be mediated through a direct inhibition of the K$^+$ channels by mGluR5. Several groups have shown that group-I mGluR agonists depolarize neurons by inhibiting voltage-independent K$^+$ conductances termed $I_{K,leak}$ and $I_{K,M}$. Group-I mGluRs can also excite neurons by activating non-selective cation conductances that have not always been identified, but which include Na$^+$/Ca^{2+} exchangers and receptor-operated TRPC channels. Thus it is tempting to conclude that group-I mGluRs rather increase neuronal excitability, whereas group-II and group-III mGluR agonists induce a K$^+$ conductance-based hyperpolarization. However, the physiological situation may be more complex than that, as the increase in intracellular Ca^{2+} that results from mGluR1 stimulation can also activate Ca^{2+}-sensitive K$^+$ channels in cerebellar neurons (**Figure 12.7c**) and this was shown to result in reduced cell excitability.

Modulation of Ca^{2+} channels by mGluRs also bridges changes in neuronal excitability. Ca^{2+} channels help generating action potential upstroke in certain neurons and also mediate the Ca^{2+} influx required for synaptic transmission (**Figure 12.6**). Furthermore, Ca^{2+} influx at various locations within neurons results in the activation of multiple signal-transduction pathways. In general, group-II and group-III mGluRs inhibit Ca^{2+} currents, and the specific subtype of Ca^{2+} current varies depending on the preparation. In hippocampal CA3 pyramidal neurons, group-I mGluR agonists also inhibit N-type Ca^{2+} currents. This effect would involve a membrane delimited mechanism, such as direct binding of G-protein $\beta\gamma$-subunits to the pore-forming subunit of Ca^{2+} channels. In cerebellar neurons, group-I mGluR agonist can trigger a direct functional coupling between ryanodine receptors and L-type Ca^{2+} channels upon voltage activation of the channels (**Figure 12.7a,b**).

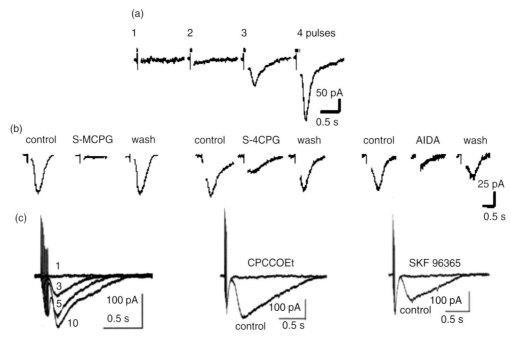

FIGURE 12.10 **MGluR1 mediates slow EPSCs in hippocampal CA3 pyramidal cells and cerebellar Purkinje cells.**

(a) Activity-dependent induction of the slow EPSC in hippocampal CA3 pyramidal cells is induced by repetitive stimulation of afferent mossy fibres. **(b)** The evoked EPSC amplitude is reduced by mGluR1 antagonists S-MCPG (500 μM), S-4CPG (100 μM) and AIDA (200 μM). Adapted from Heuss C, Scanziani M, Gahwiller BH, Gerber U (1999) G-protein-independent signalling mediated by metabotropic glutamate receptors. *Nat. Neurosci.* **2**, 1070–1077. **(c)** (Left) A similar slow EPSC can be elicited in cerebellar Purkinje cells by repetitive (100 Hz) parallel fibres stimulation, in the presence of the AMPA and GABA receptor antagonists, CNQX and gabazine respectively. (Centre and right) The slow EPSC evoked by 5 successive pulses was blocked by the mGluR1 antagonist, CPCCOEt (100 μM, centre) and TRPC1 channel antagonist, SKF96365 (30 μM; right). Note that at least three successive stimulation pulses (100 Hz) are required to induce the mGluR1-dependent slow EPSC, in both hippocampal neuron and cerebellar Purkinje cell. The amplitude of the EPSC then increased with the number of pulses. Adapted from Kim SJ, Kim YS, Yuan JP, Petralla RS, Worley PF, Linden DJ (2003) Activation of the TRPC1 cation channel by metabotropic glutamate receptor mGluR1. *Nature* **426**, 285–291.

This leads to local increase in intracellular Ca^{2+} and opening of a Ca^{2+}-sensitive K^+ channel located in close proximity to the activated Ca^{2+} channels (**Figure 12.7c,d**). These examples point out the complexity of the mechanisms by which mGluRs can control channel activity and neuronal excitability. It clearly appears that the subcellular localization of mGluRs and that of the regulated ion channels specifies the effects of these receptors on cell excitability.

12.7 METABOTROPIC GLUTAMATE RECEPTORS MEDIATE AND MODULATE SYNAPTIC TRANSMISSION

Besides altering neuronal excitability, mGluRs are also able to mediate slow excitatory postsynaptic responses. Repetitive stimulation of glutamatergic afferents in slices induce a slow EPSC in cerebellar Purkinje neurons and hippocampal CA3 pyramidal cells, that results from activation of group-I mGluRs (likely mGluR1) and opening of TRPC1 channels (**Figures 12.8** and **12.10**). In the hippocampus, but not cerebellum, this effect is mediated via mobilization of a G-protein-independent tyrosine kinase pathway.

MgluRs can also modulate synaptic transmission. In general, mGluR effects on synaptic transmission can be divided simply into presynaptic and postsynaptic (**Figure 12.6**). Presynaptically, group-II and group-III mGluRs typically act as glutamate autoreceptors, decreasing glutamate release from presynaptic nerve terminals, typically by inhibiting presynaptic N-type Ca^{2+} channels. The P/Q-type Ca^{2+} channels are also typically presynaptic and mGluR7 inhibits excitatory synaptic events through specific blockade of these channels. Conversely, in the hippocampus it appears that mGluR4 reduces frequency with no effect on amplitude of mEPSCs, indicating a reduction in presynaptic Ca^{2+}-independent release probability. Presynaptic mGluRs can also act as heteroreceptors, on GABAergic nerve terminals, where they reduce GABA release.

MGluRs can execute subtype- and location-specific postsynaptic roles and control neurotransmission in a paracrine manner. Activation of mGluR1 in cerebellar Purkinje cells induces the release of endocanabinoids, which then retrogradely acts on presynaptic type-1 cannabinoid (CB1) receptors located on both excitatory (climbing fibre) and inhibitory (GABA interneuron) axon terminals, and inhibit neurotransmitter release. In cerebellar Purkinje cells, postsynaptic mGluR1 plays a pivotal role in synaptic plasticity. Purkinje cells receive excitatory inputs from climbing fibres and parallel fibres. When these inputs are conjunctively stimulated, it results in a long-term depression (LTD) of the parallel fibre-Purkinje cell synaptic response. Based on a large body of evidence, it has been established that LTD of the parallel fibre-Purkinje cell synapse results from iGluR and group-I mGluR mobilization of voltage-gated Ca^{2+}-channel activation in the Purkinje cell. In this effect, mGluR contribution is to raise intracellular Ca^{2+} stores, which results in PKC-activation, protein kinase G-activation through nitric oxide (NO) production, and guanylyl-cyclase-activation. In the hippocampus, group-I mGluRs can also induce LTD, but pharmacological and gene deletion studies raised controversial results regarding involvement of intracellular Ca^{2+} stores. In the nucleus accumbens, the mGluR1-mediated production of endocanabinoid and retrograde inhibition of neurotransmitter release would contribute to LTD. Group-II and group-III mGluRs can also induce LTD in different regions of the brain: in the nucleus accumbens, striatum, hippocampal CA3 region.

Long-term potentiation (LTP) is characterized by a long-lasting increase in excitatory postsynaptic response, which can be induced by repetitive stimulation of afferents. In the CA1 area of the hippocampus, induction of LTP is an NMDA receptor-dependent mechanism. Consistent with the observation that group-I mGluRs potentiate NMDA currents, group-I mGluR agonists potentiate the NMDA component of EPSCs and promote LTP (see **Figure 12.8**). Furthermore, mice lacking the *mGluR5* gene are specifically deficient in the NMDA component of LTP, while having normal non-NMDA LTP.

12.8 PRE- AND POSTSYNAPTIC FUNCTIONAL ASSEMBLY OF METABOTROPIC GLUTAMATE RECEPTORS

In order to carry out their functions, mGluRs must properly localize in neurons (**Figure 12.6**). Immunological and functional studies show a predominant postsynaptic localization of group-I mGluRs, presynaptic localization of mGluR7, and pre- and postsynaptic localization of the other mGluR subtypes in the brain. Ultrastructurally, glutamatergic postsynaptic membrane contains an electron-dense zone that has been named postsynaptic density (PSD). Postsynaptic mGluRs lie within the PSD, except group-I mGluRs, which form an annulus surrounding the PSD. The presynaptic mGluRs are localized in the symmetrically positioned active zone.

Recent studies have begun to uncover the biochemical nature of the PSD. It consists of a large multimeric protein complex that comprises not only postsynaptic glutamate receptors, but also scaffolding proteins that physically link these receptors to downstream signalling molecules and cytoskeleton. Analogous to the PSD, the presynaptic active zone consists of a protein matrix that assembles receptors and channels to their signalling pathways, including those involved in neurotransmitter release. Homer, calmodulin and PDZ domain proteins have been characterized as mGluRs partners (**Figure 12.8**). For instance, the EVH domain of homer proteins can interact with a PPxxF consensus sequence that is present in the C-terminus of mGluR1a and mGluR5, but also ryanodine receptors, IP3 receptors and TRPC1 channels. The coiled-domain of homer allows multimerization of the protein. In general, PDZ interactions depend on the last three C-terminal amino acids (PDZ motif) of one partner that fits within the PDZ hydrophobic pocket (PDZ domain) of the other partner. For instance, the PDZ domain-containing protein, tamalin, was found to bind the C-terminus of mGluR1a and mGluR5, whereas the PDZ proteins PICK1 and GRIP bind to mGluR7.

Mutation/deletion and overexpression experiments have shown that these proteins play important roles in the synaptic localization, membrane targeting and function of mGluRs. As such, homer multimers can form a large multiprotein complex with mGluR1a/5, IP3/ryanodine receptors and TRPC1 channels (**Figure 12.8**). The homer1a isoform is induced following epileptic activity, acute cocaine intake or inflammatory pain, and disrupts the complex. This is functionally important as disruption of the complex can alter the intracellular Ca^{2+} signalling and induce agonist-independent/constitutive activity of mGluR1a and mGluR5. Thus, transgenic mice that constitutively express homer1a or mice that have been infused with viral vectors carrying homer1a transcript display reduced LTP and hyposensitivity to epileptogenic agents and cocaine. Conversely, preventing activity-dependent upregulation of homer1a with small interference RNA exacerbates inflammatory pain in mice. Thus homer1a blunts the physiological actions of mGluR1a/5.

Recently, homer was also found to bind the shank family of PSD proteins, whose members also contain the PPxxF motif. Shank binds to GKAP (guanylate kinase-associated protein) through its PDZ domain, thereby crosslinking the homer-group-I mGluR with the GKAP/PSD-95/NMDA receptor complex (**Figure 12.8**). One can now envision the beginnings of a large PSD complex comprising group-I mGluRs and NMDA receptors, linked by scaffolding and effector proteins. Thus, this shank-based physical coupling may provide the structural correlate for a functional cross-talk between these receptors. Group-I mGluRs can also directly interact with adenosine A1 receptor, thus adding a second-order complexity in the interplay between glutamatergic and other neurotransmitter systems. Finally, mGluR1 can directly interact with the Cav2.1 subunit of P/Q-type Ca^{2+} channels, suggesting a more direct control of these channels by glutamate.

The notion that proper functioning of receptors and channels requires their association with multiprotein complexes also applies to the presynaptic mGluR7 subtype. The C-terminal PDZ motif of this receptor can interact with the N-terminal PDZ domain of PICK1. PICK1 can dimerize through its C-terminal coiled-coil domain and interact with PKCα (**Figure 12.8b**). The integrity of this complex is required for inhibition of Ca^{2+} channels and synaptic transmission by mGluR7, probably because of the necessity for mGluR7 to be in close vicinity of its effector, PKC, to mediate its effect. MGluR7 can also block Ca^{2+} channels and neurotransmission through a separate pathway that requires interaction of the receptor with Ca^{2+}-calmodulin. Another protein called MacMARCKS can bind mGluR7 and causes the release of Ca^{2+}-calmodulin from the C-terminus of the receptor (**Figure 12.8c**), thus blunting the Ca^{2+}-calmodulin-dependent mGluR7 signalling. Collectively these examples show the importance of multiprotein complexes in the localization and function of mGluRs.

12.9 PHYSIOLOGICAL ROLES OF METABOTROPIC GLUTAMATE RECEPTOR – A STUDY OF KNOCK-OUT MODELS

A way to study the physiological role of a protein is to knock-out the corresponding gene and to characterize the phenotype of the genetically modified animal (**Table 12.1**). Such an approach, combined with pharmacological and functional studies, has been used to study the physiological roles of mGluRs. The mGluR1 knock-out mouse has no gross anatomical or basic electrophysiological abnormalities, except poly-innervation

TABLE 12.1   mGluR knock-out phenotypes

mGluR subtype knock-out	Phenotype
mGluR1	Ataxia, impaired cerebellar LTD, impaired hippocampal mossy fibre LTP
mGluR2	Impaired hippocampal mossy fibre LTD
mGluR4	Increased susceptibility to absence epilepsy, impaired paired-pulse facilitation and post-tetanic potentiation, impaired rotating rod motor learning
mGluR5	Impaired fear-conditioning, impaired behavioural effects of cocaine
mGluR6	Altered electroretinogram
mGluR7	Altered theta rhythm and impaired working memory, deficits in taste aversion and fear responses, increased epileptic seizure susceptibility
mGluR8	Increased anxiety, altered photoreceptor function

of cerebellar Purkinje cells by climbing fibres, impaired cerebellar LTD and impaired hippocampal mossy fibre LTP. These functional disorders are accompanied by deficient spatial learning and eye-blink conditioning, as well as severe motor coordination impairment. This phenotype clearly shows an important function of this receptor in cerebellar development and function, as well as in motor learning tasks.

The mGluR5 knock-out mouse displays altered synaptic functions in lateral amygdala and as expected this is accompanied by impaired fear-conditioning. The mGluR5 deletion suppresses reinforcing properties of cocaine administration, without affecting the dopamine response in the nucleus accumbens. This suggests a potential therapeutic value of mGluR5 in the treatment of cocaine addiction. The mGluR5 knock-out phenotype is different from that of the mGluR1 knock-out, although mGluR5 is functionally related to mGluR1. This observation is indeed not quite surprising, if we consider that these receptors display mirror distribution in the brain. Pharmacological studies have also shown that mGluR5 regulates extrapyramidal motor functions and may be a potential target for the treatment of Parkinson's disease. The mGluR1a/5-homer complex mediates a component of hyperalgesia and might be therapeutically targeted to prevent and treat inflammatory pain. Finally, group-I mGluRs have also been involved in temporal lobe epilepsy.

The mGluR2 knock-out mouse shows no histological changes and no alteration in basal neurotransmission. The NMDA receptor-mediated LTD was however almost abolished at the hippocampal mossy fibre pathway.

Nevertheless, the transgenic animal performs normally in the water maze learning task. Thus, mGluR2 is essential for mossy fibre hippocampal LTD and this phenomenon does not seem to be required for spatial learning.

The mGluR4 subtype is highly expressed presynaptically on thalamo-cortical neurons that are implicated in absence epilepsy. The mGluR4-deficient mouse is completely resistant to generalized absence seizures induced by GABA-A receptor antagonists and displays increased evoked glutamate release in structures that are involved in absence epilepsy (ventrobasal thalamus, nucleus reticularis thalami and cerebral cortex laminae IV–VI). This indicates a role of mGluR4 on neurotransmitter release and its possible implication in absence seizures. The mGluR4 knock-out mouse also shows altered short-term synaptic plasticity and poorly performs in rotating rod motor learning. Therefore a function of mGluR4 may be to control synaptic efficacy and to support learning of complex motor tasks.

The mGluR6 subtype is exclusively expressed on retinal ON and OFF bipolar cells. Light inhibits (hyperpolarizes) the photoreceptors (the rods and cones) and shuts down glutamate release onto the secondary retinal bipolar cells. Now, what is the sign-inverting mechanism that translates the inhibitory photoreceptor response into an excitatory signal suitable for transmission of visual information to the visual cortex? Knock-out experiments have shown that the sign inversion is mediated by mGluR6. Glutamate released from photoreceptors binds to mGluR6 and a G-protein called transducin is activated, which causes phosphodiesterase to breakdown cyclic GMP. This triggers the closing of cyclic GMP-gated cation-selective channels, resulting in hyperpolarization. Light triggers depolarization of the bipolar cell and transmission of visual information by inhibiting glutamate release and shut-down of this pathway.

The mGluR8 knock-out mouse shows increased anxiety-related behaviour, suggesting a role of this receptor in response to a novel stressful environment. The mGluR8 subtype is ubiquitously expressed in the brain, but also in the retina where it is located on rod photoreceptors. Its activation leads to a decrease in glutamate release, probably via inhibition of voltage-gated Ca^{2+} channels. Therefore its function would be to control neurotransmitter release from rod spherules. It is expected that the mGluR8-deficient mouse displays altered visual perception.

MGluR7 deletion in mice increases amplitude and power of the electroencephalographic theta rhythm. This change is accompanied by deficits in working memory, conditioned taste aversion and fear-conditioned response. The mGluR7-deficient mouse also shows increased epileptic seizure susceptibility.

The mGluR7 subtype therefore plays a crucial role in neuronal excitability and specific cognitive functions.

12.10 SUMMARY

- 'Metabotropic' glutamate receptors were first hypothesized based on glutamate-stimulated, G-protein-dependent PI hydrolysis in neuronal preparations.
- mGluR1 was first cloned using expression cloning. Subsequently, mGluR2 to mGluR8 were cloned using homology-based techniques.
- mGluRs can be divided into three groups based on sequence homology and signalling:

  Group I, mGluR1 and 5 couple to Gq protein and stimulate IP hydrolysis.
  Group II, mGluR2 and 3, as well as *Group III*, mGluR4, 6, 7 and 8, couple to Gi/Go proteins, inhibit AC and activate GIRK channels.

- mGluRs are large and complex proteins made of an extracellular ligand-binding domain connected through a cystein-rich region to a seven transmembrane domain activating G-proteins.
- Structure-function studies revealed a dimeric functioning of these proteins, agonist binding resulting in a change in the relative position of the two protomers, switching the transmembrane domains into their active state.
- High-throughput screening approaches identified allosteric modulators of mGluRs, that either inhibit or facilitate receptor activation by acting in the seven transmembrane domains. These compounds are very selective, and have the advantage of maintaining the normal biological activity of the receptor *in vivo*.
- mGluRs can control neuronal excitability by modulating K^+ and Ca^{2+} conductances via a large variety of intracellular messengers. These include mainly G-protein $\beta\gamma$-subunits, Ca^{2+} ions and protein kinases.
- Group-I mGluR function is down-regulated by PKC-mediated heterologous desensitization. This allows for negative feedback as well as for crosstalk with other neurotransmitter systems.
- mGluRs act as presynaptic auto- and heteroreceptors, reducing the release of many neurotransmitters. The major mechanism for presynaptic reduction of transmitter release is inhibition of Ca^{2+} channels and reduced release probability.
- Postsynaptically, mGluRs have a dual action on synaptic transmission. They can induce slow excitatory synaptic responses and/or modulate the

efficacy of the fast synaptic transmission mediated by ionotropic glutamate receptors. For example, mGluRs are involved in long-term depression (LTD) and long-term potentiation (LTP) of excitatory synaptic transmission, in cerebellum and hippocampus.

• The synaptic localization and functions of mGluRs depend on multiprotein complexes that include the receptors themselves and various scaffolding, signalling and effector molecules.

FURTHER READING

Aiba A, Kano M, Chen C et al. (1994) Deficient cerebellar long-term depression and impaired motor learning in mGluR1 mutant mice. Cell 79, 377–388.

Ango F, Prezeau L, Muller T, Tu JC, Xiao B, Worley PF, Pin JP, Bockaert J, Fagni L (2001) Agonist-independent activation of metabotropic glutamate receptors by the intracellular protein Homer. Nature 411, 962–965.

Charpak S, Gahwiler BH, Do KQ, Knopfel T (1990) Potassium conductances in hippocampal neurons blocked by excitatory amino-acid transmitters. Nature 347, 765–767.

Conn PJ and Pin J-P (1997) Pharmacology and functions of metabotropic glutamate receptors. Ann. Rev. Pharmacol. Toxicol. 37, 205–237.

Dhami GK and Ferguson SS (2006) Regulation of metabotropic glutamate receptor signalling, desensitization and endocytosis. Pharmacol. Ther. 111, 260–271.

El Far O and Betz H (2002) G-protein coupled receptors for neurotransmitter amino acids: C-terminal tails, crowded signalosomes. Biochem. J. (Review) 365(Pt 2), 329–336.

Fagni L, Worley P, Ango F (2002) Homer as both a scaffold and transduction molecule. STKE (Review) (137):RE8.

Gereau RW and Conn PJ (1995) Multiple presynaptic metabotropic glutamate receptors modulate excitatory and inhibitory synaptic transmission in hippocampal area CA1. J. Neurosci. 15, 6879–6889.

Gereau RW and Heinemann SF (1998) Role of protein kinase C phosphorylation in rapid desensitization of metabotropic glutamate receptor 5. Neuron 20, 143–151.

Gomeza J, Joly C, Kuhn R et al. (1996) The second intracellular loop of metabotropic glutamate receptor 1 cooperates with other intracellular domains to control coupling to G-proteins. J. Biol. Chem. 271, 2199–2205.

Goudet C, Binet V, Prezeau L, Pin J-P (2004) Allosteric modulators of class-C G-protein coupled receptors open new possibilities for therapeutic application. Drug Discovery Today: Ther. Strat. 1, 125–133.

Goudet C, Gaven F, Kniazeff J, Vol C, Liu J, Cohen-Gonsaud M, Acher F, Prézeau L, Pin J-P (2004) Heptahelical domain of metabotropic glutamate receptor 5 behaves like rhodopsin-like receptors. Proc. Natl. Acad. Sci. (USA) 101, 378–383.

Jia Z, Lu Y, Henderson J et al. (1998) Selective abolition of the NMDA component of long-term potentiation in mice lacking mGluR5. Learn. Mem. 5, 331–343.

Kim E and Sheng M (2004) PDZ domain proteins of synapses. Nat. Rev. Mol. Cell. Biol. (Review) 4, 833–841.

Kniazeff J, Bessis AS, Maurel D, Ansanay H, Prezeau L, Pin J-P (2004) Closed state of both binding domains of homodimeric mGlu receptors is required for full activity. Nat. Str. Mol. Biol., 11, 706–713.

Kunishima N, Shimada Y, Tsuji Y, Sato T, Yamamoto M, Kumasaka T, Nakanishi S, Jingami H, Morikawa K (2000) Structural basis of glutamate recognition by a dimeric metabotropic glutamate receptor. Nature 407, 971–977.

Pagano A, Rüegg D, Litschig S, Stoehr N, Stierlin C, Heinrich M, Floersheim P, Prézeau L, Carroll F, Pin J-P, Cambria A, Vranesic I, Flor PJ, Gasparini F, Kuhn R (2000) The non-competitive antagonists 2-Methyl-6-(phenylethynyl)pyridine and 7-hydroxyiminocyclopropan [b]chromen-1a-carboxylic acid ethyl ester interact with overlapping binding pockets in the transmembrane region of group I metabotropic glutamate receptors. J. Biol. Chem. 275, 33750–33758.

Pin J-P and Acher F (2002) The metabotropic glutamate receptors: structure, activation mechanism and pharmacology. Cur. Drug Targets – CNS & Neur. Dis. 1, 297–317.

Pin J-P, Joly C, Heinemann SF, Bockaert J (1994) Domains involved in the specificity of G-protein activation in phospholipase C-coupled metabotropic glutamate receptors. EMBO J. 13, 342–348.

Pin J-P, Kniazeff J, Liu J, Binet V, Goudet C, Rondard P, Prézeau L (2005) Allosteric functioning of dimeric class C G-protein coupled receptors. FEBS J. 272, 2947–2955.

Sladeczek F, Pin J-P, Recasens M et al. (1985) Glutamate stimulates inositol phosphate formation in striatal neurons. Nature 317, 717–719.

Smitt PS, Kinoshita A, Leeuw BD et al. (2000) Paraneoplastic cerebellar ataxia due to autoantibodies against a glutamate receptor. New Engl. J. Med. 342, 21–27.

Swartz KJ and Bean BP (1992) Inhibition of calcium channels in rat CA3 pyramidal neurons by a metabotropic glutamate receptor. J. Neurosci. 12, 4358–4371.

Takahashi K, Tsuchida K, Taneba Y et al. (1993) Role of the large extracellular domain of metabotropic glutamate receptors in agonist selectivity determination. J. Biol. Chem. 258, 19341–19345.

Tateyama M, Abe H, Nakata H, Saito O, Kubo Y (2004) Ligand-induced rearrangement of the dimeric metabotropic glutamate receptor 1alpha. Nat. Struct. Mol. Biol. 11, 637–642.

Tateyama M and Kubo Y (2006) Dual signalling is differentially activated by different active states of the metabotropic glutamate receptor 1. Proc. Natl Acad. Sci. USA 103, 1124–1128.

Tu JC, Xiao B, Naisbitt S et al. (1999) Coupling of mGluR/homer and PSD-95 complexes by the Shank family of postsynaptic density proteins. Neuron 23, 583–592.

Tu JC, Xiao B, Yuan JP et al. (1998) Homer binds a novel proline-rich motif and links group 1 metabotropic glutamate receptors with IP3 receptors. Neuron 21, 717–726.

CHAPTER
13

Somato-dendritic processing of postsynaptic potentials I: Passive properties of dendrites

Neurons of the mammalian central nervous system receive many afferents which contact different parts of their somato-dendritic arborization. When these afferents are activated, if their combined effect is depolarizing enough, they trigger the firing of sodium action potentials in the postsynaptic neuron. Classically, it is accepted that these action potentials are generated at a central point in the neuron, at the level of the initial segment of the axon (action potential generating zone; see Section 4.4.3 and **Figure 13.1**).

Action potentials are the response of the postsynaptic neuron. This response may be simple, consisting of a single action potential. In this case it can be described by a single characteristic: its latency. However, the postsynaptic response is generally more complex, consisting of several action potentials. It can then be described by several parameters: the latency of the first action potential, the duration of the response, the frequency of the action potentials that compose the response and the overall form – the pattern, or configuration – of the response (see **Figure 14.1**).

The events that lead to a postsynaptic response can be separated into several stages. When the afferent synapses are activated, an excitatory or inhibitory current is generated at the subsynaptic membrane, as a result of activation of receptor channels by the neurotransmitter(s). These postsynaptic currents propagate through the dendrites to the soma and to the initial segment of the postsynaptic neuron. In the course of their propagation, the postsynaptic currents summate. If the sum of the postsynaptic currents is sufficient to depolarize the membrane of the initial segment as far as the threshold potential for activation of the voltage-sensitive

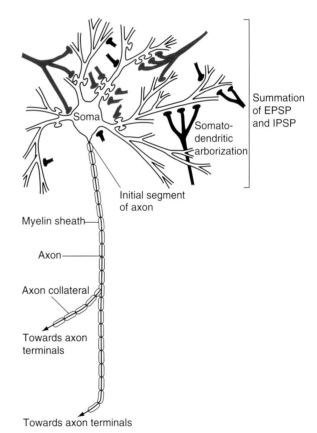

FIGURE 13.1 **Schematic of a neuron and some of its afferents.** Afferent fibres establish synaptic contacts on spines and dendritic branches which are situated at different distances from the soma of the postsynaptic neuron. When these afferents are activated, the depolarizing or hyperpolarizing postsynaptic currents are conducted towards the soma and initial segment of the axon. It is at this level that the response of the postsynaptic neuron is generated. The response is then conducted along the axon and its collateral branches.

sodium channels, a response is triggered in the postsynaptic neuron (**Figure 13.1**).

However, we will see in the following chapters that the presence of a postsynaptic response and the characteristics of this response are not the result only of the integration of different currents of synaptic origin over the somato-dendritic tree. In fact the response of the postsynaptic neuron is the result of two types of currents: currents across receptor channels in the postsynaptic membrane evoked by neurotransmitters and currents across voltage-sensitive channels present in the non-synaptic membrane. Currents of the first type are generated strictly at the postsynaptic membrane and their presence and duration is determined essentially by the interaction between the transmitter and the receptor channel and the intrinsic properties of the receptor channel. Currents of the second type are generated at the non-synaptic membrane (dendritic, somatic or at the initial segment) by voltage changes resulting from currents of synaptic origin, or from currents generated during the first action potential that is fired. The voltage-gated channels responsible for this second type of current are different from those of the sodium action potential and are generally activated in the sub-threshold range of membrane potentials. The duration of these subliminal voltage-gated currents is determined by the gating properties of the corresponding channels.

This chapter looks at the conduction and the summation of synaptic currents (first type of currents) over the dendritic tree. The characteristics of the diverse non-synaptic, subliminal currents (second type of currents) together with their role in the pattern of the postsynaptic discharge will be studied in the following chapters.

13.1 PROPAGATION OF EXCITATORY AND INHIBITORY POSTSYNAPTIC POTENTIALS THROUGH THE DENDRITIC ARBORIZATION

Excitatory and inhibitory postsynaptic potentials result, respectively, from depolarizing or hyperpolarizing currents through channels opened by neurotransmitters (receptor channels) in the postsynaptic membrane. These currents are generated over the somato-dendritic tree, at sites more or less distant from the soma (distal dendritic sites or proximal dendritic sites). Once generated, the postsynaptic currents propagate the length of the dendrites to the soma. For a long time it was thought that postsynaptic currents propagated passively and decrementally along the dendrites: passively because dendrites do not generate action potentials, the propagation of the signal depending only

on the cable properties of the dendrite; and decrementally because the signal attenuates as it propagates, owing to the leakage properties of the membrane. From this it would be expected that depolarizations evoked by distal excitatory synapses would be smaller in amplitude at the soma and would have a longer risetime than depolarizations evoked by proximal synapses.

In fact, it seems that propagation is not always passive and not always decremental. There may be at least two types of propagation of postsynaptic currents through dendrites:

- a passive and decremental propagation, which implies an attenuation of distal postsynaptic currents;
- a passive but only slightly decremental propagation, which occurs where the cable properties of the dendrite are very good and involve no attenuation, or a weak attenuation, of distal postsynaptic currents.

These two alternatives are treated in Sections 13.1.2 and 13.1.3.

13.1.1 The complexity of synaptic organization (Figure 13.1)

Presynaptic complexity

A presynaptic afferent axon gives off many axon terminals (terminal boutons or 'en passant' terminals). In this way, it generally establishes several synaptic contacts with the postsynaptic neuron. In addition, the postsynaptic neuron receives synapses coming from many other presynaptic axons. It is thus possible to distinguish several levels of complexity in postsynaptic potentials:

- the postsynaptic potential resulting from the activity of a single synaptic bouton: miniature potential;
- the postsynaptic potential representing the sum of postsynaptic potentials generated by synaptic boutons coming from the same presynaptic axon: unitary postsynaptic potential;
- the postsynaptic potential representing the sum of all the postsynaptic potentials generated at all the active synaptic boutons: composite postsynaptic potential.

Postsynaptic complexity

Different dendritic postsynaptic regions (spines, branches and main trunks) are not equivalent. The diameter of dendritic trunks is greater than that of branches, particularly distal branches. Thus different dendritic compartments have different resistances (note that $R = \rho l/s$, ρ being the resistivity, l the length and s the cross-section of the dendrite). This means that spines with a neck, or a very small diameter pedicle, have a high resistance. Consequently, synaptic currents generated at different points do not give the same potential change: for the same inward current I, the amplitude of the resulting postsynaptic depolarization (V_{EPSP}) will be greater for the dendritic regions where the resistance $r_m = 1/g_m$ is large ($V_{EPSP} = I_{EPSP}/g_m$).

Complexities of the propagation of postsynaptic action potentials

Postsynaptic potentials (EPSPs and IPSPs) propagate along the dendrites to the action potential initiation zone, which is generally situated in the initial region of the axon (initial segment). Depending on the cable properties of the dendrites, the postsynaptic potentials can change their characteristics (amplitude, risetime) during their propagation.

13.1.2 Passive decremental propagation of postsynaptic potentials

'Decremental' means that the postsynaptic potentials attenuate as they propagate. This implies that the postsynaptic potentials are not regenerated at each point along the dendrites, as is the action potential as it travels along the axon. This passive propagation depends on the cable properties of the dendrite. In order to estimate quantitatively the modifications of postsynaptic potentials in the course of their conduction, a theoretical model of the passive properties of membrane potential changes was first established by Wilfred Rall from data obtained on the squid giant axon. Thus a postsynaptic potential conducted with decrement (i) reduces in amplitude, and (ii) has a risetime (rt) which gets longer as it is propagated along the dendrites (**Figure 13.2a**).

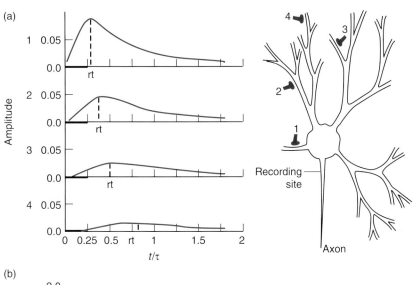

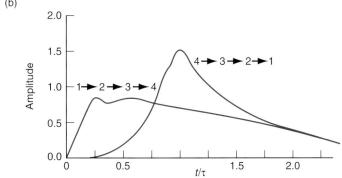

FIGURE 13.2 Theoretical model of decremental conduction of excitatory postsynaptic potentials (EPSP) along dendrites.
(a) Four EPSPs numbered 1 to 4 are generated at the instant t between $t = 0$ and $t = 0.25$ ms (black bar in simulation diagram on left), at different sites within the dendritic tree (schematic drawing on right). At the site of generation, these EPSPs are identical in amplitude and duration. After conduction along the dendrites, their shapes are different (theoretical recordings at the level of the soma, simulation diagram on left). It can be observed that the further away the site of generation of the EPSP (case 4), the smaller is its amplitude and the longer is its risetime (rt) when it arrives at the level of the soma (compare the theoretical recordings 1 to 4). **(b)** Theoretical model of the linear summation of EPSP (see text for explanation). From Rall W (1977) *Handbook of Physiology*, vol. 1, part 1, Bethesda, MA: American Physiological Society, with permission.

The reduction in amplitude of the postsynaptic current as it gets further from the generation site is due to the fact that the current flows not only longitudinally along the dendrite but also transversely across the channels that are open in the dendritic membrane potential. This 'leak' of ions towards the extracellular medium results in a reduction in the postsynaptic current and a consequent reduction in the amplitude of the postsynaptic potential. Thus, the fewer the number of channels open in the dendritic membrane, the higher will be the value of r_m, the better will be the cable properties of the dendrite and the less will be the reduction in amplitude of postsynaptic potentials of distal origin.

The increase in the risetime of the postsynaptic potentials is due to the fact that part of the postsynaptic current serves to charge the capacity of each unit of membrane along the dendrite. The consequence of this is a change in the time course of the postsynaptic current: as it gets further from its point of generation, its risetime becomes longer (it can also be said that the speed of rising becomes slower).

13.1.3 Passive and non-decremental propagation of postsynaptic potentials

This type of propagation means that postsynaptic potentials are conducted passively along the dendrites but, because of the good cable properties of the dendritic arborization, they are almost unattenuated as they propagate. Thus, in the model of the synapse of Ia afferent fibres with spinal motoneurons, it has been shown that the unitary EPSPs evoked by the activity of afferent fibres and recorded in the soma have very similar amplitudes even though their risetimes may be different; i.e. when they are generated at different distances from the soma. This implies that, in this model, there must be local dendritic mechanisms that allow an almost non-attenuating conduction of the distal postsynaptic potentials.

13.2 SUMMATION OF EXCITATORY AND INHIBITORY POSTSYNAPTIC POTENTIALS

13.2.1 Linear and nonlinear summation of excitatory postsynaptic potentials

In general many excitatory synaptic afferents converge on a single neuron. At each excitatory synapse that is activated, there is an inward current of positive charges. When the membrane potential is not held at a fixed value, this inward current of positive charges depolarizes the postsynaptic membrane: this is the postsynaptic potential, or EPSP (see, for example, the current clamp recordings of the synaptic response to glutamate in Chapter 10).

A unitary EPSP (meaning one caused by the activation of a single afferent fibre; Section 13.1.1) cannot trigger action potentials. EPSPs generated in isolation are too small in amplitude to depolarize the membrane of the initial segment to the threshold potential for the opening of voltage-sensitive Na^+ channels. However, if many EPSPs generated at different sites in the dendritic arborization arrive more or less simultaneously at the level of the initial segment, the probability that they will generate action potentials becomes much greater. This is due to the fact that the EPSPs summate.

Linear summation of excitatory postsynaptic potentials

The term 'linear summation' means that the composite EPSP (see Section 13.1.1) resulting from the activity of several excitatory synapses has an amplitude that is equal to the geometric sum of the different EPSPs contributing to it. This is true when the EPSPs are generated at sites that are sufficiently far or isolated from one another to avoid interactions between them (on different dendritic branches, or on different dendritic spines, for instance).

A postsynaptic neuron generally receives many excitatory synapses at different points on its somato-dendritic arborization (see **Figures 13.1** and **13.2a**). These EPSPs summate as they propagate, in a temporo-spatial manner. To grasp this phenomenon, it must be understood that the EPSPs generated at different sites in the dendritic arborization and conducted to the initial segment of the axon can arrive spread out in time. The offset between the EPSPs will depend on the distances between the generation sites and on the respective times at which they were generated. The examples demonstrated here are based on theoretical calculations of the cable properties of dendrites. These data give a qualitative understanding of the phenomena of summation but do not constitute a real experimental demonstration.

Let us consider the example of four EPSPs of the same amplitude, generated at different sites in the dendritic arborization, at times such that their arrivals at the initial segment are offset in time. **Figure 13.2b** shows the 'composite EPSP' obtained in two cases of arrival sequences. In the case $1 \rightarrow 2 \rightarrow 3 \rightarrow 4$, the four EPSPs are generated at the same time t; but since some are generated at more distal sites, their arrivals at the initial segment are staggered, the most proximal arriving first and the most distal arriving last. In the case $4 \rightarrow 3 \rightarrow 2 \rightarrow 1$, the most distal EPSPs are generated well before the proximal EPSPs, so that the distal EPSPs arrive before the proximal EPSPs. The theoretical results show that in the first case, in which the proximal EPSPs occur first and are followed by the more distal EPSPs, the 'composite EPSP' has a short latency, a long duration and a small amplitude; while in the second case, in which the distal EPSPs arrive before the proximal EPSPs, the 'composite EPSP' has a long latency and a large amplitude (**Figure 13.2b**).

Nonlinear summation of excitatory postsynaptic potentials

The term 'nonlinear summation' means that the 'composite EPSP' has an amplitude that is not equal to the geometric sum of the different EPSPs contributing to it. This occurs, for instance, when two EPSPs are generated at the same site or at sites that are close.

Let us take the example of two excitatory synapses whose neurotransmitter is glutamate and which are situated close together on the same dendritic segment (**Figure 13.3**), supposing that the membrane potential of the dendritic segment is V_m. When synapse 1 is active alone, EPSP$_1$ is recorded, due to the excitatory postsynaptic current I_1, such that $I_1 = g_{\text{cations}} (V_m - E_{\text{cations}})$, whose amplitude is $V_{\text{EPSP1}} = I_1/g_m$, where g_m is the membrane conductance (**Figure 13.3a**). When synapse 2 is active alone, EPSP$_2$ is recorded at level 2, due to the postsynaptic current I_2, such that $I_2 = g_{\text{cations}} (V_m - E_{\text{cations}})$, whose amplitude is $V_{\text{EPSP2}} = I_2/g_m$ (**Figure 13.3b**). If we suppose that when the two EPSPs are generated separately, $V_{\text{EPSP1}} = V_{\text{EPSP2}}$, what is the amplitude of the 'composite EPSP' when the two synapses are active at the same time?

When synapse 1 is activated first, EPSP$_1$ is recorded in the postsynaptic element and will be conducted passively to neighbouring regions (**Figure 13.3a**). At time $t + \Delta t$, EPSP$_1$ arrives at the postsynaptic element 2. The

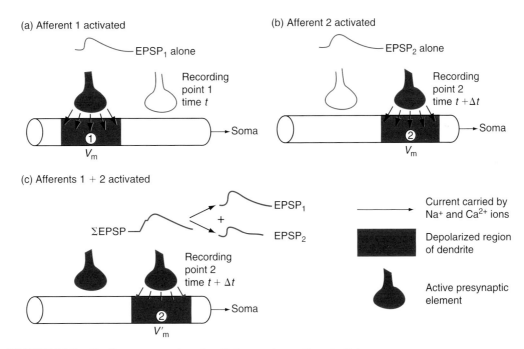

FIGURE 13.3 **Nonlinear summation of excitatory postsynaptic potentials.**
Suppose that there are two excitatory synapses situated close together on the same dendritic segment. **(a)** When afferent 1 is activated at time t, a depolarization of the postsynaptic membrane 1 is recorded at time t (EPSP$_1$ alone). This depolarization propagates in the two directions away from 1. **(b)** When afferent 2 is activated, at time $t + \Delta t$, a depolarization of the postsynaptic membrane 2 is recorded (EPSP$_2$ alone). **(c)** When the two afferents 1 and 2 are activated as before, but together, one at time t and the other at time $t + \Delta t$, a depolarization of the postsynaptic membrane 2 (ΣEPSP) is recorded at time $t + \Delta t$, which does not correspond to the geometric sum EPSP$_1$ alone + EPSP$_2$ alone, since EPSP$_2$' has an amplitude which is smaller than EPSP$_2$ (see text for explanation).

membrane of the postsynaptic element 2 is then at a potential V_m' which is more positive than V_m (**Figure 13.3c**). If at this moment $(t + \Delta t)$ synapse 2 is active, the postsynaptic current I_2' will be smaller than if it had taken place independently from I_1 because the electrochemical gradient of Na^+ and Ca^{2+} ions is reduced. $I_2' = g_{cations}(V_m' - E_{cations})$ and $I_2' < I_2$ because $(V_m' - E_{cations}) < (V_m - E_{cations})$. The 'composite EPSP' will have an amplitude less than the geometric sum $EPSP_1 + EPSP_2$ (**Figure 13.3c**).

This is also the case when a single excitatory synapse is activated repetitively by the arrival of high-frequency presynaptic action potentials. When an excitatory postsynaptic current is generated before the preceding current has ended, it has a smaller amplitude because the postsynaptic membrane is depolarized. Thus, during high-frequency activation, successive excitatory postsynaptic potentials have amplitudes that are smaller and smaller.

13.2.2 Linear and nonlinear summation of inhibitory postsynaptic potentials

When inhibitory synapses are active they cause, in the postsynaptic membrane, an outward postsynaptic current of positive charges (carried by K^+ ions) or an inward current of negative charges (carried by Cl^- ions) which hyperpolarizes the membrane: this is the inhibitory postsynaptic potential, or IPSP.

Linear summation of IPSPs is symmetrically the same as linear summation of EPSPs. Nonlinear summation of IPSPs is symmetrically the same as nonlinear summation of EPSPs.

13.2.3 The integration of excitatory and inhibitory postsynaptic potentials partly determines the configuration of the postsynaptic discharge

In order for an action potential to be triggered at the initial segment, the membrane of the initial segment must be depolarized to the threshold potential for the opening of voltage-sensitive Na^+ channels. It is also necessary for this depolarization to have a relatively rapid risetime so that the Na^+ channels do not inactivate during the depolarization. The characteristics of depolarization of the initial segment (amplitude, duration, risetime) result partly from the summation of excitatory and inhibitory postsynaptic potentials.

Integration of depolarizing (excitatory) postsynaptic potential with hyperpolarizing (inhibitory) postsynaptic potential

A hyperpolarizing postsynaptic potential is due to a current whose reversal potential is more negative than

the resting membrane potential of the cell. This type of inhibition is generally due to the opening of K^+ channels ($GABA_B$-type inhibition). Since the equilibrium potential of K^+ ions is more negative than the resting membrane potential, the opening of K^+ channels gives rise to an outward current (an exit of positive charges) and to a hyperpolarization of the membrane; i.e. an IPSP. If this IPSP is concomitant with an EPSP, it will reduce the amplitude of the EPSP. This type of summation of EPSP and IPSP is summarized in **Figure 13.4**.

Integration of depolarizing (excitatory) postsynaptic potential and silent (inhibitory) postsynaptic potential

A silent postsynaptic potential is due to a current whose reversal potential is close to the resting potential of the cell. Generally, this is caused by a current of Cl^- ions through $GABA_A$ channels (see Chapter 9). When the equilibrium potential of Cl^- ions is close to the membrane resting potential, the opening of Cl^- channels does not reveal a hyperpolarizing current at the resting potential (from which comes the term 'silent' for this inhibition). However, when the membrane is depolarized by an EPSP, the inhibition is no longer silent, but becomes hyperpolarizing and results in a reduction or even a complete suppression of the EPSP (**Figure 13.5**).

Integration of depolarizing (excitatory) postsynaptic potential and depolarizing inhibitory postsynaptic potential

A depolarizing inhibitory postsynaptic potential is due to a synaptic current whose reversal potential is more positive than the resting potential of the membrane but more negative than the threshold for the opening of the Na^+ channels of the action potential. This is generally due to the opening of Cl^- channels in cells in which the reversal potential for Cl^- ions is situated between the resting potential and the threshold potential for the opening of the Na^+ channels of the action potential. Thus, when the membrane is at its resting potential, this Cl^- current causes a slight depolarization of the membrane, but does not trigger action potentials. When the membrane is depolarized (by an EPSP) above the inversion potential of Cl^+ ions, this current causes a hyperpolarization of the membrane and an inhibition of the EPSP.

13.3 SUMMARY

Several types of inhibition appear over the length of the somato-dendritic arborization and these limit the effect of excitatory synapses. The opening or non-opening

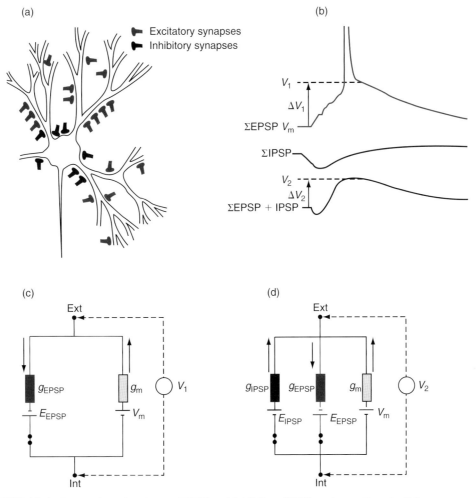

FIGURE 13.4   Integration of excitatory (EPSP) and inhibitory (IPSP) postsynaptic potentials.
(a) Suppose that on a dendritic tree, there are glutaminergic excitatory synapses which are situated distally, and GABA$_B$-type inhibitory synapses which are situated proximally, and that all of these are active at the same instant t. **(b)** If only the excitatory synapses are active, a depolarization, a composite EPSP (ΣEPSP) will be recorded at the soma which corresponds to the linear and nonlinear summation of all the different EPSPs (top trace). We will suppose that the ΣEPSP has an amplitude that is sufficient to trigger an action potential (upper trace). If only the inhibitory synapses are active, a hyperpolarization, a composite IPSP (ΣIPSP) will be recorded at the soma which corresponds to the linear and nonlinear summation of all the different IPSPs (middle trace). When all these different synapses are activated at the same time t, a depolarization preceded by a hyperpolarization, a composite PSP, will be recorded at the soma, corresponding to the sum of the different synaptic potentials (ΣEPSP + ΣIPSP) (bottom trace). In this case the amplitude of the depolarization is no longer sufficient to trigger an action potential. **(c)** Electrical equivalent of the membrane at the level of the initial segment, for an EPSP alone. **(d)** Electrical equivalent for the membrane when an EPSP and an IPSP summate. The currents I_{EPSP} and I_{IPSP} are opposite and subtract from one another. By comparing with (c), it is observed that I_{EPSP} in (c) is greater than $I_{EPSP} + I_{IPSP}$ in (d), and $\Delta V_1 > \Delta V_2$.

of the Na$^+$ channels of the action potential, and in consequence the generation of action potentials which will constitute the response of the postsynaptic neuron, are the result of this summation of excitatory and inhibitory postsynaptic potentials. However, the characteristics of the response of the postsynaptic neuron are determined not only by the amplitude and duration of the depolarization of synaptic origin but also by the characteristics of the membrane of the initial segment, also known as 'input–output' characteristics.

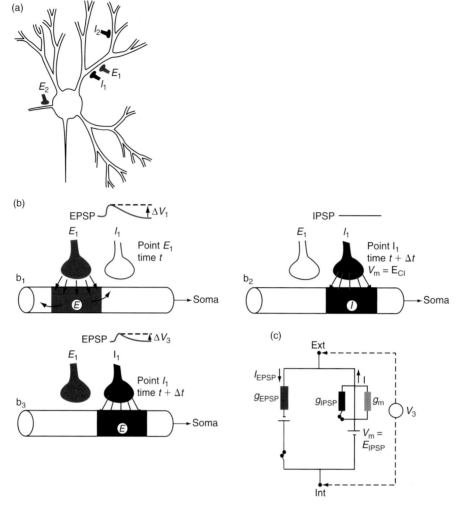

FIGURE 13.5   Role of silent inhibition.

(a) This diagram shows two synapses, one glutamatergic with postsynaptic AMPA receptors (E_1) and the other GABAergic with $GABA_A$ postsynaptic receptors (I_1), situated close to one another on the same dendritic segment, such that the inhibitory synapse is closer to the soma than the excitatory synapse. **(b)** When the excitatory synapse is excited alone, an EPSP of ΔV_1 in amplitude is recorded (b_1). When the inhibitory synapse is activated alone, no change in potential is recorded because $V_m = E_{Cl}$(b_2). When both synapses are activated, the EPSP which propagates towards the soma is reduced in amplitude (amplitude ΔV_3), or even cancelled out. This type of inhibition is selective because it only acts on excitatory synapses that are situated distally. **(c)** Electrical equivalent of the membrane at the dendritic segment. If this is compared with Figure 13.4c, it can be seen that $\Delta V_3 < \Delta V_1$ because $g_m + g_{IPSP} > g_m$.

FURTHER READING

Buhl EH, Halasy K, Somogyi P (1994) Diverse sources of hippocampal unitary inhibitory postsynaptic potentials and the number of release sites. *Nature* **368**, 823–828.

Cauller LJ and Connors BW (1992) Functions of very distal dendrites. In: McKenna TM, Davis J, Zornetzer SE (eds), *Single Neuron Computation*, New York: Academic Press.

Larkum ME, Launey T, Dityatev A, Lüscher HR (1998) Integration of excitatory postsynaptic potentials in dendrites of motoneurons of rat spinal cord slice cultures. *J. Neurophysiol.* **80**, 924–935.

Miles R, Toth K, Gulyas AI, Hajos N, Freund TF (1996) Differences between somatic and dendritic inhibition in the hippocampus. *Neuron* **16**, 815–823.

Rall W (1977) Core conductor theory and cable properties of neurons. In: Brookhart JM, Mountcastle VB, Kandel ER, Geiger SR (eds), *Handbook of Physiology*, vol. 1, part 1, Bethesda, MD: American Physiological Society.

Redman SJ (1973) The attenuation of passively propagating dendritic potentials in a motoneuron cable model. *J. Physiol.* **234**, 637–664.

Shepherd GM (1994) The significance of real neuroarchitectures for neural network simulations. In: Schwartz EL (ed.), *Computational Neuroscience*, New York: Oxford University Press.

Spruston N and Johnston D (1992) Perforated patch clamp analysis of the passive membrane properties of three classes of hippocampal neurons. *J. Neurophysiol.* **67**, 508–528.

Spruston N, Jaffe DB, Johnston D (1994) Dendritic attenuation of synaptic potentials and currents: the role of passive membrane properties. *Trends Neurosci.* **17**, 161–166.

CHAPTER

14

Subliminal voltage-gated currents of the somato-dendritic membrane

Not all neurons respond in the same way when they are activated by a depolarizing current pulse or when they are hyperpolarized: it can be said that they do not have the same pattern of firing (**Figure 14.1**). When depolarized, spinal motoneurons may respond with a low-frequency regular discharge, certain pyramidal neurons of the hippocampus with a burst of action potentials followed by a long silence, neurons of the inferior olive nucleus with an irregular sustained activity of bursts of action potentials. Conversely, when hyperpolarized, hippocampal neurons become silent whereas thalamic and subthalamic neurons discharge bursts of action potentials. It should be noted that the term 'firing pattern' is not equivalent to the term 'discharge frequency', except in cases where the response consists of action potentials generated at a regular frequency. In this case only, the mean value of the discharge frequency is sufficient to describe the response of the neuron. In other cases, the mean frequency value has no significance.

Neurons, as pointed out by Llinas (1988), are not interchangeable; i.e. a neuron cannot be functionally replaced by one of another type even if their synaptic connectivity, type of afferent neurotransmitters and receptors to these neurotransmitters are identical. The activity of a neuronal network is related not only to the *excitatory and inhibitory interactions* among neurons but also to their *intrinsic electrical properties* as well. The 'personality' of a neuron is defined by its input–output characteristics; i.e. its firing pattern (output) in response to a depolarization or a hyperpolarization (input).

Input–output characteristics are the result of a rich repertoire of ionic currents other than those of the action potentials. These currents, inward or outward, are called *subliminal voltage-gated currents* because they are activated at voltages sub-threshold to that of action

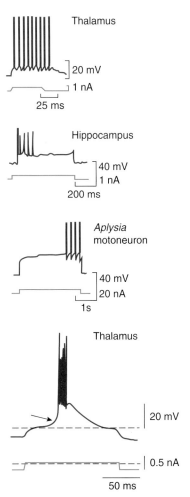

FIGURE 14.1 Different neuronal responses to a depolarizing current pulse.
From top to bottom: thalamo-cortical neuron, pyramidal neuron of the hippocampus, motoneuron innervating the ink gland of *Aplysia*, and again a thalamo-cortical neuron but from a more hyperpolarized membrane potential.

potentials. They are located either in the dendritic or the somatic membrane or both. In many experiments, recordings are performed at the level of the soma and the question may remain as to where these currents are generated: at the level of the dendrites, the soma, the initial segment of the axon?

14.1   OBSERVATIONS AND QUESTIONS

Figure 14.1 shows the different responses of central neurons recorded in current clamp mode, in response to a depolarizing current pulse and at different membrane potentials.

- What are the currents activated during the depolarizing current pulse that stop the firing of hippocampal neurons after 100–200 ms? Are these currents activated by action potentials?
- What are the currents activated by the depolarizing current pulse that delay the firing of motoneurons innervating the ink gland of *Aplysia*?
- What are the currents that make a thalamic neuron fire in the bursting mode? Why are they activated at a hyperpolarized membrane potential?

To answer these questions, the main subliminal currents must be first explained individually (this chapter). Their influence on postsynaptic potentials and firing patterns are explained in Chapters 15–17.

14.2   THE SUBLIMINAL VOLTAGE-GATED CURRENTS THAT DEPOLARIZE THE MEMBRANE

To depolarize and excite a membrane at rest, currents have to be inward and to be turned on at potentials more negative than the threshold for the opening of voltage-gated Na^+ channels of the action potential. Three types of inward subliminal currents will be explained: the persistent Na^+ current (I_{NaP}), the transient Ca^{2+} current (I_{CaT}) and the hyperpolarization-activated cationic current (I_h).

14.2.1   The persistent inward Na^+ current, I_{NaP}

I_{NaP} is a slowly inactivating current. It is thus called 'P' for persistent in comparison with the fast-inactivating Na^+ current of the action potential (I_{Na}). I_{NaP} is activated at potentials of about 10 mV negative to I_{Na}; i.e. at sub-threshold potentials. I_{NaP} is present in many vertebrate central neurons, and in particular in Purkinje cells.

I_{NaP} was first described by Prince and co-workers in hippocampal (1979) and neocortical (1982) neurons in slices, as an increase in slope resistance at potentials 10–15 mV positive to resting potential (the I_{NaP}/V relation showed an inward rectification). In the presence of TTX in the extracellular medium or when external Na^+ is replaced by the impermeant ion choline, this inward rectification disappears.

Structure of the main channel subunit

To determine whether the I_{NaP} channel is a non-inactivating subtype of the classical fast-inactivating Na^+ channel, or is a different channel, the Na^+ channel α-subunit transcripts expressed in Purkinje cells are studied using single-cell reverse transcription-PCR (RT-PCR). Purkinje cells have been chosen because the two different voltage-gated Na^+ currents are recorded from them: one responsible for the fast depolarization phase of action potentials and a second responsible for the TTX-sensitive prolonged potential plateau. mRNA transcripts for two α-subunits have been found. $Na_v1.6$ (also named cerIII) is suggested to be responsible for I_{NaP}; the other, $Na_v1.1$ (rat brain I), is responsible for I_{Na} (see **Figure 4.6**). This result favours the hypothesis that I_{NaP} does not result from a Na^+ channel with multiple gating states but corresponds to a different channel.

Gating properties and ionic nature

In whole-cell recordings, in the presence of Cd^{2+} (200 μM) and K^+ channel blockers (20 mM TEA, 2 mM Cs^+), a slowly inactivating Na^+ current is recorded in response to incremental depolarizing steps between −60 and −50 mV from a holding potential of −90 mV (**Figure 14.2a**). Note that at more depolarized levels, the fast-inactivating Na^+ current superimposes on the slow one. I_{NaP} has a fast activation time of 2–4 ms, and once it is evoked it is present for several hundreds of milliseconds and is totally and reversibly blocked by TTX (1 μM) (**Figure 14.2b**). This observation, together with the lack of effect of extracellular Co^{2+} on this current, suggest that it is an inward current which uses Na^+ as the charge carrier.

The I/V relation gives an estimated reversal potential of $+49.1 \pm 1.3$ mV (**Figure 14.2c**). The voltage dependence of activation is determined by applying different voltage steps from a holding potential $V_H = -90$ mV. The normalized peak current amplitude is plotted against the test potential to give the activation curve. In suprachiasmatic neurons, I_{NaP} is half-activated at $Vm = -43$ mV, a potential at least 10 mV more negative than $V_{1/2}$ of I_{Na} (**Figure 14.2d**). To study the inactivation properties of I_{NaP}, the membrane is

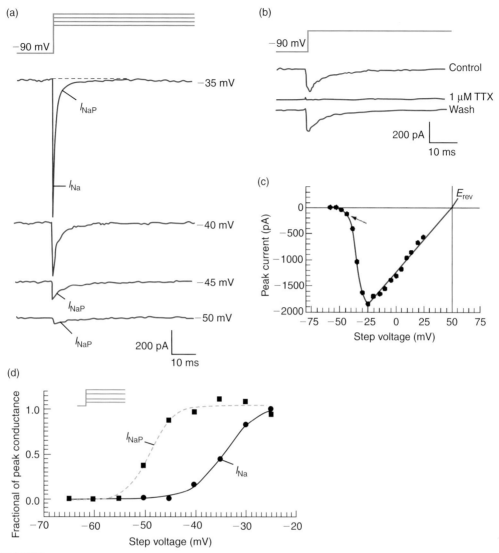

FIGURE 14.2 **Activation properties of the slowly inactivating Na⁺ current, I_{NaP}.**
The activity of neurons of the suprachiasmatic nucleus is recorded in slices (whole-cell configuration, voltage clamp mode). **(a)** Incremental steps from a holding potential of -90 mV. **(b)** Reversible block by TTX. **(c)** I/V plot for the peak I_{NaP} measured from records in (a). **(d)** Activation curve ($V_{1/2} = -43$ mV). Adapted from Pennartz CMA, Bierlaagh MA, Geurtsen AMS (1997) Cellular mechanisms underlying spontaneous firing in rat suprachiasmatic nucleus: involvement of a slowly inactivating component of sodium current. *J. Neurophysiol.* **78**, 1811–1825, with permission.

clamped at different holding potentials and the current in response to a depolarizing step to -50 mV is recorded. A plot of peak current amplitude against holding potential gives a measure of the voltage dependence of inactivation. In these neurons, I_{NaP} is half inactivated at $V_{1/2} = -68$ mV (**Figure 14.3**).

Pharmacology

I_{NaP} is sensitive to TTX at micromolar doses in the extracellular solution and to internal QX-314 (a derivative of lidocaine) injected into the intracellular medium.

These toxins also block I_{Na}, which is a problem when studying the role of I_{NaP} on the pattern of discharge of a neuron. There is no selective blocker of I_{NaP}. To study I_{NaP} in isolation, Ca^{2+} and K^+ channels must be blocked by Cd^{2+} or Co^{2+} or Ni^{2+} ($200\,\mu M$ to $1\,mM$), 4-AP ($1\,mM$), TEA ($10\,mM$), Ba^{2+} ($1\,mM$). In these conditions I_{Na} will be also recorded at some voltage steps but is easily recognized by its high amplitude and fast inactivation.

In summary, I_{NaP} is a TTX-sensitive Na⁺ current that activates at around -60 to -50 mV. I_{NaP} can be distinguished from the Na⁺ current of action potential, I_{Na}, by its low threshold of activation and its slow inactivation.

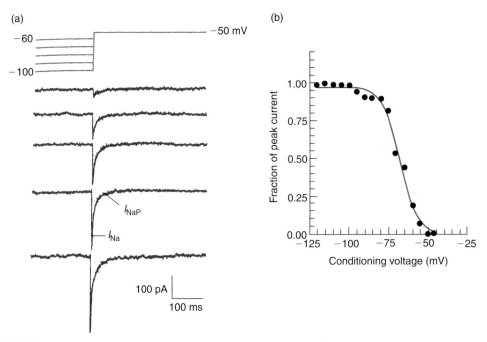

FIGURE 14.3 **Inactivation properties of the slowly inactivating Na$^+$ current, I_{NaP}.**
The activity of neurons of the suprachiasmatic nucleus is recorded in slices (whole-cell configuration, voltage clamp mode). **(a)** Currents recorded in response to a fixed step to -50 mV from incremental holding potentials. **(b)** Inactivation curve ($V_{1/2} = -68$ mV). Adapted from Pennartz CMA, Bierlaagh MA, Geurtsen AMS (1997) Cellular mechanisms underlying spontaneous firing in rat suprachiasmatic nucleus: involvement of a slowly inactivating component of sodium current. *J. Neurophysiol.* **78**, 1811–1825, with permission.

14.2.2 The low-threshold transient Ca^{2+} current, I_{CaT}

The ability of neurons to fire low-threshold Ca^{2+} spikes suggested the existence of a low-voltage-activated Ca^{2+} current (I_{CaT}). I_{CaT} is called 'T' for transient since once activated it rapidly inactivates. It was originally described by Carbone and Lux in 1984 but its existence had first been suggested by Eccles and co-workers in 1964 to explain their observation of a period of enhanced excitability following membrane hyperpolarization in some central neurons.

Structure of the main channel subunit

The Ca$_v$3 family of α_1-subunits (see **Figure 5.2b**) conducts T-type Ca^{2+} currents, which are activated and inactivated more rapidly and at more negative membrane potentials than other Ca^{2+} current types. Three Ca$_v$3 different α_1-subunits have been described: Ca$_v$3.1 (α_{1G}), Ca$_v$3.2 (α_{1H}) and Ca$_v$3.3 (α_{1I}). Their primary structure shows the presence of four homologous domains, each containing six putative transmembrane segments (S1 to S6) and a P loop between segments 5 and 6. In contrast to the α_1-subunits of high-voltage-activated channels (L, N, P), T channel α_1-subunits contain a large extracellular loop located between S5 and the P loop. Ca$_v$3 α_1-subunits are only distantly related to these homologs with less than 25% amino acid sequence identity.

Single-channel conductance

The activity of chick dorsal root ganglion cells in culture is recorded in patch clamp (cell-attached patch). The patch pipette contains 110 mM of BaCl$_2$ instead of 2 mM of CaCl$_2$ for the following reasons: Ca^{2+} channels are permeable to Ba^{2+}, Ba^{2+} does not inactivate Ca^{2+} channels, Ba^{2+} does not activate Ca^{2+}-dependent channels such as Ca^{2+}-activated K$^+$ channels. Ba^{2+} current, at such high concentration of charge carrier, 110 mM, through Ca^{2+} channels has in general a large amplitude and is more easy to study. Moreover, when Ba^{2+} is the only external cation, Na$^+$ currents are not recorded.

The membrane is held at a hyperpolarized potential ($V_H = -80$ mV) to keep T channels non-inactivated. When depolarizing voltage steps of small amplitude ($30-60$ mV) are applied to the patch, unitary inward Ba^{2+} currents are recorded. These unitary inward currents have a small amplitude (**Figure 14.4b**) and the unitary conductance γ_T is very low compared with that

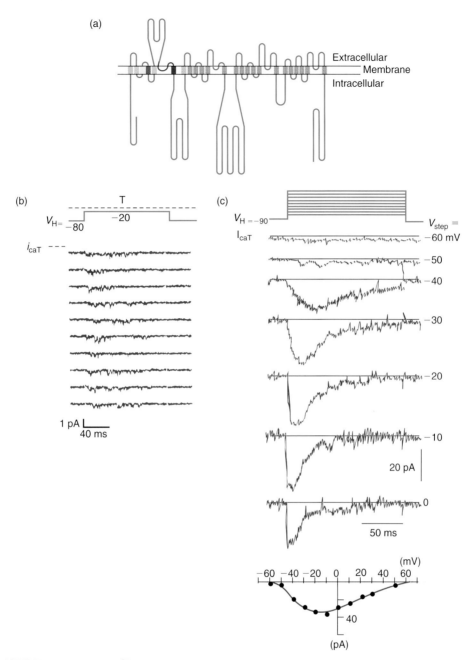

FIGURE 14.4 T-type Ca^{2+} channel and current, I_T.
(a) Predicted topology of the rat α_{1I} subunit. **(b)** Activity of a single T channel recorded from dorsal root ganglion cell bodies in patch clamp (cell-attached patch). Mean i_{CaT} amplitude at -20 mV is -0.62 ± 0.03 pA. **(c)** I/V relation of the whole-cell T current from a freshly plated (5 h) dissociated rat hippocampal neuron recorded in 10 mM of external Ca^{2+}. Part (a) adapted from Lee JH, Daud AN, Cribbs LL *et al.* (1999) Cloning and expression of a novel member of the low-voltage-activated T-type calcium channel family. *J. Neurosci.* **19**, 1912–1921, with permission. Part (b) adapted from Nowycky MC, Fox AP, Tsien RW (1985) Three types of neuronal calcium channels with different calcium agonist sensitivity. *Nature* **316**, 440–443, with permission. Part (c) adapted from Yaari Y, Hamon B, Lux HD (1987) Development of two types of calcium channels in cultured mammalian hippocampal neurons. *Science* **235**, 680–682, with permission.

of high-threshold-activated Ca^{2+} channels: it varies between 5 and 9 pS in 110 mM of Ba^{2+}. At physiological concentrations of external Ca^{2+} (2 mM) the expected conductance would be around 1 pS. Single-channel T currents are present at the beginning of the depolarizing step and then disappear though the membrane is still depolarized consistent with a rapid inactivation of T channels. T channels are more resistant to rundown of activity following patch excision or whole-cell recording than are L-type Ca^{2+} channels.

Gating properties and ionic nature

The voltage-dependence of the T current is studied in whole-cell recordings. In order to study the I/V relation of the T current in isolation, the other Ca^{2+} currents present in the cell, such as the high-threshold L-, N- and P-type Ca^{2+} currents (see Chapter 5) are pharmacologically blocked. However, in some preparations, T channels predominate and can be studied in the absence of blockers. For example, during development, embryonic hippocampal neurons in culture first express T-type Ca^{2+} channels and then, with neurite extension, also express high-threshold Ca^{2+} channels.

To study the activation properties of I_{CaT}, the membrane potential is held at $-90\,mV$. The I/V relation shows that I_{CaT} activates in response to depolarizations positive to $-55\,mV$ and is maximal around $-10\,mV$ when recorded in the presence of $10\,mM$ of external Ca^{2+} (**Figure 14.4c**). The voltage-dependence of activation is determined by applying different voltage steps from a holding potential $V_H = -105\,mV$ (**Figure 14.5b**). The normalized peak current amplitude is plotted against the test potential (**Figure 14.5c**). In chick sensory neurons, it is half-activated at $-51\,mV$ (in $10\,mM$ of external Ca^{2+}). Note that the rate of activation is highly voltage-dependent.

During a 150 ms depolarizing pulse to $-10\,mV$, I_{CaT} rapidly inactivates (in 50 ms): it is transient (**Figure 14.4c**). To study the inactivation properties of I_{CaT}, the membrane is clamped at different holding potentials amplitude and the current in response to a depolarizing step to $-35\,mV$ is recorded (**Figure 14.5a**). A plot of peak current against holding potential (**Figure 14.5c**) gives a measure of the voltage-dependence of inactivation. In chick sensory neurons, I_{CaT} is half-inactivated at $-78\,mV$ (in $10\,mM$ of external Ca^{2+}).

The removal of inactivation (de-inactivation) of the T current is time-dependent. This is studied in lateral geniculate cells *in vitro* (in single-electrode voltage clamp) with the following protocol. The membrane is held at $-55\,mV$ to inactivate completely the T current. A voltage step to $-95\,mV$ of variable duration is then applied. On stepping back to $-55\,mV$, the peak amplitude of the T current is measured (I) and compared with its maximum amplitude (I_{max}). At $35°C$, 500–600 ms at $-95\,mV$ are needed to totally remove inactivation. The removal of inactivation of the T current is also voltage-dependent: at potentials close to $-55\,mV$ the de-inactivation is less complete than at more hyperpolarized potentials.

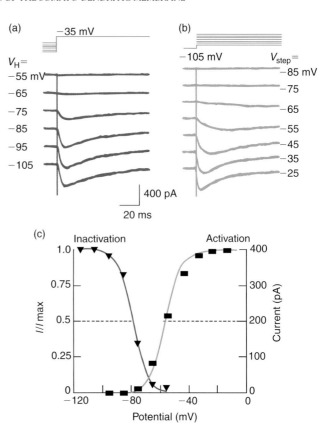

FIGURE 14.5 Voltage-dependence of activation and inactivation of the T current.
Recordings in $5–10\,mM$ of external Ca^{2+}: **(a)** inactivation; **(b)** activation. See text for explanation. **(c, left)** I_{max} is the maximal peak current amplitude obtained at $V_H = -105\,mV$. **(c, right)** I_{max} is the maximal peak current amplitude obtained at $V_{step} = -35$ to $-20\,mV$. Adapted from Fox AP, Nowycky MC, Tsien RW (1987) Kinetic and pharmacological properties distinguishing three types of calcium currents in chick sensory neurones. *J. Physiol.* **394**, 149–172, with permission.

Pharmacology

There are no highly specific antagonists or toxins for the T-type Ca^{2+} channels. Low concentrations of the inorganic cation Ni^{2+} ($20–50\,\mu M$) strongly depress I_{CaT}. Ethosuximide and amiloride have been also reported to reduce I_{CaT} in some preparations. Specific toxins acting at HVA channels are inefficient in T channels. To study I_{CaT} in isolation, Na^+ and K^+ channels must be blocked by TTX ($1\,\mu M$), 4-AP ($1\,mM$), TEA ($10\,mM$), Ba^{2+} ($1\,mM$) and HVA Ca^{2+} channels by their specific blockers (nifedipin, ω-conotoxin GVIA, ω-agatoxin).

In summary, T-type Ca^{2+} current is an inward current carried by Ca^{2+} ions; the activation threshold is around resting membrane potential (-50 to $-60\,mV$), $10–20\,mV$ negative to spike threshold. It is named 'T' for transient (owing to fast inactivation) and also for tiny (owing to small single-channel conductance). T-type current is a low-threshold-activated (LVA) Ca^{2+} current that can be distinguished from the high-threshold-activated (HVA)

L-, N- and P-type Ca^{2+} currents by the following criteria: it is activated after small depolarizations of the membrane (low voltage of activation), is transient, has a tiny single-channel conductance, and closes slowly upon repolarization of the membrane (generating a slow deactivation tail current). It is totally inactivated at potentials close to the resting potential and is de-inactivated during a transient hyperpolarization of the membrane. Therefore, it is fully activated by a depolarization only when the membrane potential has been previously maintained at a potential more hyperpolarized than resting membrane potential.

14.2.3 The hyperpolarization-activated cationic current, I_h, I_f, I_Q

I_h has an unusual voltage-dependence since it is activated upon hyperpolarization of the membrane beyond resting membrane potential. For this reason it has several names, 'h' for hyperpolarization, 'f' for funny (in the sinoatrial node of the heart) and 'Q' for queer, in some early studies in view of its odd electrophysiological behaviour and its undefined functional significance. It was originally observed by Ito and co-workers in 1962 in cat motoneurons as a non-ohmic behaviour of the I/V relation in the hyperpolarizing direction.

Structure of the main channel subunit

The I_h channel is a family of channels whose name is HCN: *h*yperpolarization-activated, *c*yclic *n*ucleotide-modulated channels. There are four known HCN subunit isoforms, HCN1–4, which combine to form tetrameric channels. HCN channels belong with the cyclic-nucleotide-gated (CNG) channels (CNG) to the superfamily of K^+ channels. They contain the conserved motifs of K^+ voltage-gated channels including the S1–S6 segments, a charged S4 voltage sensor and a pore-lining P loop (**Figure 14.6a**). In addition, all family members contain a conserved cyclic nucleotide-binding (CNB) domain in their carboxy-terminus. This domain is homologous to the CNB domain of protein kinases and of CNG channels (see **Table 3.1**), showing that the gating of I_h channels is directly regulated by cyclic nucleotides such as cAMP or cGMP.

Gating properties and ionic nature

The type of voltage clamp experiment that allows one to record I_h is called a *relaxation experiment*. The activity of thalamic neurons in slices is recorded under two-electrode voltage clamp (**Figure 14.6b**). A 1.5 s hyperpolarizing voltage step to $-90\,mV$ while the membrane potential is held at $V_H = -60\,mV$ evokes an instantaneous inward current (I_{leak}) followed by a slowly developing inward

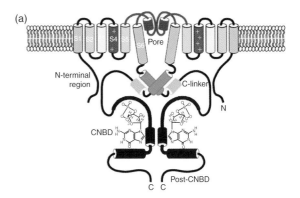

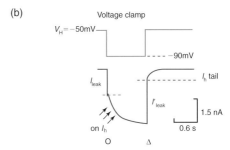

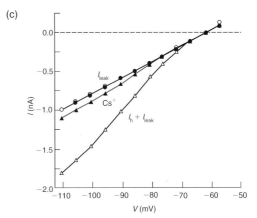

FIGURE 14.6 The hyperpolarization-activated H channel and current Ih.

(a) Membrane topology of HCN channels. Two of the four subunits that form a HCN channel are shown with the transmembrane segments (S1–S6), the pore region (S6 and pore loop) and the voltage sensor (S4) indicated by positive charges. The C-terminal region contains the cyclic nucleotide–binding domain (CNBD, *blue*, shown with cGMP bound). (b) The activity of a thalamic neuron recorded in slices (intracellular recording, voltage clamp mode). Relaxation experiment. The extracellular medium contains 14.5 mM of K^+ and 116 mM of Na^+. (c) I/V relation obtained from experiment in Figure 14.7c: instantaneous leak current is plotted as circles, steady-state current ($I_{leak} + I_h$), measured at the end of the voltage steps, is plotted as triangles. Open symbols represent currents under control conditions and filled symbols are currents after application of Cs^+. Part (a) adapted from Craven KB, Zagotta WN (2006) CNG and HCN channels: two peas, one pod. *Ann. Rev. Physiol.* **68**, 375–401, with permission. Part (c) adapted from McCormick DA, Pape HC (1990) Properties of a hyperpolarization-activated cation current and its role in rhythmic oscillation in thalamic relay neurones. *J. Physiol.* **431**, 291–318, with permission.

current which shows no inactivation with time (on I_h). I_{leak} reflects the leak current through channels open at $V_H = -50$ mV (mostly K^+ channels). The following slow inward current reflects the slow opening of I_h channels. When the membrane is then repolarized to -60 mV at the end of the voltage step, an instantaneous current (I' leak) is recorded followed by an inward tail current (tail I_h). The instantaneous current reflects the leak current through channels open at -90 mV. The tail current reflects the kinetics of closure (deactivation) of I_h channels. I' is larger than I_{leak} because at -90 mV not only the leak channels are open but also the H (I_h) channels that have been opened by the hyperpolarization.

By varying the amplitude of voltage steps, I_h is shown to activate at between -45 and -60 mV and to be half-activated at around -75 and -85 mV (**Figure 14.6c**). I_h reverses at around -50 to -30 mV depending on the

neuronal type. Decreasing the external Na^+ concentration from 153 to 26 mM reduces the amplitude of I_h (**Figure 14.7a**). When Na^+ ions are totally replaced by the non-permeant cation choline, I_h disappears almost completely (since E_K is around -90 mV). Similarly, raising the external K^+ concentration from 2.5 mM to 12.5 mM enhances I_h (not shown). These results indicate that I_h is carried by both Na^+ and K^+ ions, which is consistent with the extrapolated reversal potential. H channels are in fact four times more permeable to K^+ than to Na^+.

Pharmacology

Application of cAMP to the internal surface of an inside-out patch induces a reversible increase in the magnitude of the inward current during a step to -100 mV (**Figure 14.7b**). This effect of cAMP is due to a

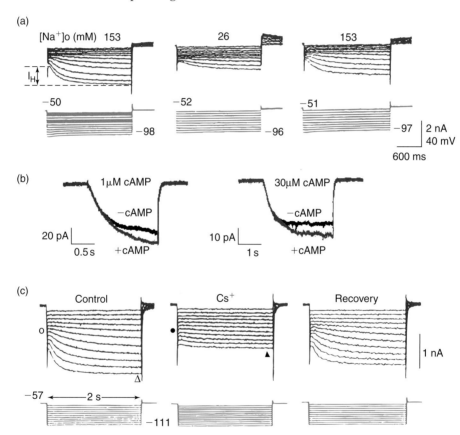

FIGURE 14.7 Ionic selectivity and pharmacology of I_h.
The activity of thalamic neurons is recorded in slices (intracellular recording, voltage-clamp mode). **(a)** Effect on I_h of changing the extracellular concentration of Na^+ ions. A family of I_h currents (upper traces) are evoked by stepping membrane potential to hyperpolarized potentials from $V_H = -50$ mV (bottom traces), in control condition (153 mM, left and right) and during reduced Na^+ concentration (26 mM, centre). It should be noted that there is no change in the baseline current. **(b)** The mBCNG-1 channel is expressed in *Xenopus* oocytes. The I_h current is evoked in inside-out macropatches by a hyperpolarizing step to -100 mV in the absence or presence of cAMP in the bath. **(c)** Similar experiment as in (a), in control conditions and in the presence of Cs^+. Parts (a) and (c) adapted from McCormick DA, Pape HC (1990) Properties of a hyperpolarization-activated cation current and its role in rhythmic oscillation in thalamic relay neurones. *J. Physiol.* **431**, 291–318, with permission. Part (b) adapted from Santoro B, Liu DT, Yao H *et al.* (1998) Identification of a gene encoding a hyperpolarization-activated pacemaker channel of brain. *Cell* **93**, 717–729, with permission.

positive shift in the steady-state activation curve of I_h current by 2 to 10 mV. I_h is completely blocked in the presence of 1–3 mM of Cs^+ in the extracellular solution (**Figures 14.6c** and **14.7c**). A bradycardic agent ZM 227189 (10–100 μM; Zeneca) selectively blocks I_h. In contrast, I_h is insensitive to external Ba^{2+}, TTX, TEA and 4-AP that effectively block Na^+ or K^+ channels. To study I_h in isolation, Ca^{2+}, Na^+ and K^+ channels must be blocked by Cd^{2+} or Co^{2+} or Ni^{2+} (200 μM to 1 mM), TTX (1 μM), 4-AP (1 mM), TEA (10 mM), Ba^{2+} (1 mM).

In summary, I_h is carried by Na^+ and K^+ ions and is a voltage- and time-dependent current. It has a slow time course of activation upon hyperpolarization and is inactived at depolarized potentials where action potentials are firing. I_h is directly modulated by internal cyclic nucleotides such as cAMP or cGMP. It is reversibly decreased by 1–3 mM of extracellular Cs^+.

14.3 THE SUBLIMINAL VOLTAGE-GATED CURRENTS THAT HYPERPOLARIZE THE MEMBRANE

The common characteristic of these subliminal currents is to be outward, carried by K^+ ions and turned on at potentials more negative than the threshold for the opening of voltage-gated Na^+ channels of the action potential. Four types of outward K^+ currents will be explained: the early K^+ currents (I_A, I_D), the K^+ currents activated by intracellular Ca^{2+} (I_{KCa}), the muscarine sensitive K^+ current (I_M) and the inward rectifier K^+ current (I_{KIR}). All K channels operate as complexes comprised of pore forming α-subunits plus a number of associated ancillary subunits. α-subunits tetramerize to form a central pore.

14.3.1 The rapidly inactivating transient K^+ current: I_A or I_{Af}

I_A is a K^+ current which rapidly activates and inactivates in response to depolarizing steps from holding potentials negative to the resting membrane potential. It was originally described in 1971 by Connor and Stevens, in molluscan neurons, and termed 'A current'.

Structure of the main channel subunit

The I_A channel belongs to the family of *Shal* K^+ channels (or K_v4, they are gated by transmembrane voltage, hence the *Kv* nomenclature). They have *six putative membrane* spanning segments designated S1 to S6, flanked by intracellular domains of variable length and a pore domain P between S5 and S6. The Shal-type family in mammals is comprised of three distinct genes: $K_v4.1$, $K_v4.2$ and $K_v4.3$.

Gating properties and ionic nature

The activity of medullary or hippocampal neurons in culture is recorded in voltage clamp mode in a medium containing TTX (to block Na^+ currents) and TEA (to block the K^+ currents of the delayed rectification), Cd^{2+} (200 μM) to depress Ca^{2+}-dependent currents, carbachol (50 μM) to block I_M (see Section 14.3.4) and Cs^+ (1–3 mM) to block I_h. When the membrane potential is maintained at −70 mV, depolarizing voltage steps (of 10 to 58 mV amplitude) evoke an outward current whose amplitude increases with the amplitude of the depolarization (**Figure 14.8a**). This current activates rapidly (within milliseconds) and then inactivates rapidly and exponentially during the step. This is the 'A current', I_A. Inactivation of I_A is time- and voltage-dependent. It appears after a few milliseconds, which makes I_A short-lasting. In this respect I_A

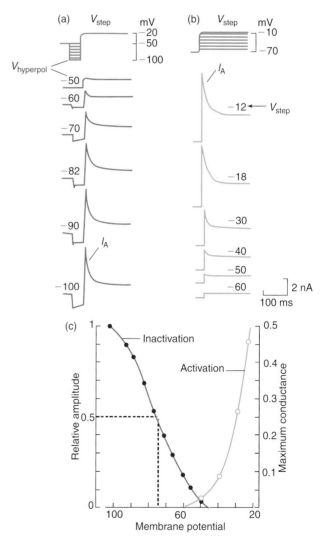

FIGURE 14.8 **Activation–inactivation properties of I_A.**
From Segal M, Rogawski MA, Barker JL (1984) A transient potassium conductance regulates the excitability of cultured hippocampal and spinal neurons. *J. Neurosci.* **4**, 604–609, with permission.

resembles more the I_{Na} of the action potential than the current of the delayed rectification, I_{KDR}. The current plateau which follows the peak of the I_A current represents the sum of the leak current and of the outward currents which are not blocked by TEA.

When a hyperpolarizing voltage step ($V_{hyperpol}$) *of* varying amplitude is now applied during the 50 ms before the depolarizing voltage step (V_{step}) to -20 mV (**Figure 14.8b**), it can be seen that I_A is inactivated when the membrane potential is maintained at a value more positive than -50 mV; at these membrane potentials I_A cannot be activated by a depolarization. Also, I_A is de-inactivated when the membrane potential is maintained at a value more negative than -50 mV and can then be activated by depolarizations to membrane potentials positive to -50 mV and its amplitude increases with the difference $V_{hyperpol} - V_{step}$. Activation and inactivation curves (**Figure 14.8c**) have been constructed from the data in traces (b) and (a), respectively.

Pharmacology

I_A is blocked by application of 4-aminopyridine (4-AP, 1–3 mM) in the extracellular medium. It is insensitive to TEA, Cs^+ and Ba^{2+}. Thus there are two ways of blocking I_A, either by depolarizing the membrane above -50 mV, or by applying 4-aminopyridine. To study I_A, Na^+ and Ca^{2+} channels must be blocked by TTX (1 µM), Cd^{2+} or $^{2+}$ or Ni^{2+} (200 µM to 1 mM).

In summary, I_A is a fast-inactivating (in the order of milliseconds) K^+ current. The threshold potential for its activation is situated at around -60 to -45 mV; i.e. at a value slightly more negative than the threshold potential for the inward Na^+ current of the action potential. I_A can be fully activated by a depolarization only when the membrane potential has been previously maintained at a potential more hyperpolarized than -60 mV (I_A is inactivated at resting membrane potential). It has characteristics which distinguish it from the K^+ currents of the delayed rectification: it is a rapidly activating and inactivating K^+ current and has pharmacological properties different from that of the delayed rectifier currents.

14.3.2 The slowly inactivating transient K^+ current, I_D or I_{As}

In addition to I_A, a K^+ current that also activates rapidly (within milliseconds) but slowly inactivates (over seconds) was first described by Storm (1988) in hippocampal pyramidal cells. It was termed 'D current' because it delays the cell firing.

Gating properties

The activity of the pyramidal cells of the hippocampus is intracellularly recorded in brain slices. Voltage

clamp experiments are performed in the presence of external Cd^{2+} (200 µM) to depress Ca^{2+}-dependent currents, carbachol (50 µM) to block I_M (see Section 14.3.4) and Cs^+ (1–3 mM) to block I_h. When, in such conditions, a depolarizing voltage step to -26 mV is applied from a holding potential $V_H = -80$ mV, an outward current consisting of two components is recorded: a fast-inactivating component, sensitive to high doses of 4-aminopyridine (4-AP, 1–3 mM) and a slowly inactivating component sensitive to low doses (40 µM) of 4-AP (**Figure 14.9a**). The first component corresponds to I_A and the second one is termed I_D.

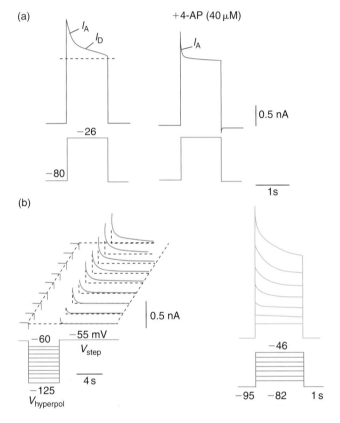

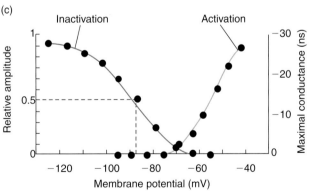

FIGURE 14.9 Activation–inactivation properties of I_D.
From Storm JF (1988) Temporal integration by a slowly inactivating k$^+$ current in hippocampal neurons. *Nature* **336**, 379–381, with permission.

In order to construct the activation–inactivation curves of I_D, the same protocol as that explained for I_A is applied (**Figure 14.9b, c**). They show that I_D is half-inactivated at -88 mV, and is inactivated when the membrane potential is maintained at a value more positive than -50 mV (at these membrane potentials I_D cannot be activated by a depolarization). Further, I_D is de-inactivated when the membrane potential is maintained at a value more negative than -60 mV (it can then be activated by depolarizations to membrane potentials more positive than -70 mV). Therefore, I_D contrasts with I_A in having a threshold for both activation and inactivation 10–20 mV more negative (compare **Figures 14.8c** and **14.9c**).

Pharmacology

I_D is much more sensitive to 4-aminopyridine than I_A, the latter requiring 1–3 mM for a block and the former 30–40 μM. I_D is insensitive to TEA and Cs^+ ions. Thus there are two ways of blocking I_D, either by depolarizing the membrane above -70 mV or by applying 40 μM of 4-aminopyridine in the extracellular medium. To study I_D, Na^+ and Ca^{2+} channels must be blocked by TTX (1 μM), Cd^{2+} or Co^{2+} or Ni^{2+} (200 μM to 1 mM).

In summary, I_D has characteristics which distinguish it from the other K^+ currents. It inactivates more slowly (in the order of seconds) than the transient current I_A and activates more rapidly and at more negative potentials than the currents of the delayed rectification I_{KDR}. It has also different pharmacological properties.

14.3.3 The K^+ currents activated by intracellular Ca^{2+} ions, I_{KCa}

I_{KCa} currents are outward K^+ currents sensitive to the intracellular concentration of Ca^{2+} ($[Ca^{2+}]_i$). In vertebrate neurons, these currents may be more or less sensitive to voltage, but for all of them an increase of the intracellular Ca^{2+} concentration is a necessary prerequisite to their activation. $[Ca^{2+}]_i$ increase may be the result of Ca^{2+} entry through voltage-dependent Ca^{2+} channels opened by depolarization, or Ca^{2+} entry through cationic receptor-channels largely permeable to Ca^{2+} ions such as the NMDA-type glutamate receptors, or Ca^{2+} release from intracellular stores. I_{KCa} have been explained in Chapter 5.

14.3.4 The K^+ current sensitive to muscarine, I_M

I_M is a depolarization-activated K^+ current originally described in frog sympathetic neurons by Brown and Adams in 1980, and studied since in a variety of other vertebrate neurons. I_M was so-called because it is inhibited by muscarinic acetylcholine receptor agonists such as the alkaloid muscarine. It is therefore under the control of muscarinic cholinergic receptors.

Structure of the main channel subunit

Potassium M channels are composed of the subunits of the K_v7 (KCNQ) gene family, principally $K_v7.2$ and $K_v7.3$. These subunits have a predicted structure similar to the Shaker family of K^+ channels with 6 transmembrane domains (S1–S6), a single pore (P)-loop that forms the selectivity filter of the pore, a positively charged fourth transmembrane domain (S4) that acts as a voltage sensor, but they possess a C-terminus that is longer than that of most of the other voltage-gated K^+ channels (**Figure 14.10a**). M current is decreased by stimulating $G_{q/11}$-coupled receptors, for example M_1 muscarinic acetylcholine receptors. The activated G-protein does not appear to gate the M channels directly but instead induces their closure by an indirect mechanism.

Gating properties and ionic nature

I_M can be recorded in response to two types of voltage clamp protocols (in the presence of TTX). In the first protocol the membrane potential is stepped from -60 to -30 mV. At -60 mV, most of the M channels are closed. A step to -30 mV reveals, superimposed on the leak current (measured from the response to a symmetrical hyperpolarizing step to -90 mV), a slowly developing outward current due to the slow opening of M channels in response to membrane depolarization (**Figure 14.10b**).

The second type of protocol involves relaxation experiments (**Figure 14.10c**). The membrane is held at $V_H = -30$ mV, a potential at which M channels remain open and contribute a steady outward current. A negative step to -60 mV causes an instantaneous inward current (I_{leak}) followed by an inward current which slowly develops (I_M). When the membrane is repolarized to -30 mV, at the end of the voltage step, an instantaneous inward current (I'_{leak}) is recorded followed by a slow outward tail current (tail I_M).

The explanation of the recordings of **Figure 14.10c** is the following. When the membrane is stepped from 30 to -60 mV, all the M channels open at -30 mV do not close immediately or at the same time. There is an instantaneous diminution of the outward current (recorded as an instantaneous inward current, I_{leak}). It represents the current through channels open at $V_H = -30$ mV; i.e. M channels and 'leak channels'. It therefore depends on the number of channels open at -30 mV. Then there appears an exponential diminution of outward current or slow inward relaxation (I_M) which reflects the kinetic of M channels closure in

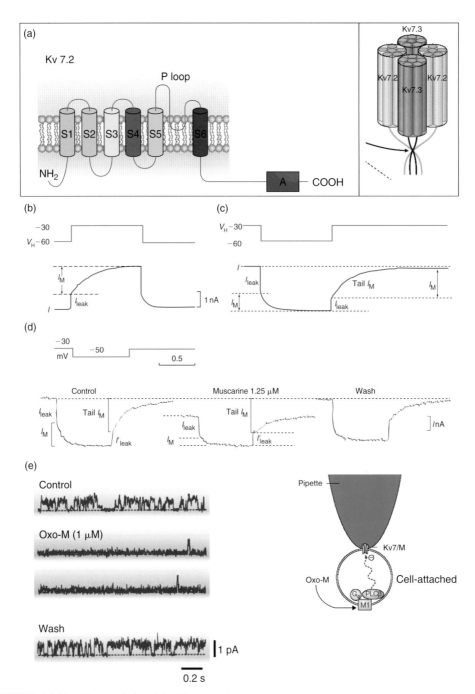

FIGURE 14.10 Characteristics of the M current, I_M.

(a) Membrane topology of Kv7/KCNQ channel subunits. They have a conventional Shaker-like K+ channel structure with a long intracellular carboxy-terminal tail. Four such subunits make up a functional Kv7 channel. All five (Kv7.1–Kv7.5) Kv7 channel subunits can form homomeric channels, whereas the formation of heteromers is restricted to certain combinations. (b–e) M currents are recorded from a sympathetic neuron in voltage clamp mode. (b, c) See text. (d) Effect of muscarine on I_M. The membrane is held at −30 mV and stepped to −50 mV. Muscarine evokes an inward current (the current trace is lower): it reduces the steady outward current through M channels open at −30 mV. In contrast, the baseline level attained at −60 mV, at the end of the command step, remains the same: muscarine does not produce an inward current at voltages where the M channels are normally shut. Inward and outward relaxations are largely depressed. (e) Shows remote signalling by muscarinic acetylcholine receptors. Bath application of the muscarinic agonist oxotremorine-methiodide (Oxo-M) strongly depresses unitary M currents in sympathetic neurons. Channel activity recovers fully after Oxo-M is washed out. Parts (a and e) from Delmas P, Brown DA (2005) Pathways modulating neural KCNQ/M (Kv7) potassium channels. *Nat. Rev. Neurosci.* **6**, 850–862. (b and c) adapted from Brown D, Adams PR (1980) Muscarinic suppression of a novel voltage sensitive K+ current in a vertebrate neuron. *Nature* **283**, 673–676; and Adams PR, Brown DA, Constanti A (1982) M currents and other potassium currents in bullfrog sympathetic neurones. *J. Physiol.* **330**, 537–572. Part (d) adapted from Adams PR, Brown DA, Constanti A (1982) Pharmacological inhibition of the M-current. *J. Physiol.* **332**, 223–262; all with permission.

response to the hyperpolarizing step to $-60\,mV$. When the membrane is stepped back to $-30\,mV$, the M channels do not open instantaneously. There is a first instantaneous outward current (I'_{leak}) which represents the current through channels open at $-60\,mV$ ('leak channels' only since M channels are closed). This instantaneous outward current is smaller than the fast inward one recorded from -30 to $-60\,mV$. It clearly indicates that M channels had closed in response to the preceding step from -30 to $-60\,mV$ (when the ohmic current is smaller in response to the same ΔV, it means that the membrane conductance is smaller). Then appears an outward tail current (tail I_M) through the M channels which slowly open again in response to the depolarization. The time course of the outward current evoked by the return to the steady holding potential can be distorted by the presence of other K^+ currents activated by this protocol (I_{KDR}, I_A, I_D and I_{KCa}), particularly when the membrane is stepped back to V_H from a potential more negative than $-70\,mV$. The M channels seem to be fully closed at membrane potentials more negative than $-60\,mV$, because in response to hyperpolarizing commands from $-60\,mV$ ($V_H = -60\,mV$) only ohmic (passive) current is recorded (not shown). With increasing step commands the inward relaxation reverses in direction at step potentials between -70 and $-100\,mV$. This reversal potential shifts to a more positive value on raising external K^+ concentration. Thus I_M is largely a K^+ current.

Pharmacology

Muscarinic agonists decrease I_M as shown in **Figure 14.10d**. The consequent loss of the steady outward current under muscarine generates a steady inward current at $-30\,mV$ (difference between baseline control and muscarine) and a step to $-50\,mV$ now reveals mostly the leak current (I_{leak}) showing that most of M channels are already closed.

In single-channel patch clamp recordings from sympathetic neurons, stimulating muscarine receptors with the muscarinic agonist oxotremorine-methiodide (Oxo-M) reversibly closes M channels inside the patch. These findings indicate that inhibition of M-channel activity involves a molecule that is capable of diffusing into (or out of) the membrane region circumscribed by the patch electrode (**Figure 14.10e**).

In summary, I_M is activated by depolarizations to membrane potentials positive to $-60\,mV$. It activates slowly (within hundreds of milliseconds) and does not inactivate. I_M differs from the delayed rectifier current I_{KDR} involved in spike repolarization by having a 40 mV more negative activation threshold and slower kinetics. It differs from I_A and I_D transient K^+ currents because it

does not inactivate. It is turned off by agonists of metabotropic receptors coupled to $G_{q/11}$ proteins, for example M_1 muscarinic acetylcholine receptors.

14.3.5 The inward rectifier K^+ current, I_{KIR}

I_{KIR} is a K^+ current that was originally described in skeletal muscle fibres by Sir Bernard Katz (1949). It is a Ba^{2+}-sensitive K^+ current with an I/V relation showing rectification when the current is inward (at potentials more hyperpolarized than $-90\,mV$). I_{Kir} is modulated by G-proteins.

Structure of the main channel subunit

These channels form a new channel-gene superfamily: inwardly rectifying K^+ channels possess only two putative transmembrane segments, which correspond to transmembrane regions S5 and S6 of the voltage-gated K^+ channels with six transmembrane domains (such as K_v channels) (**Figure 14.11a**).

Gating properties and ionic nature

The activity of the neurons of the accumbens nucleus is recorded intracellularly. The resting potential is very negative, around $-85\,mV$. In single-electrode voltage clamp, when the current evoked by hyperpolarizing steps is recorded in the presence of varying concentrations of external K^+, a series of I/V curves is obtained (**Figure 14.11b**). These curves increase in steepness between -50 and $-120\,mV$. Thus the permeability to K^+ is high when the current is inward ($V - E_K$ is negative) and low when the current is outward ($V - E_K$ is positive). In other words, inwardly rectifying channels conduct more efficiently when the membrane is negative to E_K.

Increasing the external K^+ concentration increases the slope of the I/V curve (i.e. the conductance) and shifts the reversal potential to more positive values, thus showing that K^+ ions participate in the current responsible for the inward rectification. Rectification in native channels is due in part to voltage-dependent block by cytoplasmic Mg^{2+} ions and polyamines.

Pharmacology

Ba^{2+} ($30–100\,\mu M$) causes an inward current at the resting potential (due to the blockade of the outward I_{Kir} present at rest) and decreases I_{Kir} at all potentials tested. Ba^{2+} thus linearizes the I/V curves (**Figure 14.11c**).

In summary, I_{Kir} is activated at resting membrane potential and thus maintains potential close to E_K. Owing to rectification, I_{Kir} has a low amplitude in the outward direction. It is modulated by G-proteins via the activation of synaptic metabotropic receptors.

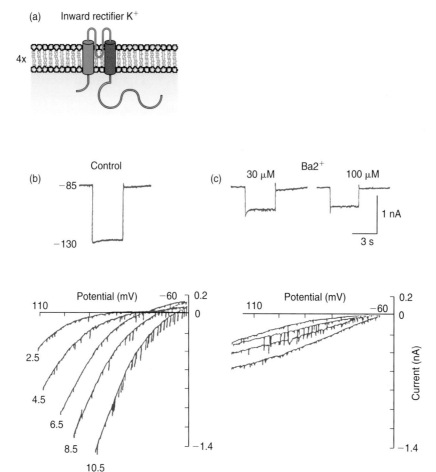

FIGURE 14.11   Characteristics of the inward rectifier current, I_{Kir}.
(a) Membrane topology of a single subunit of an inward rectifier K$^+$ channel. **(b–c)** Currents are recorded from neurons of the nucleus accumbens, in single-electrode voltage clamp mode. Inward current recorded in response to a hyperpolarizing step to -130 mV from a holding potential of -85 mV in control external K$^+$ concentration (2.5 mM) (b, top traces). I/V relations in five different external K$^+$ concentrations (b, bottom traces). The amplitude of the current has been measured at the end of 3–5 s steps, at steady state. Same current as in (b) recorded in the presence of 30 and 100 μM of Ba^{2+} (c, top traces). I/V relations as in (b) but in the presence of 30 μM of Ba^{2+} (c, bottom traces). Part (a) adapted from Swartz KJ (2004) Towards a structural view of gating in potassium channels. *Nature Rev. Neurosci.* **5**, 905–916. Parts (b–c) adapted from Uchimura N, Cherubini E, North A (1989) Inward rectification in rat nucleus accumbens. *J. Neurophysiol.* **62**, 1280–1286, with permission.

14.4 CONCLUSIONS

Comparison between inward and outward voltage-gated currents is shown in **Figure 14.12**. Subliminal currents are on the left and supraliminal currents on the right. I_{CaT} has voltage properties similar to those of I_A and I_D but with opposite functions. I_h and I_M have symmetrical voltage properties and opposite functions.

Depending on their location, subliminal currents are activated by different signals. When located in dendrites they are activated by a depolarization (EPSP) or a hyperpolarization (IPSP) of synaptic origin. When located in the soma–initial segment membrane, they are activated by the first action potential generated or

by the hyperpolarization that follows an action potential (after spike hyperpolarization).

Depending on their location, subliminal voltage-gated currents also have different roles. When present in the dendritic membrane they may boost or counter-act EPSPs or IPSPs (see Chapter 15), but when present in the soma–initial segment membrane they underlie intrinsic firing patterns, modulate synaptically driven firing patterns or participate in network oscillations (see Chapters 17, 19 and 20). When present in the whole neuronal membrane, subliminal currents that are activated around rest and that do not rapidly inactive (I_h, I_M, I_{KIR}) also determine resting membrane potential (see Chapter 3).

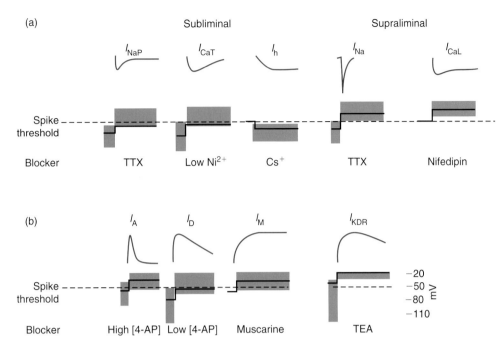

FIGURE 14.12 Comparison between subliminal and supraliminal voltage-gated currents.
(a) Inward and **(b)** outward voltage-gated currents. Subliminal currents are on the left and high-threshold-activated currents are on the right. Currents (top traces) are shown in response to voltage steps (solid bottom traces). The voltage ranges of activation and inactivation are shown in shaded green. The effective blocking agents are indicated below. Part (b) adapted from Storm JF (1988) Temporal integration of a slowly inactivating K$^+$ current in hippocampal neurons. *Nature* **336**, 379–381, with permission.

FURTHER READING

Araki T, Ito M, Oshima T (1962) Potential changes produced by application of current steps in motoneurones. *Nature* **191**, 1104–1105.

Baldwin TJ, Tsaur ML, Lopez GA *et al.* (1991) Characterization of a mammalian cDNA for an inactivating voltage-sensitive K$^+$ channel. *Neuron* **7**, 471–483.

Birnbaum SG, Varga AW, Yuan LL *et al.* (2004) Structure and function of Kv4-family transient potassium channels. *Physiol. Rev.* **84**, 803–833.

Carbone E and Lux HD (1984) A low voltage-activated, fully inactivating Ca^{2+} channel in vertebrate sensory neurons. *Nature* **310**, 501–502.

Connor JA and Stevens CF (1971) Prediction of repetitive firing behaviour from voltage-clamp data on an isolated neurone soma. *J. Physiol.* **213**, 31–53.

Crill WE (1996) Persistent sodium current in mammalian central neurons. *Ann. Rev. Physiol.* **58**, 349–362.

Hotson JR, Prince DA, Schwartzkroin PA (1979) Anomalous inward rectification in hippocampal neurons. *J. Neurophysiol.* **42**, 889–895.

Huguenard JR (1996) Low-threshold calcium currents in central nervous system neurons. *Ann. Rev. Physiol.* **58**, 329–348.

Kay AR, Sugimori M, Llinas R (1998) Kinetic and stochastic nature of a persistent sodium current in mature guinea pig cerebellar Purkinje cells. *J. Neurophysiol.* **80**, 1167–1179.

MacKinnon R (2003) Potassium channels. *FEBS Letters* **555**, 62–65.

Matsuda H, Saigusa A, Irisawa H (1987) Ohmic conductance through the inwardly rectifying K$^+$ channel and blocking by internal Mg^{2+}. *Nature* **325**, 156–158.

Miller AG and Aldrich RW (1996) Conversion of a delayed rectifier K$^+$ channel by three amino acids substitution. *Neuron* **16**, 853–858.

Pape HC (1996) Queer current and pacemaker: the hyperpolarization-activated cation current in neurons. *Ann. Rev. Physiol.* **58**, 299–327.

Perez-Reyes E, Cribbs LL, Daud A *et al.* (1998) Molecular characterization of a neuronal low-voltage-activated T-type calcium channel. *Nature* **391**, 896–900.

Song WJ (2002) Genes responsible for native depolarization-activated K$^+$ currents in neurons. *Neurosci. Res.* **42**, 7–14.

Standen NB and Stanfield PR (1978) A potential- and time-dependent blockade of inward rectification in frog skeletal muscle fibres by barium and strontium ions. *J. Physiol. (Lond.)* **280**, 169–191.

15

Somato-dendritic processing of postsynaptic potentials II. Role of subliminal depolarizing voltage-gated currents

Chapter 13 showed that in a dendritic tree in which currents propagate passively, EPSPs are attenuated in amplitude and slowed in time course as they spread to the soma due to passive properties of the dendrites. In summary, the influence of an EPSP on neuronal output (firing) that depends on its ability to depolarize the axon, depends upon the initial size and shape of the synaptic response, as well as how the cable properties of the dendritic tree filter the response as it spreads from the synapse to the site of action potential generation. As a consequence, excitatory or inhibitory synapses located on distal dendrites should be less efficient in depolarizing or hyperpolarizing the soma–initial segment region threshold. In other words, the combination of the large variation in synaptic distance and the cable-filtering properties of dendrites should, in theory, cause the amplitude and temporal characteristics of functionally similar inputs to be highly variable at the final integration site.

Although theoretical analyses have predicted such a clear location-dependent variability of synaptic input, there is now considerable evidence to indicate that the shape of EPSPs may be relatively independent of synapse location (e.g. in pyramidal neurons of the CA1 region of the hippocampus, **Figure 15.1**). The ability to simultaneously record synaptic activity from several locations on the same neuron (distal or proximal dendrites, soma) and the advent of imaging techniques with high spatial and temporal resolution now give the opportunity to understand the real dendritic processing of synaptic information. The hypothesis was then formulated: if some dendrites do not behave as simple

cables, is it because their membrane express voltage-gated channels? Experiments were thus designed to answer the following questions:

- Are subliminal voltage-gated currents present in dendritic membranes?
- Are they distributed uniformly over soma–dendritic membranes, or are the electrogenic properties in dendrites fundamentally different from those in the soma?
- How do these dendritic or somatic currents shape excitatory postsynaptic potentials and affect input summation?

To address these questions, the models used are brain regions organized in layers such as the neocortex or the hippocampus. In such regions, dendritic recordings are much easier than in other structures since the dendritic layer is easily recognizable from the somatic layer. To answer the first two questions, the activity of subliminal voltage-gated channels must be recorded in patches of dendritic membrane. Then, the amplitude of subliminal currents recorded from similar-sized patches of dendritic and somatic membranes must be compared. Experiments are thus performed in brain slices to allow outside-out, dendrite-attached or soma-attached recordings. For technical reasons, dendritic recording is limited to dendritic branches with a diameter greater than 1 μm. To answer the third question the EPSPs evoked at different locations of the dendritic tree are simultaneously recorded in whole-cell somatic and dendritic patch clamp configuration combined or not to imaging techniques.

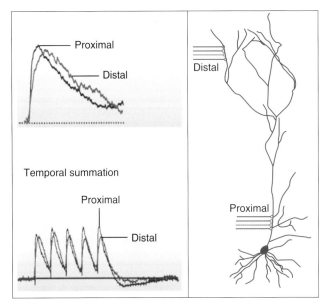

FIGURE 15.1 Synaptic integration in CA1 pyramidal neurons is independent of location.
Upper traces: The average unitary excitatory postsynaptic potentials (EPSPs) recorded at the soma of a neuron receiving distal and proximal input. Somatic EPSP amplitude is similar in spite of location differences. Lower traces: The amount of temporal summation at the soma is the same for a 50 Hz train of stimuli applied to distal (~300 μm) or to proximal Schaffer collateral inputs (~50 μm). (b) A CA1 pyramidal neuron showing the location of proximal and distal synaptic inputs across the dendritic arbor. Adapted from Magee JC (2000) Dendritic integration of excitatory synaptic input. *Nature Neuroscience Reviews* **1**, 181–190, with permission.

15.1 PERSISTENT Na⁺ CHANNELS ARE PRESENT IN SOMA AND DENDRITES OF NEOCORTICAL NEURONS; I_{NaP} BOOSTS EPSPs IN AMPLITUDE AND DURATION

The persistent Na⁺ current I_{NaP} is a TTX-sensitive Na⁺ current that activates below spike threshold and slowly inactivates (see Section 14.2.1). To record Na⁺ current in isolation, the solution bathing the extracellular face of the patch contains NaCl. TEACl is added in the external solution or CsCl in the internal solution to strongly reduce outward K⁺ currents. Voltage-gated Ca²⁺ currents (and therefore Ca²⁺-activated currents) are blocked by substitution of Mn²⁺ for Ca²⁺ in the extracellular medium. Persistent Na⁺ current is identified by inward polarity, its low threshold of activation (10–15 mV negative to I_{Na}), and its blockade by TTX. It can be easily differentiated from the Na⁺ current of action potentials (I_{Na}) as the latter activates at a higher threshold, has a far greater amplitude and inactivates much faster.

15.1.1 Persistent Na⁺ channels are present in the dendrites and soma of pyramidal neurons of the neocortex

Dendritic recordings

Single Na⁺ channel activity is recorded in patch clamp from dendrites of acutely isolated pyramidal neurons of the neocortex (dendrite-attached recordings). Depolarizing voltage steps to −60/+10 mV are applied from a holding potential of −100 mV. They evoke Na⁺ channel openings in all the recorded patches (**Figure 15.2a**). The most prominent activity in multichannel patches consists of early, short-lived openings clustered within the first few milliseconds that correspond to I_{Na}. In addition, a different Na⁺ channel activity consisting of prolonged or late openings is recorded (I_{NaP}). When many consecutive traces are averaged, this persistent channel activity is able to produce sizable net inward current even for 500 ms. The I/V relationship of the persistent component (**Figure 15.2b**) is not different from that of the macroscopic current recorded in the same cells.

Somatic recordings

Pyramidal neurons of the neocortex are loaded with the Na⁺-sensitive, membrane-impermeant fluorescent dye SBFI (sodium benzofuram isophthalate). A slow depolarizing ramp is applied through the somatic intracellular electrode to a final depolarization (around −50 mV) that is known to activate I_{NaP} and is subthreshold for action potential initiation. **Figure 15.3a** (top trace, arrow) shows the sub-threshold inward current evoked by the depolarizing ramp (middle trace), that totally disappears in the presence of TTX (**Figure 15.3b** top trace). During activation, a TTX-sensitive increase of intracellular Na⁺ concentration is observed in the soma (**Figure 15.3a, b**, bottom traces) as well as in the proximal part of the apical dendrite (not shown). This strongly suggests that I_{NaP} channels are present in the somatic membrane. However, this experiment does not allow one to localize the exact site(s) of I_{NaP} generation since the rise of Na⁺ in the proximal dendrite can result from Na⁺ diffusion from the soma, and vice versa.

15.1.2 Are dendritic persistent Na⁺ channels activated by EPSPs? Where does I_{NaP} boost EPSP amplitude, in the dendrites, in the soma?

Activation of dendritic I_{NaP} by local synaptic inputs is tested by simultaneous whole-cell dendritic and somatic recordings (in current clamp mode) made from the same pyramidal neuron of the neocortex. Dendritic EPSPs can be evoked either by (i) stimulation of afferents

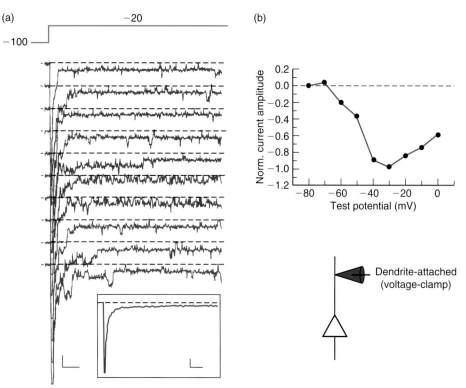

FIGURE 15.2   Persistent Na⁺ channel activity in dendrites of cortical pyramidal neurons.
(a) Na⁺ channel currents evoked by a 50 ms depolarizing pulse. The current traces shown are consecutive sweeps (scale bar 2 pA and 5 ms). *Insets*: Ensemble average current obtained from 20 consecutive sweeps (scale bar 2.5 pA and 5 ms). **(b)** Voltage-dependence of the persistent component of ensemble average currents obtained as in (a). The plot is normalized to the absolute value of its peak amplitude. Adapted from Magistretti J, Ragsdale DS, Alonso A (1999) Direct demonstration of persistent Na⁺ channel activity in dendritic processes of mammalian cortical neurones. *J. Physiol. (Lond.)* **521**, 629–636, with permission.

or by (ii) intradendritic current injection (simulated EPSP). These EPSPs must be sub-threshold (they must not trigger Na⁺ action potentials), in order to only evoke the low-threshold, TTX-sensitive, persistent Na⁺ current (I_{NaP}) and not the voltage-gated Na⁺ current of action potentials. The role of I_{NaP} on these EPSPs is then deduced by studying the effect of TTX on the amplitude and duration of EPSPs. However, since TTX affects postsynaptic Na⁺ channels as well as presynaptic ones, it affects synaptic transmission. Such a study therefore needs to bypass synaptic transmission by using only simulated EPSPs (protocol (ii)). Voltage change during an EPSP is simulated by dendritic current injection with a time course similar to that of an excitatory postsynaptic current (EPSC). With this aim, EPSCs are first recorded and then simulated. Simulated EPSPs generated by dendritic current injections are recorded both at their site of generation (dendritic site) and in the soma. Bath and local applications of TTX are used to determine whether I_{NaP} is involved in the amplification of simulated EPSPs and to localize where EPSPs are amplified, locally in dendrites or in the soma region.

When the amplitude of EPSPs recorded at the soma is greater than 5 mV, bath application of TTX causes a substantial 29 ± 2% reduction in the peak amplitude and a 53 ± 5% reduction of the EPSPs surface (when V_{rest} = −65 mV) (**Figure 15.4a**). At which site does a synaptic signal experience amplification while it travels from the dendrites to the axon initial segment? The site of EPSP amplification is tested by local application of TTX to either the site of simulated EPSP generation in the dendrites or to the somatic region. Local application of TTX is achieved by pressure ejection of TTX from a patch pipette, the tip of which is placed close to either the dendritic recording site or the soma. To minimize the spread of TTX, a low concentration of TTX (100 nM) is used. That this TTX application does in fact block dendritic Na⁺ channels is verified by its ability to reduce the amplitude of backpropagated Na⁺ action potentials (see Chapter 16). I_{NaP} seems to be mostly located in the soma as shown in **Figure 15.4b**: local application of TTX to the dendritic recording site has little or no effect on the simulated EPSP, but when applied to the soma TTX reduced the somatic EPSP peak amplitude and integral.

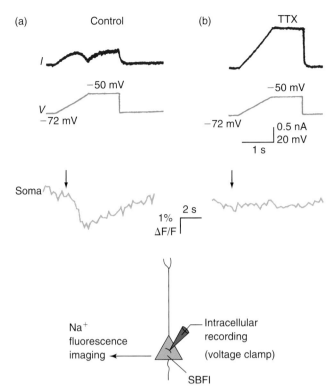

FIGURE 15.3 *I_{NaP} activation and the resultant SBFI fluorescence changes in the soma of a pyramidal neuron of the neocortex.* **(a)** Intracellular recording (voltage clamp mode) of the current evoked by a depolarizing ramp from −72 mV to a 1 s constant step at −50 mV (top traces). The corresponding decrease of SBFI fluorescence in the soma is shown in the bottom trace. The arrow indicates time of voltage clamp. A decrease of SBFI fluorescence reflects an increase of intracellular Na$^+$ concentration. **(b)** The same experiment in the presence of 1 μM of TTX in the bath. Adapted from Mittman T, Linton SM, Schwindt P, Crill W (1997) Evidence of persistent Na$^+$ current in apical dendrites of rat neocortical neurons from imaging of Na$^+$-sensitive dye. *J. Neurophysiol.* **78**, 1188–1192, with permission.

Therefore, in the neocortex, EPSPs activate I_{NaP} and in turn, I_{NaP} boosts EPSPs in amplitude and duration. This amplification mainly occurs in the somatic region. However, assuming a resting membrane potential of −65/−70 mV, to activate I_{NaP}, EPSPs must depolarize the membrane by 5–15 mV. Only summed EPSPs can reach this amplitude.

15.2 T-TYPE Ca^{2+} CHANNELS ARE PRESENT IN DENDRITES OF NEOCORTICAL NEURONS; I_{CaT} BOOSTS EPSPs IN AMPLITUDE AND DURATION

The T-type Ca^{2+} current is an amiloride- and Ni^{2+}-sensitive Ca^{2+} current which activates below spike threshold, inactivates rapidly with time and is totally inactivated at −40 mV (see Section 14.2.2). In order to

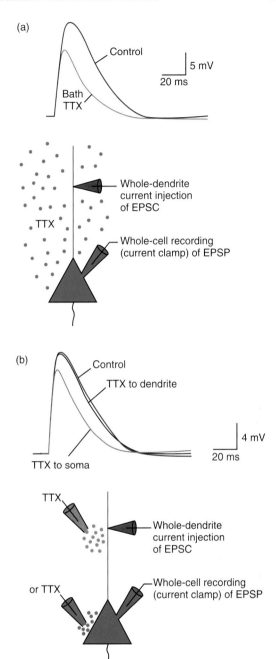

FIGURE 15.4 **Effect of bath or local applications of TTX on the amplitude and duration of simulated dendritic EPSPs (pyramidal neurons of the neocortex).**
EPSPs are generated (simulated EPSPs) by injection of an exponentially rising and falling voltage waveform into the current clamp input of the amplifier (dendritic current injections are performed 330 μm away from the soma). EPSPs are recorded from the soma (in whole-cell configuration and current clamp mode) after propagation in the dendritic tree. **(a)** Effect of bath application of TTX on the simulated EPSP recorded from the soma at resting membrane potential (−65 mV). **(b)** Effect of TTX locally applied near the dendritic or somatic recording sites, on simulated EPSPs recorded at resting membrane potential (−63 mV). Adapted from Stuart G and Sakmann B (1995) Amplification of EPSPs by axosomatic sodium channels in neocortical pyramidal neurons. *Neuron* **15**, 1065–1076, with permission.

record Ca^{2+} current in isolation, the solution bathing the extracellular face of the patch contains Ba^{2+} (110–120 mM) as the charge carrier and TEACl and TTX for blocking K^+ and Na^+ currents, respectively. T-type Ca^{2+} current is identified by inward polarity, unitary current amplitude, activation–inactivation characteristics, sensitivity to Ni^{2+} or amiloride, and insensitivity to dihydropyridines (L-type blocker), ω-conotoxins or funnel web toxin (N-type and P-type blockers).

Are the pharmacological tools used in all the above experiments sufficiently selective to allow the conclusion that the observed Ca^{2+} current is I_{CaT}? The answer is no. These experiments do not exclude some partial contribution of a dendritic R-type Ca^{2+} current (I_{CaR}) which is also sensitive to Ni^{2+} and amiloride. However, since I_{CaR} is a high-threshold-activated current – it activates at higher depolarized potentials than I_{CaT} – it has been considered in the following experiments that the Ca^{2+} current activated in dendrites by step depolarizations or by EPSPs is I_{CaT}.

15.2.1 T-type Ca^{2+} channels are present in dendrites of pyramidal neurons of the hippocampus

The activity of a patch of apical dendritic membrane is recorded in the dendrite-attached configuration (voltage clamp mode). In response to depolarizing steps to -15 mV from a hyperpolarized potential (-85 mV, to deinactivate T channels), channel openings are recorded. They occur mostly at the beginning of the depolarizing step, are of small unitary current amplitude (**Figure 15.5a**), and the i_T/V plot gives a unitary conductance γ_T of 7–11 pS (**Figure 15.5b**). They are sensitive to Ni^{2+} and amiloride (not shown). These data, together with the activation–inactivation characteristics (**Figure 15.5c**) reveal that T-type Ca^{2+} channels are present within the apical dendrite of pyramidal neurons. They are similar in basic characteristics to T-type Ca^{2+} channels recorded from many neuronal soma (compare with **Figure 14.5**).

15.2.2 Dendritic T-type Ca^{2+} channels are activated by EPSPs; in turn, I_{CaT} boosts EPSPs amplitude

Activation of dendritic I_{CaT} by local synaptic inputs is tested by simultaneous dendrite-attached and whole-cell somatic recordings from the same pyramidal neuron of the CA1 region. Sub-threshold EPSPs are evoked by Schaffer collateral stimulation. These EPSPs must be sub-threshold (they must not trigger Na^+ action potentials), in order to evoke only the low-threshold, Ni^{2+}-sensitive, Ca^{2+} current (I_{CaT}) and not

the high-voltage-activated Ca^{2+} currents such as the L-, N- or P/Q-type currents. EPSPs are recorded from the soma (in current clamp mode) after propagation in the dendritic tree. Single-channel T-type Ca^{2+} currents are recorded from the patch of dendritic membrane (in voltage clamp mode). If channel openings only occur during EPSPs, they are considered to have been triggered by it.

In response to Schaffer collateral stimulation, the activity of single Ca^{2+} channels is recorded (**Figure 15.6a**). These single-channel currents are not recorded when the Ca^{2+} channel blocker $CdCl_2$ (0.5 mM) is present in the pipette (not shown). Single-channel openings are most often observed near the peak or falling phases of the EPSPs. EPSP-activated channel openings display small unitary current amplitude and slope conductance ($\gamma = 9 \pm 1.6$ pS) characteristic of T-type dendritic channels (see **Figure 15.5**). EPSPs with a peak amplitude of 10 mV (at the somatic recording site) are necessary for activation of T-type dendritic Ca^{2+} channels. When a 4 s hyperpolarizing prepulse is applied 400 ms before synaptic stimulation, the open probability (p_o) of T-type Ca^{2+} channels is increased in a voltage-dependent manner (**Figure 15.6b**). This suggests that a large proportion of the T-type Ca^{2+} channel population is inactivated at resting potential. Therefore, membrane hyperpolarization (as during IPSPs), by allowing channel deinactivation, is necessary for maximal channel activation by EPSPs. Thus, the contribution of LVA Ca^{2+} channels to EPSP amplitude would be particularly enhanced for EPSPs occurring after hyperpolarizing IPSPs.

Another way to address the question of the activation of T-type Ca^{2+} current by EPSPs is to measure Ni^{2+}-sensitive intradendritic $[Ca^{2+}]$ increase during sub-threshold EPSPs. Whole-cell recordings in the soma are performed in conjunction with high-speed fluorescence imaging (FURA-2). To measure changes in intracellular Ca^{2+} concentration the fluorescent indicator FURA-2 is included in the pipette solution. Detectable increases in Ca^{2+} concentration are observed in response to as few as two consecutive synaptic stimulations (50 Hz) but a short train of five stimuli provides a very reproducible increase above baseline ($2.2 \pm 0.5\%$ $\Delta F/F$) (**Figure 15.7a**).

The rise in $[Ca^{2+}]_i$ continues throughout the course of the synaptic stimulation and begins to decay back to baseline several milliseconds after the end of the EPSP train. It thus appears that sub-threshold stimulations of sufficient amplitude result in a transient elevation of intradendritic $[Ca^{2+}]$. $[Ca^{2+}]_i$ transients are localized primarily to the area of the synaptic input (not shown). The localized nature of these $[Ca^{2+}]_i$ signals implies that the largest changes in $[Ca^{2+}]_i$ occur in the dendrites where the synaptic input is located and that this signal

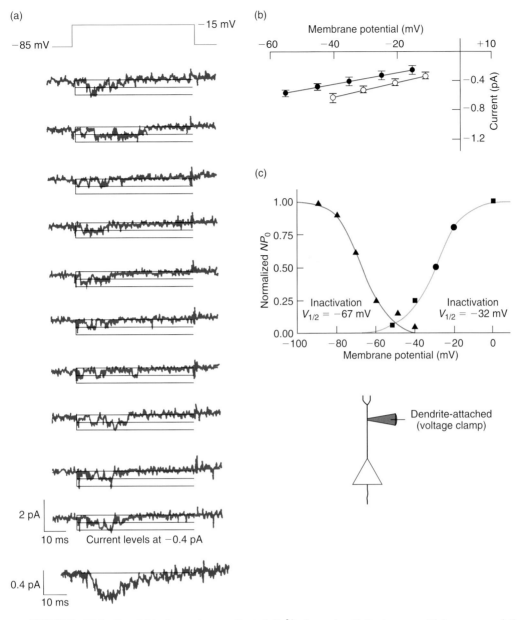

FIGURE 15.5 **Dendritic low-voltage-activated Ca²⁺ channel activity in pyramidal neurons of the hippocampus.**
(a) Consecutive sweeps of T-type Ca²⁺ channel activity recorded from a dendrite-attached patch (voltage clamp mode) in response to 60 ms depolarizing steps to -15 mV ($V_H = -85$ mV). Bottom trace is the ensemble average (104 sweeps) demonstrating significant inactivation during the 60 ms depolarizing step (110 mM of Ba²⁺ in the recording solution). **(b)** i_T/V plot of T-type Ca²⁺ channel activity. Unitary current amplitude is plotted as a function of membrane potential for patches recorded with either 20 mM (●) or 110 mM (○) Ba²⁺ as charge carrier. The slope (unitary conductance) γ_T is between 7 pS (20 mM of Ba²⁺) and 11 pS (110 mM of Ba²⁺). **(c)** Representative steady-state activation (■) and inactivation (▲) plots for dendritic LVA Ca²⁺ channels recorded in 20 mM of Ba²⁺. Adapted from Magee JC and Johnston D (1995) Characterization of single voltage-gated Na⁺ and Ca²⁺ channels in apical dendrites of rat CA1 pyramidal neurons. *J. Physiol. (Lond.)* **487**, 67–90, with permission.

attenuates as it approaches the soma. Through which types of dendritic Ca²⁺ channels are Ca²⁺ ions entering the cell; or does this [Ca²⁺]ᵢ increase result from the release of intradendritic Ca²⁺ stores? For the first

hypothesis, the candidates are Ca²⁺-permeable receptor channels (NMDA or AMPA receptors; see Sections 10.2 and 10.4) and voltage-gated ²⁺ channels. Application of APV (50 μM), a specific antagonist of NMDA channels,

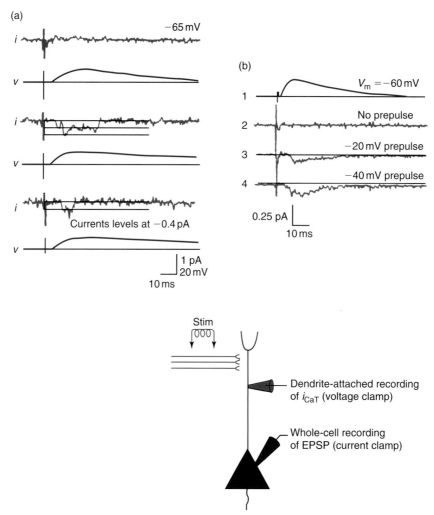

FIGURE 15.6 Synaptic activation of LVA Ca²⁺ channels in hippocampal CA1 pyramidal neurons.
Sub-threshold EPSPs are evoked by Schaffer collateral stimulation and are recorded from the soma (in current clamp mode) after propagation in the dendritic tree. **(a)** Consecutive sweeps of dendrite-attached recordings (voltage clamp mode) with the patch held at −65 mV showing Ca²⁺ channel activity recorded at the dendritic site (*i*, top traces) and of sub-threshold EPSPs (*v*, bottom traces) recorded at the somatic site (whole-cell configuration). **(b)** Hyperpolarizing prepulses (not shown) increase the activation of T-type Ca²⁺ channels by an EPSP. Ensemble average of 50 consecutive current traces without prepulse (2), ensemble average of 60 consecutive current traces after a 4 s prepulse of −20 mV (3), and ensemble average of 60 consecutive traces after a 4 s prepulse of −40 mV (4). The patch is returned to a holding potential that is 10 mV depolarized from resting potential 400 ms before synaptic stimulation in order to evoke an EPSP of similar amplitude (1) in all trials. Adapted from Magee JC and Johnston D (1995) Synaptic activation of voltage-gated channels in the dendrites of hippocampal pyramidal neurons. *Science* **268**, 301–304, with permission.

has very little effect on the sub-threshold Ca²⁺ signals as long as the EPSP amplitude is maintained constant. 50 μM of APV is therefore included in the bath solution for the remainder of the experiment. In contrast, membrane hyperpolarization to around −100 mV during synaptic stimulation prevents the synaptically-induced rise in intradendritic Ca²⁺ concentration, indicating that Ca²⁺ entry is voltage-dependent. All these data demonstrate that [Ca²⁺]ᵢ signals result from Ca²⁺ influx but not through NMDA or AMPA receptors

(antagonists of AMPA receptors cannot be tested since they would cancel the EPSP which is an AMPA-mediated EPSP). This influx is then likely to occur through voltage-gated ion channels. The primary candidate is the T-type Ca²⁺ channel. The effect of 50 μM of Ni²⁺ is therefore tested. When bath is applied, such a concentration of Ni²⁺ produces a 54 ± 5% block of the synaptically-induced influx of Ca²⁺ and this block is completely reversible with washout of Ni²⁺ (**Figure 15.7b**).

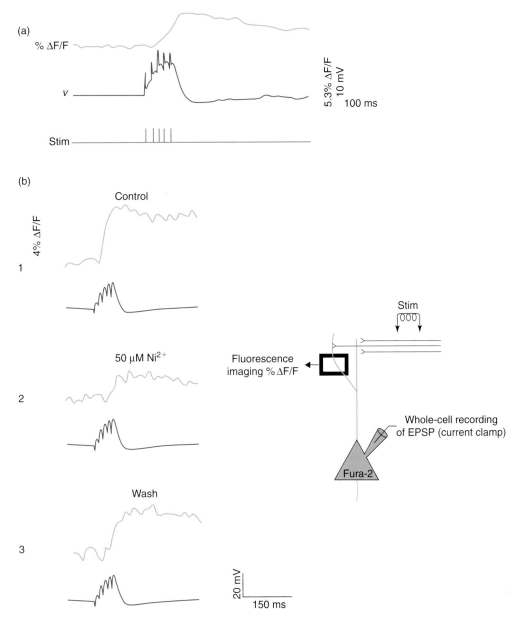

FIGURE 15.7 Sub-threshold EPSPs cause a localized, Ni²⁺-sensitive elevation of intradendritic Ca²⁺ concentration.
Sub-threshold EPSPs are evoked by stimulation of afferents close to the dendrite under study. **(a)** Time course of percentage change in FURA-2 fluorescence in a dendrite ($\%\Delta F/F$, top trace) evoked by a short train of five EPSPs and somatic voltage recordings (V) of the five EPSPs (whole-cell configuration, current clamp mode, bottom trace). The fluorescence trace is from the region delimited by the small black frame on the schematic representation of the FURA-2 loaded neuron. **(b)** Localized percent change in FURA-2 fluorescence ($\%\Delta F/F$, top trace) induced by a short train of five EPSPs and somatic voltage recordings of the five EPSPs (bottom traces) in the absence (1), presence (2) and 20 min after washing (3) of $50\,\mu$M of NiCl$_2$. The somatic recording of EPSPs (bottom traces) is unaffected by Ni²⁺ application. All traces in the figure are averages of five consecutive sweeps. Adapted from Magee JC, Christofi G, Miyakawa H *et al.* (1995) Subthreshold synaptic activation of voltage-gated Ca²⁺ channels mediates a localized Ca²⁺ influx into the dendrites of hippocampal pyramidal neurons. *J. Neurophysiol.* **74**, 1335–1342, with permission.

What are the consequences of this local intradendritic Ca²⁺ increase? Does dendritic I_{CaT} boost EPSPs? To address this question, EPSPs are evoked far out on the apical dendrite and their shape is recorded at the soma with the dendritic I_{CaT} active or partially suppressed by local pressure application of I_{CaT} blockers such as Ni²⁺ or amiloride. EPSPs are evoked by afferent fibre stimulation at a frequency of 0.2 Hz and are

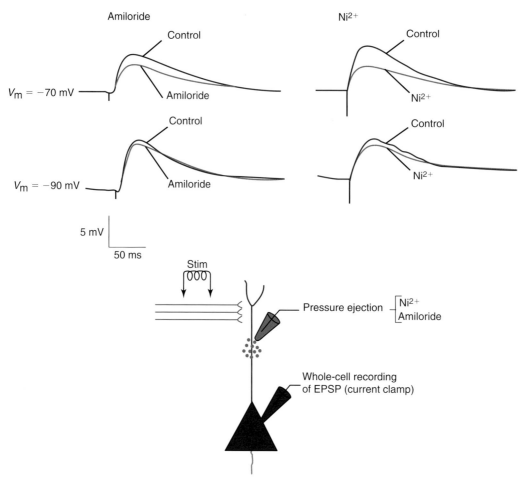

FIGURE 15.8   EPSPs in hippocampal pyramidal dendrites are amplified by an amiloride- and Ni²⁺-sensitive Ca²⁺ current.
EPSPs are evoked by stimulation of afferent fibres in the outer stratum radiatum. EPSPs are recorded from the soma (whole-cell configuration, current clamp mode) at two different membrane potentials (-70 and -90 mV) adjusted by current injection through the whole-cell pipette. Superimposed traces of averaged EPSPs ($n = 50$) recorded before and during local dendritic application of amiloride (50 µM, left) or Ni²⁺ (5 µM, right) show that both drugs reduce EPSPs recorded at -70 mV but do not significantly reduce them at -90 mV. Adapted from Gillessen T and Alzheimer C (1997) Amplification of EPSPs by low Ni²⁺ and amiloride-sensitive Ca²⁺ channels in apical dendrites of rat CA1 pyramidal neurons. *J. Neurophysiol.* **77**, 1639–1643, with permission.

recorded at the level of the soma (whole-cell configuration). To visualize the approximate spread of Ni²⁺ (5 µM) or amiloride (50 µM) in the tissue, both drugs are dissolved in 2% food colour solution. In control experiments, dendritic pressure application of food colour solution alone produces negligible reductions of EPSP amplitude. To study the role of dendritic I_{CaT} with minimum contamination by somatic I_{CaT}, the membrane potential is set to -70 mV at the soma and stimulation amplitude is adjusted to obtain EPSP peak amplitudes at the soma of 7 mV on average. Under these conditions somatic EPSP amplitude should be too small to activate LVA Ca²⁺ channels. Dendritic

amiloride application reduces EPSP amplitude by $27 \pm 2\%$ and Ni²⁺ application reduces it by $33 \pm 2.9\%$ (**Figure 15.8**, top traces). The effects of both amiloride and Ni²⁺ reverse within 15–20 min of drug washout. Hyperpolarization of the membrane to -90 mV attenuates the effect of both antagonists (**Figure 15.8**, bottom traces). However, any of the observed effects can be due to a presynaptic action of Ni²⁺ or amiloride.

In order to check this, EPSPs are recorded extracellularly near the dendrite, at the level of afferent stimulation in control conditions and in the presence of blockers. Bath application of both drugs, at a concentration 10 times higher than that achieved during focal application around the apical dendrite, fails to impair synaptic transmission. Thereby this excludes any presynaptic action of Ni^{2+} or amiloride. Therefore, in CA1 pyramidal neurons, EPSPs activate a T-type Ca^{2+} current that can indeed alter the weight of EPSPs. This amplification occurs in dendritic regions. However, as for persistent Na^+ channels, assuming a resting membrane potential of $-65/-70\,mV$, to activate I_{CaT}, EPSPs must depolarize the membrane by $5-10\,mV$. Only summed EPSPs can reach this amplitude.

15.3 THE HYPERPOLARIZATION-ACTIVATED CATIONIC CURRENT I_H IS PRESENT IN DENDRITES OF HIPPOCAMPAL PYRAMIDAL NEURONS; FOR EPSPs, DENDRITIC I_H DECREASES THE CURRENT TRANSMITTED FROM THE DENDRITES TO THE SOMA

The hyperpolarization-activated cation current I_h is a Cs^+-sensitive current turned on by hyperpolarization. It is inward at potentials more hyperpolarized than its reversal potential (around -50 to $-30\,mV$), and does not inactivate (see Section 14.2.3). Therefore, I_h is active at resting membrane potential and deactivates during membrane depolarization. To record I_h, the solution bathing the extracellular face of the membrane and the pipette solution contain control concentrations of Na^+, K^+ and Ca^{2+} ions.

15.3.1 H-type cationic channels are expressed in dendrites of pyramidal neurons of the hippocampus

The basic biophysical properties and the subcellular distribution of I_h are investigated in cell-attached configuration, voltage clamp mode. Long duration (1–3 s) hyperpolarizing steps evoke inward currents from cell-attached macropatches obtained from both the soma and apical dendritic regions but with different amplitudes (**Figure 15.9a**). These inward currents are slowly activating, non-inactivating and slowly deactivating (as seen on tail currents; see the inset of **Figure 15.9d** and Appendix 5.2). Currents begin to activate near $-60\,mV$ and steady-state current amplitude increases in an approximately linear manner with membrane hyperpolarization up to $-140\,mV$ (**Figure 15.9b**). Inclusion of 5 mM of Cs^+ in the external recording solution totally blocks the current, thus showing that it is an I_h current (**Figure 15.9c**).

The steady-state current amplitude at $-130\,mV$ progressively increases with distance away from the soma

(soma: $8.9 \pm 1.6\,pA$, $n = 21$; dendrite 300–350 μm away from the soma: $62.3 \pm 8.5\,pA$, $n = 14$) (the mean dendritic length is 500 μm). The mean current can be converted to mean current density (per μm^2) by normalizing to a 5 μm^2 patch area. It is $1.8 \pm 0.3\,pA\,\mu m^{-2}$ at the soma as compared with a density of $12.5 \pm 1.7\,pA\,\mu m^{-2}$ recorded from dendrites located 300–350 μm away from the soma. Therefore the density of I_h increases over 6-fold in 350 μm towards distal dendrites. Even with these elevated I_h densities, absolute I_h density is quite small compared with other dendritic channel densities, K^+ channels in particular. In pyramidal cells of the cortex, where recordings have been performed up to 800 μm from the soma this density increased more than 10 times (**Figure 15.10**).

In conclusion, the ionic selectivity (data not shown), voltage ranges of activation and kinetics of activation and inactivation as well as the sensitivity to external Cs^+ all fall within the ranges reported for a wide variety of central and peripheral I_h in neurons, as well as in cardiac cell types. Moreover, a 6- to 13-fold increase in I_h density is found across the somato-dendritic axis.

15.3.2 Dendritic H-type cationic channels are activated by IPSPs; in turn, Ih decreases EPSPs amplitude

The impact of I_h channels on the shape and propagation of sub-threshold voltage signals is determined by using simultaneous whole-cell current clamp recordings from both the soma and dendrites. EPSPs are simulated by dendritic current injection. Under control conditions, current injections in the dendritic compartment result in EPSP-shaped voltage transients, the amplitude and kinetics of which are filtered significantly as they propagate from the dendritic injection site to the recording somatic site (**Figure 15.11a**). When the amplitude of the simulated EPSP is $8.0 \pm 0.5\,mV$ at the dendritic recording site, it becomes $3.0 \pm 0.2\,mV$ at the somatic recording site. When the EPSP duration is $15 \pm 0.8\,ms$ at the dendritic site, it becomes $39 \pm 2\,ms$ at the somatic site. Repetitive dendritic current injections are also given to mimic repetitive synaptic inputs. These events are filtered similarly by dendritic arborizations (**Figure 15.11b**). When the peak amplitude is $24 \pm 3\,mV$ at the dendritic site, it becomes $8 \pm 1\,mV$ at the somatic site. When the duration is $136 \pm 1\,ms$ at the dendritic site, it becomes $154 \pm 2\,ms$ at the somatic site. H channel blockade with external Cs^+ increases single EPSP amplitude by $7 \pm 4\%$ at the dendritic site and by $10 \pm 2\%$ at the somatic site. It also increases EPSP duration by $10 \pm 2\%$ at the dendritic site and by $38 \pm 9\%$ at the somatic site (**Figure 15.11a**). For repetitive EPSPs, the presence of external

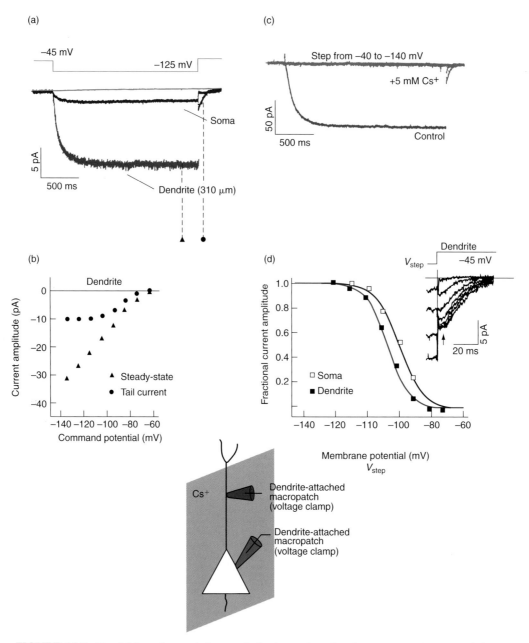

FIGURE 15.9 **Dendritic and somatic hyperpolarization-activated cation current I_h in pyramidal neurons of the hippocampus.**
(a) In a dendrite-attached macropatch located in the apical dendrite (310 μm from the soma), hyperpolarizing steps to −125 mV (VH = −45 mV) evoke inward currents that are larger than those recorded from the soma with similar-sized pipettes. **(b)** I/V plots for steady-state inward current measured 900 ms after the start of the step (▲) and for inward tail current measured 5 ms after the end of the step (●). **(c)** Blockade by 5 mM of external Cs⁺ of the inward current evoked by a hyperpolarizing step to −140 mV. **(d)** Activation curves generated from the tail currents (inset). The dendritic curve (V1/2 = −89 mV) is shifted 6 mV hyperpolarized with respect to the somatic curve (V1/2 = −83 mV). Command potentials (V_{step}) are given in 10 mV increments from −65 to −135 mV. Adapted from Magee JC (1998) Dendritic hyperpolarization-activated currents modify the integrative properties of hippocampal CA1 pyramidal neurons. *J. Neurosci.* **18**, 7613–7624, with permission.

Cs⁺ increases the amplitude by 22 ± 5% at the dendritic site and by 42 ± 8% at the somatic site. For the duration the increase is 3 ± 1% at the dendritic site and 7 ± 2% at the somatic site (**Figure 15.11b**). Therefore,

for single EPSPs, the amount of amplitude attenuation occurring between the dendrites and soma is the same in the presence or absence of I_h. In contrast, for repetitive EPSPs, I_h reduced by a factor of 2 the peak

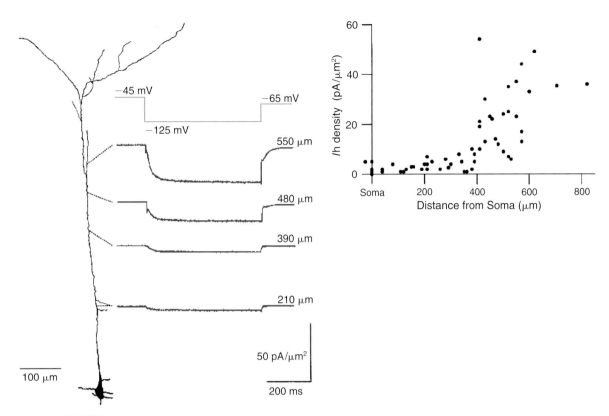

FIGURE 15.10 I_h is situated on the distal apical dendrite.
(a) Consecutive cell-attached patches along the apical dendrite of a layer V pyramidal neuron at different distances from the soma using a high-K$^+$ pipette solution. Hyperpolarizing voltage commands to approximately -125 mV resulted in the activation of tiny I_h currents at distances smaller than 400 μm from the soma, while at more distal recording sites a large I_h current flow could be induced. After the cell-attached recordings, the cell was filled with biocytin by going to whole-cell mode and the resting membrane potential measured. (b) The I_h current densities from 60 cell-attached recordings were plotted against their distance from the soma. While on the basal dendrites, the soma, and the apical dendrite <400 μm nearly no I_h currents could be found, more distal recordings <820 μm showed a marked nonlinear increase in I_h density. Adapted from Berger T, Larkum ME, Luscher HR (2001) High I_h channel density in the distal apical dendrite of layer V pyramidal cells increases bidirectional attenuation of EPSPs. *J. Neurophysiol.* **85**, 855–868, with permission.

amplitude reached during the train. The increase of amplitude of somatic EPSPs under conditions of I_h blockade is mostly the result of an elevation in effective input membrane resistance (due to I_h channels closure).

In conclusion, dendritic I_h decreases the amount of current transmitted from the dendrites to the soma in particular for summed EPSPs. Also, since I_h density is 6- to 13-fold more elevated in distal dendrites, the absolute effectiveness of distal synaptic inputs (i.e. the total charge transferred from synapse to soma) is reduced by the increasingly large I_h conductance. Notably in pyramidal cortical neurons of layer V, the high I_h channel density in the apical tuft increases the electrotonic distance between this distal dendritic compartment and the somatic compartment in comparison

to a passive dendrite. These findings suggest that integration of synaptic input to the apical tuft and the basal dendrites occurs spatially independently.

15.4 FUNCTIONAL CONSEQUENCES

15.4.1 Amplification of distal EPSPs by I_{NaP} and I_{CaT} counteracts their attenuation owing to passive propagation to the soma; it also favours temporal summation versus spatial summation

The subliminal voltage-gated I_{NaP} and I_{CaT} present in the dendrites of pyramidal neurons of the neocortex or the hippocampus boost the effects of local EPSPs

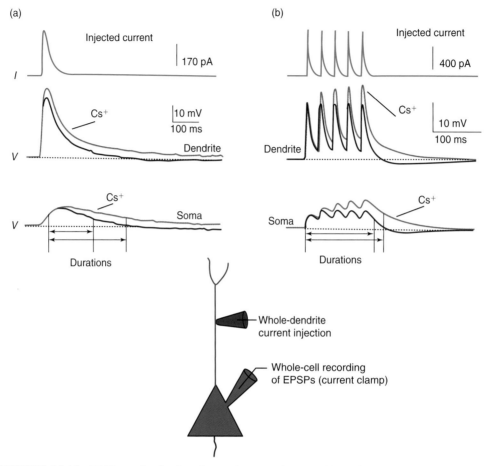

FIGURE 15.11 EPSP amplitude, duration and summation are all regulated by I_h in hippocampal pyramidal neurons.
EPSPs are generated by injection of an exponentially rising and falling voltage waveform into the current clamp input of the amplifier (dendritic current injections are performed 250 μm away from the soma). Simultaneous whole-cell recordings are performed from the dendrite and the soma of the same pyramidal neuron. (a) A single current injection into the dendritic electrode produces an EPSP-shaped transient, the amplitude and duration of which is increased in the presence of 3 mM of external Cs$^+$. (b) Repetitive current injections produce a train of EPSP-shaped voltage transients, the peak amplitude and duration of which are also increased in the presence of 3 mM of external Cs$^+$. Adapted from Magee JC (1998) Dendritic hyperpolarization-activated currents modify the integrative properties of hippocampal CA1 pyramidal neurons. *J. Neurosci.* **18**, 7613–7624, with permission.

by acting as either voltage or current amplifiers. This could be one solution to overcome the passive decay of EPSPs en route to the soma. This argument, however, is somewhat somatocentric; i.e. the emphasis is put on how dendrites amplify events so that they are bigger in the soma. An alternative viewpoint is that these channels are more important for dendritic interactions in the immediate vicinity of the synaptic inputs. For example, multiple EPSPs occurring on the same branch and within a narrow time should activate voltage-gated channels more strongly than a single EPSP and produce a much bigger response than would occur if EPSPs were on separate branches.

15.4.2 Activation of dendritic I_{CaT} generates a local dendritic [Ca^{2+}] transient

Under physiological conditions, an EPSP-evoked [Ca^{2+}]$_i$ transient would occur mainly after the summation of a number of unitary EPSPs and thus would represent the integrated result of dendritic activity at a given moment and at a given location. A number of possible physiological functions for dendritic [Ca^{2+}]$_i$ transients exists. Intracellular Ca^{2+} may activate biochemical pathways, Ca^{2+}-induced Ca^{2+} release as well as Ca^{2+}-activated K$^+$ currents present in the dendritic membrane. Such outward current would change the shape of the EPSP. Ca^{2+} is also implicated in postsynaptically induced forms of plasticity such as long-term potentiation or depression (see Chapter 18).

15.4.3 Activation of dendritic I_h, I_{CaT} and I_{NaP} alter the local membrane resistance and time constant

This in turn will influence both spatial and temporal summation of EPSPs. Moreover, I_h participates to resting potential, so that different densities of I_h from the soma could lead to different resting potentials in the dendrites. Moreover, I_h is deactivated by membrane depolarization as a result of EPSPs. This will produce an increase of membrane resistance. In contrast, activation of I_{NaP} and I_{CaT} by EPSPs will produce a decrease of membrane resistance.

15.5 CONCLUSIONS

We can now answer the three questions asked in the introduction:

- Subliminal depolarizing voltage-gated currents are present in the dendritic membrane of some CNS neurons. These are the currents I_h, I_{CaT} and I_{NaP}. The transient K^+ current I_A is also present in dendrites but this has not been studied here.
- They are not distributed uniformly over somato-dendritic membranes. In pyramidal neurons of the neocortex, I_{NaP} seems more efficient at the soma. In contrast, in pyramidal neurons of the hippocampus and of the layer V of the cortex, I_h density increases from soma to distal dendrites.
- I_{NaP} and I_{CaT} activation boost EPSP amplitude and duration. Moreover I_{CaT} activation induces a transient and local increase of intradendritic Ca^{2+} concentration. In contrast, it is I_h deactivation which generates an outward current that by producing an hyperpolarization shorten the duration of local EPSPs and attenuates their temporal summation.

It seems that although dendritic voltage-gated subliminar currents may have a limited impact on the amplitude of unitary synaptic input they actively shape repetitive synaptic potentials of larger amplitude. Finally, one must keep in mind that the state of voltage-gated channels, closed, open or inactivated, depends on the history of the membrane. If a segment of dendritic membrane has been previously depolarized before synaptic activity, the voltage-gated channels present in this segment of dendritic membrane will be already inactivated and will not play a role. There is therefore a dynamic aspect in the active properties of dendrites.

FURTHER READING

Gulledge AT, Kampa BM, Stuart GJ (2005) Synaptic integration in dendritic trees. *J Neurobiol.* **64**, 75–90.

Hausser M, Spruston N, Stuart GJ (2000) Diversity and dynamic of dendritic processing. *Science* **290**, 739–744.

Isomura Y, Fujiwara-Tsukamoto Y, Imanishi M *et al.* (2002) Distance-dependent Ni^{2+}-sensitivity of synaptic plasticity in apical dendrites of hippocampal CA1 pyramidal cells. *J. Neurophysiol.* **87**, 1169–1174.

Johnston D, Magee JC, Colbert CM, Christie BR (1996) Active properties of neuronal dendrites. *Annu. Rev. Neurosci.* **19**, 165–186.

Judkewitz B, Roth A, Hausser M (2006) Dendritic enlightment: using patterned two-photon uncaging to reveal the secrets of the brain's smallest dendrites. *Neuron* **50**, 180–183.

Lipowsky R, Gillessen T, Alzheimer C (1996) Dendritic Na^+ channels amplify EPSPs in hippocampal CA1 pyramidal cells. *J. Neurophysiol.* **76**, 2181–2190.

Magee JC (1999) Dendritic I_h normalizes temporal summation in hippocampal CA1 neurons. *Nature Neurosci.* **2**, 508–514.

Magee JC and Cook EP (2000) Somatic EPSP amplitude is independent of synapse location in hippocampal pyramidal neurons. *Nature Neurosci.* **3**, 895–903.

Markram H and Sakmann B (1994) Calcium transients in dendrites of neocortical neuron evoked by single subthreshold excitatory postsynaptic potentials via low-voltage-activated calcium channels. *Proc. Natl Acad. Sci. USA* **91**, 5207–5211.

Mouginot D, Bossu JL, Gähwiler BH (1997) Low-threshold Ca^{2+} currents in dendritic recordings from Purkinje cells in rat cerebellar slice cultures. *J. Neurosci.* **17**, 160–170.

Oviedo H and Reyes AD (2002) Boosting of neuronal firing evoked with asynchronous and synchronous inputs to the dendrite. *Nat Neurosci.* **5**, 261–266.

Schwindt PC and Crill WE (1995) Amplification of synaptic current by persistent sodium conductance in apical dendrite of neocortical neurons. *J. Neurosci.* **74**, 2220–2224.

16

Somato-dendritic processing of postsynaptic potentials III. Role of high-voltage-activated depolarizing currents

Dendrites of neurons of the mammalian central nervous system (CNS) have long been considered as electrically passive structures which funnel postsynaptic potentials to the soma and axon initial segment, the site of action potential initiation. However, the recording of dendritic action potentials (at first with intracellular electrodes) from dendrites of some neurons of mammalian central nervous system (**Figure 16.1**), indicated that these dendrites express high-threshold-activated Na$^+$ or Ca^{2+} channels. This led to the suggestion that, in these neurons, synaptic integration is not solely governed by (passive) cable properties of dendrites. In the previous chapter we studied the role of subliminal voltage-gated currents in the shaping of postsynaptic potentials. This chapter will examine the roles of high-voltage-activated currents.

Dendritic events have recently come under direct experimental scrutiny by the use of dendritic patch recordings and by the advent of imaging techniques with high spatial and temporal resolution. This permitted the design of experiments to answer the following questions:

- Are high-voltage-activated (HVA) currents present in the dendritic membrane of some CNS neurons?
- Are they distributed uniformly over the soma-dendritic membrane so that electrogenic properties in dendrites are not fundamentally different from that in soma?
- How do the currents affect synaptic potentials and input summation?

This chapter looks at experiments performed on four types of central neurons, on pyramidal neurons of the neocortex or hippocampus, dopaminergic neurons of substantia nigra pars compacta, and Purkinje cells of the cerebellar cortex.

16.1 HIGH-VOLTAGE-ACTIVATED Na$^+$ AND/OR Ca^{2+} CHANNELS ARE PRESENT IN THE DENDRITIC MEMBRANE OF SOME CNS NEURONS, BUT ARE THEY DISTRIBUTED WITH COMPARABLE DENSITIES IN SOMA AND DENDRITES?

One way to answer the above questions is first to identify the activity of HVA channels in patches of dendritic membrane and then to compare the amplitude of the HVA current recorded from similar-sized patches of dendritic and somatic membranes. Experiments are performed in brain slices and recordings are either in the outside-out or dendrite-attached configuration. For technical reasons this type of patch recording is limited to dendritic branches with a diameter greater than 1 μm. In order to record Na$^+$ channels, the solution bathing the extracellular face of the patch contains NaCl; and in order to strongly reduce outward K$^+$ currents, TEACl is added in the external solution or CsCl in the internal solution. Na$^+$ channel activity is identified by inward current polarity, voltage-dependent

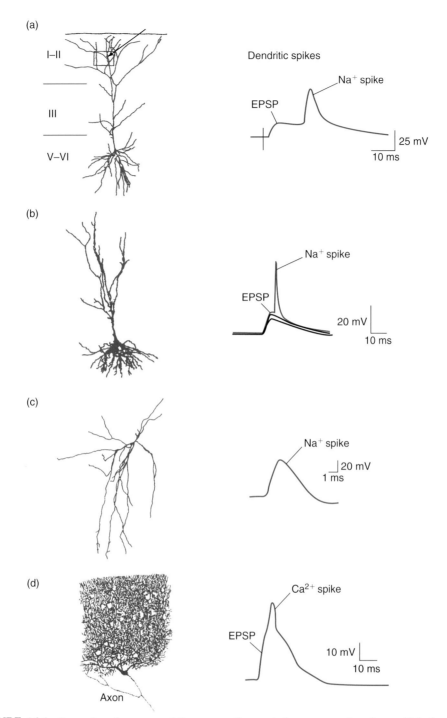

FIGURE 16.1 **Examples of neurons of the mammalian central nervous system from which dendritic spikes are recorded.**

Drawing of neurons (left) with the corresponding recording of dendritic spikes (right) evoked by afferent stimulation. **(a)** Pyramidal neuron of the neocortex. **(b)** Pyramidal neuron of the hippocampus. **(c)** Dopaminergic neurons of the substantia nigra. **(d)** Purkinje cell of the cerebellar cortex. Part (a) adapted from Seamans JK, Gorelova N, Yang CR (1997) Contribution of voltage-gated Ca^{2+} channels in the proximal versus distal dendrites to synaptic integration in prefrontal cortical neurons. *J. Neurosci.* **17**, 5936–5948, with permission. Part (b) drawing by Taras Pankevitch and Roustem Khazipov and adapted from Tsubokawa H and Ross WN (1996) IPSPs modulate spike backpropagation and associated $[Ca^{2+}]_i$ changes in dendrites of hippocampal CA1 pyramidal neurons. *J. Neurophysiol* **76**, 2896–2906, with permission. Part (c) drawing by Jérôme Yelnik and adapted from Häusser M, Stuart G, Racca C, Sakmann B (1995) Axonal initiation and active dendrite propagation of action potentials in substantia nigra neurons. *Neuron* **15**, 637–647, with permission. Part (d) adapted from Callaway JC, Lasser-Ross N, Ross WN (1995) IPSPs strongly inhibit climbing fiber-activated $[Ca^{2+}]_i$ increases in the dendrites of cerebellar Purkinje neurons. *J. Neurosci.* **15**, 2777–2787, with permission.

channel gating, unitary current amplitude, its blockade by TTX and the lack of effect of Cd^{2+}.

16.1.1 High-voltage-activated Na^+ channels are present in some dendrites

Pyramidal neurons of the hippocampus

In every dendrite-attached patch, Na^+ channel activity is consistently found and more than a single channel is always recorded (**Figure 16.2a**). Na^+ channels are opened by depolarizations of about 15 mV from rest. The i_{Na}/V relationship shows that Na^+ channels have a unitary conductance γ_{Na} of 15 pS and a unitary current that reverses at $E_{rev} = +54$ mV (**Figure 16.2b**). This value is close to the calculated Nernst equilibrium potential assuming an intracellular Na^+ concentration of 10 mM (extracellular concentration is 110 mM). These data, together with the activation–inactivation characteristics of dendritic I_{Na} (**Figure 16.2c**) reveal that HVA Na^+ channels present in the apical dendrites of pyramidal neurons have basic characteristics similar as HVA Na^+ channels recorded in many soma (compare with **Figure 4.8**) with a difference concerning the inactivation properties (slow inactivation and slow recovery from inactivation for dendritic channels). This explains the decrease of amplitude of repetitive dendritic action potentials.

Pyramidal neurons of the neocortex

Outside-out macropatches of dendritic membrane are excised at different distances from the soma, up to 500 μm. In response to step depolarizations applied through the recording electrode, an inward current that rapidly inactivates and totally disappears in the presence of TTX in the bath is recorded (**Figure 16.3a**). This is observed whether patches are excised from proximal or more distal dendrites. Moreover, this TTX-sensitive Na^+ current has a similar amplitude in patches taken from dendritic or somatic membranes, thus suggesting a similar somatic and dendritic density of Na^+ channels in both membranes.

Dopaminergic neurons of the pars compacta of the substantia nigra

Outside-out patches are excised from somatic or dendritic membranes. In an attempt to maintain constant patch membrane area, all recordings are made with pipettes of similar size. Multichannel TTX-sensitive sodium currents are recorded from both patches. The average peak Na^+ current in somatic patches is 5.0 ± 1.3 pA and that in dendritic patches is 3.6 ± 1.6 pA.

Again, in these neurons, there is a similar density of sodium channels in dendritic and somatic membranes.

Purkinje cells of the cerebellar cortex

The situation in these cells is fundamentally different. As suggested by the absence of large-amplitude Na^+ action potentials in dendrites of Purkinje cells, there is an extremely low-amplitude, TTX-sensitive Na^+ current in outside-out macropatches excised from dendrites (1.9 ± 0.4 pA) compared with that in patches excised from the soma (12.4 ± 1.5 pA) (**Figure 16.3b**). In fact, sodium-channel density steeply declines in dendrites with distance from the soma. These results were confirmed by experiments in Purkinje cells loaded with the Na^+ indicator SBFI (sodium benzofuran isophthalate). During Na^+ spikes, changes in the intracellular Na^+ concentration were detected only in soma and not in dendrites. In these cells, Na^+ channels are distributed non-uniformly over the somatic and dendritic membrane.

Conclusions

In pyramidal neurons of the neocortex or the hippocampus and in dopaminergic neurons of the substantia nigra there is a similar density of TTX-sensitive Na^+ channels in somatic and dendritic membranes up to several hundreds of μm from the soma. In contrast, there is a low density of Na^+ channels in the dendritic membrane of Purkinje cells. It must be pointed out that most CNS dendrites contain a low density of HVA Na^+ channels. Pyramidal neurons of the neocortex or the hippocampus and substantia nigra neurons are exceptions.

16.1.2 Dendritic Na^+ channels are opened by EPSPs and the resultant Na^+ current boosts EPSPs in amplitude and duration

Pyramidal neurons of the hippocampus

Activation of dendritic voltage-gated channels by local synaptic inputs is tested by simultaneous dendrite-attached and whole-cell somatic recordings from the same pyramidal neuron of the CA1 region (**Figure 16.4**). Sub-threshold EPSPs are evoked by Schaffer collateral stimulation (stim) and are recorded from the soma (in current clamp mode) after propagation in the dendritic tree. Single Na^+ channel currents are recorded from the patch of dendritic membrane (in voltage clamp mode). If channel openings occur only during EPSPs, they are considered to have been triggered by it.

When EPSPs of 15–20 mV amplitude and below action potential threshold are evoked, they consistently

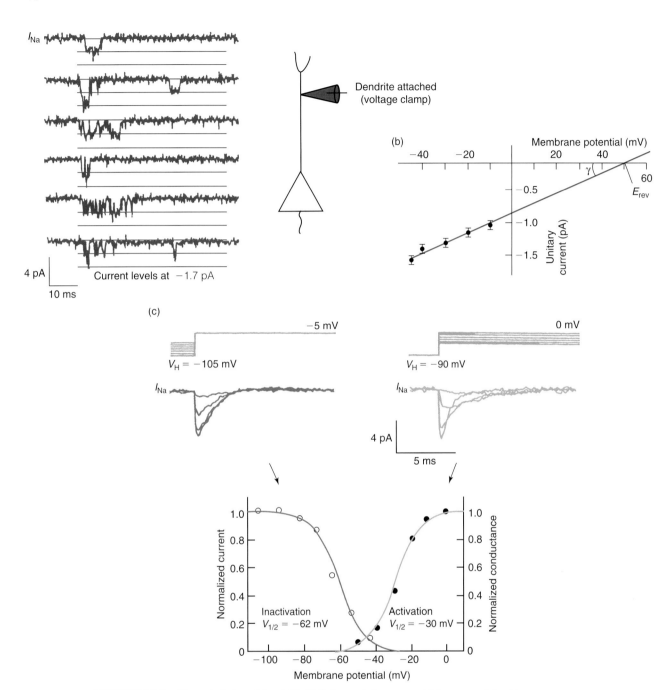

FIGURE 16.2 **Characteristics of dendritic Na⁺ channels in a pyramidal neuron of the hippocampus.**
(a) Consecutive sweeps showing Na⁺ channel openings (dendrite-attached configuration, voltage clamp mode) in response to step depolarizations to –40 mV (V_H = –70 mV). Most of the channel openings occur at the beginning of the step but there are some late reopenings. (b) Current–voltage plot of Na+ channel activity. Unitary current amplitude from a total of 27 patches is plotted as a function of membrane potential. Bars are standard error of the mean (SEM). The slope indicates a unitary conductance γ of 15 pS and the extrapolated reversal potential E_{rev} is +54 mV. (c) Dendritic Na⁺ channel steady-state activation and inactivation characteristics. Activation is tested by applying depolarizing steps to –65 to 0 mV from V_H = –90 mV. Inactivation is tested by applying a depolarizing step to –5 mV from a V_H varying from –105 to –45 mV. The representative steady-state activation (black circle) and inactivation (open circle) plots for dendritic Na⁺ channels indicate that they are half-activated at $V_{1/2}$ = –30 mV and half-inactivated at $V_{1/2}$ = –62 mV. Adapted from Magee JC and Johnston D (1995) Characterization of single voltage-gated Na⁺ and Ca²⁺ channels in apical dendrites of rat CA1 pyramidal neurons. *J. Physiol. (Lond.)* **487**, 67–90, with permission.

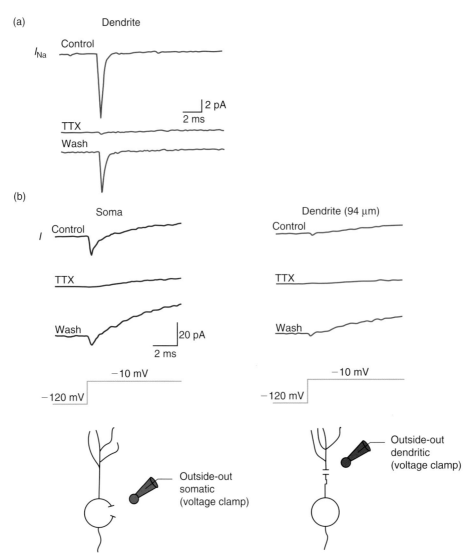

FIGURE 16.3   **TTX-sensitive inward currents in dendrites of pyramidal neurons of the neocortex and Purkinje cells.**
(a) Neocortex. Rapidly inactivating inward current evoked by a depolarizing step to –10 mV (V_H = –90 mV) in an outside-out dendritic macropatch excised from the apical dendrite of a layer V pyramidal neuron (439 μm from the soma) (control). This current is reversibly blocked in the presence of 500 nM of TTX in the external solution. **(b)** Purkinje cells. Voltage-activated currents evoked by a depolarizing step to –10 mV (V_H = –120 mV) in outside-out macropatches excised from either the soma (left) or dendrite (right, 94 μm from the soma) of Purkinje cells using similar-sized patch pipettes. A rapidly inactivating inward current followed by an outward current that is more prominent in the somatic membrane are recorded (control). Rapidly inactivating currents in both somatic and dendritic patches are reversibly blocked by the presence of 500 nM of TTX in the extracellular medium (TTX). Part (a) adapted from Stuart G and Sakmann B (1994) Active propagation of somatic action potentials into neocortical pyramidal cell dendrites. *Nature* **367**, 69–72, with permission. Part (b) adapted from Stuart GJ and Häusser M (1994) Initiation and spread of sodium action potentials in cerebellar Purkinje cells. *Neuron* **13**, 703–712, with permission.

activate Na$^+$ channels located within the CA1 dendrite-attached patch (**Figure 16.4a**). Most channel openings are near the peak of the EPSP, but occasional openings are encountered during either the rising or the falling phase of the EPSP. With TTX present in the pipette, there is no EPSP-associated channel activity in dendritic patches (not shown). Are these Na$^+$ channels different from that opened by step depolarizations as

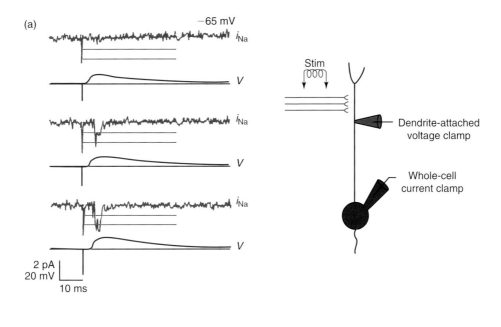

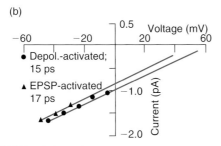

FIGURE 16.4 **Na⁺ channel openings evoked by sub-threshold EPSPs in dendrites of hippocampal pyramidal neurons.**

(a) Consecutive sweeps of unitary Na⁺ currents recorded from a dendrite-attached patch (i_{Na}) and simultaneous whole-cell somatic recordings of the EPSPs (V) evoked by Schaffer collateral stimulation (bottom traces). **(b)** i_{Na}/V plots of unitary Na⁺ current evoked in the same patch by depolarizing steps (depol-activated) or by EPSPs (EPSP-activated). Adapted from Magee JC and Johnston D (1995) Synaptic activation of voltage-gated channels in the dendrites of hippocampal pyramidal neurons. *Science* **268**, 301–304, with permission.

shown in **Figure 16.2**? Plots of unitary current amplitude versus approximate membrane potential (in cell-attached recordings, membrane potential can only be approximately evaluated) show that the slope unitary conductance γ (16 ± 1 pS) and reversal potential E_{rev} (+56 ± 1 mV) of EPSP-activated channel openings are similar to those calculated from step depolarization in the same patches (**Figures 16.2b** and **16.4b**). This suggests that Na⁺ channels opened by EPSPs are the same as those opened by intracellular depolarization.

Activation of dendritic HVA Na⁺ channels may elevate EPSP amplitude and prolong their duration. However, due to the high voltage of activation of Na⁺ channels, only summed EPSPs of 10–20 mV amplitude can activate them (assuming that V_{rest} is close to −70/−60 mV). Only in these conditions will dendritic

HVA Na⁺ channels increase both the strength and duration of excitatory synaptic inputs and thus enhance the efficacy of more distal and widely distributed synaptic contacts.

16.1.3 Dendritic Na⁺ channels are opened by backpropagating Na⁺ action potentials

The question concerning the role of dendritic Na⁺ channels in dendritic Na⁺ action potentials is the following: do dendritic Na⁺ channels allow the initiation of Na⁺ action potentials in dendrites in response to synaptic activity (**Figure 16.5a**), or do dendritic Na⁺ channels only allow active backpropagation of Na⁺

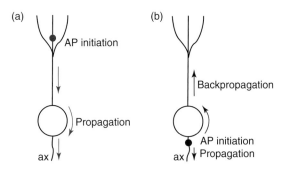

FIGURE 16.5   Schematic drawings of two hypotheses concerning the site of dendritic Na⁺ spike initiation.
(a) In response to dendritic EPSPs, Na⁺ action potential (AP) is locally initiated in dendrites (black point) and then actively propagates to the soma-initial segment and along the axon. (b) In response to dendritic EPSPs, Na⁺ action potential is first initiated at the soma-initial segment (black point) and then actively propagates along the axon and backpropagates (actively or passively) into the dendritic tree.

action potentials first initiated in the axon hillock region (**Figure 16.5b**)? To further explain the latter hypothesis it must be assumed that, in general, Na⁺ action potentials, once they have been initiated, actively (i.e. in a regenerative manner) propagate along the axon (orthodromic propagation; see Section 4.4) and at the same time passively propagate into the soma and dendritic tree (passive backpropagation). Therefore the question is not 'do Na⁺ action potentials backpropagate in the dendritic tree?' but rather 'do they backpropagate *actively* in the dendritic tree of neurons that contain HVA Na⁺ channels in their dendrites?'

Pyramidal neurons of the neocortex

Simultaneous whole-cell recordings (current clamp mode) are made from the soma and apical dendrite of the same pyramidal neuron in slices *in vitro*. To confirm that the recorded dendrite and soma belong to the same neuron, the cell is simultaneously filled from the somatic and the dendritic recording sites with different coloured fluorescent dyes present in the recording pipettes. In response to suprathreshold synaptic stimulation, action potential initiation occurs first at or near the soma (**Figure 16.6a**). Simultaneous recordings obtained from the soma and axon hillock of the same cell further show that initiation occurs first in the axon, possibly as a result of differences in geometry between soma and axon, as well as possible differences in the density, distribution and properties of voltage-activated Na⁺ channels in these structures.

Action potentials have also been observed to be generated in distal dendrites of neocortical pyramidal neurons in response to stimulation of afferents. However, these Na⁺ spikes attenuate as they spread to

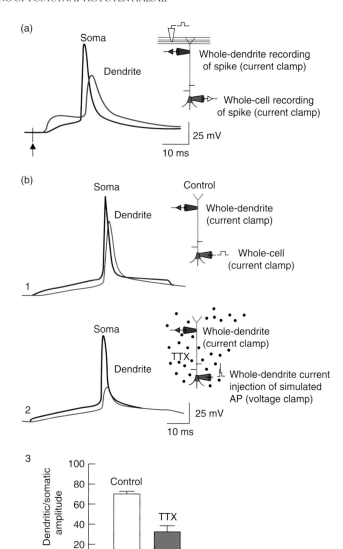

FIGURE 16.6   Site of initiation of Na⁺ action potential and its active backpropagation into the dendritic tree of pyramidal neurons of the neocortex.
(a) Na⁺ action potential evoked by distal synaptic stimulation in layer I and simultaneously recorded from the soma and a dendrite (dendritic recording is 525 μm from the soma). (b) Comparison of active and passive propagation of Na⁺ action potential in the apical dendrite studied with simultaneous somatic and dendritic recordings. b₁: An action potential is evoked in the soma by a depolarizing current pulse (200 pA, soma). It propagates in the apical dendrite where it is recorded (dendrite, 310 μm from the soma). b₂: A simulated action potential waveform is injected in the soma in the presence of 1 μM of TTX. The somatic voltage response (soma) propagates passively in the dendrites where it is recorded at the same location as in b₁ but in the presence of TTX (dendrite). The simulated somatic action potential is recorded later at the soma with a second somatic recording pipette (soma). b₃: Histogram of the average amplitude of dendritic action potentials recorded as in b₁ (open column) and of dendritic responses recorded as in b₂ (black column). Data are expressed as a percentage of the response recorded at the soma ± SEM; dendritic recordings 165–470 μm from the soma. Adapted from Stuart G and Sakmann B (1994) Active propagation of somatic action potentials into neocortical pyramidal cell dendrites. *Nature* **367**, 69–72, with permission.

the soma and axon. As a consequence, Na$^+$ action potentials are always initiated in the axon before the soma even when synaptic activation is intense enough to initiate dendritic regenerative potentials. Na$^+$ action potentials then actively propagate along the axon. Do they also actively backpropagate in the dendritic tree (**Figure 16.5b**)?

To investigate whether voltage-activated Na$^+$ channels aid the backpropagation of somatic action potentials into the dendrites, the internal Na$^+$ channel blocker QX-314 is included in the dendritic patch pipette during simultaneous somatic and dendritic recording. Following establishment of the dendritic patch, dendritic action potentials are observed to decrease progressively in amplitude before any change is observed in the amplitude or time course of the somatic action potential. This suggests that dendritic Na$^+$ channels boost the amplitude of dendritic action potentials as they backpropagate into the dendritic tree.

To compare the expected attenuation of dendritic action potentials in the presence and absence of Na$^+$ dendritic channels, TTX is applied in the bath. Since action potentials can no longer be evoked in this condition by current injection in the soma (Na$^+$ channels are blocked), a voltage command simulating an action potential is applied at the soma. The amount of attenuation is compared with that of action potentials evoked in the soma in the absence of TTX in the extracellular medium (**Figure 16.6b**, 1,2). On average, from these experiments, the amplitude of evoked dendritic action potentials is 70% of that of somatic action potentials, whereas in the presence of TTX it represents only 30% (**Figure 16.6b**, 2,3). These results show unequivocally that there is a regenerative (active) backpropagation of somatic action potentials in the dendrites of layer-V pyramidal neurons via the activation of TTX-sensitive dendritic Na$^+$ channels.

Dopaminergic neurons of the substantia nigra pars compacta

Simultaneous whole-cell recordings (current clamp mode) are made under visual control from the soma and dendrite of nigral dopaminergic neurons in slices. First, the site of action potential initiation is determined. In many dopaminergic cells, action potential is observed to occur first at the dendritic recording site (**Figure 16.7a**) and in some cases it is observed to occur first at the soma. To visualize the neuron recorded, the somatic pipette is filled with a biocytin-containing solution. Morphological examination of biocytin-filled neurons shows that in every case where the action potential is observed to occur first in the dendrite, the axon of the neuron is found to emerge from the

dendrite from which the recording had been made (**Figure 16.7a**). In 76% of dopaminergic neurons, the axon is found to emerge from a dendrite sometimes as far as 240 μm from the soma. When the action potential is observed to occur first in the soma, the axon is found to originate either from the soma or from a dendrite other that the one from which the dendritic recording had been made. Finally, in cases where the action potential appears to be simultaneous at the somatic and dendritic recording site, the axon is found to emerge from the dendrite in between the two recording pipettes. These findings indicate that the site of action potential initiation is always the axon hillock.

To determine whether Na$^+$ dendritic channels support the regenerative backpropagation of Na$^+$-dependent action potentials, the amplitude of action potentials evoked in control extracellular solution is compared with that of a voltage waveform (simulated action potential) injected in the soma in the presence of TTX in the bath. In all such experiments, the attenuation of a simulated action potential waveform injected in the soma in the presence of TTX is greater than that of the action potential evoked by somatic current injection in the absence of TTX (**Figure 16.7b**). These results suggest that dendritic Na$^+$ channels support the *active* backpropagation of Na$^+$ action potentials in the dendritic tree of nigral dopaminergic neurons.

The emergence of the axon from a dendrite rather than from the soma may have interesting consequences. It reverses the normal direction of propagation of the action potential in that the action potential will travel from the dendritic tree toward the soma. Consequently, the dendrite bearing the axon will experience the action potential before it spreads into the soma and other dendrites. In these neurons the final site of integration prior to the axon will not be at the soma but rather in the dendrites at the point where the axon emerges. This suggests that synapses made on the axon-bearing dendrite will be in an electrotonically privileged position (the concept of a 'privileged dendrite').

Purkinje cells

In these cells, Na$^+$ action potentials are initiated in the axon and decrease markedly with increasing distance from the soma, as shown with simultaneous somatic and dendritic recordings. On average, the amplitude of somatic Na$^+$ action potentials is 78.1 $\pm$ 7.6 mV whereas that of dendritic Na$^+$ action potentials is only a few millivolts at distances greater than 100 μm from the soma (**Figure 16.8a**). This strongly suggests that, in these neurons, Na$^+$ action potentials spread *passively* into the dendritic tree. In fact, in the presence of TTX, a

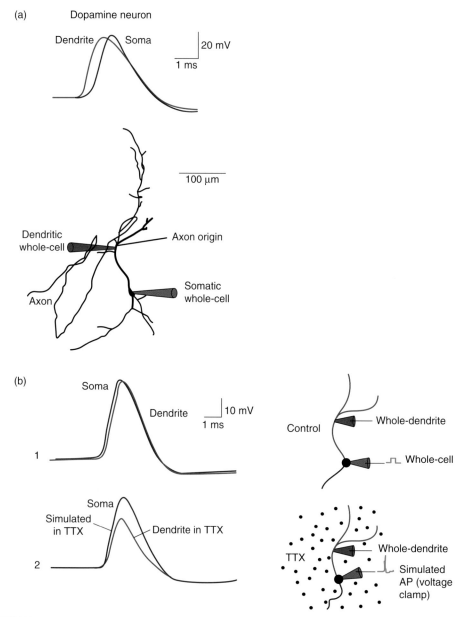

FIGURE 16.7   **Site of initiation of Na$^+$ action potential and its active backpropagation in the dendritic tree of dopaminergic neurons of the substantia nigra.**
(a) Spontaneous Na$^+$ action potential recorded simultaneously at the soma and dendrite (top) and the morphological reconstruction of the filled recorded neuron (below) with the location of the somatic and dendritic pipettes. The axon origin is indicated. The action potential is observed to occur first at the dendritic recording site, 195 μm from the soma; the axon of this cell emerges from the dendrite from which the dendritic recording is made (215 μm from the soma). (b) b$_1$: An action potential is evoked in the soma by a depolarizing current pulse (200 pA, soma). It propagates in the dendrite where it is recorded (dendrite, 100 μm from the soma). b$_2$: A simulated action potential waveform is injected in the soma in the presence of 1 μM of TTX. The somatic voltage response (soma) propagates passively in the dendrites where it is recorded at the same location as in b$_1$ but in the presence of TTX (dendrite). The simulated somatic action potential is recorded later at the soma with a second somatic recording pipette (soma). Adapted from Häusser M, Stuart G, Eacca C, Sakmann B (1995) Axonal initiation and active dendritic propagation of action potentials in substantia nigra neurons. *Neuron* **15**, 637–647, with permission.

simulated somatic action potential waveform attenuates in a similar manner as the synaptically evoked action potential (**Figure 16.8b**). This represents a striking contrast with neocortical layer-V pyramidal cells

or nigral dopaminergic neurons in which somatic Na$^+$ action potentials *actively* backpropagate into the dendrites. This marked attenuation of Na$^+$ action potentials is consistent with the observed low Na$^+$

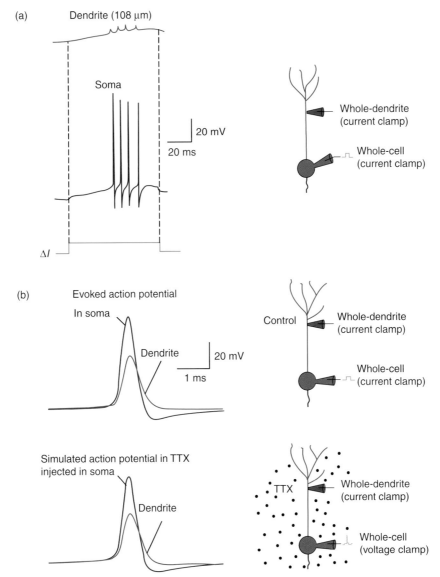

FIGURE 16.8 **Passive propagation of Na$^+$ action potentials in the dendritic tree of Purkinje cells.**
(a) Simultaneous recordings at the soma and dendrite (108 μm from the soma) of a train of Na$^+$ action potentials evoked by a somatic long depolarizing current pulse (100 pA). **(b)** b$_1$: An action potential is evoked in the soma by a depolarizing current pulse (soma). It propagates in the dendrite where it is recorded (dendrite, 47 μm from the soma). b$_2$: A simulated action potential waveform is injected in the soma in the presence of 1 μM of TTX. The somatic voltage response propagates passively in the dendrites where it is recorded at the same location as in b$_1$ but in the presence of TTX (dendrite). The simulated somatic action potential is recorded later at the soma with a second somatic recording pipette (soma). Adapted from Stuart GJ and Häusser M (1994) Initiation and spread of sodium action potentials in cerebellar Purkinje cells. *Neuron* **13**, 703–712, with permission.

current density in the dendrites of Purkinje cells compared with that found in the soma.

Conclusions

When TTX-sensitive voltage-gated Na$^+$ channels are present in high density in the dendritic membrane, they allow *active* backpropagation of Na$^+$ action

potentials in the dendritic tree. This is, for example, the case with dendrites of pyramidal neurons of the neocortex and hippocampus and of dopaminergic neurons of the substantia nigra. In contrast, the active backpropagation of Na$^+$ action potentials does not exist in dendrites that have in their membrane a low density of Na$^+$ channels, like Purkinje cells of the cerebellum. In this latter case, which is in fact the general case, Na$^+$

action potentials backpropagate *passively* (with decrement) in the dendritic tree.

16.2 HIGH-VOLTAGE-ACTIVATED Ca^{2+} CHANNELS ARE PRESENT IN THE DENDRITIC MEMBRANE OF SOME CNS NEURONS, BUT ARE THEY DISTRIBUTED WITH COMPARABLE DENSITIES IN SOMA AND DENDRITES?

16.2.1 High-voltage-activated Ca^{2+} channels are present in some dendrites

In order to record Ca^{2+} channels in isolation, the solution bathing the extracellular face of the patch contains Ba^{2+} as the charge carrier and TEACl and TTX for blocking K$^+$ and Na$^+$ currents, respectively. Ca^{2+} channel activity is identified by inward current polarity, voltage-dependent channel gating, unitary current amplitude, single-channel behaviour and its blockade by Cd^{2+}.

Purkinje cells of the cerebellar cortex

To determine whether the dendrites of Purkinje cells contain HVA Ca^{2+} channels, dendrite-attached patch recordings are performed in slices. Patches always show the activity of several channels, thus suggesting a tight clustering of Ca^{2+} channels in the dendritic membrane. **Figure 16.9a** shows the I/V relationship of a multichannel inward current carried by 10 mM of Ba^{2+} and evoked by a voltage ramp from −80 to +80 mV applied to a dendrite-attached macropatch. This dendritic Ba^{2+} current activates at −35 mV and is maximal around 0 mV. This HVA current is insensitive to the presence in the pipette solution of the specific blocker of N-type $^{2+}$ channels, ω-conotoxin GVIA (ωCgTx) and to the L-type channel opener (Bay K 8644). To test whether it is a P/Q-type Ca^{2+} current, a specific blocker, the funnel web spider toxin (FTX), is applied. Owing to the patch configuration, drugs must either be included in the pipette or be superfused over the cell before dendrite-attached recording, and a population of dendrite-attached recordings in control conditions is compared with the same number of recordings in the presence of the Ca^{2+} channel blocker. Funnel web toxin is the only drug that blocks the dendritic Ca^{2+} current (**Figure 16.9b**), thus showing that the dendritic Ba^{2+} current recorded is carried through P/Q-type Ca^{2+} channels. Their characteristics are close to that of P/Q-type channels recorded in Purkinje cells somata (see Section 5.2.2).

Pyramidal neurons of the hippocampus

In contrast to the above findings, in pyramidal neurons of the hippocampus there is a heterogeneous distribution of different types of Ca^{2+} channels within the soma-proximal dendritic trunks and more distal dendrites. Recordings of single Ca^{2+} channels in dendrite-attached patches show that L-type Ca^{2+} channels (sensitive to dihydropyridines) are observed at fairly high density only in the first 50 μm from the soma and at extremely low density in more distal dendritic patches where mainly Ni^{2+}-sensitive, T-type Ca^{2+} channels are present. Therefore HVA Ca^{2+} channels in these cells would be confined to the soma and very proximal dendrites.

Conclusions

Dendrites of Purkinje cells contain a high density of HVA Ca^{2+} channels of the P/Q type. In contrast, HVA Ca^{2+} channels are present at low density in dendrites of pyramidal neurons of the neocortex or the hippocampus. It must be pointed out that in most CNS dendrites there is a low density of HVA Ca^{2+} channels. Purkinje cells are exceptions.

16.2.2 High-voltage-activated Ca^{2+} channels of Purkinje cell dendrites are opened by climbing fibre EPSP; this initiates Ca^{2+} action potentials in the dendritic tree of Purkinje cells

The cerebellar Purkinje cells receive two kinds of excitatory inputs, a single powerful climbing fibre (CF) and many thousands of small parallel fibres (PF). The CF synapse arises from an axonal projection from the inferior olive, a brainstem nucleus. The CF synapse is composed of around 300 synaptic contacts located on the largest dendritic branches (thick and smooth dendrites) and on the smaller spiny dendrites (see **Figures 6.8 and 6.9**). The pioneering work of Llinas and Nicholson (1976) with intradendritic recordings showed that activation of this single distributed synapse evokes a large, all-or-none EPSP surmounted with one or two Ca^{2+}-dependent action potentials (complex spike; **Figure 16.10a**). They are Ca^{2+} spikes since they disappear in the presence of Cd^{2+}. They are characterized by a rather slow onset, but a large amplitude at dendritic level which fluctuates between 30 and 60 mV. Their time course is much longer than that of Na$^+$ spikes. Their threshold is lower at the dendritic level: depolarizations at around 10 mV are sufficient to generate Ca^{2+} dendritic spikes at the dendritic level while 20 mV depolarizations are required at the somatic level to evoke them. This complex spike then evokes bursts of Na$^+$ action potentials in the axon.

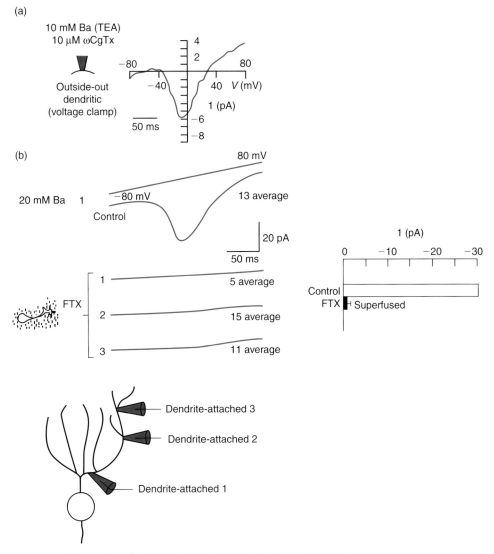

FIGURE 16.9   P-type Ca^{2+} channel current in dendrites of Purkinje cells.
(a) In the presence of 10 μM of ω-CgTx added to the 10 mM of Ba^{2+} pipette solution, currents are evoked in an outside-out macropatch of dendritic membrane by a depolarizing voltage ramp from –80 to +80 mV. The I/V plot shows that the evoked inward current peaks at –9 mV and activates at –44 mV. **(b)** Currents carried by 20 mM of Ba^{2+} evoked by voltage ramps (from –80 to +80 mV) in dendrite-attached macropatches in different conditions (left). Top: Averaged current in control conditions. Lower traces: Funnel web toxin (FTX) is first applied in the extracellular medium, then the patch is performed. Three different dendrite-attached patch recordings are shown (the approximate positions of the recording pipettes are indicated). The averaged currents recorded show the absence of inward Ba^{2+} current (right). Adapted from Usowicz MM, Sugimori M, Cherksey B, Llinas R (1992) P-type calcium channels in the somata and dendrites of adult cerebellar Purkinje cells. *Neuron* **9**, 1185–1199, with permission.

The climbing fibre-evoked EPSP underlying this complex spike can be uncovered in dendritic recordings by evoking a simultaneous IPSP by stimulation of interneurons (**Figure 16.10b**). Climbing fibre EPSP has a lower amplitude and a longer duration than the complex spike. This suggests that activation of a dendritic voltage-gated depolarizing current(s) amplifies the CF EPSP. Many data suggest that this dendritic depolarizing current is a P-type Ca^{2+} current. First, P channels are present in the dendritic membrane (see Section 16.2.1). Second, the complex spike is accompanied by a transient rise in intracellular Ca^{2+} concentration which is most prominent at dendritic locations (**Figure 16.10c**).

Modelling of the complex spike shows the currents underlying the CF-evoked complex spike and their sequence of activation (**Figure 16.11**). These currents are the large CF synaptic inward current (resulting from the summation of around 300 unitary synaptic

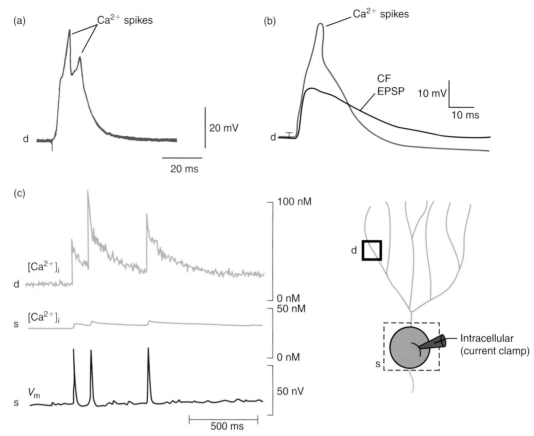

FIGURE 16.10   **P-type Ca²⁺ current activated by climbing fibre EPSP in dendrites of Purkinje cells.**
(a) Intradendritic recording of the synaptically evoked climbing fibre response that is surmounted by two Ca²⁺ spikes (intracellular recording in current clamp mode). **(b)** Intradendritic recording at resting potential of the climbing fibre EPSP showing a 2–3 ms-wide Ca²⁺ spike and of the climbing fibre EPSP recorded during a concomitant IPSP (CF EPSP) (intracellular recording in current clamp mode). **(c)** Time course of [Ca²⁺]ᵢ recorded at a dendritic (d, top trace) and somatic (s, middle trace) site during spontaneous climbing fibre responses (s, bottom trace) recorded with simultaneous microfluorometric measurements of cytosolic free calcium concentration and intracellular (intrasomatic) electrophysiological recordings (current clamp mode). Part (a) adapted from Llinas R and Sugimori M (1980) Electrophysiological properties of *in vitro* Purkinje cell dendrites in mammalian cerebellar slices. *J. Physiol. (Lond.)* **305**, 197–213, with permission. Part (b) adapted from Callaway JC, Lasser-Ross N, Ross WN (1995) IPSPs strongly inhibit climbing fiber-activated [Ca²⁺]ᵢ increases in the dendrites of cerebellar Purkinje neurons. *J. Neurosci.* **15**, 2777–2787, with permission. Part (c) adapted from Knöpfel T, Vranesic I, Staub C, Gähwiler BH (1991) Climbing fiber response in olivo-cerebellar slice cultures: II. Dynamics of cytosolic calcium in Purkinje cells. *Eur. J. Neurophysiol.* **3**, 343–348, with permission.

currents through glutamate AMPA receptors). CF-induced synaptic inward current depolarizes the dendritic membrane and thus activates P-type Ca²⁺ channels over large regions of the dendrites. The resulting Ca²⁺ current is responsible for almost all of the resulting additional depolarization and for dendritic Ca²⁺ spikes. Ca²⁺ spikes are generated at multiple sites along the dendritic tree, which explains why the CF EPSP recorded in a dendrite is sometimes surmounted by more than one Ca²⁺ spike. In turn, Ca²⁺ entry increases intradendritic Ca²⁺ concentration and thus activates the Ca²⁺-activated K⁺ outward current in the whole dendritic tree. This repolarizes the

complex spike. In conclusion, activation of the dendritic P-type Ca²⁺ current boosts the amplitude of the CF EPSP, and by activating K⁺ current leads to a faster repolarization of the EPSP (**Figure 16.10b**).

16.2.3   Dendritic high-voltage-activated Ca²⁺ channels are opened by backpropagating Na⁺ action potentials

Pyramidal neurons of the hippocampus

To test whether dendritic HVA Ca²⁺ channels are activated by sub-threshold EPSPs or by higher amplitude

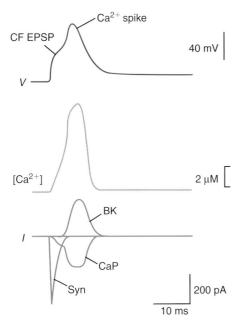

FIGURE 16.11 Dendritic currents underlying the climbing fibre-evoked EPSP in Purkinje cells.
(a) Modelling of the CF response recorded in current clamp mode in a dendrite (top trace) and the underlying $[Ca^{2+}]_i$ transient (middle trace) and currents (*I*, bottom traces). The underlying currents are the synaptic glutamatergic current (Syn) which generates the climbing fibre EPSP, depolarizes the dendritic membrane and thus activates the dendritic P-type Ca^{2+} current (CaP) which further depolarizes the dendritic membrane, amplifies the EPSP and generates a Ca^{2+} spike (shown in top trace). The resultant $[Ca^{2+}]_i$ increase (middle trace) activates the BK current (BK, bottom trace) which rapidly repolarizes the membrane. Adapted from De Schutter E and Bower JM (1994) An active membrane model of the cerebellar Purkinje cell: II. Simulation of synaptic response. *J. Neurophysiol.* **71**, 401–419, with permission.

depolarizations such as backpropagating Na^+ action potentials, simultaneous dendrite-attached (voltage clamp mode) and whole-cell somatic (current clamp mode) recordings are performed in the same neuron (**Figure 16.12a**). Excitatory postsynaptic potentials are evoked by Schaffer collateral stimulation. Na^+ spikes are evoked by intrasomatic injection of a depolarizing current pulse. EPSPs and spikes are recorded from the soma while channel openings are recorded from the dendritic patch of membrane. Ca^{2+} channel activity present in the dendritic patch is first recorded and Ca^{2+} channels are classified as HVA or LVA (low-voltage-activated, also called 'subliminal') channels. Two types of Ca^{2+} channels are encountered regularly on dendrites greater than 100 μm from the soma, essentially the LVA T-type Ca^{2+} channels and less frequently the HVA L-type Ca^{2+} channels. Only the T-type channels are opened in response to sub-threshold EPSPs.

Instead, somatically generated action potentials or trains of suprathreshold synaptic stimulation are required for HVA channel activation (**Figure 16.12a**).

Dendritic HVA channel openings are observed during and after the repolarization phase of somatically generated Na^+ action potentials. This strongly suggests that dendritic HVA Ca^{2+} channels are opened by Na^+ action potentials backpropagating into the dendrites. Openings occur tens of milliseconds after the action potential. This provides an influx of Ca^{2+} throughout an extended portion of the dendritic tree (defined by the extent of action potential propagation). The spatial domain of the effects of these HVA Ca^{2+} channels will therefore be much more extensive compared with a local opening of Ca^{2+} channels by EPSPs. Thus, the HVA Ca^{2+} channels in the CA1 apical dendrites may modify synaptic strength over broad areas of the dendrites (see Chapter 18).

Conclusions

When HVA Ca^{2+} channels are present with a high density in the dendritic membrane, they allow *generation and propagation* of Ca^{2+} action potentials in dendrites. This is, for example, the case with dendrites of Purkinje cells. In contrast, Ca^{2+} action potentials do not exist in dendrites that contain in their membrane a low density of HVA Ca^{2+} channels, such as pyramidal neurons of the hippocampus.

16.3 FUNCTIONAL CONSEQUENCES

16.3.1 Amplification of distal synaptic responses by dendritic HVA currents counteracts their attenuation due to passive propagation to the soma

High-voltage-activated Na^+ and Ca^{2+} channels opened by EPSPs boost the effect of local synaptic inputs by acting as either voltage or current amplifiers. This could be one solution to overcome the passive decay of EPSPs, en route to the soma. This argument is somewhat somatocentric; i.e. the emphasis is on how dendrites might amplify events so that they are bigger in the soma. An alternative viewpoint is that these channels are more important for dendritic interactions in the immediate vicinity of synaptic inputs. For example, multiple EPSPs occurring on the same branch and within a narrow time should activate voltage-gated channels more strongly than a single EPSP and produce a much bigger response than would occur if EPSPs were on separate branches.

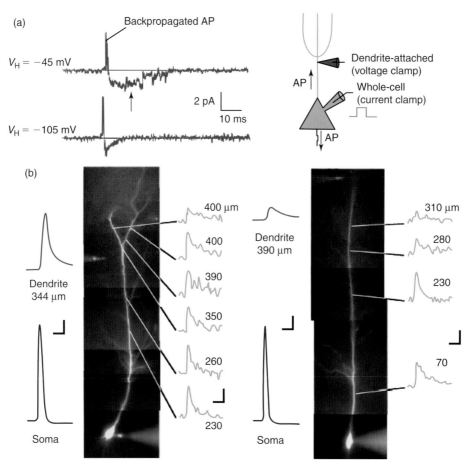

FIGURE 16.12 Activation of dendritic HVA Ca^{2+} channels by backpropagated Na+ action potential in hippocampal pyramidal neurons. Distal dendritic calcium influx is correlated with the efficacy of action potential backpropagation.

(a) An action potential is evoked in the soma by a depolarizing current pulse. It backpropagates in the dendrites and is recorded at a dendritic site as a capacitative current at two different holding potentials (backpropagated AP). When the dendritic patch is held 20 mV more depolarized than resting potential (–45 mV), numerous openings of channels are observed following the action potential (arrow). In contrast, when the dendritic membrane is held at –105 mV, the backpropagated action potential does not evoke channel openings. **(b)** Spike-induced [Ca^{2+}]$_i$ transients in a FURA-2 loaded neuron. Single action potentials (e.g. *left*) that propagate efficiently to the distal dendrite trigger robust calcium influx (expressed as the relative change in fluorescence, DF/F) in both proximal and distal dendritic compartments. *Right*: A different pyramidal neuron filled with Fura-2 exhibits weak action-potential backpropagation in the distal dendrites. The associated calcium influx shows significant attenuation in distal dendritic regions. Physiology scale bars: 20 mV, 1 ms. Imaging scale bars: 5% DF/F, 300 ms. Part (a) adapted from Magee JC and Johnston D (1995) Synaptic activation of voltage-gated channels in the dendrites of hippocampal pyramidal neurons. *Science* **268**, 301–304, with permission. Part (b) adapted from Golding NL, Kath WL, Spruston N (2001) Dichotomy of action-potential backpropagation in CA1 pyramidal neuron dendrites. *J. Neurophysiol.* **86**, 2998–3010.

16.3.2 Active backpropagation of Na$^+$ spikes in the dendritic tree depolarizes the dendritic membrane, with multiple consequences

Most of the consequences of the presence of Na$^+$ spikes (large amplitude depolarizations) in the dendrites are still hypotheses. The only well-demonstrated one is the opening of dendritic Ca^{2+} channels

and the consequent increase in intradendritic Ca^{2+} concentration. Such an increase will have by itself other consequences.

A retrograde signal that activates voltage-sensitive dendritic Ca^{2+} channels

In the hippocampus, Na$^+$ action potentials open dendritic Ca^{2+} channels, leading to a widespread influx of Ca^{2+} in the dendrites. In order to localize and quantify the increase of intradendritic Ca^{2+} concentration resulting from backpropagated Na^+ action potentials, pyramidal neurons of the hippocampus are loaded with FURA-2 and a train of action potentials is evoked by somatic depolarization through the whole-cell recording electrode. The evoked Ca^{2+} influx is thus visualized in the dendrite under fluorescence observation. Then to identify the type of Ca^{2+} channel involved and its localization along the dendrite, the same experiment is repeated in the presence of specific Ca^{2+} channel blockers. Finally a control experiment is performed in the absence of extracellular Ca^{2+}. Results show that $[Ca^{2+}]_i$ transients are largest in the proximal dendrites and smaller changes occur in more distal dendritic regions (**Figure 16.12b**).

One particular role for an intradendritic increase of Ca^{2+} concentration is found in dopaminergic neurons of the substantia nigra. $[Ca^{2+}]_i$ increase triggers transmitter release from dendrites (a *presynaptic* effect). In these cells, clusters of synaptic vesicles containing dopamine are present in dendrites that behave in certain sites as presynaptic elements. Dendritic release of dopamine is Ca^{2+}-dependent and TTX-sensitive. Backpropagated action potentials may thus provide the stimulus (i.e. intradendritic $[Ca^{2+}]_i$ increase) to trigger dopamine release and evoke synaptic transmission from nigral dendrites to postsynaptic sites.

Apart from this very particular case, intradendritic $[Ca^{2+}]_i$ increase will have a *postsynaptic* effect. Intracellular Ca^{2+} activates biochemical pathways, Ca^{2+}-induced Ca^{2+} release as well as Ca^{2+}-activated K^+ currents present in the dendritic membrane. Such outward current changes the shape of EPSPs. Intracellular Ca^{2+} is also implicated in postsynaptically induced forms of plasticity such as long-term potentiation or depression (see Chapter 18).

A retrograde signal that amplifies NMDA-mediated synaptic currents

The transient depolarization due to backpropagated Na^+ spikes may relieve the voltage-dependent Mg^{2+} block of NMDA receptor channels and amplify the signal mediated by these channels (see Section 10.4).

A retrograde signal that shunts ongoing synaptic integration

Backpropagated Na^+ action potentials act as a signal to the dendritic tree that the axon has fired. This transient depolarization, by reducing the electrochemical gradient for cations, will diminish ongoing postsynaptic excitatory currents. Moreover, dendritic action potentials will open voltage-sensitive channels and

thus diminish the resistance of the dendritic membrane and shunt ongoing synaptic integration. Finally, the rise in dendritic Ca^{2+} concentration could also transiently shunt out parts of the dendritic tree by opening Ca^{2+}-activated K^+ currents.

16.3.3 Initiation of Ca^{2+} spikes in the dendritic tree of Purkinje cells evokes a widespread intradendritic $[Ca^{2+}]$ increase

As seen above, intradendritic $[Ca^{2+}]$ increase will have a *postsynaptic* effect in Purkinje cells. It activates biochemical pathways, Ca^{2+}-induced Ca^{2+} release as well as Ca^{2+}-activated K^+ currents present in the dendritic membrane. Such outward currents change the shape of EPSPs. Intracellular Ca^{2+} is also implicated in postsynaptically induced forms of plasticity such as long-term depression which has been extensively studied in Purkinje cells (see Chapter 18).

16.4 CONCLUSIONS

High-voltage-activated channels have been shown in the dendritic membrane of some CNS neurons such as pyramidal neurons of the neocortex and hippocampus, dopaminergic neurons of the substantia nigra pars compacta, and Purkinje cells of the cerebellar cortex. To answer the questions asked in the introduction:

- High voltage-gated currents are present in the dendritic membrane of some CNS neurons. These are the depolarizing currents I_{Na}, I_{CaP} and I_{CaL}.
- They are not all distributed equally over somato-dendritic membranes. I_{Na} is present at the same density in the somatic and dendritic membranes in pyramidal neurons of the neocortex or hippocampus and in dopaminergic neurons of the substantia nigra but is nearly absent in dendrites of Purkinje cells. I_{CaP} is present at the same density in the somatic and dendritic membranes in Purkinje cells. In contrast, I_{CaL} is mostly present in the somatic and very proximal dendritic membranes of pyramidal neurons of the hippocampus.
- Since I_{Na}, I_{CaP} and I_{CaL} are activated at high voltage, they can be activated only by already large summed EPSPs or by backpropagating Na^+ action potentials (for I_{CaP} and I_{CaL}).
- I_{Na} may in theory boost EPSPs but it mostly supports the active backpropagation of Na^+ action potentials in pyramidal neurons of the neocortex or hippocampus and in dopaminergic neurons of the substantia nigra in response to suprathreshold

EPSPs (these Na^+ action potentials are first initiated at the axon initial segment). I_{CaP} boosts climbing fibre EPSP and supports the initiation and active propagation of Ca^{2+} action potentials in dendrites of Purkinje cells. Direct (by EPSPs) or indirect (via dendritic Na^+ action potentials) activation of I_{CaP} and I_{CaL} induces a transient increase of intradendritic Ca^{2+} concentration that is more or less localized depending on the neuron considered.

FURTHER READING

Callaway JC and Ross WN (1997) Spatial distribution of synaptically activated sodium concentration changes in cerebellar Purkinje neurons. *J. Neurophysiol.* **77**, 145–152.

Christie BR, Eliot LS, Ito K *et al.* (1995) Different Ca^{2+} channels in soma and dendrites of hippocampal pyramidal neurons mediate spike-induced Ca^{2+} influx. *J. Neurophysiol.* **73**, 2553–2557.

Colbert CM, Magee JC, Hoffman DA, Johnston D (1997) Slow recovery from inactivation of Na^+ channels underlies the activity-dependent attenuation of dendritic action potentials in hippocampal CA1 pyramidal neurons. *J Neurosci.* **17**, 6512–6521.

Hausser M, Spruston N, Stuart GJ (2000) Diversity and dynamics of dendritic signaling. *Science* **290**, 739–744.

Johnston D, Magee JC, Colbert CM, Christie BR (1996) Active properties of neuronal dendrites. *Annu. Rev. Neurosci.* **19**, 165–186.

Lüscher HR and Larkum ME (1998) Modeling action potential initiation and back-propagation in dendrites of cultured rat motoneurons. *J. Neurophysiol.* **80**, 715–729.

Markram H, Helm PJ, Sakmann B (1995) Dendritic calcium transients evoked by single backpropagating action potentials in rat neocortical pyramidal neurons. *J. Physiol. (Lond.)* **485**, 1–20.

Markram H, Lubke J, Frotscher M, Sakmann B (1997) Regulation of synaptic efficacy by coincidence of postsynaptic APs and EPSPs. *Science* **275**, 213–215.

Mitgaard J (1994) Processing of information from different sources: spatial synaptic integration in the dendrites of vertebrate CNS neurons. *Trend. Neurosci.* **17**, 166–172.

Miyakawa H, Lev-Ram V, Lasser-Ross N, Ross WN (1992). Calcium transients evoked by climbing fiber and parallel fiber synaptic inputs in guinea pig cerebellar Purkinje neurons. *J. Neurophysiol.* **68**, 1178–1188.

Schiller J, Schiller Y, Stuart G, Sakmann B (1997) Calcium action potentials restricted to distal apical dendrites of rat neocortical pyramidal neurons. *J. Physiol. (Lond.)* **505**, 605–616.

Stuart GJ and Hausser M (2001) Dendritic coincidence detection of EPSPs and action potentials. *Nat. Neurosci.* **4**, 63–71.

Stuart G, Schiller J, Sakmann B (1997) Action potential initiation and propagation in rat neocortical pyramidal neurons. *J. Physiol. (Lond.)* **505**, 617–632.

Stuart G, Spruston N, Sakmann B, Hausser M (1997) Action potential initiation and backpropagation in neurons of the mammalian CNS. *Trend. Neurosci.* **20**, 125–131.

Vetter PR, Roth A, Hausser M (2000) Propagation of action potentials in dendrites depends on dendritic morphology. *J. Neurophysiol.* **85**, 926–937.

Williams SR and Stuart GJ (2000) Backpropagation of physiological spike trains in neocortical pyramidal neurons: implications for temporal coding in dendrites. *J. Neurosci.* **20**, 8238–8246.

Yuste R and Tank DW (1996) Dendritic integration in mammalian neurons, a century after Cajal. *Neuron* **16**, 701–716.

CHAPTER

17

Firing patterns of neurons

The electrical activity of a neuron is related not only to the excitatory and inhibitory synaptic inputs that it receives, but also to its intrinsic electrophysiological membrane properties; i.e. the subliminal voltage-gated channels present in its dendritic, somatic and initial segment membranes and activated in the near-threshold range of membrane potential. As a result, the same postsynaptic depolarizing current will trigger different firing patterns according to the neuronal cell type recorded. In brief, the firing pattern (output) of a neuron results from the integration of synaptic currents (input) and subliminal voltage-gated currents present in the somatic and dendritic membrane. This concept was stated simply by Rodolpho Llinas in 1990: 'Nerve cells are not interchangeable: a neuron of a given kind cannot be functionally replaced by one of another type even if their synaptic connectivity and the type of neurotransmitter outputs are identical.'

In this chapter we shall consider the mammalian central nervous system to demonstrate how these intrinsic electrophysiological properties determine the firing patterns of neurons. We shall study the mechanisms underlying the firing patterns of medium spiny neurons of the striatum, of inferior olivary neurons, of Purkinje cells of the cerebellar cortex, and of thalamic and subthalamic neurons.

17.1 MEDIUM SPINY NEURONS OF THE NEOSTRIATUM ARE SILENT NEURONS THAT RESPOND WITH A LONG LATENCY

The neostriatum belongs to basal ganglia. It contains several types of neurons among which medium spiny neurons are the most numerous. Medium spiny

neurons are Golgi type I neurons that use GABA as a neurotransmitter. They project to globus pallidus and substantia nigra. They receive numerous inputs. Excitatory glutamatergic inputs come from neocortical and thalamic neurons. Several thousands of these, from nearly as many different afferent neurons, impinge on medium spiny neurons. Inhibitory GABAergic inputs come from local interneurons and from neurons of other basal ganglia nuclei. *In vivo*, medium spiny neurons are silent or exhibit a low level of spontaneous activity.

17.1.1 Medium spiny neurons are silent at rest owing to the activation of an inward rectifier K^+ current

When spontaneous synaptic transmission is intact, medium spiny neurons are silent or show brief episodes (0.1–3 s) of firing separated by long periods of silence. Intracellular recordings show that even during cell silence, membrane potential abruptly shifts between two preferred levels: a hyperpolarized level (-80 to -95 mV), called the *down state*, and a near-threshold depolarized level (-40 to -50 mV) called the *up state*. In the down state, neurons are silent; in the up state, neurons are either silent or generate intermittent action potentials on top of the largest membrane fluctuations (**Figures 17.1a, b**). These two states can last from several hundred milliseconds to seconds.

The down state

This does not result from a tonic inhibitory afferent synaptic activity since it is still observed when bicuculline (the $GABA_A$ receptor antagonist) is iontophoretically applied near the recording electrode. Moreover, in the presence of blockers of synaptic transmission, such as Ca^{2+} channel blockers, membrane fluctuations

325

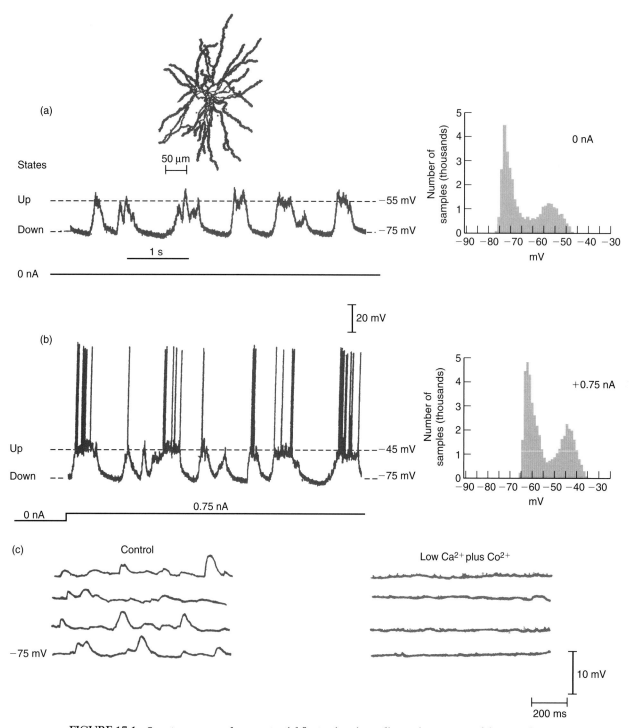

FIGURE 17.1 **Spontaneous membrane potential fluctuations in medium spiny neurons of the neostriatum.**
In vivo intracellular recordings of 'up' and 'down' states **(a)** at resting membrane potential and **(b)** during the continuous injection of a depolarizing current, of a medium spiny neuron (inset, calibration bar 50 μm). Histograms represent the time spent at various membrane potentials. The two peaks of each histogram represent the down and up states of the membrane potential. The proportion of the area of the histogram under each of the peaks represents the proportion of time spent in each state. The depolarizing current moves the peaks to the right. **(c)** *In vitro* intracellular recordings of up- and down-state transitions in control conditions and in the presence of a low external Ca^{2+} concentration (0.5 mM) and Co^{2+} (0.5 mM). In (a) and (b), intracellular electrodes are filled with 1 M of K acetate and in (c) with 2 M of KCl. Parts (a) and (b) adapted from Wilson CJ and Kawaguchi Y (1996) The origins of two state spontaneous membrane potential fluctuations of neostriatal spiny neurons. *J. Neurosci.* **16**, 2397–2410, with permission. Part (c) adapted from Calabresi P, Mercuri NB, Bernardi G (1990) Synaptic and intrinsic control of membrane excitability of neostriatal neurons: II. An *in vitro* analysis. *J. Neurophysiol.* **63**, 663–675, with permission. Drawing by Jérôme Yelnik.

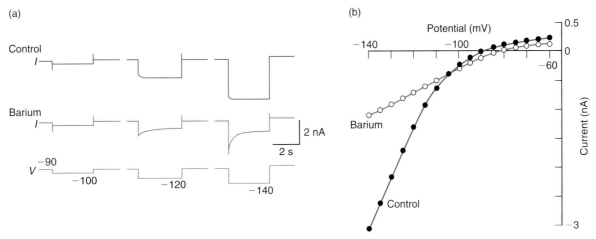

FIGURE 17.2   The inward rectification K^1 current of medium spiny neurons.
(a) Membrane current (I) evoked by hyperpolarizing steps (V) to the indicated potentials (in mV), in control and in the presence of barium (10 μM). (b) I/V relations constructed from the steady-state I currents recorded in (a). Note the reversal potential at around −100 mV. Adapted from Uchimura N, Cherubini E, North RA (1989) Inward rectification in rat nucleus accumbens neurons. *J. Neurophysiol.* **62**, 1280–1286, with permission.

disappear and membrane potential remains stable in the down state (**Figure 17.1c**). Therefore, the down state does not depend on afferent activity. The down state is attributable to the presence of a strong and rapidly activating inwardly rectifying potassium-selective current (I_{KIR}; see Section 14.3.5) (**Figure 17.2**). Therefore, the membrane of these cells is 'clamped' near the K$^+$ equilibrium potential by the K$^+$-selective inward rectifier current, that is open at rest, in the absence of afferent synaptic activity. In brief, during the down state the membrane potential is determined by I_{KIR} which dominates the other currents in the absence of strong depolarizing synaptic currents. As a result, these neurons are characterized by a low input resistance at resting membrane potential.

The up state

The up state, in contrast, absolutely requires the integrity of excitatory synaptic inputs to the neostriatal neurons. When the cortex is removed or temporarily inactivated, or when blockers of synaptic transmission are iontophoretically applied near the recording electrode, up-state transitions are abolished (**Figure 17.1c**). Similarly, up-state transitions are not recorded in neostriatal slices in which afferent input is interrupted. Up-state transitions depend on the synchronous activity of excitatory afferents arising from the cortico- and/or thalamo-striatal pathways. In brief, during the up state the membrane potential results from the interaction between strong depolarizing synaptic currents and intrinsic voltage-dependent subliminal currents, as explained below.

17.1.2 When activated, the response of medium spiny neurons is a long-latency regular discharge

The long-latency response of medium spiny neurons can be observed in response to an intracellular current pulse that mimics a depolarizing synaptic input (**Figure 17.3**). In the presence of a low dose of 4-aminopyridine (4-AP, 30–100 μM), known to preferentially block the slowly inactivating transient K$^+$ current (called I_{As} or I_D) (see Section 14.3.2), the latency of the first spike in response to a 400 ms depolarizing current pulse is largely reduced.

Voltage clamp recordings have shown that neostriatal medium spiny neurons possess at least three types of depolarization-activated K$^+$ currents. There are the two types of transient A currents; i.e. the fast- (I_{Af} or I_A) and slow- (I_{As} or I_D) inactivating, activated at sub-threshold membrane potentials (around –65 mV) and both sensitive to 4-AP (see Sections 14.3.1 and 14.3.2). There is also a non-inactivating current (I_{KDR}) available at more depolarized potentials (−20 to −30 mV) and relatively resistant to 4-AP but blocked by TEA (see Section 4.3). The importance of these voltage-dependent K$^+$ currents in opposing depolarization and firing is also indicated by the large increase in the amplitude of the up state after such currents are poisoned by intracellular injection of caesium (cells depolarize to a mean potential of − 30 mV instead of −55 mV in control solution). Moreover, caesium greatly enhances their frequency of occurrence and extends their duration.

Why does the response consist of a regular discharge with no adaptation? Adaptation (slowing of spike

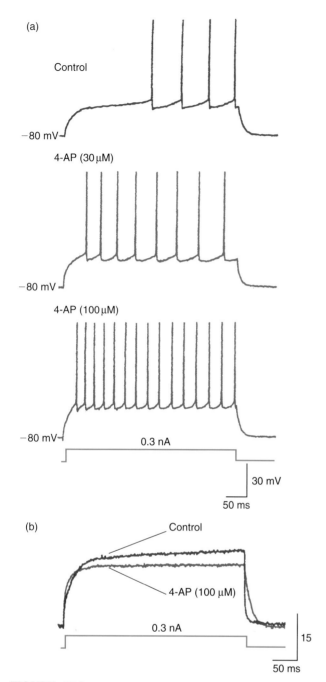

FIGURE 17.3 **The long-latency discharge of medium spiny neurons.**

(a) A suprathreshold current pulse is delivered in control conditions and in the presence of 4-AP. Between pulses, the cell membrane is hyperpolarized back to the original resting membrane potential (−80 mV). 4-AP decreases the first spike latency and increases the frequency of discharge. **(b)** Comparison of the voltage deflections produced by a sub-threshold 0.5 nA current pulse (400 ms duration) in the presence of TTX shows that 4-AP reduces the slope of the ramp potential and decreases the apparent time constant of the membrane (average of four responses). Adapted from Nisenbaum ES, Xu ZC, Wilson CJ (1994) Contribution of a slowly inactivating potassium current to the transition to firing of neostriatal spiny projection neurons. *J. Neurophysiol.* **71**, 1174–1189, with permission.

frequency inside a train) results from the progressive summation of the Ca^{2+}-activated K^+ current that underlies the slow AHP. In medium spiny neurons, this current is weak or absent.

Why do medium spiny neurons have a unique firing pattern? Rhythmic bursting currents are either suppressed in these neurons by the presence of K^+ currents at both hyperpolarized and depolarized potentials, or are absent.

Summary

In the absence of afferent synaptic activity, medium spiny neurons are in the *down state* and silent. Transition from the down state to the *up state* is determined by excitatory synaptic inputs; but the level of depolarization during up states – which in turn determines the triggering or not of action potentials as well as the latency of this discharge – results from the interaction between the depolarizing glutamatergic synaptic current (mediated by AMPA receptors) and the intrinsic subliminal voltage-gated K^+ currents that oppose depolarization.

17.2 INFERIOR OLIVARY CELLS ARE SILENT NEURONS THAT CAN OSCILLATE

The inferior olive is a brainstem nucleus whose neurons innervate and monosynaptically excite cerebellar Purkinje cells through characteristic axonal terminations known as *climbing fibres* (see **Figures 6.8** and **6.9**). They use an excitatory amino acid as a neurotransmitter, most probably glutamate.

17.2.1 Inferior olivary cells are silent at rest in the absence of afferent activity

In slices *in vitro*, extracellular and intracellular recordings reveal that inferior olivary neurons are generally silent at the resting membrane potential but can spontaneously display sequences of membrane oscillations in response to afferent synaptic activity (**Figure 17.4a**). Non-oscillating inferior olive cells have a resting membrane potential of −55 to −60 mV.

In response to stimulation of afferents or to intracellular current pulses, a typical rhythmic bursting activity is recorded with a frequency varying from 3 to 12 Hz, depending on membrane potential. Intracellular recordings in slices *in vitro* revealed that these cells have the intrinsic properties necessary to oscillate endogenously.

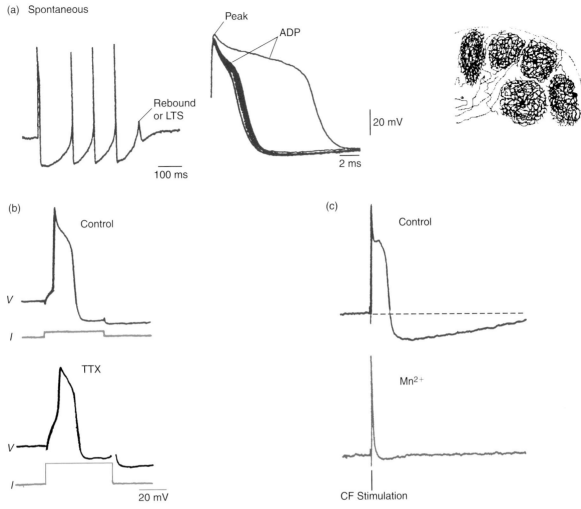

FIGURE 17.4 Complex spikes of inferior olivary neurons.
The activity of olivary neuron is intracellularly recorded under current clamp in cerebellar slices (inset represent five of these neurons). **(a)** Spontaneous low-frequency train of spikes from an olivary neuron, displayed at two different sweep speeds. The action potentials shown at left are displayed superimposed at right at a faster sweep speed. The first action potential which arises from the resting membrane potential level has a slightly higher amplitude at the peak and a rather prolonged plateau (after-spike depolarization, ADP) which is followed by an after-hyperpolarization. The rest of the spikes in the train become progressively shorter until failure of spike generation occurs (arrow, left) and the train terminates. **(b)** Effect of TTX (left) and Mn^{2+} (right) on the different parts of the complex spike evoked in two olivary neurons either by a depolarizing intracellular current pulse (left) or climbing fibre stimulation (CF, right). Part (a) adapted from Llinas R and Yarom Y (1986) Oscillatory properties of guinea-pig inferior olivary neurones and their pharmacological modulation: an *in vitro* study. *J. Physiol. (Lond.)* **376**, 163–182, with permission. Part (b) adapted from Llinas R and Yarom Y. (1981) Electrophysiology of mammalian olivary neurones *in vitro*: different types of voltage-dependent ionic conductances. *J. Physiol. (Lond.)* **315**, 549–567, with permission. Drawing by Ramon Y Cajal, 1911.

17.2.2 When depolarized, inferior olivary cells oscillate at a low frequency (3–6 Hz)

When inferior olivary cells are slightly depolarized, their response to a depolarizing current pulse is characterized by an initial fast-rising spike (1 ms duration) which is prolonged to 10–15 ms by a plateau (ADP: after-spike depolarization) on which small action potentials are sometimes superimposed (**Figure 17.4**). It is followed by a large-amplitude long-lasting (150–200 ms) after-hyperpolarization (AHP) which silences the spike-generating activity and terminates in a rebound depolarization (arrowhead). The rebound depolarization may evoke another complex action potential: these cells have oscillatory membrane properties. Owing to their difference of threshold potential of initiation, the peak and plateau are called *high-threshold spike* (HTS), whereas the rebound depolarization is called *low-threshold spike* (LTS).

Pioneering *in vitro* studies by Llinas and Yarom in 1981 described the ionic currents that underlie the endogenous oscillatory properties of single inferior olivary neurons. The analysis of the currents responsible for this discharge configuration gives the following description. To record the low- and high-threshold spikes together or the low-threshold spike in isolation, the membrane potential is respectively maintained at a depolarized potential (**Figures 17.4b** and **17.5a**) or a hyperpolarized potential (**Figure 17.5b**).

- TTX abolishes the peak of the action potential, showing that it results from the activation of the voltage-dependent I_{Na} (**Figures 17.4b** and **17.5a**).
- Ca^{2+} channel blockers decrease the after-depolarization (ADP), the small superimposed action potentials, the AHP and the rebound depolarization, but leave intact the early Na^+-dependent spike (**Figures 17.4c**

and **17.5b**). This shows that ADP, AHP and rebound depolarization are all Ca^{2+}-dependent. The plateau is the result of the activation of a high-threshold Ca^{2+} current since the depolarization required to evoke it in the presence of TTX is high (see current trace I in **Figure 17.4b**, compare control and TTX). This current, localized in the dendrites is activated by the fast Na^+-dependent action potential.

- The after-hyperpolarization (AHP) is dependent on the amplitude of the ADP (**Figure 17.5a**, compare a_1 and a_2) and is blocked by external Ba^{2+} ions. It results from the activation of the Ca^{2+}-sensitive K^+ currents (I_{KCa}).
- The rebound depolarization or low-threshold spike (LTS) is suppressed by Ca^{2+} channel blockers (**Figure 17.5b**) and is activated at sub-threshold potentials. It is due to the activation of a low-threshold Ca^{2+} current (I_{CaT}) localized at the level of the soma. This current is de-inactivated during the period of after-hyperpolarization and activated when the hyperpolarization decreases.

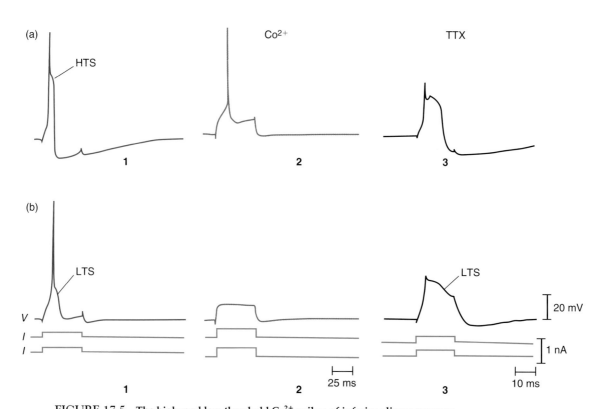

FIGURE 17.5 **The high- and low-threshold Ca^{2+} spikes of inferior olivary neurons.**
Effect of membrane potential on excitability. A depolarizing current pulse of constant amplitude evokes (**a**, 1) a high-threshold Ca^{2+} spike (HTS) at resting membrane potential and (**b**, 1) a low-threshold Ca^{2+} spike (LTS) at a more hyperpolarized potential. Note that the ADP and AHP are smaller in (b) than in (a). From left to right, effect of Co^{2+} and TTX in the same conditions. Adapted from Llinas R and Yarom Y (1981) Electrophysiology of mammalian olivary neurones *in vitro*: different types of voltage-dependent ionic conductances. *J. Physiol. (Lond.)* **315**, 549–567, with permission.

Summary

Inferior olivary neurons are silent at rest. When depolarized to the threshold potential of the voltage-sensitive Na^+ channels, a sodium action potential is generated in the soma–initial segment region and the dendritic membrane is depolarized up to the level of activation of the high-threshold Ca^{2+} channels. The entry of Ca^{2+} ions through these channels causes a dendritic calcium plateau (ADP) and then the activation of Ca^{2+}-sensitive K^+ channels. The resulting I_{KCa} hyperpolarizes the membrane (AHP). This after-spike hyperpolarization allows the de-inactivation of the T-type Ca^{2+} channels. As the amplitude of the AHP diminishes and the membrane potentials return to baseline, the low-threshold Ca^{2+} current (I_{CaT}) is activated, generates a 'low-threshold' Ca^{2+}-dependent spike, which reinitiates the cycle by activating again the Na^+/Ca^{2+} action potential (sodium spike–ADP sequence). The cycle can thus repeat itself at 3–6 Hz without any external intervention (**Figure 17.6a**).

17.2.3 When hyperpolarized, inferior olivary cells oscillate at a higher frequency (9–12 Hz)

When inferior olivary cells are slightly hyperpolarized, their response to a depolarizing current pulse is characterized by cycles of low-threshold Ca^{2+} spikes activating one or two fast Na^+ spikes and followed by a pronounced after-hyperpolarization, at a frequency of 9–12 Hz (**Figure 17.6b**). The enhancement of rhythmic oscillations with hyperpolarization suggests that a depolarizing current such as I_h may contribute to these oscillations. I_h is activated upon membrane hyperpolarization, is carried by both Na^+ and K^+ ions and has a reversal potential around −30 to −40 mV. Therefore, at hyperpolarized potentials, I_h is inward and depolarizing.

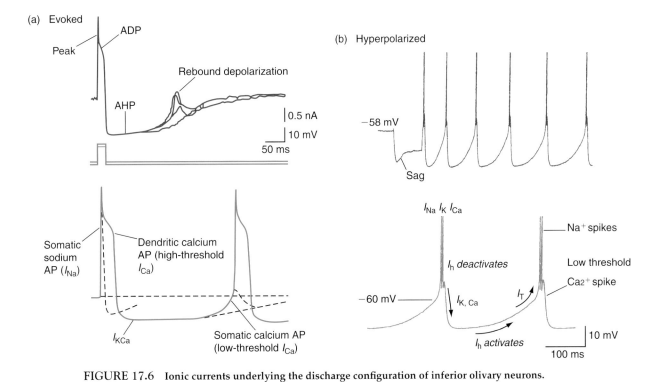

FIGURE 17.6 Ionic currents underlying the discharge configuration of inferior olivary neurons.
(a) In slightly depolarized cells, direct stimulation of the neuron by injecting a depolarizing current step evokes a sequence consisting of a TTX-sensitive action potential, followed by Ca^{2+}-dependent events, a plateau (ADP), a period of after-hyperpolarization (AHP) and a depolarizing rebound of variable amplitude (four superimposed top traces). Schematic of this discharge configuration and indication of the different currents sequentially activated (see text for explanation) (bottom trace). **(b)** Direct intracellular injection of a hyperpolarizing current pulse is associated with a depolarizing sag and the generation of a rhythmic sequence of low-threshold Ca^{2+} spikes (top trace). Schematic of this discharge configuration and indication of the different currents sequentially activated (see text for explanation) (bottom trace). Part (a) adapted from Llinas R and Yarom Y (1981) Properties and distribution of ionic conductances generating electroresponsiveness of mammalian inferior olive neurons *in vitro. J. Physiol. (Lond.)* **315**, 569–584, with permission. Part (b) adapted from Bal T and McCormick D (1997) Synchronized oscillations in the inferior olive are controlled by the hyperpolarization-activated cation current I_h. *J. Neurophysiol.* **77**, 3145–3156, with permission.

On the basis of pharmacological experiments, the following sequence of events is proposed to explain oscillations at hyperpolarized potentials. Activation of a somatic low-threshold Ca^{2+} spike which generates one or two Na^+ spikes is followed by an AHP, mediated largely by the activation of an apamin-sensitive Ca^{2+}-activated K^+ current. In addition, during the low-threshold Ca^{2+} spike, a portion of I_h is deactivated; this facilitates the generation of the AHP by allowing it to reach more negative potentials. The AHP subsequently results in two important effects: removal of inactivation of I_T and the activation of I_h. Activation of I_h depolarizes the membrane toward the threshold of activation of I_T and subsequently promotes the generation of a low-threshold Ca^{2+} spike and associated Na^+ action potentials and therefore reinitiates the oscillation (**Figure 17.6b**). In hyperpolarized olivary cells, 9–12 Hz oscillations are recorded owing to a decrease of the involvement of the high-threshold Ca^{2+} current, resulting in a shortening of the duration of the AHP.

When the membrane potential is in a region between rhythmic oscillations at hyperpolarized and depolarized membrane potentials, inferior olivary cells are silent.

17.3 PURKINJE CELLS ARE PACEMAKER NEURONS THAT RESPOND BY A COMPLEX SPIKE FOLLOWED BY A PERIOD OF SILENCE

Purkinje cells are located in the cerebellar cortex in the so-called Purkinje cell layer (see **Figure 6.8**). The dendritic tree of Purkinje cells in the rat receive about 175,000 excitatory glutamatergic synaptic contacts from parallel fibres of granule cells and around 300 from a single climbing fibre of an inferior olivary neuron. They also receive about 1500 GABAergic inputs from local interneurons. However, even when deconnected from these inputs, Purkinje cells present a tonic, single-spike, spontaneous activity – thus called *intrinsic*.

17.3.1 Purkinje cells present an intrinsic tonic firing that depends on a persistent Na^+ current

Cerebellar Purkinje neurons *in vivo* show high-frequency, regular spontaneous firing that is independent of synaptic activity since it is still recorded in cerebellar slice preparations or cultured Purkinje neurons when synaptic activity is blocked or in isolated Purkinje neurons (**Figure 17.7a**). TTX abolishes this intrinsic firing in all cells tested (**Figure 17.7b**), whereas Ca^{2+} channel blockers did not suppress it (not shown), suggesting

that it consists of Na^+-dependent spikes. How are these spikes generated in the absence of synaptic activity?

Spikes are generated by the spontaneous depolarization that, between consecutive spikes, depolarizes the membrane from the peak of the AHP to the threshold potential of the following spike. This phase of slow depolarization is called *pacemaker potential* or *pacemaker depolarization*, by analogy with pacemaker activity of cardiac cells. To identify the ionic currents that flow during spontaneous activity, previously recorded action potentials are used as voltage commands, and ionic currents during these voltage commands are recorded in voltage clamp (**Figure 17.7c**). This shows that pacemaker depolarization depends mainly on a persistent TTX-sensitive Na^+ current (I_{NaP}) (see Section 14.2.1) present in the cell body of Purkinje cells.

Another key factor allowing spontaneous firing is the lack of active K^+ currents between -70 and -50 mV which allows a high input membrane resistance during the pacemaker depolarization (note that these membrane properties are just the opposite of that of striatal medium spiny neurons; see Section 17.1.1). Thus, initially a small Na^+ current can depolarize the membrane to the threshold potential of Na^+ spikes. Moreover, the cationic I_h present in these cells may also play a role at the beginning of the pacemaker depolarization. The K^+ currents that repolarize the spikes in Purkinje neurons are notable for their very fast deactivation so that the membrane does not hyperpolarize very deeply (there is not a prominent AHP). The rapid deactivation of K^+ current also returns the input resistance to a high value within milliseconds so that the small interspike I_h and I_{NaP} can effectively depolarize the membrane for another action potential. These Na^+ spikes then passively propagate in the dendritic tree (see **Figure 17.9**).

17.3.2 Purkinje cells respond to climbing fibre activation by a complex spike

We will study the response of Purkinje cells to one of its excitatory afferents, the climbing fibres that are the axons of inferior olivary neurons. In the adult, a single climbing fibre innervates each Purkinje cell. This innervation has the following particular characteristic. The climbing fibre winds itself around the dendrites making a great number of 'en passant' boutons along its course (see **Figures 6.8** and **6.9**). These synapses are excitatory and the neurotransmitter is an excitatory amino acid, probably glutamate. The activation of a climbing fibre thus causes a massive all-or-none depolarization of the dendritic arborization and an activation of the high-threshold Ca^{2+} channels present at different points along the dendrites (see **Figures 16.10** and **17.9**).

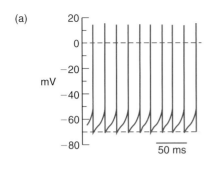

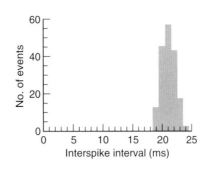

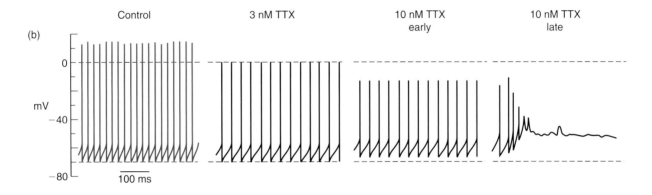

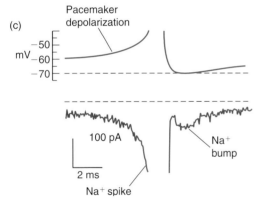

FIGURE 17.7   **The intrinsic tonic firing of isolated Purkinje cells.**
(a) Spontaneous action potentials recorded from an isolated Purkinje neuron in control conditions (left) and interspike interval histogram for the same cell (right). Dotted lines indicate -70 and $0\,mV$. (b) Spontaneous firing in control extracellular medium and in the presence of TTX as indicated. This cell continues to fire for some time in $10\,nM$ of TTX (early) before silencing and resting at $-51\,mV$ (late). (c) Kinetics of Na^+ currents evoked by the spike train protocol. The spike train in (a) is used as a command voltage (top trace) and the currents evoked are recorded in voltage clamp (bottom trace). The first $13\,ms$ are shown. The arrow indicates the bump of Na^+ current that occurs when the action potential command reaches its trough. Spike and bump Na^+ currents are sensitive to TTX (not shown). Adapted from Raman IM and Bean BP (1999) Ionic currents underlying spontaneous action potentials in isolated Purkinje neurons. *J. Neurosci.* **19**, 1663–1674, with permission.

Thus, in response to the activation of a climbing fibre, several Ca^{2+} action potentials are generated in the dendrites. These Ca^{2+} action potentials propagate passively along the dendrites, summing together and depolarizing the axon initial segment to the threshold for triggering sodium action potentials (**Figure 17.8a**). Ca^{2+} spikes force the cell to respond with a high-frequency burst of Na^+ spikes at the level of the soma and axon. Afferent information coming from the inferior olive is thus amplified.

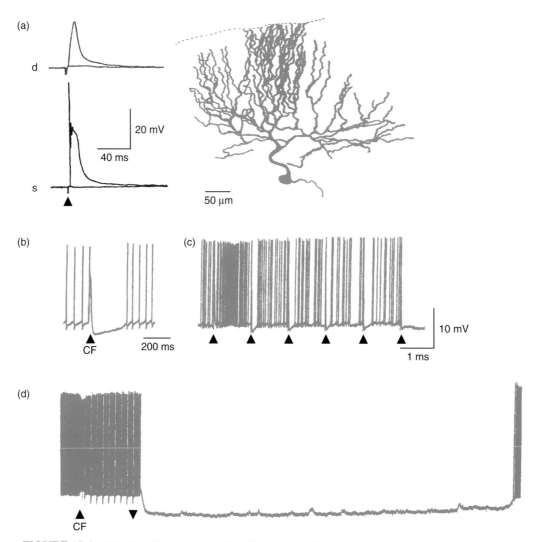

FIGURE 17.8   Climbing fibre response of Purkinje cells and its after-effect.
(a) All-or-none dendritic (d, top) and somatic (s, bottom) climbing fibre response. The position of the traces relative to the drawing of the recorded Purkinje cell indicates the recording sites. **(b)** Climbing fibre (CF) response followed by a transient inactivation of spontaneous firing. **(c)** Climbing fibre stimulation at 1 Hz (arrowheads). **(d)** At a slower sweep speed, the long-lasting hyperpolarization following a train of climbing fibre stimulation at 1 Hz is shown. Adapted from Hounsgaard J and Mitgaard J (1989) Synaptic control of excitability in turtle cerebellar Purkinje cells. *J. Physiol. (Lond.)* **409**, 157–170, with permission.

The climbing fibre response is followed by a long-lasting hyperpolarization (**Figure 17.8b**) that is abolished in the presence of TEA, but unaffected by TTX, thus suggesting that it is mediated by a voltage-sensitive, Ca^{2+}-dependent K^+ current. Repeated activation of the climbing fibre gradually induces an additional hyperpolarization with a much longer time course that is accompanied by a reduction of the frequency of Na^+ spikes (**Figure 17.8c**). Therefore, climbing fibre responses are potent regulators of Purkinje cell excitability. For example, in cells with a high spontaneous firing rate, climbing fibre responses evoked at 10 Hz shift the membrane potential by 10–15 mV, to a level well below the threshold for Na^+ spikes (**Figure 17.8d**). The nonlinear membrane properties of the soma-dendritic membrane of Purkinje cells are such that only small changes in current are needed to shift the membrane potential in the depolarizing or hyperpolarizing direction.

Summary

Purkinje cells are not silent at rest, they display a tonic firing mode of Na^+-dependent action potentials that depends on the depolarizing drive of an intrinsic persistent Na^+ current. These action potentials passively backpropagate in the dendritic tree. In response to

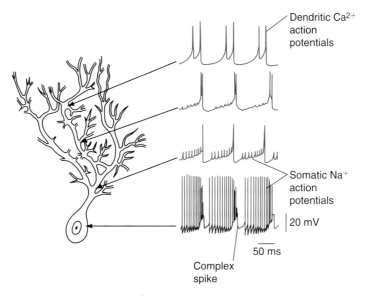

FIGURE 17.9   Integration of Na$^+$ and Ca^{2+} action potentials in Purkinje cells.
The activity of Purkinje cells is recorded intracellularly in the soma and in three different regions of the dendritic tree (current clamp mode) in cerebellar slices. Na$^+$-dependent action potentials are spontaneously evoked at the soma-axon hillock region. They passively backpropagate in the dendritic tree (note their rapid and strong diminution in amplitude). Ca^{2+}-dependent action potentials evoked at different points of the dendritic tree (in response to climbing fibre activation) propagate passively to the axon hillock region where they evoke the complex response followed by a period of cell silence. Adapted from Llinas R and Sugimori M (1980) Electrophysiological properties of *in vitro* Purkinje cell dendrites in mammalian cerebellar slices. *J. Physiol. (Lond.)* **305**, 197–213, with permission.

climbing fibre activation, a Na$^+$–Ca^{2+} spike is evoked (resulting from the activation of the high-threshold dendritic Ca^{2+} current and of the somatic Na$^+$ current). It is followed by a long-lasting inhibition of intrinsic tonic firing due to the activation of Ca^{2+}-activated K$^+$ currents. Therefore, repetitive activation of an excitatory input (climbing fibre) can lead to a long-lasting inhibition of Purkinje cells, as a result of the activation of intrinsic outward currents.

17.4 THALAMIC AND SUBTHALAMIC NEURONS ARE PACEMAKER NEURONS WITH TWO INTRINSIC FIRING MODES: A TONIC AND A BURSTING MODE

The *thalamus* relays and integrates information destined for the cerebral cortex (see **Figure 1.14**). It is formed from many nuclei which are classically separated into two groups: the specific nuclei and the non-specific nuclei, according to whether they project to a localized area of the cerebral cortex or to several functionally different areas. When recorded in brain slices *in vitro*, the thalamocortical and thalamic reticular neurons have complex intrinsic properties that allow them to display two firing patterns, a tonic one and a bursting one, depending on membrane potential (**Figure 17.10a**). Similarly, *in vivo*, during periods of slow-wave sleep, rhythmic burst firing is prevalent, whereas waking activity is dominated by the occurrence of trains of action potentials.

The *subthalamic nucleus* (STN) is part of basal ganglia. Its name comes from its localization ventral to the thalamus. STN controls the output of basal ganglia to the thalamus. It contains a homogeneous population of Golgi type I neurons that use glutamate as a neurotransmitter and project to both substantia nigra and the internal pallidal segment. Like thalamo-cortical neurons, when recorded in brain slices *in vitro*, STN neurons display two firing patterns, a tonic one and a bursting one, depending on the membrane potential (**Figure 17.10b**). In awake resting animals, STN neurons have a tonic mode of discharge; whereas during and after limb and eye movements as well as in a parkinsonian state, STN neurons discharge bursts of high-frequency spikes.

In both neuronal types, tonic firing is recorded at more depolarized potential than burst firing (**Figure 17.10**). Modulation of the membrane potential by the

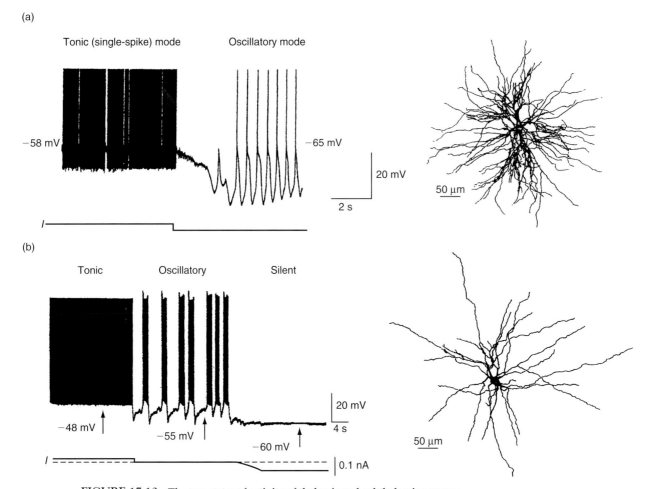

FIGURE 17.10   The two states of activity of thalamic and subthalamic neurons.
(a) The activity of a thalamocortical neuron (inset) is recorded in current clamp. When depolarized to –58 mV with intracellular injection of current, the neuron displays the tonic firing mode and switches to the oscillatory bursting mode when hyperpolarized. (b) The same protocol applied to a subthalamic neuron (inset) allows one to record the two firing modes. When the membrane is further hyperpolarized, the cell becomes silent. Part (a) adapted from McCormick DA and Pape HC (1990) Properties of a hyperpolarization-activated cation current and its role in rhythmic oscillation in thalamic relay neurons. *J. Physiol. (Lond.)* **431**, 291–318, with permission. Part (b) adapted from Beurrier C, Congar P, Bioulac B, Hammond C (1999) Subthalamic neurons switch from single-spike activity to burst-firing mode. *J. Neurosci.* **19**, 599–609, with permission. Drawings by Jérôme Yelnik.

activity of afferents thus plays an important role in the triggering of either one of the discharge configurations.

17.4.1 The intrinsic tonic (single-spike) mode depends on a persistent Na$^+$ current

Tonic activity of STN neurons recorded in slices *in vitro* is still present when blockers of synaptic transmission are added in the external medium, such as the Ca^{2+} channel blockers Co^{2+}, Cd^{2+} or Mn^{2+} (**Figure 7.11a**). This shows that the single-spike mode results from a cascade of voltage-gated currents intrinsic to the membrane. As in Purkinje neurons, spikes are generated by the spontaneous depolarization that, between consecutive spikes, depolarizes the membrane from the peak

of the AHP to the threshold potential of the following spike. TTX (1 μM) abolishes spontaneous firing, indicating that it consists of Na$^+$ spikes. Interestingly, at the onset of action of TTX, a few sub-threshold slow depolarizations that normally lead to spike firing are still observed (**Figure 17.11b**). As in Purkinje cells, this phase of slow depolarization, the pacemaker depolarization, depends mainly on the activation of a persistent TTX-sensitive Na$^+$ current (I_{NaP}) present in these neurons and which presents a voltage-dependency that allows it to be activated in the pacemaker range (**Figure 17.11c**).

The same ionic mechanism underlies tonic firing in thalamic neurons (see **Figure 17.14a**). It is important to note that in both cells the other key factor that allows

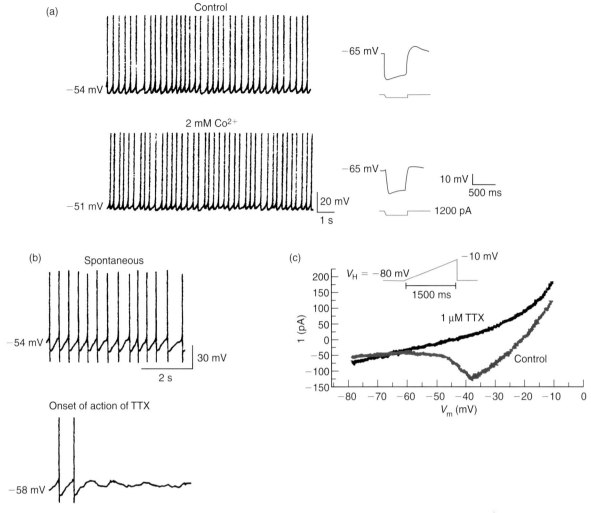

FIGURE 17.11 **Na$^+$ currents are critical for intrinsic tonic firing mode of subthalamic neurons.**
(a) Tonic activity of a STN neuron recorded in control medium and during application of Co^{2+} (left). Right traces show that the low-threshold Ca^{2+} spike evoked at the break of a hyperpolarization pulse is strongly decreased in Co^{2+} to attest that Ca^{2+} channels are effectively blocked in these conditions. **(b)** Tonic activity recorded in control medium and at the onset of TTX (1 μM) application. **(c)** Persistent Na$^+$ current recorded in whole-cell patch clamp in response to a depolarizing ramp (5 mV s^{-1}) in the absence (control) and presence of TTX. Parts (a) and (c) adapted from Beurrier C, Bioulac B, Hammond C (2000) Slowly inactivating sodium current (I_{NaP}) underlies single-spike activity in rat subthalamic nucleus. *J. Neurophysiol.* **83**, 1951–1957, with permission. Part (b) adapted from Bevan MD and Wilson CJ (1999) Mechanisms underlying spontaneous oscillation and rhythmic firing in rat subthalamic neurons. *J. Neurosci.* **19**, 7617–7628, with permission.

spontaneous firing is the weak presence of active K$^+$ currents between −70 and −50 mV which allows a high input membrane resistance. Thus, a small Na$^+$ current can depolarize the membrane to the threshold potential of Na$^+$ spikes. Moreover, in thalamic and subthalamic neurons, there is a significant contribution of I_h (see Section 14.2.3) to the resting membrane potential as shown by the hyperpolarizing effect of external Cs$^+$. This hyperpolarization is in general large enough to move the cell into the burst mode of action potential generation (see **Figure 17.14**), suggesting that the

fraction of I_h open at rest depolarizes the membrane and maintains it in a stable state where the neuron discharges in the single-spike mode.

17.4.2 The bursting mode depends on a cascade of subliminal inward currents: I_h, I_{CaT}, I_{CAN}

The burst of action potentials which rise from the peak of each slow depolarization or low-threshold spike (LTS) disappear in the presence of TTX. In contrast, LTS

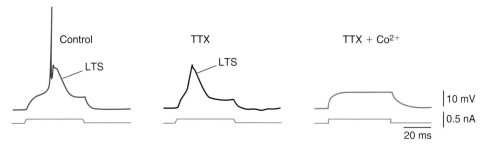

FIGURE 17.12   Thalamic oscillations depend on a low-threshold Ca²⁺ spike (LTS).
When the membrane of a thalamocortical neuron is hyperpolarized to −65 mV, a depolarizing current pulse evokes a LTS that is insensitive to TTX and abolished by Co^{2+} (1 mM). Note the presence of a TTX-sensitive Na^+ spike in control conditions. Adapted from Llinas R and Jahnsen H (1982) Electrophysiology of thalamic neurons *in vitro. Nature* **297**, 406–408, with permission.

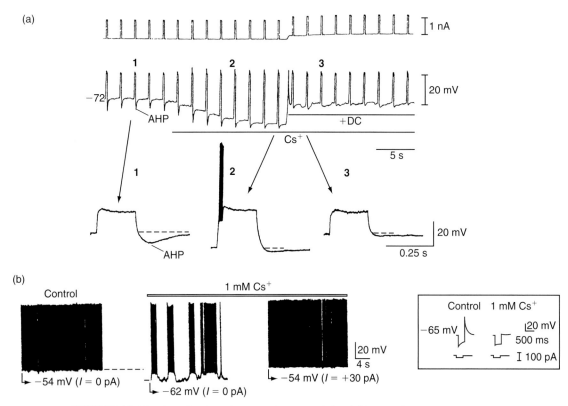

FIGURE 17.13   Contribution of I_h to resting potential and firing mode.
(a) Thalamocortical neuron. A depolarizing current pulse from resting potential (−72 mV) which does not result in a LTS (1) or the generation of action potential is applied. Cs application results in a substantial hyperpolarization of the membrane that de-inactivates the LTS thereby activating a burst of spikes (2). Compensation for the hyperpolarization with intracellular injection of current (+DC) reveals that the AHP is nearly abolished during Cs^+ (3). **(b)** Subthalamic (STN) neuron. In control conditions, at rest, a STN neuron discharges in the single-spike mode. Bath application of Cs^+ hyperpolarizes the membrane by 8 mV and shifts STN activity to burst firing mode. Continuous injection of positive current shifts the membrane potential back to the control value and to single-spike activity, though Cs^+ is still present. Concomitantly, the depolarizing sag in response to negative current pulse is strongly decreased as well as the depolarizing rebound seen at the break of pulse, to attest that I_h is strongly reduced (insets). Part (a) adapted from McCormick DA and Pape HC (1990) Properties of a hyperpolarization-activated cation current and its role in rhythmic oscillation in thalamic relay neurons. *J. Physiol. (Lond.)* **431**, 291–318, with permission. Part (b) adapted from Beurrier C, Bioulac B, Hammond C (2000) Slowly inactivating sodium current (I_{NaP}) underlies single-spike activity in rat subthalamic nucleus. *J. Neurophysiol.* **83**, 1951–1957, with permission.

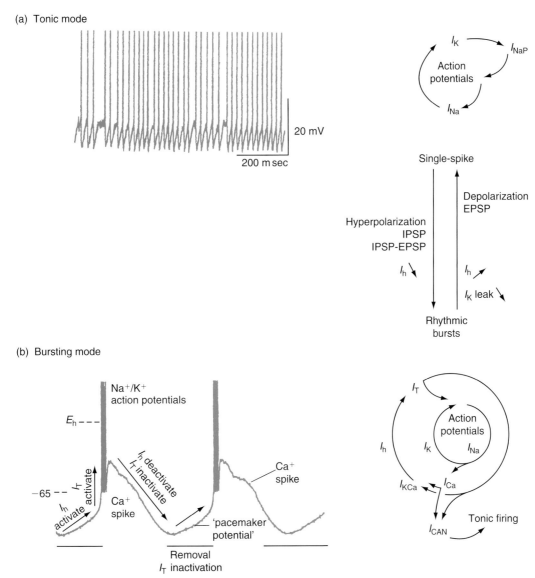

(a) Tonic mode

20 mV

200 m sec

I_K → I_{NaP}

Action
potentials

I_{Na}

Single-spike

Depolarization
EPSP

Hyperpolarization
IPSP
IPSP-EPSP

I_h

I_h
I_K leak

Rhythmic
bursts

(b) Bursting mode

E_h - - -

Na^+/K^+
action potentials

Ca^+
spike

I_h deactivate
I_T inactivate

−65 - - I_T activate
I_h activate

Ca^+
spike

'pacemaker
potential'

Removal
I_T inactivation

I_T

Action
potentials

I_h I_K I_{Na}

I_{KCa} I_{Ca}

I_{CAN} Tonic firing

FIGURE 17.14   Currents underlying the tonic and burst firing modes.
Recordings and scheme of the ionic basis of **(a)** the tonic mode and **(b)** the bursting mode of thalamic neurons.
Adapted from McCormick DA and Pape HC (1990) Properties of a hyperpolarization-activated cation current
and its role in rhythmic oscillation in thalamic relay neurons. *J. Physiol. (Lond.)* **431**, 291–318; and Bal T and
McCormick DA (1993) Ionic mechanisms of rhythmic burst firing and tonic activity in the nucleus reticularis
thalami, a mammalian pacemaker. *J. Physiol. (Lond.)* **468**, 669–691, with permission.

is not affected by TTX but disappears in the presence of Ca^{2+} channel blockers (**Figure 17.12**). This demonstrates that the fast action potentials are sodium spikes and that LTS results from a Ca^{2+} current. This slow depolarization appears only when the membrane has been previously hyperpolarized for at least 150 ms, suggesting that it results from a low-threshold-activated T-type Ca^{2+} current (I_{CaT}). This current is normally inactivated at resting membrane potential (or at potentials more depolarized than resting potential) and is de-inactivated by a transient hyperpolarization of the membrane. The low-threshold spike leads to the activation of a high-threshold Ca^{2+} current, the entry of Ca^{2+} ions (probably in the dendrites) and the activation of Ca^{2+}-sensitive K^+ currents (I_{KCa}). Each action potential is followed by a phase of after-hyperpolarization (see **Figure 17.14b**).

The hyperpolarization-activated cationic current (I_h) known to be present and activated in the oscillatory range in thalamic neurons also plays a role. For example, application of small amounts of Cs^+ hyperpolarizes the membrane and reduces the AHP. As already said, a fraction of I_h is open at rest and depolarizes the membrane. In addition, during the low-threshold Ca^{2+} spike and the generation of action potentials, a portion of I_h is deactivated (owing to the depolarization). This deactivation of I_h facilitates the generation of the AHP by allowing it to reach more negative potentials (see **Figure 17.14b**). This pronounced AHP subsequently results in two important effects: removal of inactivation of I_{CaT} and the activation of I_h. The latter in turn depolarizes the membrane potential toward the threshold for activation of I_{CaT} and subsequently promotes the generation of a low-threshold Ca^{2+} spike and associated Na^{2+}-dependent action potentials.

17.4.3 The transition from one mode to the other in response to synaptic inputs

When do thalamic and STN neurons discharge in a single spike? At resting membrane potential or at potentials more positive than rest, I_{CaT} (see Section 14.2.2) is inactivated and the regular frequency firing pattern can thus occur. In this state, an EPSP evokes a regular train of discharge.

When do thalamic and STN neurons discharge in bursting mode? Bursting mode requires that the membrane is at a potential more negative than the resting potential, so that I_{CaT} is de-inactivated and thus may be activated. Bursting state is present as long as the membrane is hyperpolarized. For example, at the break of an IPSP or in response to an EPSP evoked during or after an IPSP, a short sequence of bursts is recorded. In this case the bursting mode is transient; it is not a stable state. Unless hyperpolarized, thalamic and STN neurons discharge in single-spike mode. We see here that IPSP does not always mean inhibition of the post-synaptic neuron: when neurons have the ability to oscillate, an IPSP can evoke a burst of spikes (i.e. an excitation). This is observed, for example, in STN neurons, during and after the execution of a conditioned movement.

In vivo, thalamocortical neurons discharge in the single-spike or bursting mode, depending on the waking state of the animal: a stable bursting mode is observed during slow-wave sleep, a stable single-spike mode during waking or paradoxical sleep (see **Figure 17.15**). The transition from the electroencephalogram

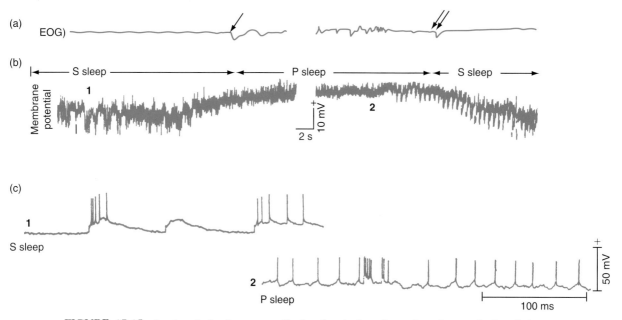

FIGURE 17.15 *In vivo*, **thalamic neurons display the single-spike or bursting mode, in relation to behavioural state.**
Simultaneous display of **(a)** eye movements (electro-oculogram, EOG) and **(b)** membrane potential of an intracellularly recorded thalamic neuron during slow-wave sleep (S sleep) and paradoxical sleep (P sleep) in an intact animal. **(b)** The neuron is already depolarized by 8 mV, when the animal enters P sleep (first eye movement, arrow). Depolarization is maintained throughout P sleep. Upon last eye movement (double arrow), membrane potential repolarizes as the animal goes back to S sleep (the trace is filtered at 0–75 Hz). **(c)** Enlarged sequences (labelled 1 and 2 under trace (b)) of spontaneous activities: 1, bursting mode during S sleep (hyperpolarized resting potential); 2, single-spike mode during P sleep (depolarized resting potential). Adapted from Hirsch J, Fourment A, Marc ME (1983) Sleep-related variations of membrane potential in the lateral geniculate body relay neurons of the cat. *Brain Res.* **259**, 308–312, with permission.

(EEG)-synchronized sleep to the waking or rapid-eye-movement (REM)-sleep state (paradoxical sleep) occurs with a progressive depolarization of thalamocortical cells and the abolition of intracellular slow oscillations (LTS) and burst firing, and the appearance of tonic activity. Such changes can be mimicked by the activation of muscarinic or glutamatergic metabotropic receptors that reduce a resting leak K^+ current in thalamocortical neurons (see **Figure 17.14**). The modulation of I_h can also play a role as seen in **Figure 17.13**. For example, activation of serotoninergic and β-adrenergic metabotropic receptors shifts the voltage-dependence of I_h to more positive membrane potentials. This reduces the ability of cells to oscillate. Together these results suggest that the release of acetylcholine, glutamate, serotonin and norepinephrine abolishes sleep-related activity in thalamocortical networks and facilitates the single-spike activity typical of the waking state.

Summary

Thalamic and subthalamic neurons can function either as relay systems or as oscillators. During oscillations afferent informations have a low probability of evoking a response. For example, when thalamic neurons are oscillating during slow-wave sleep, there is a marked diminution of responsiveness of thalamic neurons to activation of their receptive fields, 'presumably owing to the hyperpolarized state of these neurons, the interrupting effects of spontaneous thalamocortical rhythms and the frequency limitations of the burst firing mode'. Oscillations are also recorded in pathological conditions: in the STN of parkinsonian patients and in the thalamocortical networks during absence epileptic seizures. Noteworthy, during these oscillations, motor or sensory processing is unpaired.

FURTHER READING

Bal T, Von Krosigk M, McCormick DA (1994) From cellular to network mechanisms of a thalamic synchronized oscillation. In: Buzski G (ed.) *Temporal Coding in the Brain*, Berlin: Springer-Verlag.

Byrne JH (1980) Analysis of ionic conductance mechanisms in motor cells mediating inking behavior in *Aplysia californica. J. Neurophysiol.* **43**, 630–650.

Crépel F and Pénit-Soria J (1986) Inward rectification and low threshold calcium conductance in rat cerebellar Purkinje cells: an *in vitro* study. *J. Physiol. (Lond.)* **372**, 1–23.

Llinas RR (1988) The intrinsic electrophysiological properties of mammalian neurons: insights into central nervous system function. *Science* **242**, 1654–1664.

Llinas RR and Jahnsen H (1982) Electrophysiology of mammalian thalamic neurons. *Nature* **297**, 406–408.

McCormick DA and Bal T (1997) Sleep and arousal: thalamocortical mechanisms. *Ann. Rev. Neurosci.* **20**, 185–215.

CHAPTER

18

Synaptic plasticity

Synaptic responses undergo short- and long-term modifications. This chapter examines the mechanisms underlying plasticity in adult synapses. Developmental forms of plasticity are not covered here. There are two main forms of long-term changes of synaptic efficacy, long-term potentiation (LTP) and long-term depression (LTD). Moreover, there are several forms of LTP and LTD classified by their mechanisms. We have chosen two examples, the NMDA receptor-dependent LTP and the mGluR-dependent LTD. Before explaining these forms of LTP and LTD, we shall examine the meaning of 'long term' versus 'short term'.

18.1 SHORT-TERM POTENTIATION (STP) OF A CHOLINERGIC SYNAPTIC RESPONSE AS AN EXAMPLE OF SHORT-TERM PLASTICITY: THE CHOLINERGIC RESPONSE OF MUSCLE CELLS TO MOTONEURON STIMULATION

Repetitive high-frequency (>15 Hz) stimulation of the presynaptic element (motoneuron) leads to a short-term potentiation (STP) of the postsynaptic response of the muscle cell. As shown in **Figure 18.1**, successive stimulations produce in these conditions excitatory postsynaptic currents (EPSC) of greater and greater amplitudes. This phenomenon, first discovered at the neuromuscular junction, is also observed at the squid giant synapse and in mammalian afferent synapses to motoneurons.

In the squid giant synapse, synaptic facilitation has the following characteristics. When the presynaptic element repeatedly fires an increase of the postsynaptic

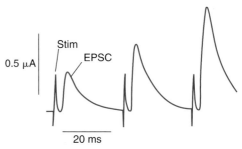

FIGURE 18.1 Presynaptic facilitation at the frog neuromuscular junction.
The activity of a frog sartorius muscle cell is recorded in normal Ringer solution (V_m = –90 mV). The motor endplate currents are evoked by repetitive stimulations (stim) of the motor nerve (2 μA intensity, 5 ms duration). The average current intensity (EPSC) in response to the first stimulation is 0.5 μA. This amplitude gradually rises following second and third stimulations. The inward currents are represented upwardly, which is unusual. Adapted from Katz B and Miledi R (1979) Estimates of quantal content during chemical potentiation of transmitter release. *Proc. R. Soc. Lond.* **B205**, 369–378, with permission.

response amplitude is observed. This increase diminishes with a time constant of the order of tens of milliseconds. Simultaneous recordings of presynaptic action potentials, presynaptic Ca^{2+} current (I_{Ca}), variations of the intracellular Ca^{2+} concentration and postsynaptic depolarization shows that the postsynaptic response amplitude increases when:

- the amplitude and length of presynaptic spikes are unchanged;
- the amplitude of the presynaptic I_{Ca} evoked by each presynaptic depolarizing pulse or action potential is constant;

- the increase of the presynaptic intracellular Ca^{2+} concentration is identical in response to each depolarizing pulse or action potential.

The increase of intracellular Ca^{2+} concentration ($[Ca^{2+}]_i$) in the presynaptic element slowly disappears, in about one second, whereas the Ca^{2+} current and the release of the neurotransmitter both last about 1 ms. Katz and Miledi, in 1965, were the first to propose that STP is due to residual Ca^{2+} ions still present in the presynaptic active zone when the second presynaptic spike occurs. The following hypothesis was proposed. Ca^{2+} ions enter the presynaptic element through voltage-gated Ca^{2+} channels opened by the depolarization. The intracellular Ca^{2+} concentration is very high at active zones at the end of the action potential. These Ca^{2+} ions act rapidly and locally on target molecules to trigger the exocytosis of synaptic vesicles with a probability p. At the same time, the Ca^{2+} ions are also buffered in the cytoplasm and are actively transported to the extracellular medium or inside the organelles (see Section 7.2.4). But a residual and quite high $[Ca^{2+}]_i$ is still present close to the presynaptic membrane for some time. This $[Ca^{2+}]_i$ value is not high enough to trigger neurotransmitter release, but added to the incoming increase of $[Ca^{2+}]_i$ accompanying the arrival of the second action potential (when the delay between the two action potentials is short) it increases neurotransmitter release probability to the second action potential, and thus causes potentiation of the postsynaptic response.

STP can also be induced by high-frequency stimulation (conditioning tetanus) of the afferent motoneuron (model of the crayfish neuromuscular junction) (**Figure 18.2a**). In that case, the postsynaptic response (EPSP) that is recorded at regular intervals after the tetanus is potentiated, and then decays to control amplitude within 1.5 s (**Figure 18.2b**, 1). In order to test the hypothesis of Katz and Miledi, a photolabile Ca^{2+} chelator, diazo-2, is injected into the presynaptic terminals. The motoneuron is penetrated at the level of an axon branch with a microelectrode containing KCl (to record presynaptic activity), the photolabile Ca^{2+} chelator diazo-2 (to chelate Ca^{2+} ions with an affinity of 150 nM after photolysis) and fluorescein (to monitor the progress of injection). First, the control STP is recorded (**Figure 18.2b**, 1). Then diazo-2 is injected into the presynaptic axon in order to test that before photolysis diazo-2 has little effect on STP since the unphotolyzed chelator has a low power to chelate Ca^{2+} ions (**Figure 18.2b**, 2). An ultraviolet flash is given after the tetanus in order to produce a chelator with 150 nM^{2+} affinity: the STP of the postsynaptic response is prevented (**Figure 18.2b**, 3).

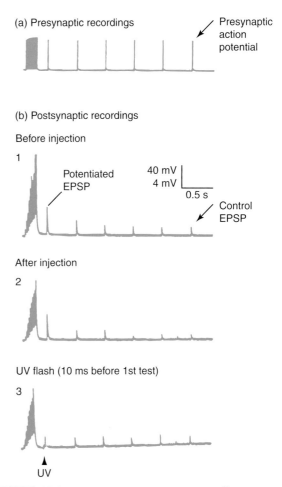

FIGURE 18.2 Rapid reduction of residual Ca^{2+} ions quickly eliminates STP.
The activity of the crayfish dactyl opener muscle cell is intracellularly recorded in current clamp mode in response to the stimulation of an axonal branch of the presynaptic motoneuron. The electrode positioned inside the presynaptic axon is filled with diazo-2 (50 mM) and fluorescein (10 mM) in KCl (3 M) in order to both stimulate the presynaptic axon and to fill it with the Ca^{2+} chelator. A conditioning tetanus (10 stimuli at 50 Hz) followed by a single stimulus at 2 Hz is applied to the axon. **(a)** Action potentials recorded from the preterminal axon branch. **(b)** The response (EPSP) of the postsynaptic muscle cell is recorded in control conditions (1), after the intracellular injection of diazo-2 (2) and after photolysis of diazo-2 by an ultraviolet flash given after the tetanus (3). Adapted from Kamiya H and Zucker RS (1994) Residual Ca^{2+} and short-term synaptic plasticity. *Nature* **371**, 603–606, with permission.

These results show that STP is due to residual-free Ca^{2+} ions following presynaptic activity. What are the molecular targets of Ca^{2+} action in short-term plasticity? Many candidates exist among vesicular, plasma membrane and cytoplasmic proteins of the presynaptic element (see Section 7.3). This identification awaits further experiments.

18.2 LONG-TERM POTENTIATION (LTP) OF A GLUTAMATERGIC SYNAPTIC RESPONSE: EXAMPLE OF THE GLUTAMATERGIC SYNAPTIC RESPONSE OF PYRAMIDAL NEURONS OF THE CA1 REGION OF THE HIPPOCAMPUS TO SCHAFFER COLLATERALS ACTIVATION

18.2.1 The Schaffer collaterals are axon collaterals of CA3 pyramidal neurons which form glutamatergic excitatory synapses with dendrites of CA1 pyramidal neurons

The hippocampus is a telencephalic structure with a rostrocaudal extension in the rat. It is composed of two closely interconnected crescent-like regions, Ammon's horn and the dentate gyrus (**Figure 18.3a**). Ammon's horn is formed by a layer of principal neurons, the pyramidal neurons, and is subdivided in three regions called CA1, CA2 and CA3 (CA for *cornu* a*mmonis*). The dentate gyrus is formed by a layer of principal neurons called *granular cells*. Numerous interneurons are present in each region (see also Chapter 19).

The pyramidal cells of CA3 have branched axons. One branch leaves the hippocampus and projects to other structures. The other branches are recurrent collaterals that form synapses with dendrites of CA1 pyramidal neurons. These collaterals run in bundles and form the Schaffer collateral pathway; and their terminals form asymmetrical synapses with the numerous spines of CA1 dendrites. These synapses are excitatory and the neurotransmitter is glutamate. Owing to the laminar organization of the hippocampal structure, it is possible to stimulate selectively the Schaffer collateral pathway (stim) and to record the evoked excitatory postsynaptic potential (EPSP) in a CA1 pyramidal soma either *in vivo* or *in vitro* (**Figure 18.3a**). EPSPs can be recorded from a single neuron (intracellular or whole-cell somatic recording) or extracellularly in the dendritic layer from a population of neurons (field EPSPs).

We shall restrict our study of long-term potentiation (LTP) to the synaptic response of CA1 pyramidal neurons to Schaffer collateral stimulation in hippocampal slices, recorded in the presence of bicuculline (an antagonist at $GABA_A$ receptors) in order to prevent the participation of GABAergic inhibitory responses resulting from interneuron activation.

18.2.2 Observation of the long-term potentiation of the Schaffer collateral-mediated EPSP

The pioneering observation was made by Bliss and Lomo, in 1973, that high-frequency stimulation of Schaffer collaterals in the rat hippocampus *in vivo* produces an increase of the amplitude of the Shaffer collateral-mediated EPSP recorded from the postsynaptic pyramidal neuron (**Figure 18.3b**). The exact protocol is the following. A single stimulus applied repeatedly at a very low frequency (0.02–0.03 Hz) to Schaffer collaterals evokes stable control EPSPs in the postsynaptic pyramidal neuron. These control responses can be averaged to give a mean control EPSP. Control EPSPs are largely mediated by non-NMDA receptors since they are nearly completely abolished by CNQX (not shown). A tetanic stimulation is then applied to the Schaffer collateral pathway through the same stimulating electrode (one train of 1 s duration, composed of 50–100 stimuli at 100 Hz). After this tetanus, the same single stimulus applied repeatedly at the same very low frequency, through the same stimulating electrode, now evokes 'post-tetanic' EPSPs of larger amplitude: the EPSP is potentiated. Since this potentiation lasts from minutes to hours it is a long-term potentiation (LTP).

Is LTP restricted to the synapses that have been tetanized?

Two EPSPs evoked in one pyramidal neuron in response to the stimulation of two different Schaffer collaterals inputs are recorded (**Figure 18.4a**). When only one input (S_1) is tetanized, the response evoked by a single shock at S_1 is potentiated (LTP of $EPSP_1$) whereas the response evoked in the same pyramidal neuron at S_2 ($EPSP_2$) is not potentiated: LTP is synapse-specific. In other words, when generated at one set of synapses by repetitive activation, LTP does not normally occur in other synapses on the same cell.

18.2.3 Long-term potentiation (LTP) of the glutamatergic EPSP recorded in CA1 pyramidal neurons results from an increase of synaptic efficacy (or synaptic strength)

LTP of the Schaffer collateral-mediated EPSP can have several origins. We shall study some of the hypotheses one by one.

Does LTP result from a non-specific change of postsynaptic cell excitability?

To test this hypothesis, a pulse of depolarizing current is injected directly into the pyramidal cell. The response of the membrane is the same before and after the tetanus, at all potentials tested. Therefore, LTP does not result from a change of the total resistance of the postsynaptic membrane. It is also not due to a persistent reduction of the inhibitory GABAergic responses

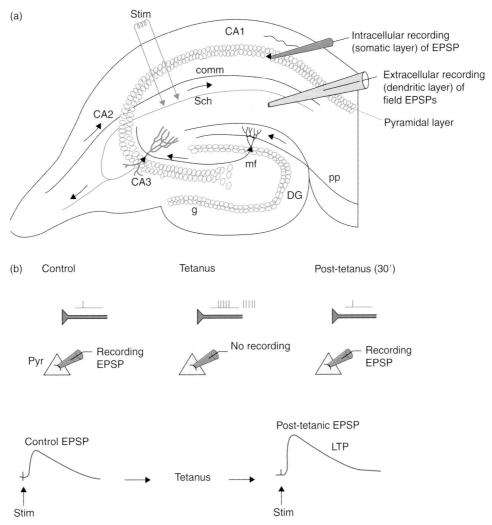

FIGURE 18.3 **Tetanic LTP in the hippocampus is induced by high-frequency stimulation of afferent fibres.**
(a) Coronal section of the rat hippocampus showing the major excitatory connections. CA1, CA2, CA3 regions of the hippocampus; DG, dentate gyrus layer composed of granular cells (g) which send their axons (mossy fibres, mf) to CA3 pyramidal dendrites. CA3 pyramidal cells send axon collaterals, called Schaffer collaterals (Sch), to CA1 pyramidal apical dendrites. The tetanic stimulation (stim, e.g. 1–4 trains of 10 stimulations at 100 Hz applied every 1 s) is applied to Schaffer collaterals and the AMPA-mediated postsynaptic response is recorded intracellularly in the soma of a pyramidal cell (EPSP or EPSC) and/or extracellularly in the layer of CA1 pyramidal dendrites. comm, commissural fibres; pp, perforant path. **(b)** A CA1 pyramidal neuron represented upside down compared with its position in the coronal section and the afferent Schaffer collaterals. The AMPA-mediated EPSP evoked by a single stimulation of Schaffer collateral (one vertical bar) is intracellularly recorded in the presence of bicuculline. After a tetanic stimulation of the Schaffer collaterals (shown as high-frequency bars), a potentiation of the glutamatergic EPSP evoked by a single stimulation is recorded. Adapted from Kauer JA, Malenka RC, Nicoll RA (1988) Persistent postsynaptic modification mediates long-term potentiation in the hippocampus. *Neuron* **1**, 911–917, with permission.

since it is still observed in the presence of bicuculline, a GABA$_A$ receptor antagonist.

Does LTP result from an increase in the number of stimulated axons?

One way to answer this question is to record simultaneously the presynaptic action potentials (afferent volley) and the postsynaptic response. This is possible with extracellular recordings at the level of apical dendrites of pyramidal cells. A single stimulus applied repeatedly at a low frequency (0.02–0.03 Hz) to the Schaffer collaterals evokes a stable 'control' field EPSP recorded by an extracellular electrode placed in the dendritic field of CA1 pyramidal neurons (**Figures 18.3a** and **18.4b**, left and middle). A field EPSP

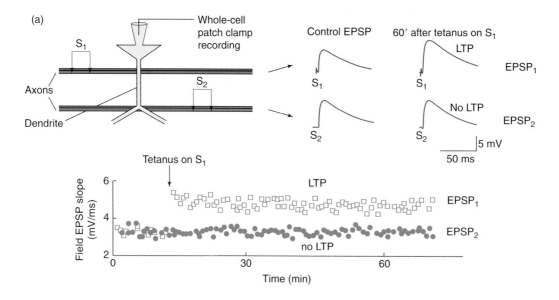

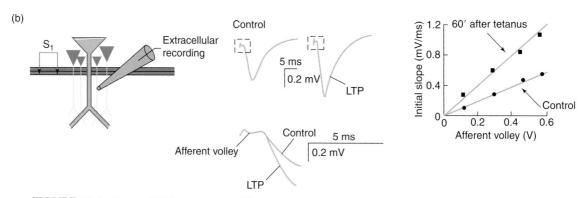

FIGURE 18.4   Tetanic LTP is synapse specific.
(a) The postsynaptic responses (control $EPSP_1$ and $EPSP_2$) of a single pyramidal neuron are intracellularly recorded in current clamp mode (whole-cell patch) in response to stimulations S_1 and S_2. Then, stimulus S_1 is tetanized but not stimulus S_2. Sixty minutes after the tetanus on S_1, $EPSP_1$ and $EPSP_2$ are recorded again. The diagram illustrates the time course of the initial slope of $EPSP_1$ and $EPSP_2$ before and after the tetanus on S_1. **(b)** Extracellular recording of the response of a population of CA1 pyramidal neurons to stimulation (S_1) of afferent Schaffer collaterals. The stimulation S_1 evokes an afferent volley (the extracellular recording of presynaptic action potentials in all stimulated afferent axons) and a field EPSP (the extracellular recording of the postsynaptic response of pyramidal neurons). Sixty minutes after a tetanus (two trains of $100\,Hz$, $1\,s$ duration, $30\,s$ interval) applied through the same stimulating electrode, the field EPSP is recorded. Note the increased initial slope $60\,min$ after the tetanus (enlarged dotted squares). The input/output curves depict the amplitude of the afferent volley versus the initial slope of the field EPSP. Part (a) from L. Aniksztejn and (b) from H. Gozlan, personal communications.

corresponds to the response of a population of pyramidal neurons situated close to the recording electrode and connected to the stimulated axons. Its slope is proportional to the amplitude of the currents generated in the postsynaptic neurons. After the stimulating artefact, before the field EPSP develops, the afferent volley is recorded (inset).

As a control intracellular EPSP, the control field EPSP is mediated predominantly by non-NMDA receptors since it is nearly completely abolished by the bath application of CNQX (not shown), a selective antagonist of AMPA receptors. A tetanic stimulation (one train of $1\,s$ duration, composed of 50–100 stimuli at $100\,Hz$) is then applied to the Schaffer collateral pathway through the stimulating electrode. After this tetanus, the same single stimulus, again through the same stimulating electrode, now evokes a 'post-tetanic' field EPSP of larger amplitude and with a steeper initial slope than the control one: the field EPSP is potentiated (LTP). The value of the initial slope of a field EPSP (or of an intracellular EPSP) is an accurate index of the changes of the monosynaptically evoked postsynaptic excitatory response since the field EPSP (as well as the intracellular EPSP) can be composed of monosynaptic as well as polysynaptic unitary EPSPs. This potentiation of EPSP amplitude and slope is persistent: it lasts hours when recorded in the *in vitro* hippocampal slice preparation and days when induced in the freely moving animal. However, the afferent volley (the presynaptic component) is unchanged.

LTP is an increase of synaptic strength

Following a tetanic stimulation, the presynaptic component (the afferent volley) is unchanged whereas the peak amplitude and the initial slope of the postsynaptic one (field EPSPs) are potentiated (by 30% and 200%, respectively). The input/output curve depicting the initial slope of the field EPSPs versus afferent volley amplitude has a different slope before (control) and 60 minutes after the tetanus (**Figure 18.4b**, right). This result shows that potentiation of the postsynaptic response does not result from an increase of the number of stimulated axons but from a genuine increase in synaptic efficacy: the same input evokes an enhanced output.

LTP consists of two phases: induction and maintenance

LTP-generating mechanisms are classically separated into two phases: a brief induction phase (1–20 s) which occurs during tetanus, and a following expression phase; i.e. the mechanisms sustaining the persistent enhancement of synaptic efficacy. Therefore LTP is triggered rapidly (within seconds) whereas it is maintained for long periods of time (for hours in *in vitro* preparations and days *in vivo*). We will now analyze these two phases.

18.2.4 Induction of LTP results from a transient enhancement of glutamate release and a rise in postsynaptic intracellular Ca^{2+} concentration

Why is tetanic stimulation necessary to induce LTP? What does tetanic stimulation add to a single shock stimulation?

Tetanic stimulation evokes a large release of glutamate from Schaffer collateral terminals compared with a single shock. Glutamate released in synaptic clefts during tetanus binds to non-NMDA and NMDA receptor channels but also to the metabotropic glutamate receptors (receptors linked to G-proteins) present

in the postsynaptic membrane (see **Figure 18.7**). The fact that tetanic stimulation can be replaced by a pairing diagram consisting of low-frequency stimulation of Schaffer collaterals combined with the intracellular depolarization of the postsynaptic neuron suggests that one of the roles of the tetanus is to depolarize the postsynaptic membrane. This is confirmed by the following experiment. When the postsynaptic potential is hyperpolarized during the tetanic stimulation, LTP is not induced. The tetanus-induced depolarization is the large EPSP recorded during the tetanus and which results from the strong activation of AMPA receptors (see **Figures 18.5(2)** and **(4)** and **18.7**).

What does induce postsynaptic depolarization?

Several observations led to the conclusion that in CA1, induction of LTP is not just depolarization-dependent but also NMDA receptor-dependent: the application of APV, the selective antagonist of NMDA receptors, during the tetanic simulation prevents the induction of LTP (**Figure 18.5**). In contrast, antagonists of non-NMDA receptors such as CNQX, applied during the tetanus, do not prevent the induction of LTP. Therefore, induction of LTP is voltage- and NMDA receptor-dependent. These results confirm that, in the CA1 region of the hippocampus, tetanus induces a postsynaptic *depolarization* generated by the enhancement of glutamate release. This postsynaptic depolarization is a necessary prerequisite for LTP induction, because it allows the activation of NMDA receptors.

What is the role of postsynaptic NMDA receptor activation?

To explain the APV-sensitivity of LTP, the following hypothesis is proposed. The postsynaptic depolarization evoked by the tetanic stimulation allows the activation of postsynaptic NMDA receptors and the subsequent Ca^{2+} entry into the postsynaptic element. In fact, during tetanic stimulation an increase of $[Ca^{2+}]_i$ is observed in the dendrites of the postsynaptic pyramidal cells as visualized with a fluorescent calcium-sensitive dye. When this transient elevation of $[Ca^{2+}]_i$ is prevented by the intracellular injection of a Ca^{2+} chelator agent (BAPTA) in the recorded pyramidal cell before the tetanus or by a strong postsynaptic depolarization which decreases the driving force for Ca^{2+} entry, LTP is not observed or is reduced. A simultaneous extracellular recording of the field EPSP shows that LTP is, however, generated in the other stimulated cells (which were not injected with BAPTA or depolarized). These results indicate that an increase of $[Ca^{2+}]_i$ is essential for the induction of LTP.

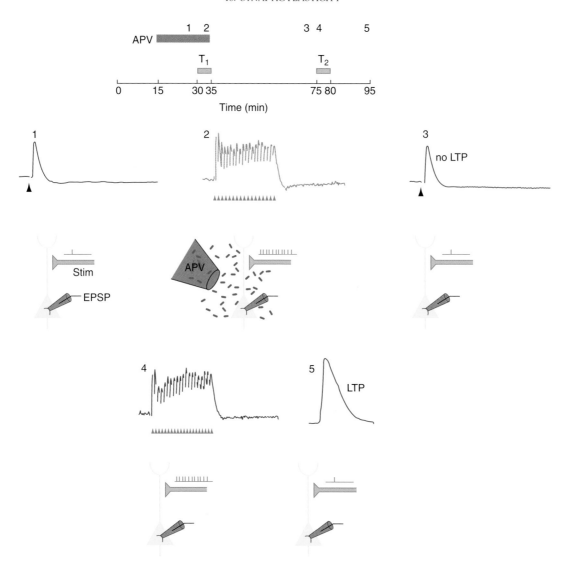

FIGURE 18.5 **NMDA receptor activation is required for LTP induction.**
An intracellular glutamatergic EPSP is evoked by Schaffer collateral stimulation (1). D-APV (20 μM, black bar)
is applied before and during the tetanus (T_1, 2). LTP is not induced since the EPSP recorded one hour after
wash of APV (3) has the same peak amplitude as the control one (1). A second tetanus (T_2, 4) is applied in the
absence of D-APV; the EPSP is now potentiated (LTP, 5). Note that APV evokes only a small change of the
depolarization of the membrane during the tetanus (compare 2 and 4). T_1 and T_2 are identical periods of
tetanic stimulation composed of 10–12 high-frequency trains at 30 s interval. Each train comprised 20 stimulations at 100 Hz. Adapted from Collingridge GL, Herron CE, Lester RAJ (1988) Frequency-dependent
N-methyl-D-aspartate receptor-mediated synaptic transmission in rat hippocampus. *J. Physiol.* **399**, 301–312,
with permission.

*For how long must $[Ca^{2+}]_i$ remain increased in the
postsynaptic element to trigger LTP?*

The duration for which $[Ca^{2+}]_i$ must remain
elevated to induce LTP was tested by injecting into
the recorded neuron a photosensitive Ca^{2+} chelator.
This compound, diazo-4, has a low affinity for Ca^{2+}
($K_D = 89 \mu M$) which can be suddenly (in 100–400 μs)
increased ($K_D = 0.55 \mu m$) when a UV flash inducing

the photolysis of its diazo-acetyl groups is applied to
the cell (**Figure 18.6a**). Thus, introduction of diazo-4
into a cell does not affect ambient Ca^{2+} levels before
the application of UV light. The manipulation of the
delay between the LTP inducing tetanus and photolysis of diazo-4 allows one to determine the minimum
duration of postsynaptic $[Ca^{2+}]_i$ increase necessary to
induce LTP. When Ca^{2+} is chelated by diazo-4 photolysis 2.5 s or more after the tetanus, LTP is still induced

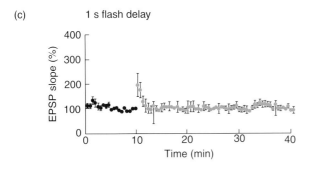

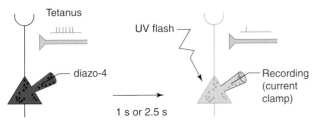

FIGURE 18.6 Photolysis of diazo-4, 1 s after the start of the tetanus, prevents LTP.
The activity of CA1 pyramidal cells is recorded in current clamp mode (whole-cell configuration) in hippocampal slices. The whole-cell electrode contains diazo-4, a Ca^{2+} chelator (1–2.5 mM). **(a)** Structure of diazo-4 before and after photolysis. **(b)** diazo-4 is photolyzed 2.5 s or 4 s following the start of the tetanus (stimuli given at 100–200 Hz for 1 s, from time 10 min). Even after this short delay, LTP of the glutamatergic EPSP is induced ($n = 8$). **(c)** Photolysis of diazo-4 immediately at the end of the 1 s duration tetanus (given at time 10 minutes) prevents the induction of LTP of the glutamatergic EPSP ($n = 5$). In the same experiments, LTP of the field (extracellular) EPSP is observed (not shown). Adapted from Malenka RC, Lancaster B, Zucker RS (1992) Temporal limits on the rise in postsynaptic calcium required for the induction of long-term potentiation. *Neuron* **9**, 121–128, with permission.

(**Figure 18.6b**). In contrast, if the UV flash follows the 1 s duration tetanus without delay, the induction of LTP is prevented (**Figure 18.6c**). Therefore an increase of $[Ca^{2+}]_i$ lasting at most 2.5 s (1 s during the tetanus plus 1.5 s after) is sufficient for LTP induction.

The hypothetical model for LTP induction

The following model is proposed to explain the induction of LTP in the CA1 region of the hippocampus.

Before tetanus

A single shock evokes the release of glutamate from the stimulated terminal. Glutamate activates postsynaptic non-NMDA and metabotropic glutamate receptors. NMDA receptors, owing to Mg^{2+} block, are weakly activated and contribute little to the basal EPSP (**Figure 18.7a**).

During tetanus

The high-frequency stimulation (tetanus) activates a certain number of afferent axons (**Figure 18.7b**). This enhances the release of glutamate from the stimulated terminals, thus evoking a postsynaptic depolarization due to the inward current through postsynaptic non-NMDA (AMPA) receptors and probably also the activation of metabotropic glutamate receptors (mGluR). Activation of AMPA receptors depolarizes the postsynaptic elements (spines) to the point where the Mg^{2+} blockade of the NMDA receptors is removed, thus allowing the influx of Ca^{2+} ions into the spines through NMDA channels and a further depolarization of the membrane. The depolarization of synaptic origin can also bring the postsynaptic membrane to the threshold for dendritic voltage-gated Ca^{2+} channel activation allowing an additional Ca^{2+} entry. The short-lasting (few seconds) rise in $[Ca^{2+}]_i$ resulting from NMDA and Ca^{2+} channel activation provides the necessary trigger for the subsequent events: activation of Ca^{2+}-dependent protein kinases and other Ca^{2+}-dependent processes, which lead to the expression of LTP; i.e. a persistent increase in synaptic efficacy (see maintenance or expression, Section 18.2.5).

Metabotropic glutamate receptors are modulators that regulate the threshold of induction of NMDAR-dependent LTP

In the presence of t-ACPD, a selective agonist of metabotropic glutamate receptors (mGluRs), a subthreshold tetanus (which alone triggers only short-term potentiation), now generates a LTP (**Figure 18.8**). This effect is blocked by APV, a selective antagonist of

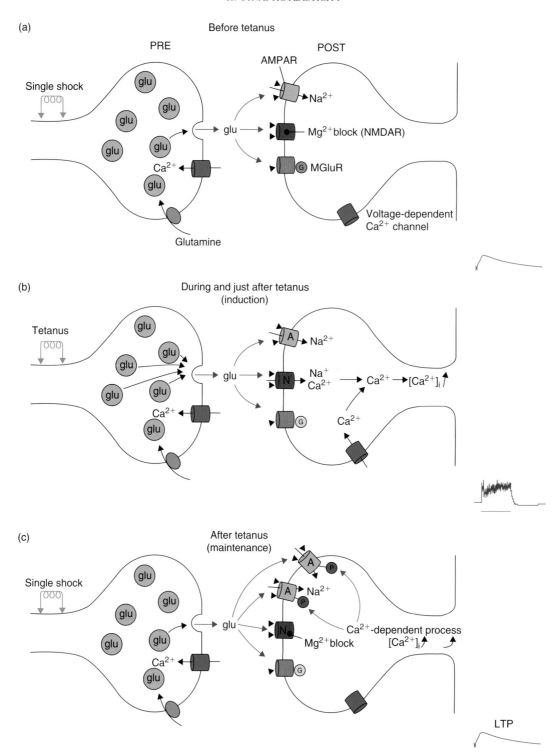

FIGURE 18.7 Schematic on the role of NMDA receptors and intracellular Ca²⁺ ions in the induction and maintenance of LTP.
See text for explanation.

NMDA receptors and by protein kinase C (PKC) inhibitors. It indicates that activation of metabotropic glutamate receptors reduces the threshold of LTP induction, an effect mediated by NMDA receptors and protein kinase C. This effect is specific to mGluRs since application in similar conditions of agonists of iGluRs, AMPA or NMDA, in addition to the sub-threshold tetanus fails to trigger LTP. In control situations (in the

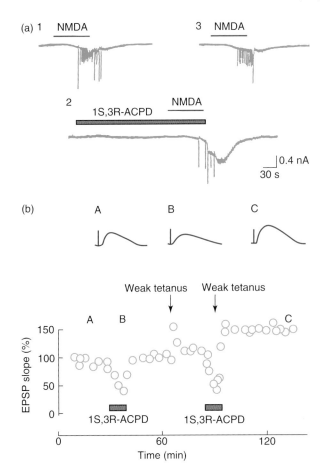

FIGURE 18.8 mGluRs activation potentiates NMDA-mediated currents and facilitates LTP induction of AMPA-mediated EPSP. (a) The activity of a CA1 pyramidal neuron is intracellularly recorded in single-electrode voltage clamp mode in slices. The external solution contains TTX to block synaptic activity and K^+ channel blockers and the intracellular electrode is filled with CsCl. Bath application of NMDA ($10\,\mu M$, 90 s) evokes an inward current (1) with rapid inward voltage-gated Ca^{2+} currents evoked in unclamped regions of the neuronal membrane. Bath application of 1S,3R-ACPD ($50\,\mu M$, 4 min), a mGluR agonist, before and during NMDA application ($10\,\mu M$, 90 s) potentiates the NMDA-evoked current (2). This effect is reversible since 5 minutes after washing NMDA ($10\,\mu M$, 90 s) evokes an inward current (3) of similar amplitude to the one observed in control (1). **(b)** The activity of a CA1 pyramidal neuron is intracellularly recorded in current clamp mode in slices. The diagram shows the amplitude of the initial slope of the AMPA-mediated EPSP recorded in response to Schaffer collaterals stimulation. Bath application of 1S,3R-ACPD ($50\,\mu M$, 2 min) reversibly depresses the EPSP (compare B with A). A sub-threshold tetanic stimulation of Schaffer collaterals (weak tetanus: stimuli at 50 Hz for 0.5 s) induces a short-term potentiation of the EPSP (trace not shown). The same weak tetanus given during bath application of 1S,3R-ACPD ($50\,\mu M$, 2 min) now induces a long-term potentiation of the EPSP (C). Part (a) adapted from Ben Ari Y and Aniksztejn L (1995) Role of glutamate metabotropic receptors in long-term potentiation in the hippocampus. *Sem. Neurosci.* **7**, 127–135, with permission. Part (b) adapted from Aniksztejn L, Otani S, Ben Ari Y (1992) Quisqualate metabotropic receptors modulate NMDA currents and facilitates induction of LTP through protein kinase C. *Eur. J. Neurosci.* **4**, 500–505, with permission.

absence of tetanus) a link between mGluRs, protein kinase C and NMDA receptors is suggested by numerous experiments.

In intracellular recordings of pyramidal neurons of the CA1 region of the hippocampus, the mGluR agonist t-ACPD enhances the current generated by NMDA but not by AMPA applications, in the presence of TTX (to block action potentials and therefore network activity) and K^+ channels blockers.

This effect is blocked by the intracellular injection of a protein kinase C inhibitor. The intracellular injection of protein kinase C enhances the NMDA receptor-mediated current.

- Protein kinase C phosphorylation sites are present on NMDA receptors.
- In oocytes transfected with cDNAs coding for mGluRs and NMDA receptors, the mGluR agonist t-ACPD increases NMDA currents, an effect blocked by protein kinase C inhibitors.
- In a wide range of cell types, kinases and phosphatases modulate rapidly and reversibly NMDA receptor activity.

These results suggest that the activation of postsynaptic mGluRs enhances (via protein kinase C) the postsynaptic NMDA receptor-mediated current (activated by the release of glutamate evoked by a sub-threshold tetanus applied to the afferents). This enhancement of NMDA receptor-mediated response, together with the activation of AMPA receptors, induces LTP of the glutamatergic AMPA receptor-mediated response.

18.2.5 Expression of LTP (also called maintenance) involves a persistent enhancement of the AMPA component of the EPSP

Owing to the Mg^{2+} block of NMDA receptors, the control glutamatergic EPSP recorded in CA1 pyramidal neurons (in the presence of physiological concentrations of Mg^{2+}) is mainly mediated by non-NMDA receptors (**Figure 18.7a**) since it is negligibly affected by APV (the selective antagonist at NMDA receptors) and nearly completely blocked by CNQX. The same analysis of the relative contribution of NMDA and non-NMDA receptors was applied to the potentiated EPSP after a tetanus.

The EPSC (postsynaptic excitatory current) evoked in CA1 pyramidal neurons in response to Schaffer collaterals stimulation is recorded with patch clamp techniques (whole-cell patch) at two different holding potentials. At $V_H = -80\,mV$, the EPSC is mainly mediated by AMPA receptors owing to the Mg^{2+} block of

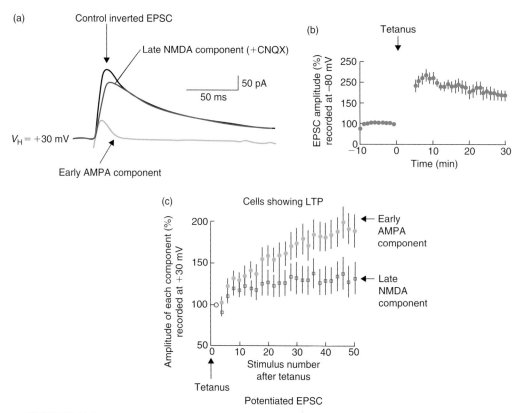

FIGURE 18.9 Differential enhancement of the non-NMDA and NMDA components in LTP.
The excitatory postsynaptic current (EPSC) evoked by Schaffer collaterals stimulation is recorded in a CA1 pyramidal neuron in hippocampal slices with patch clamp techniques (whole-cell patch). **(a)** At $V_H = +30\,mV$, the EPSC is inverted (control). Application of the non-NMDA selective antagonist (CNQX) selectively reduces the early component of the current, leaving the late component (NMDA receptor-mediated) unaffected. The subtracted record (sub = control – CNQX insensitive) illustrates the time course of the AMPA receptor-mediated component (CNQX-sensitive). **(b)** The EPSC peak amplitude (expressed as a percentage of the control) is plotted against time, before and after the tetanus applied to evoke LTP ($V_H = -80\,mV$, except during the tetanus). The total EPSC is clearly potentiated by the procedure. **(c)** The EPSC peak amplitude, expressed as a percentage of the control, is measured from just after the induction of LTP to 30 min after ($V_H = +30\,mV$). The early (CNQX-sensitive) AMPA receptor-mediated component is clearly potentiated while the late (CNQX-insensitive) NMDA receptor-mediated component is not significantly potentiated. Recordings in (b) and (c) are from the same cell; to obtain the curves in (b) and (c), the membrane potential is continuously shifted from –80 to +30 mV. Adapted from Perkel DJ and Nicoll RA (1993) Evidence for all or none regulation of neurotransmitter release: implications for long-term potentiation. *J. Physiol.* **471**, 481–500, with permission.

NMDA receptors at this hyperpolarized potential. In contrast, at $V_H = +30\,mV$ the control EPSC (which is inverted since the reversal potential of the glutamate response is 0 mV) is mixed and mainly mediated by NMDA receptors as shown by the small effect of CNQX (**Figure 18.9a**). The early rising phase of the EPSC is mainly mediated by AMPA receptors, while the current measured 100 ms after the stimulation is mainly mediated by NMDA receptors.

The recorded cell is subjected to a procedure which induces LTP. After the tetanus, the membrane potential is returned to –80 mV and the test stimulation of Schaffer collaterals is regularly applied to verify that the EPSC is now potentiated (LTP has been induced)

(**Figure 18.9b**). This potentiated EPSC is also recorded at +30 mV in order to evaluate the amplitudes of the early AMPA and the late NMDA components. The early component (AMPA-mediated) approximately doubles while the late component (NMDA-mediated) remains rather stable (**Figure 18.9c**). Therefore LTP of the glutamatergic response, in this experiment, is primarily mediated by an enhancement of the AMPA component of the synaptic current.

This differential effect of the tetanus can be explained by:

- an increase in the density of AMPA receptors in the synaptic cleft (clustering);

- a change in the properties of AMPA receptors (affinity, unitary current amplitude);
- an increase in the effective spread of synaptic current from dendritic spines into dendrites (a change of diameter of the neck of the spines for example).

This differential enhancement of the two components of the EPSC favours the hypothesis that *expression* of LTP requires postsynaptic mechanisms and does not result exclusively from a presynaptic mechanism, such as a persistent enhancement of glutamate release. If the expression of LTP resulted only from a presynaptic mechanism, a similar increase of both components of the EPSC should have been observed (assuming that AMPA and NMDA receptors are co-localized in the same postsynaptic membrane). As shown above, LTP *induction* clearly requires postsynaptic events: activation of NMDA receptors, increase of postsynaptic $[Ca^{2+}]_i$. Therefore, if LTP expression were presynaptic, that would imply that some message must be sent from the postsynaptic spines to the presynaptic elements. This retrograde messenger would be generated postsynaptically and would trigger a sustained enhanced release of glutamate by the presynaptic element. The identity of any such messenger remains elusive. It appears now safe to state that the major mechanism for the expression of LTP involves a postsynaptic mechanism (see below).

The Ca^{2+} signal is translated into an increase in synaptic strength by biochemical pathways

What are the biochemical pathways activated by intradendritic Ca^{2+} increase that are key components absolutely required for translating the Ca^{2+} signal into an increase in synaptic strength? Amongst the kinases, the Ca^{2+} calmodulin-dependent protein kinase II (CaMKII) plays a crucial role:

- CaMKII is found in high concentrations in the postsynaptic density in spines, near postsynaptic glutamate receptors.
- Injection of inhibitors of CaMKII in the postsynaptic cell, or genetic deletion of a critical CaMKII subunit, block the ability to generate LTP.
- When autophosphorylated on Thr(286), the activity of CaMKII is no longer dependent on Ca^{2+}-calmodulin. This allows its activity to continue long after the Ca^{2+} signal has returned to baseline.
- Replacement of endogenous CaMKII by a form of CaMKII containing a Thr286 point mutation (by the use of genetic techniques) blocks LTP.
- Finally, CaMKII directly phosphorylates the AMPA receptor *in situ*.

Several other protein kinases have been suggested to contribute to LTP, including protein kinase C (PKC) since its selective inhibition by intracellular injection of a PKC inhibitory peptide (PKCI) prevents LTP induction.

Expression of LTP involves the phosphorylation and persistent upregulation of AMPA receptors

A persistent enhancement of the AMPA component of the EPSP could result from a persistent modification in the function *and/or* the number of postsynaptic AMPA receptors (see **Figure 18.7c**). The former hypothesis implies either an increase in unitary current amplitude or an increase in the probability of opening of the AMPA channel. Such a change in receptor function generally involves a phosphorylation by a serine or a threonine kinase. In fact, induction of LTP specifically increases the phosphorylation of Ser831 of the GluR1 subunit, an effect that is blocked by a CaMKII inhibitor (AMPA receptors in CA1 pyramidal cells are heteromers composed primarily of GluR1 and GluR2 subunits). Moreover, genetic deletion of GluR1 subunit prevents the generation of LTP in CA1 pyramidal cells.

In agreement with the second hypothesis is the physiological and anatomical evidence of a rapid and selective upregulation of AMPA receptors after the induction of LTP. For example, when the GluR1 subunit of AMPA receptors is tagged with green fluorescent protein (GFP) and transiently expressed in hippocampal CA1 neurons, its distribution can be observed with a two-photon laser scanning microscope (see Appendix 5.1). After a tetanus a rapid delivery of tagged receptors to dendritic spines is observed. The increased number of AMPA receptors in the plasma membrane at synapses is achieved via activity-dependent changes in AMPA receptors trafficking. However, the source of of the AMPARs that are delivered to synapses during LTP is still unknown as are the detailed mechanisms that deliver and retain AMPARs in the postsynaptic density.

18.2.6 Multiple ways to induce LTP, multiple forms of LTP and multiple ways to block LTP induction

Although synchronous activation of a number of presynaptic fibres by high-frequency stimulation is the most reliable way to evoke LTP (**Figure 18.10a**), LTP can be also evoked *in vitro* by a combination of low-frequency stimulation of presynaptic afferents and postsynaptic injection of a depolarizing current pulse to activate NMDA receptors and voltage-gated Ca^{2+} channels (**Figure 18.10b**); and by a combination of low-frequency stimulation of presynaptic afferents and

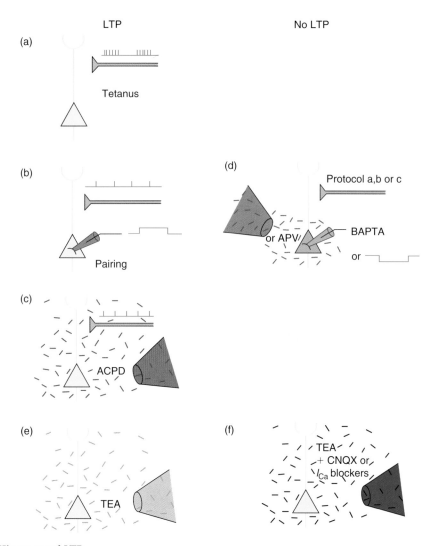

FIGURE 18.10 Hippocampal LTP.
(a)–(d) Tetanic LTP and (e, f) TEA-induced LTP. The multiple ways of induction are shown in (a)–(c) and (e), and blockade of induction in (d) and (f).

bath application of a selective agonist at metabotropic glutamate receptors (**Figure 18.10c**). All these forms of LTP induction are blocked by bath application of APV, an antagonist at NMDA receptors, by intracellular injection of a Ca^{2+} chelator (BAPTA), or by intracellular injection of a hyperpolarizing current pulse during tetanic stimulation (**Figure 18.10d**).

Another form of LTP is induced by bath application of K^+ channel blockers, such as tetraethylammonium chloride (TEA), that depolarizes the presynaptic elements and enhances transmitter release (**Figure 18.10e**). This form of LTP also requires a rise of the postsynaptic intracellular Ca^{2+} concentration. This rise is produced by the entry of Ca^{2+} ions through voltage-gated Ca^{2+} channels activated by the depolarization resulting from the closure of K^+ channels by TEA. In contrast to tetanus LTP, TEA-induced LTP is not synapse-specific since all the synapses are activated by

bath application of TEA. Moreover, TEA-induced LTP is NMDA receptor-independent. It is blocked by bath application of CNQX (an antagonist at AMPA receptors) or of Ca^{2+} channel blockers, or by intracellular injection of a Ca^{2+} chelator (**Figure 18.10f**).

The observation that a rise of the intracellular Ca^{2+} concentration is a necessary prerequisite for LTP induction raises the possibility that a wide range of physiological or pathological processes known to evoke a rise of $[Ca^{2+}]_i$ would trigger long-lasting changes of synaptic efficacy. Both seizures, which generate synchronized giant paroxysmal activity, and anoxic–ischaemic episodes which generate LTP of *NMDA receptor*-mediated EPSPs (anoxic LTP), are in fact associated with $[Ca^{2+}]_i$ rises and long-lasting changes of synaptic efficacy. In such cases, LTP of excitatory synaptic transmission may participate in the pathological consequences of these insults.

18.2.7 Summary: principal features of LTP in the Schaffer collateral–pyramidal cell glutamatergic transmission

- LTP is a long-lasting phenomenon, persisting for hours *in vitro* and days or weeks in the intact animal.
- LTP is synapse-specific.
- LTP results from an increase in the synaptic response without changes in the number of stimulated presynaptic axons.
- LTP does not result from a persistent change of postsynaptic cell excitability.
- LTP does not result from a persistent reduction of the inhibitory GABAergic responses.
- LTP consists of two phases: a brief induction phase (1–20 s) followed by the expression phase; i.e. the mechanisms sustaining the persistent enhancement of synaptic efficacy.
- *Induction* of LTP is voltage- and NMDA-dependent. It requires depolarization of the postsynaptic membrane (resulting from AMPA receptors activation) *and* activation of NMDA receptors which leads to Ca^{2+} entry into the postsynaptic spine and postsynaptic $[Ca^{2+}]_i$ increase. The NMDA receptor acts as a detector of coincident activity in the postsynaptic cells as it opens efficiently only when glutamate is released from the presynaptic terminal and the postsynaptic cell is strongly depolarized (to relieve Mg^{2+} block). Channel opening produces a rise in Ca^{2+} that is largely restricted to the dendritic spine onto which the active synapse terminates. This Ca^{2+} elevation is both necessary and sufficient for LTP induction.
- *Maintenance* of LTP, at least during the initial phase, results from the triggering of a Ca^{2+}-dependent cascade of events leading to postsynaptic modifications of AMPA receptor function and density, and to persistent enhancement of the synaptic glutamatergic response (LTP).

18.3 THE LONG-TERM DEPRESSION (LTD) OF A GLUTAMATERGIC RESPONSE: EXAMPLE OF THE RESPONSE OF PURKINJE CELLS OF THE CEREBELLUM TO PARALLEL FIBRE STIMULATION

Purkinje cells represent the single output neurons of the cerebellar cortex. Each of them receives two distinct excitatory inputs, one from parallel fibres (axons of granule cells) and the other from a climbing fibre (axons of the contralateral inferior olive cells). These two types of inputs display distinct characteristics. A single climbing fibre terminates on each Purkinje cell. This powerful one-to-one excitatory input makes multiple synapses on

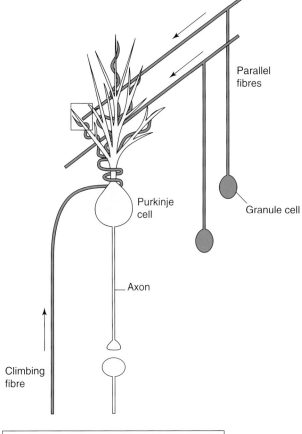

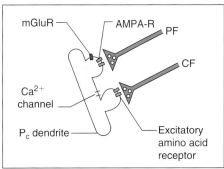

FIGURE 18.11   Simplified neural circuit in the cerebellar cortex. Inset shows a more detailed view of the synaptic contacts between a parallel or a climbing fibre terminal and the Purkinje cell dendrite. AMPA-R, AMPA receptor; mGluR, metabotropic glutamate receptors; VDCC, voltage-dependent calcium channels; CF, climbing fibre; PF, parallel fibre; Pc, Purkinje cell. Adapted from Daniel H, Levenes C, Crépel F (1998) Cellular mechanisms of cerebellar LTD. *Trend. Neurosci.* **9**, 401–407, with permission.

the soma and proximal dendrites of the Purkinje cell (**Figure 18.11**; see also **Figures 6.8** and **6.9**). In contrast, many parallel fibres converge on each Purkinje cell but each fibre makes few synapses on each Purkinje cell.

The putative neurotransmitter at parallel-fibre and climbing-fibre synapses is glutamate. Fast excitatory synaptic transmission at these synapses is mediated

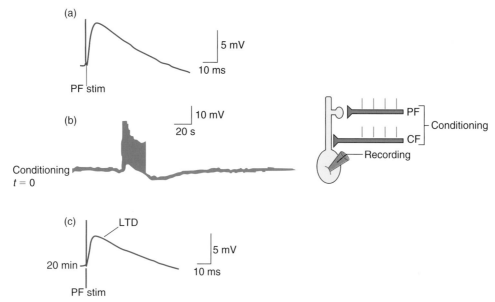

FIGURE 18.12   LTD of the parallel-fibre-mediated EPSP.
The activity of a Purkinje cell is intracellularly recorded in current clamp mode in a cerebellar slice. **(a)** The EPSP in response to PF stimulation (the stimulating electrode is placed in the superficial molecular layer) is recorded in the presence of picrotoxin (40 μM) to block IPSPs mediated by local interneurons. **(b)** CF and PF are then stimulated conjointly at 4 Hz for 25 s. To stimulate climbing fibres, a second electrode is placed in the white matter. **(c)** Twenty minutes after the end of conditioning stimulation, the EPSP recorded in response to PF stimulation is still depressed in amplitude. Adapted from Sakurai M (1990) Calcium is an intracellular mediator of the climbing fibre in induction of cerebellar long-term depression. *Proc. Natl Acad. Sci. USA* **87**, 3383–3385, with permission.

entirely by non-NMDA ionotropic glutamatergic receptors, since both synapses lack NMDA receptors in the adult – in marked contrast to most other neurons in the brain (**Figure 18.11**, inset). In addition, parallel-fibre/Purkinje-cell synapses also bear mGluR1 receptors known to be coupled to phospholipase C, activation of which leads to production of inositol trisphosphate (IP$_3$) and diacylglycerol (DAG).

The dual arrangement of the two excitatory synaptic inputs raises the question of the role of the powerful input (climbing fibre) on the weaker input (parallel fibres). The coactivation of climbing-fibre and parallel-fibre inputs induces a persistent decrease in the efficacy of the parallel-fibre/Purkinje-cell synapse. This decrease of efficacy is called *long-term depression* (LTD). With the experimental advantages of *in vitro* brain slices and culture preparations, cerebellar LTD constitutes a simple model to study activity-dependent changes confined to excitatory synapses.

18.3.1 The long-term depression of a postsynaptic response (EPSC or EPSP) is a decrease of synaptic efficacy

Ito and co-workers (1982) were the first to demonstrate in the rabbit cerebellum *in vivo* that conjunctive

stimulation of the afferent climbing and parallel fibres leads to an LTD of synaptic transmission at parallel-fibre/Purkinje-cell synapses. In other terms, LTD is the attenuation of the Purkinje cell response to parallel fibres after the conjunctive stimulation of parallel and climbing fibres.

The activity of a Purkinje cell is intracellularly recorded in current clamp mode in rat cerebellar slices. The stimulation of parallel fibres evokes an EPSP resulting from the activation of postsynaptic AMPA receptors by glutamate released from the stimulated terminals since it is totally blocked by CNQX, a selective AMPA receptor antagonist. After recording this control parallel-fibre-mediated EPSP for several minutes, parallel fibres are then stimulated in conjunction with the climbing fibre at low frequency, 4 Hz for 25 s. After this conjunctive stimulation, the same stimulation of parallel fibres as in the control now evokes a smaller EPSP (**Figure 18.12**). The parallel-fibre-mediated EPSP stays attenuated for the rest of the recording session. It is a long-term depression (LTD).

The persistent decrease of the parallel-fibre-mediated synaptic response can be also studied in another *in vitro* preparation, a culture of Purkinje cells, granule cells and an inferior olivary explant (**Figure 18.13a**). The parallel-fibre-mediated postsynaptic excitatory

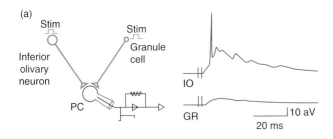

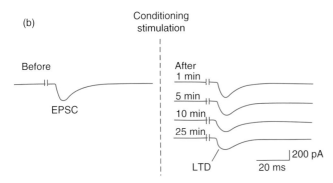

FIGURE 18.13 (Cerebellar LTD is observed when a single parallel fibre is stimulated.

(a) The activity of a Purkinje cell is recorded in patch clamp (whole-cell patch) in co-cultures of rat cerebellar Purkinje cells (PC), granule cells and an explant of inferior olivary neurons, to record the evoked postsynaptic current (EPSC). The conditioning stimulation consists of the conjunctive stimulation of a single granule cell (GR) and the inferior olivary explant (IO) at 2 Hz for 20 s while the Purkinje cell membrane is recorded in current clamp mode. **(b)** The Purkinje cell membrane is held at $V_H = -50$ 1 min mV in voltage clamp mode and the response (EPSC) to the activation of a single granule cell is recorded before and 1, 5, 10 and 25 min after the conditioning stimulation. Adapted from Hirano T (1990) Depression and potentiation of the synaptic transmission between a granule cell and a Purkinje cell in rat LTD 20 ms cerebellar culture. *Neurosci. Lett.* **119**, 141–144, with permission.

current (EPSC) is first recorded in voltage clamp (**Figure 18.13b**). The repetitive conjunctive stimulation of a single granule cell (whose axon is a parallel fibre) and the inferior olivary explant (which sends an axon, the climbing fibre, to the recorded Purkinje cell) is then applied while the Purkinje cell activity is recorded in current clamp mode. When switching back to voltage clamp mode, the parallel-fibre-mediated EPSC (in response to granule cell stimulation) is persistently decreased. This *in vitro* preparation allows the stimulation of a single presynaptic granule cell before and after LTD induction. Therefore, it can be demonstrated that LTD is observed though the number of parallel fibres stimulated before and after the conditioning stimulation is identical (a depressed EPSC or EPSP could in fact result from a decrease in the number of stimulated axons).

18.3.2 Induction of LTD requires a rise in postsynaptic intracellular Ca^{2+} concentration and the activation of postsynaptic AMPA receptors

As already studied in Sections 16.2.2 and 17.3, the response of a Purkinje cell to the activation of its afferent climbing fibre is an all-or-none response composed of an initial depolarization, an overshooting action potential and following depolarizing humps. Since the activation of the afferent climbing fibre potently activates the voltage-gated Ca^{2+} channels present in the membrane of Purkinje dendrites (**Figure 18.11**, inset), it was supposed that the consequent rise in intradendritic Ca^{2+} concentration played a role in LTD. In fact, *in vivo* experiments have shown that hyperpolarization of the membrane by the activation of stellate cells during co-stimulation of parallel and climbing fibres prevents the occurrence of LTD.

$[Ca^{2+}]_i$ rises during co-stimulation

In order to simultaneously record the synaptic responses and the intracellular Ca^{2+} concentration, the activity of a Purkinje cell is recorded in patch clamp (whole-cell patch) in the presence of a fluorescent calcium dye, FURA-2, injected into the cell (**Figure 18.14**; see also Appendix 5.1). First, the control EPSC in response to parallel fibre stimulation is recorded in the Purkinje cell. Then, parallel fibres and climbing fibres are co-stimulated in phase at a low frequency (1–4 Hz; dotted line). Five minutes after this conditioning stimulation, the response to the same parallel fibre stimulation recorded from the same Purkinje cell begins to decrease and stays attenuated thereafter. Cerebellar LTD is associated with an increase of Ca^{2+} concentration in Purkinje cell dendrites during the conditioning stimulation.

LTD of the response to parallel fibre is not observed when the parallel fibres are stimulated alone at 1–4 Hz; what adds the climbing fibre stimulation?

Voltage-gated Ca^{2+} channels located in the membrane of Purkinje cell dendrites are activated by the climbing-fibre-mediated EPSP. This suggests that the resulting increase of intradendritic Ca^{2+} concentration is a necessary prerequisite for LTD induction. This hypothesis is tested by hyperpolarizing the Purkinje cell membrane during the co-stimulation or by injecting of a Ca^{2+} chelator into the Purkinje cell before the co-stimulation (**Figure 18.15**), or by removing the external Ca^{2+} ions, in order to prevent the rise of intradendritic Ca^{2+} concentration: all these procedures block LTD

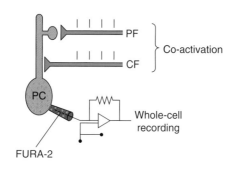

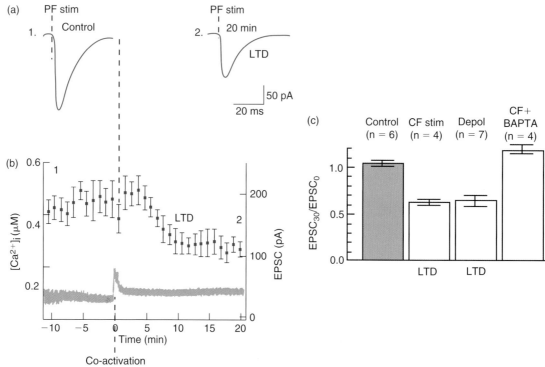

FIGURE 18.14 **An increase of intracellular Ca²⁺ concentration is observed during LTD induction.**
The activity of a Purkinje cell is recorded in patch clamp (whole-cell patch) in a thin slice of rat cerebellum. The patch pipette also contains FURA-2 in order to record on-line the intracellular Ca²⁺ concentration. **(a)** The excitatory postsynaptic current (EPSC) recorded in voltage clamp in response to parallel fibre stimulation (PF stim, 1 Hz) is recorded before (control) and 20 min after the conditioning stimulation (co-activation: conjunctive stimulation of parallel and climbing fibres while the Pc membrane is recorded in current clamp mode). **(b)** Time course of changes in parallel-fibre-mediated EPSC amplitude (top curve) and in [Ca²⁺]ᵢ (bottom curve). The conditioning stimulation (given at time 0) induces a LTD of the EPSC (with a delay) and an immediate transient rise of [Ca²⁺]i. Note that the stimulation of parallel fibres before the conditioning stimulation does not induce significant changes of [Ca²⁺]ᵢ. **(c)** Average changes in PF-mediated EPSC expressed as the ratio of EPSC amplitude before (EPSC₀) and 30 min (EPSC3₀) after co-activation, in four different conditions: no co-activation (control), co-activation (CF stim), pairing (depol) and co-activation in the presence of BAPTA (CF + BAPTA) in the patch electrode. From Konnerth A, Dreessen J, Augustine GJ (1992) Brief dendritic calcium signals initiate long-lasting synaptic depression in cerebellar Purkinje cells. *Proc. Natl Acad. Sci. USA* **89**, 7051–7055, with permission.

induction. Along the same lines, climbing fibre stimulation can be replaced by direct intracellular depolarization of the Purkinje cell which evokes Ca²⁺ spikes. This is called the 'pairing protocol' (**Figure 18.16a**). In conclusion, LTD of synaptic transmission at parallel-Purkinje cell synapses is triggered by a rise of intracellular Ca²⁺ concentration resulting from Ca²⁺ entry in Purkinje cell dendrites through voltage-gated ²⁺ channels opened by the membrane depolarization during co-stimulation of climbing and parallel fibres.

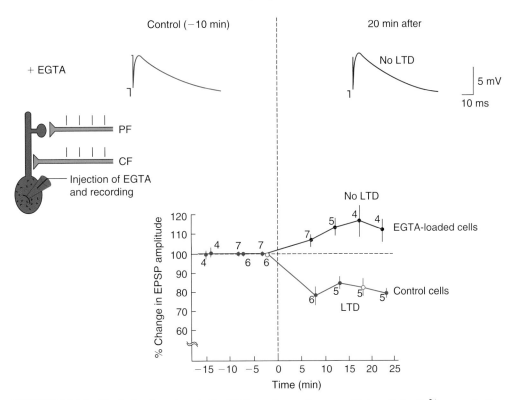

FIGURE 18.15 The induction of cerebellar LTD requires an increase of intracellular Ca²⁺ concentration.
The activity of a Purkinje cell is intracellularly recorded (current clamp mode) in a guinea pig cerebellar slice. The amplitude of the EPSP recorded in response to parallel fibre stimulation is recorded before and after the conditioning stimulation (conjunctive stimulation of PF and CF at 4 Hz for 25 s) in control cells (white circles). The same experiment is performed after intracellular injection of the Ca²⁺ chelator EGTA into the recorded Purkinje cells (black circles). The respective averaged EPSPs recorded in the presence of EGTA are shown in the insets. The time 0 represents the end of conjunctive stimulation. The values at each plotted point represent the number of cells recorded. Adapted from Sakurai M (1990) Calcium is an intracellular mediator of the climbing fibre in induction of cerebellar long-term depression. *Proc. Natl Acad. Sci. USA* **87**, 3383–3385, with permission.

LTD of the response to parallel fibre is not observed when the climbing fibre is stimulated alone at 1–4 Hz; what adds the parallel fibre stimulation?

The glutamate released from parallel fibre terminals activates the non-NMDA receptors present in the post-synaptic membrane (ionotropic AMPA receptors and metabotropic glutamate receptors, mGluR1; **Figure 18.11**, inset). AMPA receptors mediate the excitatory response (EPSP or EPSC) evoked by parallel fibre stimulation since it is totally blocked by the application of CNQX, a selective antagonist of this class of receptors. In order to test the role of non-NMDA receptors in LTD induction, CNQX is bath applied during or after a pairing protocol (direct Purkinje cell depolarization with parallel fibre stimulation). The blockade of non-NMDA receptors during the pairing protocol prevents LTD induction while it has no effect after the pairing protocol (once LTD is induced) (**Figures 18.16b, c**).

In order to test the role of the non-NMDA receptors in LTD induction, parallel fibre stimulation can also be replaced by external application of agonists at non-NMDA receptors. Parallel fibre stimulation during the conditioning stimulus can be replaced by the application on the Purkinje cell dendrites of glutamate or quisqualate (agonists on *both* AMPA and metabotropic receptors, or a solution containing *both* AMPA and an agonist of metabotropic receptors) (see **Figures 18.21c, d**). The activation of AMPA receptors alone by AMPA or the application of NMDA are ineffective. This is in keeping with the recent demonstration that antibodies directed against mGluR1 subunit blocks LTD induction in cultured Purkinje cells. The final demonstration of the participation of mGluR to LTD induction in acute cerebellar slices has been given recently by showing that LTD of the parallel-fibre-mediated EPSP is markedly impaired in knockout mice lacking mGluR1 (see **Figure 18.21f** and Section 12.4). In conclusion, the activation of the parallel fibres during the conjunctive or pairing stimulation allows the release of glutamate and the activation of both

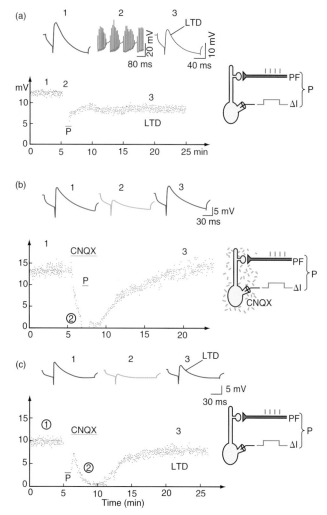

FIGURE 18.16 **Induction of cerebellar LTD requires the activation of postsynaptic AMPA receptors.**
The activity of a Purkinje cell is recorded in patch clamp (whole-cell patch, current clamp mode) in cerebellar thin slices. The PF-mediated EPSP is evoked during a hyperpolarizing current pulse before and after the pairing in order to test the variation of membrane resistance during the experiment. **(a, top traces)** A control EPSP is recorded in response to parallel fibre stimulation (1). After the pairing (P) – i.e. intracellular depolarizing pulses to evoke Ca^{2+} spikes in conjunction with parallel fibre stimulation (2) – a LTD of the parallel-fibre-mediated EPSP is observed (3). Note the change in calibrations between 1, 3 and 2. **(a, bottom trace)** Plot of the EPSP amplitude against time. **(b)** The same experiment as in (a) but in the presence of CNQX (4 μM) in the bath before, during and after pairing (P). During CNQX application the parallel-fibre-mediated EPSP (2) is completely blocked since it is mediated by AMPA receptors. **(c)** The same experiment as in (b) but CNQX is bath applied after pairing (P). Part (a) adapted from Hémart N, Daniel H, Jaillard D *et al.* (1995) Receptors and second messengers involved in long-term depression in rat cerebellar slices *in vitro*: a reappraisal. *Eur. J. Neurosci.* **7**, 45–53, with permission.

AMPA and metabotropic glutamatergic postsynaptic receptors.

These, with the concomitant rise in intracellular calcium concentration, are possibly the necessary and

sufficient processes for LTD induction, since the conditioning stimulation can be replaced by a direct depolarization of the Purkinje cell membrane to activate voltage-dependent Ca^{2+} channels (to mimic climbing fibre stimulation) and the concomitant application of agonists of AMPA and metabotropic receptors (to mimic parallel fibre stimulation) (see **Figure 18.21d**).

18.3.3 The expression of LTD involves a persistent desensitization of postsynaptic AMPA receptors

The fact that co-activation of Purkinje cells by climbing fibre stimulation and iontophoretic application of glutamate on Purkinje cell dendrites induces a long-lasting decrease of the response to this agonist led Masao Ito to postulate that LTD of parallel-fibre-mediated EPSP or EPSC is due to a long-term desensitization of ionotropic glutamate receptors of Purkinje cells (a desensitized state is a state where the probability of the channel opening is very low). This would explain the decrease in synaptic efficacy.

In Purkinje cells in cerebellar slices, a pairing procedure known to induce LTD of the synaptic response induces a long-lasting decrease of the response to iontophoretic application of glutamate (or quisqualate, not shown) but not of aspartate (**Figure 18.17**). This suggests that LTD of synaptic transmission between parallel fibres and Purkinje cells is accompanied by LTD of the responsiveness of Purkinje cells to glutamate or quisqualate, whereas that to aspartate is unaffected. The observed decrease in efficacy of glutamate or quisqualate in activating Purkinje cells could involve a desensitization of AMPA receptors. What are the mediators between Ca^{2+} entry and the long-term changes of AMPA receptors?

18.3.4 Second messengers are required for LTD induction

The metabotropic receptors mGluR1 are abundantly expressed in Purkinje cells. These receptors are coupled to phospholipase C and their activation leads to the formation of inositol trisphosphate (IP3) and diacylglycerol (DAG). Moreover, Ca^{2+}-dependent PKC is also expressed abundantly in Purkinje cells. Therefore, during the conditioning stimulus, the Ca^{2+}-dependent kinases such as protein kinase C can be activated by both the increase of intracellular Ca^{2+} concentration due to climbing fibre activation (see **Figure 18.14b**) and the activation of mGluR1 by glutamate released from parallel fibres. The role of protein kinase C in LTD is tested by injecting into the recorded Purkinje cell a selective inhibitor of protein kinase C (PKC 19-36) before the conditioning stimulus (**Figure 18.18**). In

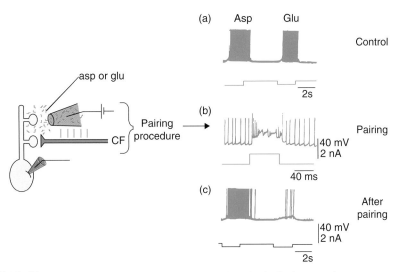

(a) Asp   Glu

Control

2s

(b)

Pairing

40 mV
2 nA

40 ms

(c)

After
pairing

40 mV
2 nA

2s

FIGURE 18.17 The postsynaptic glutamate response is selectively depressed.
The response of a Purkinje cell to iontophoretic application of glutamate (glu) or aspartate (asp) is intracellularly recorded (current clamp mode, $V_m = -65\,mV$) in cerebellar slices. **(a)** Glutamate or aspartate are alternatively ejected in the dendritic field of the recorded Purkinje cell. They both evoke a transient membrane depolarization which reaches the firing level. **(b)** The conditioning stimulation used to induce LTD consists of climbing fibre stimulation (2–4 Hz) paired for 1 min with the ejections of glutamate and aspartate at 2 min intervals. **(c)** Twenty minutes after the pairing procedure, the response to glutamate is selectively depressed (the response to aspartate is left unaffected). Adapted from Crépel F and Krupa M (1988) Activation of protein kinase C induces a long-term depression of glutamate sensitivity of cerebellar Purkinje cells: an *in vitro* study. *Brain Res.* **458**, 397–401, with permission.

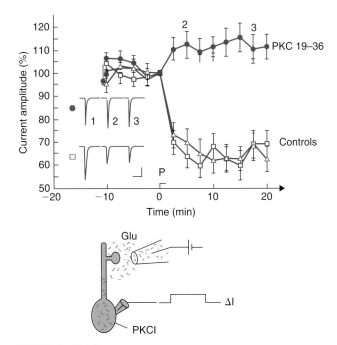

FIGURE 18.18 Protein kinase C inhibition prevents LTD induction.
The activity of a Purkinje cell is recorded in patch clamp (whole-cell patch) in the presence of the selective PKC inhibitor, PKC 19-36 (black circles) or a non-inhibitory control peptide (triangle) or the intracellular solution only (square) in the patch pipette. The control EPSC evoked by parallel fibre stimulation is recorded for 10 min and the conditioning stimulation is applied at $t = 0$. It consists of the conjunctive application of glutamate and intracellular depolarization. LTD of the EPSC is not observed in cells dialyzed with PKC 19-36. Scale bars: 100 pA, 2 s. Adapted from Linden DJ and Connor JA (1991) Participation of postsynaptic PKC in cerebellar long-term depression in culture. *Science* **254**, 1656–1659, with permission.

such conditions, LTD is not induced. Moreover, selective expression of a PKC inhibitor in Purkinje cells in transgenic mice leads to a complete blockade of LTD induction, supporting the hypothesis that activation of PKC is necessary for LTD induction.

Antibodies directed against mGluR1 as well as mGluR1 knockout mice were used to demonstrate the role of these metabotropic receptors in LTD induction: in both preparations long-term depression at the parallel-fibre/Purkinje-cell synapse is absent. These preparations also permitted testing of the possible role of internal Ca^{2+} stores: is the combination of direct activation of IP_3-sensitive Ca^{2+} stores in Purkinje cell dendrites and a conventional pairing protocol in mGluR1-deficient mice (mGluR1-/-), sufficient to rescue LTD in the cerebellum of these animals by bypassing the disrupted mGluR1 (**Figure 18.19**)? Caged-IP_3 and the fluorescent Ca^{2+}-sensitive dye fluo-3 are present in the whole-cell pipette. The recording session starts 30–45 minutes after whole-cell 'break in' to allow diffusion of the compounds in the dendrites of the recorded Purkinje cell *in vitro* (in cerebellar slices). Control parallel-fibre EPSPs are recorded in control conditions (trace 1). Pairing (simultaneous depolarization of the Purkinje cell and parallel-fibre stimulation) is first performed in the absence of photolysis of caged-IP_3. It induces only a transient depression of the EPSP (trace 2). A second pairing is then performed with concomitant photolysis of caged-IP_3. The transient intracellular Ca^{2+} increase in response to UV flash is visualized by the change of fluorescence of fluo-3 (bottom inset,

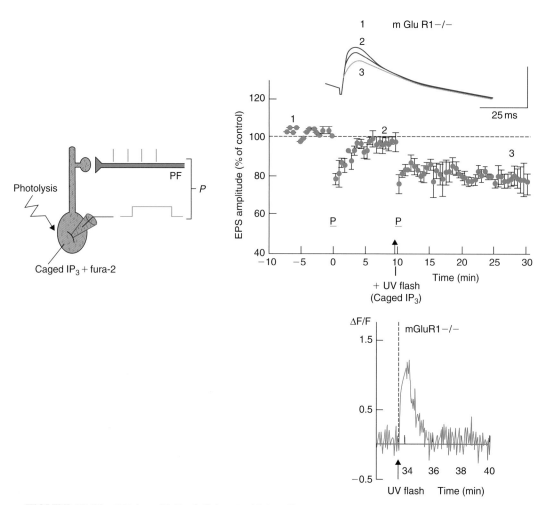

FIGURE 18.19   LTD in mGluR1-deficient Purkinje cells.
The central plot represents the normalized amplitude of PF-mediated EPSPs against time before and after two successive pairing protocols (P), first at $t = 0$ in the control condition and then at $t = 10$ minutes, combined with photolysis of caged IP3. Each point is the mean $\pm$ SEM of separate experiments in four cells. The top inset represents superimposed averaged EPSPs recorded at the indicated times. The bottom inset represents the Ca-induced fluorescence change evoked by photorelease of caged IP3 in a FURA-2-loaded cell. Adapted from Daniel H, Levenes C, Fagni L *et al.* (1999) Inositol-1,4,5-trisphosphate-mediated rescue of cerebellar long-term depression in subtype 1 metabotropic glutamate receptor mutant mouse. *Neuroscience* **91**, 1–6, with permission.

$\Delta F / F$). After such pairing, parallel-fibre EPSPs are depressed by $76.2 \pm 8.2\%$ even 20 minutes after the pairing period (trace 3). The same protocol in the presence of the inhibitory PKC 19-36 peptide fails to induce LTD (not shown). This demonstrates that the impairment of LTD in mGluR1-deficient Purkinje cells is caused by the lack of functional mGluR1 preventing the second-messenger cascade activation. It also suggests that the combination of Ca^{2+} influx through voltage-gated Ca^{2+} channels (in response to Purkinje cell membrane depolarization) and Ca^{2+} release from IP3-sensitive Ca^{2+} stores is capable of restoring LTD in mGluR1 knockout mice.

The hypothesis is the following. During the conditioning stimulus, the formation of diacylglycerol (DAG) following activation of mGluR1, together with the cytosolic Ca^{2+} increase due to the activation of voltage-gated channels (and perhaps the release of Ca^{2+} ions from internal stores due to the formation of IP3), leads to the activation of protein kinase C. This, with other second-messenger cascades, would lead directly or indirectly to phosphorylation of AMPA receptors and activation of their transition to a stable desensitized state and thus to LTD (**Figure 18.20**).

18.3.5   The different ways to induce or block cerebellar LTD

Long-term depression of the response of a Purkinje cell to parallel fibre activation can be *induced* by (**Figures 18.21a–d**):

- conjunctive stimulation of the afferent parallel and climbing fibres;

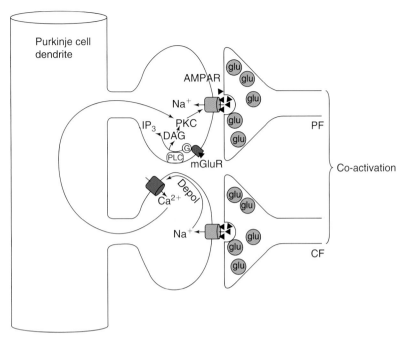

FIGURE 18.20   Schematic of some of the putative mechanisms of cerebellar LTD induction.

- conjunctive stimulation of the parallel fibres and intracellular injection of a depolarizing current (which evokes Ca^{2+} spikes) into the Purkinje cell;
- conjunctive iontophoretic application of glutamate, quisqualate or AMPA + t-ACPD to the Purkinje cell dendrites and stimulation of its afferent climbing fibre;
- conjunctive iontophoretic application of glutamate or quisqualate or AMPA + t-ACPD to the Purkinje cell dendrites and intracellular injection of depolarizing current (which evokes Ca^{2+} spikes) into the Purkinje cell.

Long-term depression of the response of a Purkinje cell to parallel fibre activation can be *blocked* by (**Figures 18.21e–g**):

- intracellular injection of a Ca^{2+} chelator into the Purkinje cell or injection of a hyperpolarizing current into the Purkinje cell during the conjunctive stimulation or the pairing protocol;
- bath application of CNQX or the lack of mGluR1 in the cerebellum and notably in Purkinje cell membrane (mGluR1 gene-deficient mice are obtained by disrupting the mGluR1 gene);
- intracellular injection in the Purkinje cell or bath application of an inhibitor of protein kinase C before the conditioning stimulus.

18.3.6 Summary: principal features of LTD in parallel-fibre/Purkinje cell glutamatergic transmission

- LTD results from a decrease of the parallel-fibre-mediated EPSC or EPSP without changes in the number of afferent axons stimulated: it is a depression of the synaptic efficacy.
- LTD is a very long-lasting phenomenon since it persists for the duration of the experiment, up to several hours.
- LTD is input specific: it is restricted to those parallel-fibre synapses activated at the same time as climbing fibres.
- LTD is associated with a large increase of Ca^{2+} concentration in Purkinje cell dendrites which occurs during the conjunctive stimulation of parallel and climbing fibres. Climbing-fibre synapses are very powerful and their activation leads to a large rise in intracellular Ca^{2+} that is permissive for LTD.
- A key signal that distinguishes active from inactive parallel-fibre synapse, and which is required to trigger LTD, is activation of group 1 mGluRs. Induction of cerebellar LTD requires activation of mGluR1.
- LTD is expressed as a depression of AMPA-mediated current at the parallel-fibre/Purkinje cell synapses activated at the same time as climbing

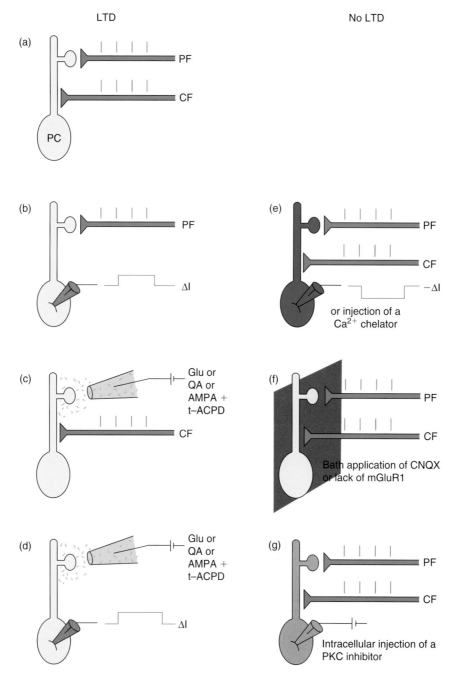

FIGURE 18.21   Cerebellar LTD.

The different ways of **(a)** induction or **(b)** blockade of induction.

fibres. It results from the long-term desensitization of AMPA receptors which requires the activation of protein kinase C.

FURTHER READING

Aiba A, Kano M, Chen C, Stanton ME *et al.* (1994) Deficient cerebellar long-term depression and impaired motor learning in mGluR1 mutant mice. *Cell* **79**, 377–388.

Bredt DS and Nicoll RA (2003) AMPA receptor trafficking at excitatory synapses. *Neuron* **40**, 361–379.

Carroll RC, Lissin DV, Zastrow von M *et al.* (1999) Rapid redistribution of glutamate receptors contributes to long-term depression in hippocampal cultures. *Nature Neurosci.* **2**, 454–460.

Conquet F, Bashir ZI, Davies CH *et al.* (1994) Motor deficit and impairment of synaptic plasticity in mice lacking mGluR1. *Nature* **372**, 237–243.

Crépel F, Audinat E, Daniel H *et al.* (1994) Cellular locus of the nitric oxide-synthase involved in cerebellar long-term depression induced by high external potassium concentration. *Neuropharmacology* **33**, 1399–1405.

De Zeeuw CI, Hansel C, Bian F *et al.* (1998) Expression of a protein kinase C inhibitor in Purkinje cells blocks cerebellar LTD and adaptation of the vestibuloocular reflex. *Neuron* **20**, 495–508.

Hartell NA (2001) Receptors, second messengers and protein kinases required for heterosynaptic cerebellar long-term depression. *Neuropharmacology* **40**, 148–161.

Ito M, Sakurai M, Tongroach P (1982) Climbing fibre induced depression of both mossy fibre responsiveness and glutamate sensitivity of cerebellar Purkinje cells. *J. Physiol.* **324**, 113–134.

Lev-Ram V, Makings LR, Keitz PF *et al.* (1995) Long-term depression in cerebellar Purkinje neurons results from coincidence of nitric oxide and depolarization induced Ca^{2+} transients. *Neuron* **15**, 407–415.

Linden DJ, Dickinson MH, Smeyne M, Connor JA (1991) A long-term depression of AMPA currents in cultured cerebellar Purkinje neurons. *Neuron* **7**, 81–89.

Lisman J, Schulman H, Cline H (2002) The molecular basis of CaMKII function in synaptic and behavioral memory. *Nat. Rev. Neurosci.* **3**, 175–190.

Malenka RC and Bear MF (2004) LTP and LTD: an embarrassment of riches. *Neuron* **44**, 5–21.

Schuman EM, Dynes JL, Steward O (2006) Synaptic regulation of dendritic mRNAs. *J. Neuroscience* **26**, 7143–7146.

Shi SH, Hayashi Y, Petralia RS *et al.* (1999) Rapid spine delivery and redistribution of AMPA receptors after synaptic NMDA receptors activation. *Science* **284**, 1811–1816.

Shigemoto R, Abe T, Nomura S *et al.* (1994) Antibodies inactivating mGluR1 metabotropic glutamate receptor block long-term depression in cultures Purkinje cells. *Neuron* **12**, 1245–1255.

APPENDIX 18.1 DEPOLARIZATION-INDUCED SUPPRESSION OF INHIBITION (DSI): AN EXAMPLE OF SHORT-TERM PLASTICITY AT GABAERGIC SYNAPSES

Inhibitory synaptic responses also undergo short- and long-term modifications. This appendix examines the mechanisms underlying short-term plasticity at inhibitory synapses.

18.1.1 Observation of depolarization-induced suppression of inhibition

This form of short-term plasticity at GABAergic synapses was first reported more than a decade ago in the cerebellum. The pioneering observation was made by Alain Marty and collaborators in 1991: the direct depolarization of cerebellar Purkinje cell induces a decrease of the amplitude of spontaneous inhibitory postsynaptic currents (IPSCs). The experimental protocol is the following (**Figure A18.1**). Evoked and spontaneous inhibitory postsynaptic currents (IPSCs) are recorded from Purkinje cells *in vitro* using the whole-cell patch clamp technique in the continuous presence of ionotropic glutamate receptors antagonists. The IPSCs are entirely blocked by bath application of bicuculline, showing that they are mediated by $GABA_A$ receptors.

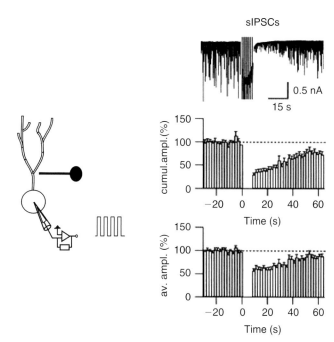

FIGURE A18.1 **Effects of postsynaptic depolarization on the spontaneous synaptic activity in Purkinje cells recorded in the presence of blockers of glutamatergic synaptic currents: evidence for depolarization-induced suppression of inhibition.**
Whole-cell recording of cerebellar Purkinje cell *in vitro*. The top trace shows a typical DSI protocol. Spontaneous GABA-IPSCs are recorded in the presence of glutamatergic receptor antagonists. After a control period, the Purkinje cell is depolarized 8 times to 0 mV for 100 msec, at 1 Hz; the trace shows the consequent dramatic inhibition of sIPSCs and, thereafter, their recovery phase over 60 sec. In this cell, four such protocols were averaged, yielding 79.60.5% inhibition in the sIPSCs cumulative amplitude (cumul. ampl.), calculated over the first 10 sec after the end of the pulse train. Decreases in the frequency and in the average amplitude (av. ampl.) contributed in equal measure to this reduction: 53.64.3 and 55.24.3%, respectively. The two bottom graphs show the time course of the DSI of cumulative and average amplitudes for *n* = 14 experiments performed with this protocol. Adapted from Diana MA and Marty A (2003) Characterization of depolarization-induced suppression of inhibition using paired interneuron–Purkinje cell recordings. *J Neurosci.* **23**, 5906–5918.

After a stable control period a train of depolarizing voltage steps (8 pulses from −70 mV to +20 mV, 100 msec duration) are applied through the recording pipette. This train of pulses leads to a transient decrease of the amplitude of evoked and spontaneous GABAergic IPSCs. The inhibition of GABAergic activity is quantified as a function of time by counting the sum of the amplitudes of spontaneous IPSCs during sampling intervals of 2 sec. A similar effect is induced by a single depolarizing pulse of 0.1 to 1 sec duration. This phenomenon, termed depolarization-induced suppression of inhibition (DSI), develops within seconds following the voltage pulses and lasts tens of

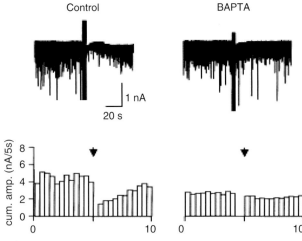

FIGURE A18.2 DSI requires a postsynaptic rise of intracellular Ca^{2+} concentration.
Whole-cell recording of a cerebellar Purkinje cell. Top traces show the effect of repeated depolarizing voltage pulse on spontaneous IPSCs in control cell (left) and in cells loaded with the fast Ca^{2+} chelator BAPTA (right). Similar depolarization has no effect on spontaneous IPSCs. Adapted from Glitsch M, Parra P, Llano I (2000) The retrograde inhibition of IPSCs in rat cerebellar Purkinje cells is highly sensitive to intracellular Ca^{2+}. *Euro. J. of Neurosci.* **12**, 987–993.

seconds. DSI is not restricted to cerebellar $GABA_A$ synapses since it is also observed for hippocampal $GABA_A$ synapses following the depolarization of CA1 pyramidal neurons. Experiments performed in either preparation contributed to the description of DSI that follows.

18.1.2 DSI requires a postsynaptic rise of intracellular calcium concentration

It was quickly recognized that an increase in intracellular Ca^{2+} concentration, through the activation of voltage-dependent Ca^{2+} channels (VDCCs) is necessary and sufficient to trigger DSI.

- When the postsynaptic neuron is loaded with the fast Ca^{2+} buffer BAPTA, the depolarization fails to trigger DSI. Moreover, release of Ca^{2+} ions via flash photolysis of caged Ca^{2+} is sufficient to induce DSI (Figure A18.2). These observations show that an increase in $[Ca^{2+}]_i$ is required to trigger DSI.
- Fura2 Ca^{2+} measurements combined with whole-cell patch clamp recordings of cerebellar Purkinje cells indicates that the extent of DSI is a function of the peak postsynaptic intracellular Ca^{2+} concentration ($[Ca^{2+}]_i$).
- DSI does not occur when Ca^{2+} ions are removed from the extracellular solution, showing that an influx of Ca^{2+} is necessary to induce DSI.

- DSI is observed in the presence of D-APV, a selective antagonist of NMDA receptors but not in the presence of blockers of voltage-dependent Ca^{2+} channels.

18.1.3 DSI results from a transient suppression of GABA release

What are the mechanisms underlying the transient decrease of GABAergic synaptic activity? DSI can be accounted for by alterations in the density or properties of postsynaptic $GABA_A$ receptors, or by a presynaptic decrease of transmitter release.

The presynaptic nature of DSI was rapidly established on the basis of the following experiments. The amplitude of miniature GABAergic IPSCs or the postsynaptic response to GABA applications are not affected by the DSI protocol. These observations strongly suggest that the postsynaptic sensitivity to GABA is not affected. In contrast, the widely used indicators of presynaptic modifications are affected. Thus, during DSI the percentage of synaptic failures increase and the frequency of miniature IPSCs decrease, indicating a decrease in release probability.

18.1.4 Endocannabinoid as a retrograde messenger in DSI

Altogether these observations indicate that the postsynaptic $[Ca^{2+}]_i$ rise triggered by the depolarization of the target (recorded) neuron leads to a transient suppression of GABA release from presynaptic terminals. It was thus hypothesized that a retrograde signal is sent from the postsynaptic neuron to the presynaptic inhibitory terminals. To approach this question, the effect of depolarizing voltage pulses is analyzed in paired recordings of neighbouring Purkinje cells separated by 25–150 μm. Simultaneous recordings from two neighbouring Purkinje cells revealed that a depolarization applied to one cell results in the transient suppression of inhibition in both recorded cells. Thus, DSI is not restricted to the stimulated neuron but can spread to its neighbours.

The following step was to identify the retrograde messenger of DSI. In some studies it was proposed that glutamate is released by the postsynaptic cell and by acting on metabotropic glutamatergic receptors (mGluRs) inhibits the release of GABA. However, in 2001 Wilson and Nicoll demonstrate that endocannabinoids and the subsequent activation of type 1 cannabinoid receptors (CB1Rs) are the retrograde signal of DSI in the hippocampus. A similar conclusion was reached about the cerebellum.

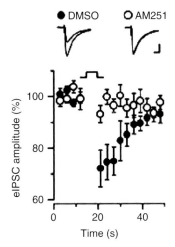

FIGURE A18.3 Activation of the cannabinoid receptor CB1 is required to trigger the depolarization-induced suppression of inhibition.
Whole-cell recording of hippocampal pyramidal cells. Hippocampal slices are pre-incubated in DMSO alone (●) or in the presence of a CB1 receptor antagonist (AM251) dissolved in DMSO (○). In the presence of the CB1R antagonist, a 5 sec depolarizing step results in little or no suppression of evoked IPSCs (eIPSCs), while in control DMSO experiment, the same depolarization leads to a robust reduction. Inserts show average eIPSCs for the 10 s before and the 10 s after the application of the depolarizing step. Scale bars: 200 pA, 20 ms. Adapted from Wilson RI and Nicoll RA (2001) Endogenous cannabinoids mediate retrograde signalling at hippocampal synapses. *Nature* **410**, 588–592.

This conclusion is supported by the following observations:

- CB1 receptor agonist induces a presynaptic inhibition of GABAergic synaptic postsynaptic currents similar to that occurring during DSI. Moreover, this effect is only observed at synapses that have the ability to express DSI. In CB1-insensitive GABAergic synapses, likely lacking CB1 receptors, DSI is not observed. DSI is abolished by the application of CB1 receptor antagonists (Figure A18.3) and not observed in CB1 receptor knock-out mice. These observations clearly show that the activation of CB1Rs is required to trigger DSI.
- Photolysis of caged Ca^{2+} inside a pyramidal neuron triggers DSI that is completely abolished by CB1R antagonist. This observation demonstrates that a postsynaptic rise of $[Ca^{2+}]_i$ leads to the release of endocannabinoids that activate presynaptic CB1Rs and inhibit GABA release.

- The retrograde messenger, like CB1 receptor, use a G-protein signalling pathway to induce DSI. In the rat hippocampus DSI is blocked by pre-treatment with pertussis toxin or with N-ethylmaleimide (NEM), which block pertussis toxin sensitive G-proteins. However, bath applied NEM and pertussis toxin have access to both presynaptic and postsynaptic sites. To determine the locus of the relevant G-protein, the postsynaptic cell is loaded with a non-hydrolyzable GTP analog. In these cells, DSI is not affected by NEM or pertussis toxin, indicating that only the activation of presynaptic G-proteins is required for DSI induction.
- Loading the postsynaptic target neuron with botulinum toxin that blocks membrane fusion and vesicular release does not prevent DSI. This observation is consistent with the mode of release of endocannabinoids, anandamide and 2-AG, that exit the cell by diffusion across the membrane or by passive transport but not by vesicular fusion.

18.1.5 Functional significance of DSI

A train of action potentials in the Purkinje cells or the activation of their afferent glutamatergic climbing fibre induces DSI of afferent GABA$_A$ activity, indicating that this form of short-term plasticity might play a role in controlling neuronal excitability *in vivo*. Similarly, a single Ca^{2+} spike is sufficient to produce DSI in the hippocampus. These observations indicate that DSI modulates the strength of GABAergic synapses *in vivo*.

FURTHER READING

Diana M and Marty A (2004) Endocannabinoid-mediated short-term synaptic plasticity: depolarization-induced suppression of inhibition (DSI) and depolarization-induced suppression of excitation (DSE). *British J. Pharmacol.* **142**, 9–19.

Llano I, Leresche N, Marty A (1991) Calcium entry increases the sensitivity of cerebellar Purkinje cells to applied GABA and decreases inhibitory synaptic currents. *Neuron* **6**, 565–574.

Piomelli D (2003) The molecular logic of endocannabinoid signaling. *Nature Rev. Neurosci.* **11**, 873–874.

Pitler TA and Alger BE (1992) Postsynaptic spike firing reduces synaptic GABA responses in hippocampal pyramidal cells. *J. Neurosci.* **12**, 4122–4132.

Vincent P and Marty A (1993) Neighboring cerebellar Purkinje cells communicate via retrograde inhibition of common presynaptic interneurons. *Neuron* **11**, 885–893.

CHAPTER

19

The adult hippocampal network

The hippocampus is part of the limbic system which mediates emotions and aspects of learning and memory. In the rat, it is a rostro-caudal structure (**Figure 19.1**) whereas in primates it is strictly localized in the temporal lobe.

The hippocampus is composed of two interconnected crescent-like regions (**Figures 19.2a,b**): the Ammon's horn (cornu ammonis) and the dentate gyrus (DG, also called fascia dentata). On a coronal section, Ammon's horn of the rat can be further subdivided into two regions, CA1 and CA3 (CA for cornu ammonis) (**Figure 19.2b**). In humans, two other subdivisions exist, CA2 and CA4. CA2 lies between CA1 and CA3 whereas CA4 is inserted in the dentate gyrus. Ammon's horn and dentate gyrus both contain a layer of principal neurons that are projection neurons (Golgi type I): the pyramidal cells and the granular cells, respectively. Numerous local interneurons (Golgi type II) are present in each region. Principal cells use an

excitatory amino acid as a neurotransmitter whereas interneurons use GABA.

19.1 OBSERVATIONS AND QUESTIONS

What is a network?

In nuclei of the central nervous system, various types of neurons are generally present: projection neurons (Golgi type I) whose axons project to neurons located outside the nucleus, and local interneurons (Golgi type II) whose axons project to neurons located inside the nucleus. These neuronal types are connected to each other (intrinsic connections): they form a network. Each network receives afferents from neurons located in other nuclei (extrinsic connections).

Are networks completely different from one nucleus to another or are there some fundamental principles of organization?

In the neocortex and hippocampus, Golgi type I neurons are the pyramidal cells, in the cerebellar cortex they are the Purkinje cells, and in the striatum they are the medium spiny neurons. Pyramidal cells are glutamatergic whereas Purkinje cells and medium spiny neurons are GABAergic. Aside from these principal cells, a large variety of local GABAergic interneurons are present in all these nuclei.

Does the precise knowledge of intrinsic and extrinsic connections as well as the firing patterns of neurons allow us to explain how network oscillations are generated?

In the hippocampus of the freely moving rat, several types of oscillations are recorded from populations of

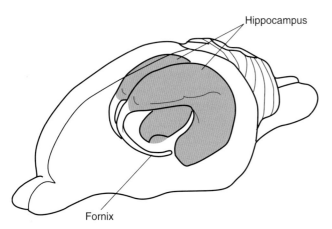

FIGURE 19.1 Schematic of the localization of the two hippocampi inside a rat brain.

368

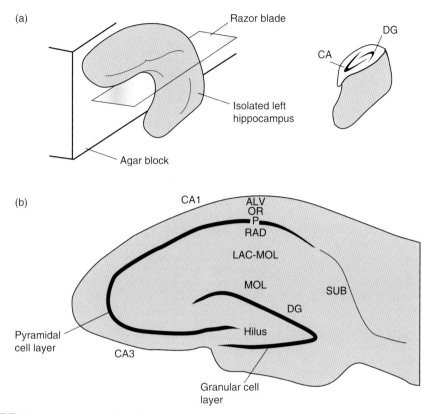

FIGURE 19.2 **Structure of the rat hippocampus**.
(a) Schematic of the slice preparation protocol. **(b)** Bright-field photomicrograph of a transverse section of the hippocampus stained with the Nissl method which stains neuronal somata and proximal dendrites (due to the presence of the Golgi apparatus). gr, granular cell; pyr, pyramidal cell; SUB, subiculum. See the text for further explanations. Adapted from Ishizuka N, Weber J, Amaral DG (1990) Organization of intrahippocampal projections originating from CA3 pyramidal cells in the rat. *J. Comp. Neurol.* **295**, 580–623, with permission.

neurons *in vivo* (extracellular recordings). For example, a rhythmic slow activity called 'theta' (5–10 Hz) is recorded during exploratory behaviour, such as sniffing, rearing and walking, and the paradoxical phase of sleep; once the animal stays still, or during consummatory behaviours or slow-wave sleep, intermittent sharp waves (SPW) are recorded in the dendritic layer of CA1–CA3 (**Figure 19.3**). These oscillations are network oscillations.

Neuronal oscillations have two main origins. They can be *intrinsic* to the neuron when they result from the activation of a cascade of currents intrinsic to the membrane, as described for thalamic and subthalamic neurons in Section 17.4. They can be *extrinsic* when they result from the activity of a group of interconnected neurons, from the activity of their synapses, as is the case for hippocampal neurons.

The aim of the present chapter is to give a description of the adult hippocampal network and to explain how it can generate oscillations.

19.2 THE HIPPOCAMPAL CIRCUITRY

19.2.1 Ammon's horn

Ammon's horn is a curved structure. It has a laminar organization with five layers. Owing to its U-shape, the layers (and pyramidal cells) are upside down in CA1 compared with CA3 (**Figures 19.2b** and **19.6b**).

Principal cells are called the pyramidal cells; they use an excitatory amino acid, probably glutamate, as a neurotransmitter

The principal cells, the pyramidal cells, have their soma aligned in a thin layer called the *pyramidal cell layer* (**Figure 19.2b**). The name of these cells comes from the clear pyramidal shape of their dendritic tree (**Figure 19.4a**). The cell body has a diameter of 20 μm. Three main dendritic trunks emerge from the cell body, one apical and two basal. Apical dendrites extend in

(a)

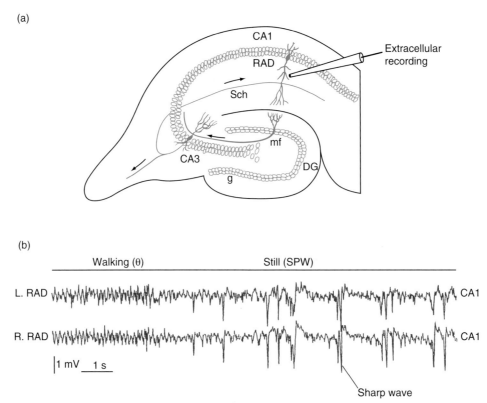

(b)

FIGURE 19.3 **Extracellular field recordings of hippocampal oscillations in a freely moving rat.**
An extracellular recording electrode is implanted in the stratum radiatum of the CA1 region of each hippocampus, the left (L) and the right (R). During exploratory activity (walking), regular theta waves are recorded (θ);
during immobility, large monophasic sharp waves (SPW) are recorded. Note the bilaterally synchronous
nature of SPW. Adapted from Buzsàki G (1989) Two-stage model of memory trace formation: a role for 'noisy'
brain states. *Neuroscience* **31**, 551–570, with permission.

the stratum radiatum (so called because of the radial
organization of apical dendrites from all pyramidal
cells) and arborize in stratum lacunosum moleculare.
Basal dendrites ramify in stratum oriens. Dendrites
have numerous spines. In CA3, the proximal part of
apical dendrites of pyramidal cells present giant spines
(thorny excrescences; **Figure 19.4b**) that are the postsynaptic elements of the synapses with granular cells of
dentate gyrus. Axons of pyramidal cells run in stratum
alveus where they emit numerous collaterals before
leaving the hippocampus.

*Several types of inhibitory interneurons innervate
pyramidal neurons; they use GABA as a
neurotransmitter, and their cell body is located in the
four more internal layers of CA*

The activity of pyramidal cells is modulated not
only by extrinsic afferences but also by intrinsic ones
coming from local inhibitory interneurons. Four main
types of inhibitory interneurons have been described
in Ammon's horn, all GABAergic. They have their cell

bodies in the five layers of the CA regions which are
from the external to the internal part of the hippocampus (**Figures 19.2b** and **19.5a**, left): stratum alveole
(ALV), stratum oriens (OR), stratum pyramidale (p),
stratum radiatum (RAD) and stratum lacunosum
moleculare (LAC-MOL). Interneurons are classified
according to their site(s) of termination on pyramidal
neurons (**Figure 19.5a**, right):

- Basket cells (BC) innervate the soma and proximal
  dendrites located in stratum pyramidale and radiatum. Cell bodies of basket cells are located in
  stratum pyramidale.
- Bistratified cells (BiC) innervate both apical and
  basal dendrites on their proximal part located in
  stratum radiatum and oriens. Cell bodies of bistratified cells are located in stratum oriens/radiatum.
- Oriens-lacunosum moleculare cells (O-LMC)
  innervate distal apical dendrites located in stratum
  lacunosum moleculare. Cell bodies of O-LMC are
  located in stratum oriens.
- Axo-axonic cells (AAC), also called chandelier cells,
  innervate exclusively the axon initial segment. Cell

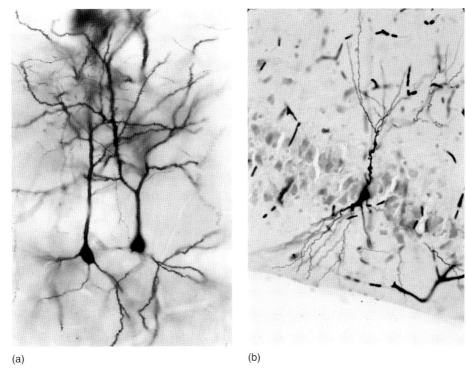

(a)   (b)

FIGURE 19.4   **Photomicrographs of stained Golgi CA1 and CA3 pyramidal neurons.**
(a) CA1; **(b)** CA3. In (b) a Nissl colouration shows the density of neuronal cell bodies in the pyramidal layer.
Photographs: (a) by Olivier Robain; (b) by Jean Luc Gaiarsa.

bodies of axo-axonic cells are located in stratum oriens.

When interneurons are activated by extrinsic afferences they participate in feedforward inhibition; when they are activated by recurrent axon collaterals of pyramidal cells they participate in feedback inhibition (**Figure 19.5b**). Some interneurons like O-LMC are involved only in feedback inhibition since they are activated only by axon collaterals from principal cells.

19.2.2 The dentate gyrus

The dentate gyrus is also a curved structure, with a U-shape and a three-layer organization (see **Figure 19.2b**).

Principal cells called the granular cells use an excitatory amino acid as a neurotransmitter

The principal cells called the granular cells have their soma densely packed in a thin layer, the granular cell layer. Somas have a small diameter (14–18 μm) and are ovoid. Dendritic trees emerge from the apical pole of somas and form the molecular layer. Axons, called mossy fibres, have a small diameter (0.5 μm) and are not myelinated. They emerge from the basal pole of

somas, divide in the hilus in numerous collaterals that contact local interneurons, and cross the hilus to make synapses with CA3 pyramidal cells.

Several types of inhibitory interneurons innervate granular cells; they use GABA as a neurotransmitter and their cell body is located in the three layers of DG

The same four types of interneurons as those found in Ammon's horn have been described in the dentate gyrus (DG). Interneurons located in stratum pyramidale in CA are located in the granular layer in DG. Similarly, interneurons located in stratum oriens of CA are in the hilus of DG and those in stratum radiatum of CA are in stratum moleculare of DG.

19.2.3 Principal cells form a tri-neural excitatory circuit

The main circuit inside the hippocampal formation involves the principal cells: granular cells of DG, pyramidal cells of CA1 and pyramidal cells of CA3. All these cells use an excitatory amino acid as a neurotransmitter. First, granular cells project on to CA3 pyramidal cells (**Figure 19.6a**). Their axon (called mossy fibres)

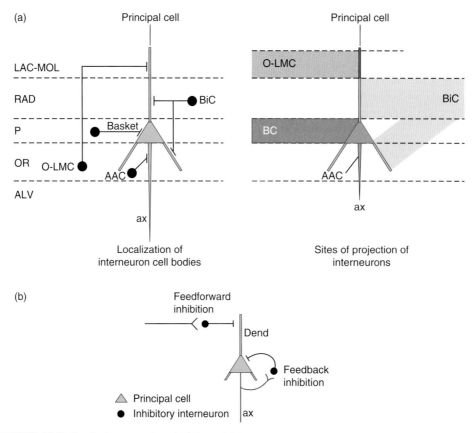

FIGURE 19.5   Intrinsic connections in CA1 and CA3.
(a) Schematic of a pyramidal cell indicating the localization of the cell bodies of the different interneurons (left) and the segregated postsynaptic domains innervated by the distinct presynaptic interneurons (right). **(b)** Illustration of feedforward and feedback inhibition. Part (a) adapted from Maccaferri G, Roberts DB, Szucs P *et al.* (2000) Cell surface domain specific postsynaptic currents evoked by identified GABAergic neurones in rat hippocampus *in vitro*. *J. Physiol.* **524**, 91–116, with permission.

terminates on the proximal portion of CA3 apical dendrites, on to giant spines. This restricted zone of projection forms the stratum lucidum (LUC), a sublayer of the radiatum that exists only in the CA3 region. Synapses between mossy fibres and dendritic spines of CA3 pyramidal cells are giant synapses (see **Figure 6.3**). In turn, CA3 pyramidal cells send axon collaterals, called the Schaffer collaterals, to CA1 pyramidal cells, on the distal part of their apical dendrites, at the level of stratum lacunosum moleculare. In coronal slices *in vitro*, all these circuits are present since they are organized in the transverse plane (**Figure 19.6b**).

In addition, pyramidal cells of the CA3 and CA1 regions emit local axon collaterals that contact local interneurons. Similarly, granular cells emit axon collaterals that locally innervate interneurons. Moreover, in CA3, pyramidal cells are connected to each other by excitatory recurrent collaterals (**Figure 19.6c**). Therefore, local circuits superimpose on the main tri-neuronal excitatory circuit (**Figure 19.7**). Local circuits are detailed below in Sections 19.3 and 19.4.

19.2.4 Extrinsic afferences to principal cells and interneurons

The two major pathways that convey afferent information to the hippocampus are the *perforant* pathway coming from entorhinal cortex and the *fornix* coming from the medial septum and anterior thalamus. Moreover, the two hippocampi are interconnected by the *commissural* pathway. These connections and their functions will not be studied here.

19.3 ACTIVATION OF INTERNEURONS EVOKE INHIBITORY GABAERGIC RESPONSES IN POSTSYNAPTIC PYRAMIDAL CELLS

To study the response of a postsynaptic pyramidal neuron to a presynaptic interneuron, the activity of these connected neurons is recorded concomitantly. Two neurons that are connected are called a pair.

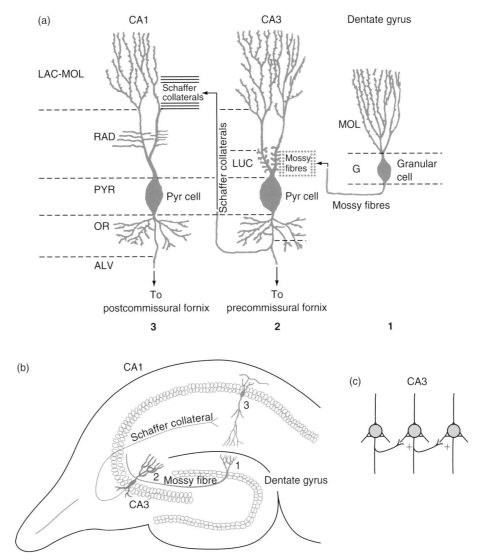

FIGURE 19.6 **The tri-neuronal circuit between principal cells.**
(a) Sites of termination of axons of principal cells on target principal cells (which are CA1 and CA3 pyramidal cells). Axons of granular cells are called mossy fibres. Axonal collaterals of CA3 pyramidal cells are called Schaffer collaterals. **(b)** The tri-neuronal circuit is organized in the transverse plane. LUC, stratum lucidum of CA3; pyr cell, pyramidal cell. **(c)** Illustration of recurrent excitation. Part (a) adapted from Altman J, Brunner RL, Bayer SA (1973) The hippocampus and behavioral maturation. *Behav. Biol.* 8, 557–596, with permission.

19.3.1 Experimental protocol to study pairs of neurons

The hippocampus is an adequate structure to study pairs of neurons. Thanks to its laminar organization, the localization of the cell bodies of the different neurons is strictly organized. For example, in slices of the rat hippocampus, when an electrode is inserted in the pyramidal layer of the CA1 region, the probability of impaling or patching the cell body of a pyramidal neuron is high and that of an interneuron (a basket cell for example) around ten times less (**Figure 19.5a**, left). Conversely, when the electrode is inserted in stratum oriens or radiatum, the probability of impaling or patching a pyramidal cell is close to 0 whereas that for an interneuron (O-LMC, AAC, BiC) is very high, close to 1.

To record the activity of an interneuron–pyramidal cell pair, one electrode (electrode 1) is placed in stratum oriens or radiatum or pyramidale to patch or impale an interneuron, and the other electrode (electrode 2) is placed in the stratum pyramidale to patch or impale a pyramidal cell (**Figure 19.8**). The activity of the interneuron is always recorded in current clamp mode and the activity of the pyramidal cell is recorded either in voltage clamp (to record the inhibitory postsynaptic current or IPSC) or current clamp (to record the

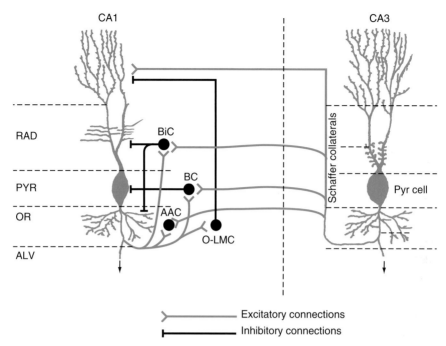

FIGURE 19.7 Schematic of the synaptic circuitry in the CA1 region of the hippocampus and afferent connections from Schaffer collaterals of CA3 pyramidal cells.
Adapted from Altman J, Brunner RL, Bayer SA (1973) The hippocampus and behavioural maturation. *Behav. Biol.* **8**, 557–596, with permission.

inhibitory postsynaptic potential or IPSP). Recordings are performed with either intracellular or whole-cell electrodes. Interneurons and pyramidal cells are identified during the recording session. To do so, spontaneous action potentials and evoked responses are recorded. Interneurons are characterized by the presence of an after-hyperpolarization following their action potentials and by their response to a long-lasting depolarizing current pulse which lacks spike frequency accommodation (**Figure 19.8** – compare recordings in (a) and (b)).

To check a connection between the interneuron and the pyramidal cell, electrode 1 is used as the stimulatory electrode and electrode 2 as the recording one. Spontaneous firing of the recorded interneuron is prevented by the continuous injection of a hyperpolarizing current. A suprathreshold square current pulse is injected into the interneuron to evoke action potentials (an example is given in **Figure 19.9**). If an IPSC or IPSP is evoked in the pyramidal cell in response to interneuron stimulation, the two neurons are thus identified as a pair of synaptically coupled neurons. If no synaptic response is recorded, electrode 1 is left in place and electrode 2 is changed (or the reverse). Another interneuron or pyramidal cell is patched (or impaled) and stimulated. When a pair is found, the study of the synaptic response can begin.

After the recording session, the type of interneurons recorded are identified on morphological criteria. To do this, electrodes 1 and 2 are filled with biocytin which

diffuses or is injected into the cell. Slices are then fixed and biocytin-filled cells are visualized by the avidin-biotiny lated horseradish peroxidase method. The dendritic tree and axonal arborization of the recorded neurons are drawn by reconstruction from serial 60 μm thick sections under a light microscope. This also allows one to check that the two neurons are connected and to count the number of contacts. Then, under electron microscopy, the type of synapses between the two neurons and the number of active zones can be precisely analyzed.

In summary, the basis for the selection of pairs of connected neurons are: (i) the presence of a short latency (monosynaptic) IPSC or IPSP in the pyramidal cell following an action potential in the putative interneuron; (ii) stable recordings from both cells for sufficient time to obtain an averaged IPSC or IPSP; and (iii) recovery of at least part of the biocytin-labelled interneuron to allow its identification.

19.3.2 Unitary inhibitory postsynaptic currents (IPSCs) evoked by different types of interneurons are all GABA$_A$-mediated but have different kinetics when recorded at the level of the soma

To study the synaptic current evoked in a pyramidal neuron in response to a single action potential in the presynaptic interneuron, the activity of both neurons is recorded in whole-cell configuration.

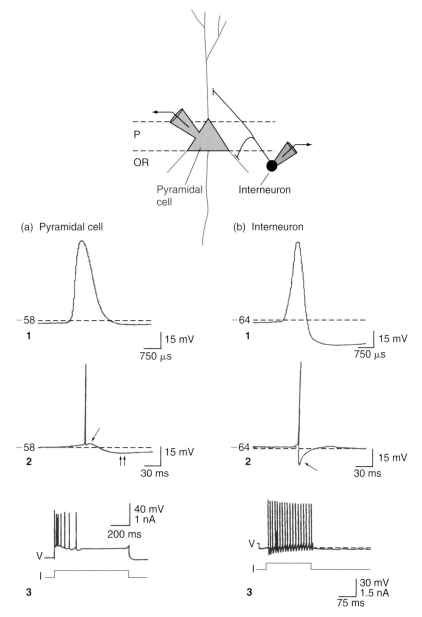

FIGURE 19.8 **Physiological characteristics that differentiate pyramidal neurons from interneurons.**
(a) Action potential of a pyramidal neuron at a fast (1) and a slow (2) timebase to show the presence of an after-spike depolarization (arrow) followed by a slow after-spike hyperpolarization (double arrow). The bottom trace (3) shows the response of a pyramidal neuron to a depolarizing current pulse. (b) Action potential of an interneuron recorded in the stratum oriens (OR) at a fast (1) and a slow (2) timebase to show the presence of a fast after-hyperpolarization (arrow). The bottom trace (3) shows the response of an interneuron to a depolarizing current pulse. Adapted from Lacaille JC and Williams S (1990) Membrane properties of interneurons in stratum oriens-alveus of the CA1 region of rat hippocampus *in vitro*. *Neuroscience* **36**, 349–359, with permission.

Whole-cell patch recordings

The intrapipette solution of electrode 1 (interneuron) is designed to allow the recording of action potentials (in mM): 130 K gluconate, 2 MgCl$_2$, 0.1 EGTA, 2 ATP, 0.3 GTP, 10 Hepes and 0.5% biocytin. Action potentials are generated in the interneuron by injection of a supra-threshold square current pulse at 0.1–1 Hz.

The intrapipette solution of electrode 2 (pyramidal cell) is designed to record GABA$_A$-mediated currents in isolation (in mM): 100 CsCl, 2 MgCl$_2$, 0.1 EGTA, 2 ATP, 0.3 GTP, 40 Hepes, 5 QX-314 and 0.5% biocytin, at a pH of 7.2. QX-314 and Cs$^+$ strongly reduce voltage-gated Na$^+$ and K$^+$ currents, respectively. Ionotropic glutamate receptors are blocked by DNQX 20 μM and D-AP5 50 μM

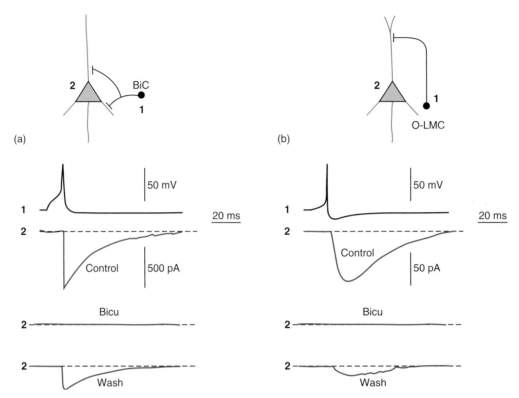

FIGURE 19.9 Unitary IPSCs (uIPSCs) evoked in pyramidal cells in response to different types of interneurons are all mediated by GABA$_A$ receptors.
Averaged uIPSC evoked in a pyramidal neuron (2) in response to a single spike in a bistratified interneuron (1a, BiC) or an oriens lacunosum moleculare interneuron (1b, O-LMC) in control conditions, in the presence of bicuculline (Bicu, 10 μM) and after partial washout of the drug (Wash). Adapted from Maccaferri G, Roberts DB, Szucs P *et al.* (2000) Cell surface domain specific postsynaptic currents evoked by identified GABAergic neurones in rat hippocampus *in vitro. J. Physiol.* **524**, 91–116, with permission.

in the bath. The synaptic current is recorded in voltage clamp mode ($V_H = -70$ mV). Internal Cl$^-$ concentration is 104 mM and the external concentration is 135 mM, which gives a reversal potential for Cl$^-$ ions close to 0 mV (recall that in physiological conditions E_{Cl} is around -70 mV; see Section 9.3.4). Therefore, at $V_H = -70$ mV, when GABA$_A$ receptors open, there is an outflow of Cl$^-$ through channels permeable to Cl$^-$ ions; it is recorded as an inward current (an inward current is by convention an inward movement of positive charges; see Section 3.3.3).

Unitary IPSCs evoked by different types of interneurons are all mediated by GABA$_A$ receptors

When a single action potential is evoked in the presynaptic interneuron, an inward current is recorded in the postsynaptic pyramidal cell (**Figure 19.9**). This inward current is totally blocked by bicuculline (Bicu, an antagonist of GABA$_A$ receptors), thus showing that it is mediated by GABA$_A$ receptors. This holds true for all the following pairs: BiC–pyr, O-LMC–pyr, BC–pyr and AAC–pyr.

These GABA$_A$-mediated currents are called *inhibitory postsynaptic currents* (IPSCs), though they are inward, because in control Cl$^-$ conditions they would be outward and thus inhibitory. An IPSC which results from a single action potential in the presynaptic neuron is called a *unitary IPSC* (uIPSC).

Proximally and distally generated unitary IPSCs have different kinetic parameters

The unitary IPSCs of **Figure 19.9** are recorded at the level of the soma of pyramidal cells. When evoked in distal dendrites by O-LMC, unitary IPSCs are passively conducted along the apical dendrite before being recorded in the soma. In contrast, unitary IPSCs evoked at the level of the soma by basket cells (BC) or at the axon initial segment by axo-axonic cells (AAC) are generated at sites close to the recording electrode. As shown in **Figure 19.10**, distally evoked unitary IPSCs have a slower time to peak (also called the rise-time) than those evoked in proximal dendrites and soma. Unitary IPSCs evoked by axo-axonic cells have

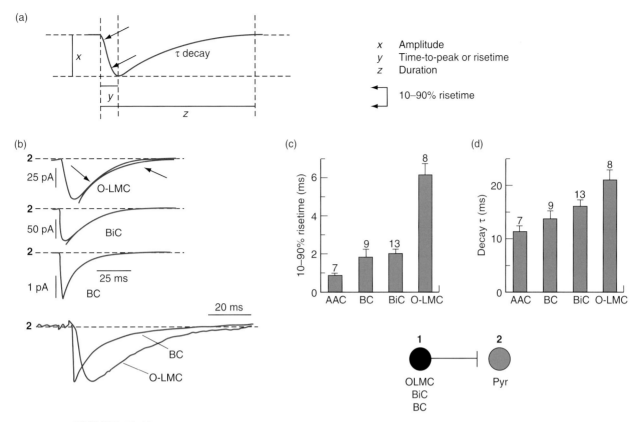

FIGURE 19.10 Kinetic parameters of unitary IPSCs evoked in a pyramidal cell in response to different types of interneurons.
(a) Definition of the parameters of postsynaptic currents. **(b)** Comparison of the risetime (or time to peak) and decay phase of three different uIPSCs. Bottom trace shows superimposed averaged uIPSCs generated by a presynaptic basket cell (BC) or oriens lacunosum moleculare cell (O-LMC). **(c)** Histogram of the risetimes (10–90%) and **(d)** of the decay (τ, time to 63% of decay) of the uIPSCs generated by different classes of presynaptic interneurons. Adapted from Maccaferri G, Roberts DB, Szucs P *et al.* (2000) Cell surface domain specific postsynaptic currents evoked by identified GABAergic neurones in rat hippocampus *in vitro. J. Physiol.* **524**, 91–116, with permission.

a fast risetime (0.8 ± 0.1 ms) whereas those evoked by basket cells have a slower risetime; and those evoked in the most distal pyramidal dendrites by oriens lacunosum moleculare cells have a very slow risetime (6.2 ± 0.6 ms). The kinetics of unitary IPSCs recorded in the soma of pyramidal cells reflect the domain of innervation: this can result from electrotonic dendritic filtering (see Section 13.1) and/or the lack of voltage clamp of the more distal locations and/or site-specific subunit composition of $GABA_A$ receptors.

In summary, in Ammon's horn, the synapses established by interneurons on the soma (BC), on proximal or distal dendrites (BiC, O-LMC) or the axon initial segment (AAC) of pyramidal cells are all inhibitory and $GABA_A$-mediated. This means that $GABA_A$ synapses are present all along the somato-dendritic tree and

axon initial segment of pyramidal cells. Depending on where they are generated on the dendritic tree, $GABA_A$-mediated IPSCs have different risetimes.

19.3.3 $GABA_A$-mediated IPSCs generate IPSPs in postsynaptic pyramidal cells

IPSCs generate transient hyperpolarizations of the postsynaptic membrane, called *inhibitory postsynaptic potentials* (IPSPs). To study the IPSPs evoked in a pyramidal neuron in response to a presynaptic interneuron in physiological conditions, the activity of both neurons is recorded with intracellular electrodes filled with 1.5 M of KCH_3SO_4 (and 2% biocytin) so as not to change the internal Cl^- concentration. In these conditions, the reversal potential for Cl^- ions is around −70 mV, which is close to that *in vivo*.

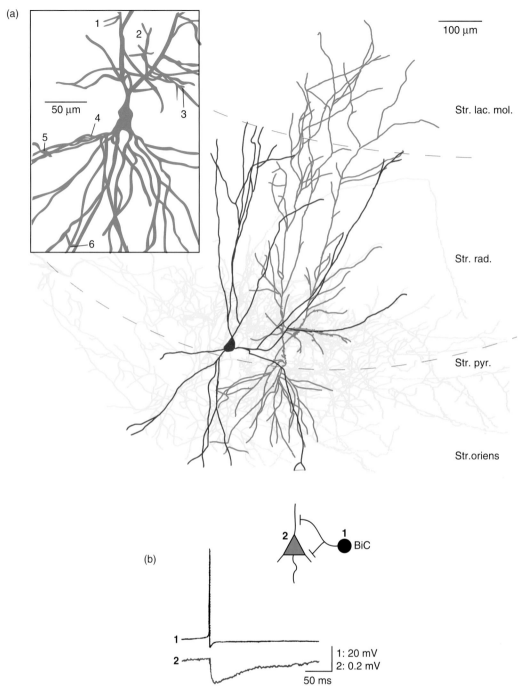

FIGURE 19.11 Unitary IPSP evoked in a CA1 pyramidal cell in response to a bistratified cell (BiC) and location of contact sites.
(a) Reconstruction of the biocytin-filled presynaptic interneuron (somato-dendritic tree in grey, axon in black) and postsynaptic pyramidal dendrites (green). The inset shows the location of the six contact sites between the GABAergic axon (black) and the postsynaptic pyramidal cell (green). (b) An action potential in BiC (1) elicits a small-amplitude, short-latency unitary IPSP in the postsynaptic pyramidal cell (2). (Trace 2 is an averaged unitary IPSP.) Adapted from Buhl EH, Halasy K, Somogyi P (1994) Diverse sources of hippocampal unitary inhibitory postsynaptic potentials and the number of synaptic release sites. *Nature* **368**, 823–828, with permission.

In **Figure 19.11**, a single spike in the presynaptic interneuron (bistratified cell) evokes a transient hyperpolarization in the postsynaptic pyramidal neuron; this is called inhibitory (IPSP) because it hyperpolarizes the membrane to a potential far from the threshold of spike initiation. It is unitary (uIPSP) since it is evoked by a single presynaptic spike. To evoke this uIPSP, the presynaptic action potential first propagates to the numerous synaptic terminals of the interneuron and evokes the release of GABA from all or some of these terminals. Therefore, a unitary IPSC or IPSP can result from the activation of one or more release sites, depending on the number of synapses established by the presynaptic interneuron on the recorded postsynaptic pyramidal neuron and on the number of active zones per synapse (see **Figure 7.2**). A study under electron microscopy then reveals the exact number of synaptic complexes since a single synaptic bouton may establish multiple synaptic complexes.

In the example of **Figure 19.11**, there are six synaptic contacts between the axon of the presynaptic basket cell and the postsynaptic pyramidal neuron. Electron microscopy shows that these six synaptic contacts correspond to six synaptic complexes (there is a single active zone per bouton). Therefore, a maximum of six release sites is responsible for this unitary IPSP. This allows one to calculate the average amplitude of an IPSP evoked by the activity of one release site only: it is around $30\,\mu V$ ($220\,\mu V$ divided by 6) which is a very small hyperpolarization. In **Figure 19.12**, the IPSP evoked by an axo-axonic cell corresponds to the activity of eight synaptic contacts between a presynaptic axo-axonic cell and a pyramidal neuron. This $GABA_A$-mediated IPSP reverses at $-78\,mV$ (range -65 to $-78\,mV$), which is close to E_{Cl}. Morphological studies show that inhibitory interneurons establish an average of 5–30 synaptic contacts with a single postsynaptic pyramidal cell.

19.3.4 $GABA_B$-mediated IPSPs are also recorded in pyramidal neurons in response to strong interneuron stimulation

The activity of a CA3 pyramidal cell is intracellularly recorded with electrodes filled with $2\,M$ of $KMeSO_4$ so as not to change the intracellular concentration of Cl^-. A stimulating electrode is placed in the hilus (it stimulates local interneurons and excitatory afferents from granular cells of the dentate gyrus). In response to the stimulation, a complex synaptic response is recorded (**Figure 19.13**): a prior EPSP (inset) followed by a biphasic IPSP (early and late). When QX-314, a derivative of lidocaine that blocks voltage-gated Na^+ currents (fast and persistent) and $GABA_B$ receptor-activated K^+ current (see Chapter 11), the late component disappears. In contrast, in the presence of bicuculline in the bath the early component disappears (not shown). This shows that the early IPSP is mediated by $GABA_A$ receptors whereas the late phase is mediated by $GABA_B$ receptors.

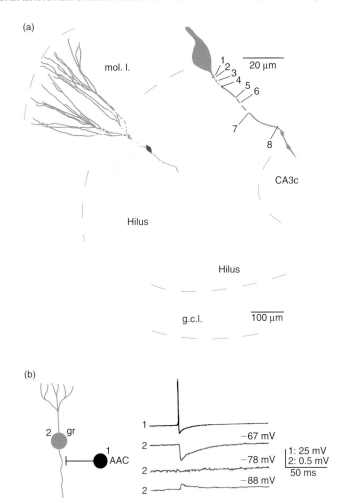

FIGURE 19.12 Unitary IPSP evoked in a granular cell in response to an axo-axonic cell (AAC) and location of contact sites.
(a) Reconstruction of the biocytin-filled presynaptic interneuron (soma in grey, axon in black) and postsynaptic granular cell (green). The top right shows the location of the eight contact sites between the GABAergic axon (black) and the axon initial segment of the postsynaptic granular cell (green). **(b)** An action potential in AAC (1) elicits a small-amplitude, short-latency unitary IPSP in the postsynaptic granular cell (2) that reverses at around $-78\,mV$ (traces 2 show averaged EPSPs). Adapted from Buhl EH, Halasy K, Somogyi P (1994) Diverse sources of hippocampal unitary inhibitory postsynaptic potentials and the number of synaptic release sites. *Nature* **368**, 823–828, with permission.

Postsynaptic $GABA_B$ receptor-mediated IPSPs are recorded only in response to a strong activation of interneurons. This effect is absent in the response to a single spike in interneurons, suggesting that a larger release of GABA in the synaptic cleft is necessary to activate postsynaptic $GABA_B$ receptors. This late IPSP prolongs the inhibition of pyramidal cells by GABAergic interneurons.

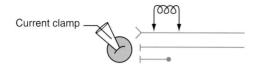

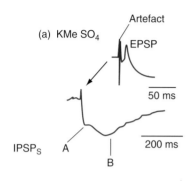

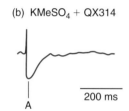

FIGURE 19.13   GABA$_A$ and GABA$_B$ receptor-mediated IPSPs in pyramidal cells.

(a) CA3 pyramidal cells respond to a hilar stimulation by a biphasic IPSP preceded by an EPSP (inset). The biphasic IPSP consist of an early (A) and a late (B) IPSP. **(b)** In the presence of QX 314 (50 mM) in the pipette solution, stimulation no longer evokes the late IPSP whereas the early one (A) is spared (as well as the EPSP, not shown). The pipette is filled with potassium methyl sulphate (KMeSO$_4$). Adapted from McLean HA, Ben Ari Y, Gaiarsa JL (1995) NMDA-dependent GABA$_A$-mediated polysynaptic potentials in the neonatal rat hippocampal CA3 region. *Eur. J. Neurosci.* **7**, 1442–1448, with permission.

19.4 ACTIVATION OF PRINCIPAL CELLS EVOKES EXCITATORY GLUTAMATERGIC RESPONSES IN POSTSYNAPTIC INTERNEURONS AND OTHER PRINCIPAL CELLS (SYNCHRONIZATION IN CA3)

To study the physiological response of a postsynaptic neuron to an action potential in the presynaptic pyramidal neuron, the activity of these connected neurons is recorded concomitantly with intracellular or whole-cell electrodes. Electrodes 1 and 2 are filled with 4% biocytin in 0.5 M of potassium acetate. In these conditions the reversal potential for Cl$^-$ ions is around −70 mV, which is close to that *in vivo*.

19.4.1 Pyramidal neurons evoke AMPA-mediated EPSPs in interneurons

EPSPs elicited in interneurons in response to a single action potential in the presynaptic pyramidal cell have a mean amplitude of 1–4 mV and a time to peak of 1.5–4 ms. They are totally blocked by CNQX, the selective blocker of AMPA receptors (not shown). They fluctuate in amplitude at all synapses examined and sometimes fail (**Figure 19.14a**). This latter observation suggests that there is a low probability of release or the existence of a few release sites. In fact, under light microscopy, in all pairs studied, a single synaptic contact is identified between the filled pyramidal cell and interneuron. Electron microscopy shows that each contact has a single active zone.

Studies are performed in the hippocampus on a large number of cells in order to obtain quantitative data: pyramidal cells are filled *in vivo* with neurobiotin, and parvalbumine-containing interneurons (basket and axo-axonic cells) are revealed by immunocytochemistry. This study confirms that each filled pyramidal axon establishes a single contact with parvalbumine-containing interneurons. By counting the boutons terminating on interneurons, it has been shown that over 1000 excitatory synapses terminate on a single inhibitory cell, suggesting that more than 1000 pyramidal cells converge on to one interneuron. This excitatory drive presumably contributes to the high frequency of the spontaneous firing of hippocampal interneurons.

Interestingly, single pyramidal cell action potentials cause inhibitory cells to fire at resting membrane potential with a probability of 0.4 (**Figure 19.15**). The mean interval between pre- and postsynaptic spikes is 2.9 ± 0.7 ms. Factors contributing to spike-to-spike transmission are the low firing threshold of inhibitory interneurons, their depolarized resting membrane potential and the large EPSCs elicited by pyramidal cells leading to large EPSPs (of the order of the millivolts) owing to the high input membrane resistance of interneurons (**Figure 19.14b**). Moreover, CA3 pyramidal cells have a tendency to discharge in bursts of several action potentials (with 5–10 ms intervals). This has as a consequence that the security of transmission is enhanced and temporal summation occurs when the interval between presynaptic spikes is shorter than the time course of EPSPs (**Figure 19.14b**).

19.4.2 EPSPs in interneurons lead to feedback inhibition of pyramidal neurons

When two pyramidal cells of the CA3 region are recorded simultaneously, firing in one pyramidal cell

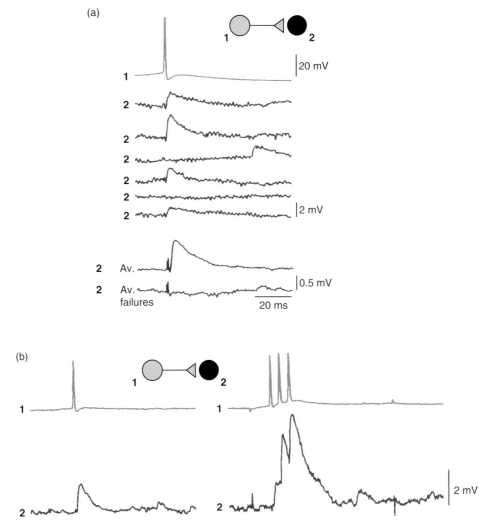

FIGURE 19.14   **Unitary EPSP evoked in an inhibitory interneuron in response to a CA3 pyramidal cell.**
(a) A single spike in the presynaptic pyramidal cell (1) evokes a unitary EPSP in the postsynaptic interneuron (2) that fluctuates in amplitude and sometimes fails. An averaged EPSP and an averaged trace of failures ($n = 38$) are shown below. Av: average. **(b)** EPSPs initiated in an inhibitory interneuron in response to a single spike (left) or a train of three spikes (right) in the presynaptic pyramidal cell (1). On the right there is a temporal summation of the EPSPs. Adapted from Miles R (1990) Synaptic excitation of inhibitory cells by single CA3 hippocampal pyramidal cells of the guinea pig *in vitro*, *J. Physiol.* **428**, 61–77, with permission.

can evoke an IPSP in the other pyramidal cell (**Figure 19.16a**). The mean IPSP latency is 3.5 ± 0.7 ms and the mean amplitude 1.9 ± 0.6 mV. Bicuculline (a GABA$_A$ receptor antagonist) as well as CNQX (an AMPA receptor antagonist) completely suppresses these evoked IPSPs. This experiment excludes the possibility that pyramidal cells establish monosynaptic inhibitory connections. It shows in contrast that these IPSPs result from a bisynaptic connection between pyramidal cells: the first synapse is glutamatergic and the second one GABAergic. There are two ways to block this IPSP: to block synaptic transmission at the first synapse with CNQX or at the second synapse with bicuculline.

19.4.3 CA3 pyramidal neurons are monosynaptically connected via glutamatergic synapses

Neighbouring pyramidal cells of CA3 are monosynaptically connected via axon collaterals (see **Figure 19.6c**). In pairs of pyramidal neurons, presynaptic pyramidal action potentials evoke EPSPs in the postsynaptic pyramidal cell (**Figure 19.17**) that are sensitive to CNQX. These EPSPs are considered to be monosynaptic on the basis of their mean latency (range $0.8–1.2$ ms) and the proportion of transmission failures (small in monosynaptic connections).

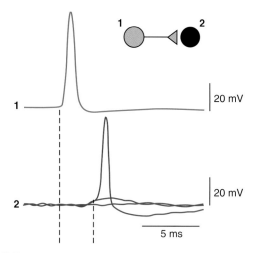

FIGURE 19.15 Spike to spike transmission at an excitatory synapse between a CA3 pyramidal neuron and a postsynaptic inhibitory interneuron.
In response to successive single spikes in a presynaptic pyramidal cell (1, only one spike is displayed), one transmission failure, one unitary EPSP and one unitary EPSP that causes postsynaptic firing are recorded in the postsynaptic interneuron (2, three superimposed traces). Adapted from Miles R (1990) Synaptic excitation of inhibitory cells by single CA3 hippocampal pyramidal cells of the guinea pig *in vitro*. *J. Physiol.* **428**, 61–77, with permission.

19.4.4 Overview of intrinsic hippocampal circuits

The main circuit is the tri-neuronal circuit between the excitatory principal cells (see **Figure 19.6a, b**). In each region, dentate gyrus, CA1 or CA3, axon collaterals of principal cells excite inhibitory interneurons which in turn inhibit other principal cells (feed forward inhibition). Interneurons are also directly activated by extrinsic excitatory afferences (feedforward inhibition). In CA3, principal cells (pyramidal cells) are monosynaptically connected via excitatory axon collaterals (see **Figure 19.6c**).

19.5 OSCILLATIONS IN THE HIPPOCAMPAL NETWORK: EXAMPLE OF SHARP WAVES (SPW)

Gyorgy Buzsaki discovered sharp waves (SPW) in 1983 when recording from CA1 stratum radiatum and a pyramidal cell layer simultaneously in the freely moving rat. When the rat is exploring, typical theta waves are present. When the animal becomes immobile or goes to sleep (slow-wave stage), large-amplitude intermittent sharp waves of 40–120 ms replace theta oscillations (see **Figure 19.3**). The frequency of the intermittent sharp waves ranges from 0.02 to 3 Hz.

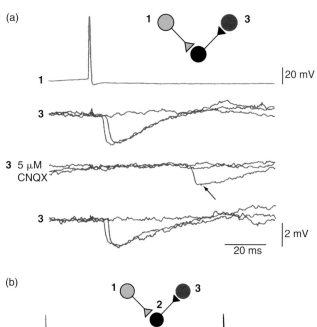

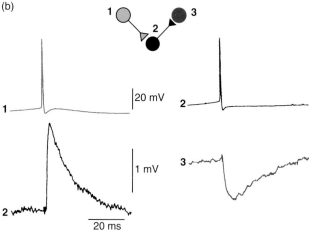

FIGURE 19.16 Feed forward inhibition between two CA3 pyramidal cells.
(a) In response to a single spike in one pyramidal cell (1), an IPSP is recorded in a neighbouring pyramidal cell (3). Upper trace 2 shows three superimposed responses to three presynaptic action potentials (trace 1, only one spike is displayed). There is one transmission failure and two IPSPs. Middle trace 3 shows the effect of CNQX in the bath: the evoked IPSPs are suppressed but spontaneous ones are still present (arrow). Bottom trace 3 shows the return to control solution. **(b)** Sequential recordings of the three connected cells. Pyramidal cell 1 activates interneuron 2 (average of uEPSP, $n = 40$) and interneuron 2 inhibits cell 3 (average of uIPSP, $n = 40$). Adapted from Miles R (1990) Synaptic excitation of inhibitory cells by single CA3 hippocampal pyramidal cells of the guinea pig *in vitro*. *J. Physiol.* **428**, 61–77, with permission.

Recordings in layers of the CA1 region with extracellular electrodes show that sharp wave amplitude is maximal in stratum radiatum, the layer where apical dendrites of CA1 pyramidal neurons extend (**Figure 19.18a, b**). The immediate cause of sharp waves in the CA1 region is the synchronous discharge of a large number of CA3 pyramidal neurons. This results in the near-simultaneous depolarization of CA1 pyramidal

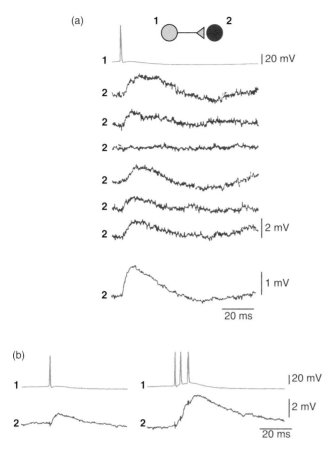

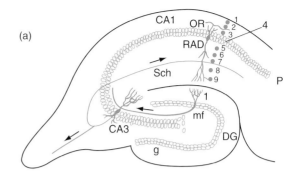

Extracellular recordings in CA1

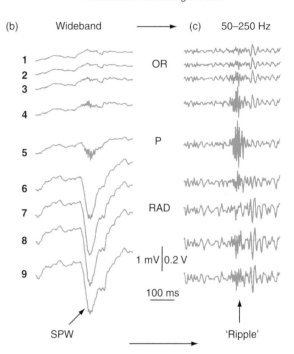

FIGURE 19.17 **Unitary EPSP evoked in a CA3 pyramidal cell in response to a CA3 pyramidal cell.**
(a) Monosynaptic unitary EPSP evoked in a postsynaptic pyramidal cell (2) in response to a single spike in the presynaptic pyramidal cell (1). The EPSP fluctuates in amplitude and sometimes fails. Bottom trace shows an averaged EPSP. (b) EPSPs initiated in a postsynaptic pyramidal cell (2) in response to a single spike (left) or a train of three spikes (right) in the presynaptic pyramidal cell (1). On the right there is a temporal summation of the EPSPs. Part (a) adapted from Miles R and Wong RKS (1986) Excitatory synaptic interactions between CA3 neurones in the guinea pig hippocampus. *J. Physiol.* **373**, 397–418, with permission. Part (b) adapted from Miles R (1990) Synaptic excitation of inhibitory cells by single CA3 hippocampal pyramidal cells of the guinea pig *in vitro. J. Physiol.* **428**, 61–77, with permission.

FIGURE 19.18 **Sharp waves (SPW) in CA1 of the awake immobile rat and the high-frequency oscillations (ripples).**
(a) The extracellular activity of a population of neurons is recorded with nine extracellular electrodes in the CA1 region. (b) Extracellular recordings show that the sharp waves are the most pronounced in the apical dendritic layer (stratum radiatum, RAD). (c) When the recordings in (b) are filtered in order to leave only the events with a frequency between 50 to 250 Hz, high-frequency oscillations that form a ripple are revealed. Ripples are particularly prominent in the pyramidal layer (p). Note that the amplitude scale is increased 5-fold between (b) and (c). Adapted from Ylinen A, Bragin A, Nadasdy Z *et al.* (1995) Sharp wave-associated high-frequency oscillation (200 Hz) in the intact hippocampus: network and intracellular mechanisms. *J. Neurosci.* **15**, 30–46, with permission.

cells via Schaffer collaterals which terminate dominantly on apical CA1 dendrites (see **Figure 19.6**) (recall that the extracellular recording of EPSPs from a population of neurons, called field EPSP, appears as a downward deflection in extracellular recordings; see **Figure 18.4b**). In brief, sharp waves represent a coherent depolarization of the apical dendrites of CA1 pyramidal neurons.

However, the synchronous discharge of a large number of CA3 pyramidal neurons also directly activates interneurons (see **Figure 19.7**). Therefore, concurrent with sharp waves, pyramidal cells and interneurons are activated synchronously (interneurons are even activated earlier than target pyramidal cells, because of their lower spike threshold). As a result, dendrites of CA1 pyramidal cells respond by summed EPSPs (see **Figure 19.14b**) to bursts of action potentials in Schaffer

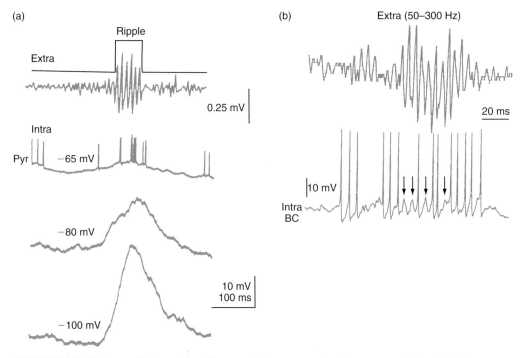

FIGURE 19.19   Intracellular activity of a pyramidal neuron and an interneuron during high-frequency oscillations (ripples).
Intracellular recording **(a)** from a CA1 pyramidal neuron and **(b)** from a CA1 basket cell (BC) during a single ripple event. In (a), membrane hyperpolarization of the pyramidal cell from −65 to −100 mV reveals a strong depolarization force during the ripple. Part (a) adapted from Ylinen A, Bragin A, Nadasdy Z *et al.* (1995) Sharp wave-associated high-frequency oscillation (200 Hz) in the intact hippocampus: network and intracellular mechanisms. *J. Neurosci.* **15**, 30–46, with permission. Part (b) adapted from Freund TF and Buzsaki G (1996) Interneurons in the hippocampus. *Hippocampus* **6**, 347–470, with permission.

collaterals and by summed IPSPs in response to bursts of action potentials in axons of interneurons. How do CA1 pyramidal cells finally respond?

Whenever a sharp wave is present in the radiatum, a 140–200 Hz field oscillation is present in the pyramidal cell layer (**Figures 19.18b, c**). This is best shown by filtering the field potential below 50 Hz (to suppress oscillations at a frequency under 50 Hz). This reveals high-frequency oscillations that form a ripple. These oscillations are most prominent in the pyramidal layer (**Figure 19.18**, trace 5). The fast field oscillation is believed to represent summed fast IPSPs in the somata of pyramidal cells brought about by the activated interneurons.

The explanation is the following. Interneurons, including basket and axo-axonic (chandelier) cells, fire together (perhaps coupled by gap junctions) at around 200 Hz and impose a series of IPSPs on the somata of pyramidal cells. Some pyramidal cells are excited through their dendrites strongly enough so that excitation can overcome somato-axonal inhibition. However, because of these series of IPSPs in the soma and axon initial

segment membrane, the spike(s) emerge at the periods where inhibition is least (i.e. out of phase with the spikes of the interneurons; **Figure 19.19**). In short, inhibition does not necessarily prevent firing but serves to time the occurrence of spikes.

Sharp waves are envisaged as an endogenous mechanism for consolidating synaptic changes and transferring information from the hippocampus to neocortex during sleep. The strong depolarization of pyramidal cell dendrites during a sharp wave burst enhances the size of fast dendritic spikes (see Section 16.3.2). The large membrane depolarization also triggers Ca^{2+} spikes which, in turn, may alter the weights of the simultaneously active nearby synapses (see Section 16.3.3).

19.6 SUMMARY

What is a neuronal network?

A neuronal network is formed by neurons from the same nucleus. It is described by the connections between these neurons, the type of synapse (glutamatergic, GABAergic, etc.), and the arrangement of the synapses on somato-dendritic trees.

Are networks completely different from one nucleus to another or are there some fundamental principles of organization?

Networks have in common a basic organization. There are always principal cells and most often interneurons (the subthalamic nucleus is, for example, devoid of interneurons whereas striatum contains many different types of these). Interneurons are always connected so as to provide feedforward and feedback inhibitions.

Afferent extrinsic connection simpinge on to principal cells and interneurons.

Does the precise knowledge of intrinsic and extrinsic connections as well as the firing patterns of neurons allow us to explain how network oscillations are generated?

Yes, it does, if the pattern of afferent activity in extrinsic afferent axons is known.

FURTHER READING

Buzsaki G (2002) Theta oscillations in the hippocampus. *Neuron.* **33**, 325–340.

Buzsaki G, Horvath Z, Urioste R *et al.* (1992) High-frequency network oscillation in the hippocampus. *Science* **256**, 1025–1027.

Csicsvari J, Hirase H, Czurko A *et al.* (1999) Oscillatory coupling of hippocampal pyramidal cells and interneurons in the behaving rat. *J. Neurosci.* **19**, 274–287.

Gloveli T, Dugladze T, Saha S *et al.* Differential involvement of oriens/pyramidale interneurones in hippocampal network oscillations *in vitro. J Physiol.* **562**, 131–147.

Kamondi A, Acsady L, Buzsaki G (1998) Dendritic spikes are enhanced by cooperative network activity in the intact hippocampus. *J Neurosci.* **18**, 3919–3928.

Maccaferri G (2005) Stratum oriens horizontal interneurone diversity and hippocampal network dynamics. *J. Physiol.* **562**, 73–80.

Maccaferri G, Lacaille JC (2003) Interneuron diversity series: hippocampal interneuron classifications–making things as simple as possible, not simpler. *Trends Neurosci.* **26**, 564–571.

Nadasdy Z, Hirase H, Czurko A *et al.* (1999) Replay and time compression of recurring spike sequences in the hippocampus. *J Neurosci.* **19**, 9497–9507.

Somogyi P and Klausberger T (2005) Defined types of cortical interneurone structure space and spike timing in the hippocampus. *J Physiol.* **562**, 9–26.

CHAPTER

20

Maturation of the hippocampal network

Developing neurons and circuits have several unique features and mechanisms that differ from those in adults. Firstly, several processes and cascades occur in the developing but seldom in the adult brain, including cell migration, differentiation, programmed cell death etc. Also, the subunit composition of ionotropic and metabotropic receptor channels or of voltage-gated ionic channels is often different in developing neurons. This chapter describes some of the sequential events that take place during the construction of the hippocampal network. It concentrates on the maturation of the main neuronal elements of the hippocampus, pyramidal neurons and interneurons, and their transmitters glutamate and GABA, which in the adult provide most of the excitatory and inhibitory drives, respectively. The properties of electrical activity that result from this maturation are described and compared with what is observed in the adult hippocampus.

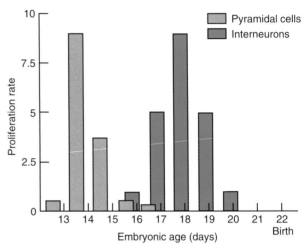

FIGURE 20.1 **Histogram showing the proliferation rate of pyramidal cells (grey) and interneurons (green) according to the embryonic age of the rat.**
Adapted from Bayer SA (1980) Development of the hippocampal region in the rat: I. Neurogenesis examined with ³H-thymidine autoradiography. *J. Comp. Neurol.* **190**, 87–114, with permission.

20.1 GABAERGIC NEURONS AND GABAERGIC SYNAPSES DEVELOP PRIOR TO GLUTAMATERGIC ONES

20.1.1 GABAergic interneurons divide and arborize prior to pyramidal neurons and granular cells

In adults, GABAergic and glutamatergic signals equilibrate in order to prevent seizures. What is the situation during brain maturation? Do interneurons and pyramidal neurons become functional at the same developmental stage or are they sequentially functional? This was investigated using the bromodeoxiuridine (BrdU) technique. When BrdU is injected systemically, it is incorporated in the DNA of cells in

the process of division. It is therefore possible to label neuronal ensembles according to their postmitotic age and determine their migration speed. This showed that GABAergic interneurons divide prior to the principal neurons (CA1–CA3 pyramidal neurons and granular cells of the dentate gyrus) (**Figure 20.1**). Thus, in the rat, interneurons divide between E13 (embryonic age, 13 days) and E17, whereas pyramidal neurons divide between E16 and E21 (there is an additional difference between CA3 and CA1 pyramidal neurons, the former reaching maturity earlier). The granule cells of the fascia dentata have a primarily postnatal division; it is estimated that over 85% of the granule cells in the rat will divide in the three-week period following birth. Interneurons are mature at an earlier stage than the bulk of principal cells.

To determine if this is also manifested by a sequential maturation of the axonal and dendritic arbors, it is possible to patch-clamp interneurons and pyramidal neurons and to inject a dye in the intracellular medium. Such studies show that interneurons indeed mature and arborize at an earlier stage than pyramidal neurons. The interneuronal circuit is therefore in a situation to exert an important modulatory role on the growth of pyramidal cells and the formation of the hippocampal network. This observation raises the following question: 'Since GABAergic interneurons are mature before glutamatergic pyramidal neurons, do these interneurons establish synapses before the glutamatergic ones on to target pyramidal neurons?'

20.1.2 GABAergic synapses are established before glutamatergic ones on to pyramidal cells

To determine the formation of GABA and glutamate synapses, the activity of pyramidal neurons is recorded in the whole-cell configuration (voltage clamp mode) in slices at an early stage – say at birth – and the properties of the postsynaptic currents (PSCs) that occur spontaneously or in response to electrical stimulation are determined. This is combined to morphological reconstruction of the recorded neurons (marked by intracellular injection of biocytin). At birth (P0, postnatal day 0), pyramidal neurons in the rat hippocampus are composed of three populations (**Figure 20.2**):

- Eighty percent of the neurons have a soma and an axon but essentially no apical or basal dendrites. These neurons are 'silent' in that no spontaneous or evoked synaptic current is recorded from them even in response to strong electrical stimuli (**Figure 20.2a**). These neurons do, however, express extra-synaptic receptors since bath applications of GABA or glutamate agonists evoke the usual currents (not shown) observed in more adult neurons, confirming (see above) that the expression of receptors precedes that of functional synapses. They are identified as neurons (and not glia) by their ability to generate spikes in response to an intracellular depolarization.

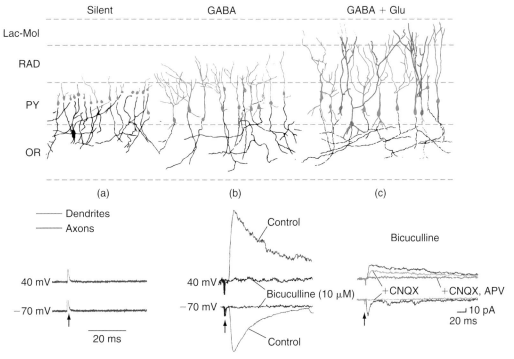

FIGURE 20.2 **Pyramidal cells at P0, grouped accordingly to their synaptic properties.**
The activity of pyramidal cells is recorded at P0 in whole-cell configuration (voltage clamp mode) in response to stratum radiatum stimulation. Recorded cells are injected with biocytin and reconstructed with a camera lucida. **(a)** Silent cells displaying no synaptic current. **(b)** Cells displaying a bicuculline-sensitive (GABA$_A$-mediated) synaptic current only. **(c)** Cells displaying bicuculline- (GABA$_A$), CNQX- (AMPA) and APV- (NMDA) sensitive synaptic currents. Arrow indicates the stimulating artefact. Adapted from Tyzio R, Represa A, Jorquera I *et al.* (1999) The establishment of GABAergic and glutamatergic synapses on CA1 pyramidal neurons is sequential and correlates with the development of the apical dendrite. *J. Neurosci.* **19**, 10372–10382, with permission.

- Ten percent of the pyramidal neurons have a bigger soma, an axon and a small apical dendrite restricted to the initial part of the stratum radiatum and no basal dendrite. In these neurons, $GABA_A$-receptor-mediated PSCs are recorded but glutamate-receptor-mediated EPSCs are absent (the synaptic response is fully abolished in the presence of bicuculline) (**Figure 20.2b**). There are only GABAergic synapses established on these neurons: they are thus of the 'GABA only' type. In these whole-cell experiments, the reversal potential of Cl^- is $0\,mV$, which explains why we speak of PSC rather than IPSC.

- Ten percent of the neurons have an extensively arborized apical dendrite that penetrates to the most distal part of the apical dendrite (lacunosum molec-ulare) and a more developed basal dendrite. In these neurons, $GABA_A$-*and* glutamate-receptor-mediated PSCs are recorded as shown by the sequential use of selective antagonists (**Figure 20.2c**). Thus, there are GABA and glutamate synapses established on these neurons: they are of the 'GABA + Glu' type.

Similar recordings in embryonic slices (E19) indicate that, at this age, virtually all pyramidal neurons (over 90%) are 'silent', instead of 80% at P0. All these results show that GABAergic synapses are established prior to glutamatergic ones on pyramidal neurons. Moreover, GABAergic synapses are formed only when the pyramidal neurons have an apical dendrite (gluta-matergic synapses are established on the pyramidal neurons when the apical dendrite reaches the stratum lacunosum moleculare). Parallel immunocytochemical data confirm that synaptic markers such as synapto-physin (**Figure 20.3**) or synapsin or markers of GABAergic terminals are first observed at birth at the level of the apical dendrites of pyramidal neurons. They are observed neither in the pyramidal layer nor in stratum oriens at this early stage.

A similar study performed in immature GABAergic interneurons indicates a similar sequence in which GABAergic synapses are formed before glutamatergic synapses. However, the sequence in interneurons occurs earlier: more than 90% of interneurons are innerv-ated at birth at a time when most pyramidal neurons are silent. Therefore, GABAergic neurons and synapses provide most of the synaptic drive at an initial stage.

20.1.3 Sequential expression of GABA and glutamate synapses is also observed in the hippocampus of subhuman primates *in utero*

Similar studies have been performed in fetal and embryonic macaque rhesus hippocampus during the second part of gestation (birth takes place around E165 in this species) in order to understand whether the GABA–Glu sequence of innervations applies also to other species, and in particular to non-human pri-mates. The monkey embryos were removed by caesar-ian intervention (between E85 and E154), the brain dissected and hippocampal slices obtained. The activ-ity of pyramidal cells is recorded in voltage clamp (whole-cell configuration) and neurons are filled with biocytin and reconstructed. As for the postnatal rat, fetal pyramidal neurons of the macaque can be divided into three populations (**Figure 20.4**):

- 'Silent' neurons, that have an axon but no den-drites, are silent in that they express no sponta-neous or evoked synaptic activity (**Figure 20.4a**). 'Silent' neurons generate sodium action potentials when depolarized (they are neurons, not glia). Bath-applied GABA or glutamate agonists evoke currents (not shown), indicating that the receptors are functional but not the synapses.

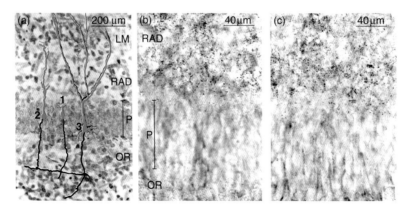

FIGURE 20.3 Presence of synaptic boutons in stratum radiatum but not in stratum pyramidale at P0.
(a) CA1 hippocampal section stained with cresyl violet (shown here as grey). Three pyramidal cells are shown: silent (middle), GABA only (left) and GABA + Glu (right), to demonstrate the distribution of the dendrites within all the layers. Other sections from the same hippocampus depict immunolabelling with (b) synaptophysin and (c) synapsin (see also Figure 20.2). The labelling is observed in stratum radiatum but not pyramidale.

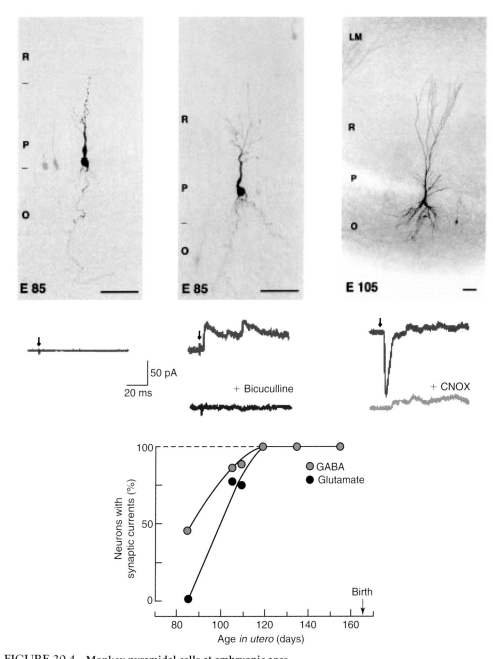

FIGURE 20.4   **Monkey pyramidal cells at embryonic ages.**
The activity of pyramidal cells is recorded in whole-cell configuration (voltage clamp mode) in response to stratum radiatum stimulation. Recorded cells are injected with biocylin and reconstructed with a camera lucida. **(a)** *From left to right*: Silent cells at E85 displaying no synaptic current, cells at E85 displaying bicuculline-sensitive (GABA$_A$-mediated) synaptic currents only, and cells at E105 displaying CNQX-(AMPA) sensitive synaptic current. Arrows indicate the stimulating artefact. **(b)** Developmental curve to depict the progressive expression at the embryonic stage of GABA and glutamate synapses. M. Esclapez and R. Khazipov, personal communication.

- Neurons that have axons and small apical dendrites express only GABA$_A$-mediated PSCs (**Figure 20.4b**). There are only GABAergic synapses established on these neurons; they are thus of the 'GABA only' type.
- Neurons that have an arborized apical dendrite as well as basal dendrites exhibit both GABA$_A$- and glutamate-mediated PSCs (**Figure 20.4c**). Thus,

there are GABA and Glu synapses established on these neurons; they are of the 'GABA + Glu' type.

This sequence is observed in both CA1 and CA3, with the only difference that the latter matures earlier. Quantification of the percentage of neurons expressing GABAergic synapses and glutamatergic synapses

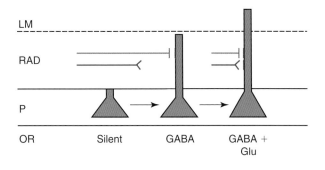

LM

RAD

P

OR   Silent   GABA   GABA +
                                                      Glu

──┤ GABAergic afferent

──< Glutamatergic afferent

FIGURE 20.5 **Schematic of the different stages of maturation of CA1 pyramidal cells at P0.**
See the caption to Figure 20.2.

indicates that in the beginning of mid-gestation over 50% of CA1 pyramidal neurons have functional GABAergic but not glutamatergic synapses (**Figure 20.4d**). In contrast, a few weeks later, one month before birth, all pyramidal neurons have both GABA and glutamate synapses. This results from the high speed of establishment of glutamatergic synapses during this period. Therefore, the sequence described for the rat hippocampus applies also to that of monkeys and probably of humans too. However, the gestation period during which the sequence is established is different: the first week postnatal in the rat, and the beginning of the second part of the gestation period in the rhesus monkey.

20.1.4 Questions about the sequential maturation of GABA and glutamate synapses

To summarize, GABAergic synapses between interneurons and the dendrites of pyramidal neurons are the first synapses to be established on pyramidal neurons of the hipppocampus (**Figure 20.5**). Glutamatergic synapses are formed at a later stage (such as synapses between two pyramidal cells or between extrinsic glutamatergic fibre tracts and pyramidal cells). This sequential expression suggests that the developmental stage of the target determines whether or not synapses will be established with presynaptic axons. This also suggests that the rules governing the formation of GABA and glutamate synapses differ, the latter requiring more mature postsynaptic targets.

These observations in turn raise the following questions:

• Does the activation of GABA$_A$ receptors in immature neurons evoke a current and a potential change identical to that in adults, or does the neonatal GABA$_A$ synaptic response have different properties?

• Is the other major inhibitory response, the metabotropic GABA$_B$-receptor-mediated IPSP, functional in immature neurons?

• What are the consequences of these developmentally regulated features on the electrical properties of the immature network? How does the immature network discharge?

20.2 GABA$_A$- AND GABA$_B$-MEDIATED RESPONSES DIFFER IN DEVELOPING AND MATURE BRAINS

20.2.1 Activation of GABA$_A$ receptors is depolarizing and excitatory in immature networks because of a high intracellular concentration of chloride

In the adult hippocampus, activation of GABA$_A$ receptors at rest evokes an inward flow of Cl$^-$ ions across the postsynaptic membrane (i.e. an outward current) that results in membrane hyperpolarization (see **Figures 9.16** and **19.9**), a decrease in membrane resistance and a shunt effect (see **Figure 9.17** and Section 9.5.3). GABA$_A$-mediated inhibition is the key element that provides the basis for the coordinated synchronized neuronal activity. Removal of this inhibition leads in the adult hippocampus to the generation of epileptiform activities. In contrast in the immature hippocampus, GABA$_A$ receptors have a totally different function – they mediate excitation (EPSPs).

GABA$_A$ receptor activation evokes a depolarization and bursts of action potentials in the immature hippocampus

To determine the properties of the GABA$_A$-mediated response in immature neurons, the activity of pyramidal cells is recorded in embryonic or early neonatal slices. One puzzling observation is that the GABA$_A$ receptor antagonist bicuculline, which in the adult generates epileptiform activity, in contrast silences ongoing activity in slices at an early developmental stage (**Figure 20.6a**), suggesting that GABA exerts an excitatory action at this stage. In keeping with this, intracellular and whole-cell recordings (current clamp mode) show that GABA$_A$ receptor activation leads to a depolarization of the membrane and the generation of sodium action potential(s). This is also observed in cell-attached recordings in which the intracellular concentration of Cl$^-$ is not altered (**Figure 20.6b**). Therefore, GABA depolarizes and excites immature neurons.

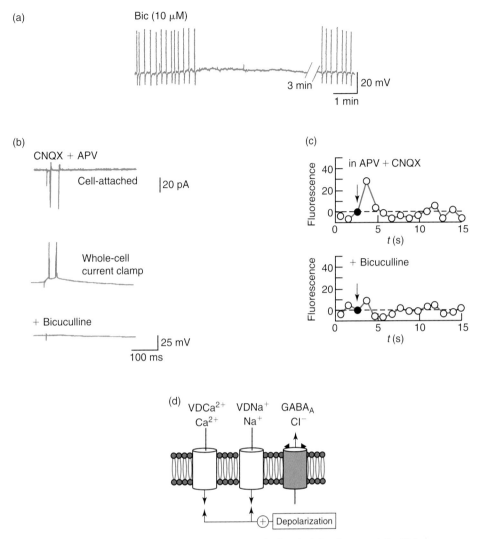

FIGURE 20.6 Synaptic activation of GABA_A receptors is depolarizing in neonatal rat hippocampus.
The activity of CA3 hippocampal neurons recorded in slices *in vitro*. **(a)** Effect of bicuculline on the spontaneous activity of a pyramidal cell recorded with K methyl sulphate intracellular electrodes at $-70\,$mV. **(b)** Response of an interneuron to stimulation in stratum radiatum in the continuous presence of CNQX ($10\,\mu$M) and APV ($50\,\mu$M). The excitatory response is recorded from the same interneuron in the cell-attached and whole-cell configuration (current clamp mode). It is totally abolished by bicuculline ($10\,\mu$M). **(c)** In the same conditions, electrical stimulation also evokes an increase of $[Ca^{2+}]_i$ thai is abolished by bicuculline, in a pyramidal neuron loaded with fluo-3. **(d)** GABA_A receptor activation leads to membrane depolarizing and thus opening of voltage-dependent (VD) Na$^+$ and Ca^{2+} channels. Part (a) adapted from Gaiarsa JL, Coradetti R, Ben-Ari Y, Cherubini E (1990) GABA mediated synaptic events in neonatal rat CA3 pyramidal neurons *in vitro*: modulation by NMDA and non-NMDA receptors, In: Ben Ari Y (ed.) *Excitatory Amino Acids and Neuronal Plasticity*, New York: Plenum Press. Part (b) adapted from Ben-Ari Y, Khazipov R, Leinekugel X *et al.* (1997) GABA_A, NMDA and AMPA receptors: a developmentally regulated 'ménage à trois'. *Trend Neurosci.* **20**, 523–529. Part (c) adapted from Khazipov R. Leinekugel X, Khalilov I *et al.* (1997) Synchronization of GABAergic interneuronal network in CA3 subfield of neonatal rat hippocampal slices. *J. Physiol. (Lond.)* **498**, 763–772; all with permission.

GABA_A-receptor-mediated depolarization evokes Ca^{2+} entry through both the voltage-gated Ca^{2+} and NMDA channels

Activation of GABA_A synapses in immature neurons leads to an increase of intracellular Ca^{2+} concentration ($[Ca^{2+}]_i$), as shown by Ca^{2+} imaging techniques (**Figure 20.6c**). In order to understand the underlying mechanism (recall that GABA_A receptor channels are not permeable to Ca^{2+} ions), two main hypotheses can be tested: the $[Ca^{2+}]_i$ increase results either from the activation of voltage-gated Ca^{2+} channels or from the activation of NMDA receptor channels.

To test the first hypothesis, synaptic GABA_A receptors are activated by stimulation of afferents in the presence of CNQX + APV, the blockers of ionotropic glutamate receptors. An increase of intracellular Ca^{2+} concentration is still observed and is blocked by the subsequent application of bicuculline (**Figure 20.6c**) or of antagonists of Ca^{2+} channels such as nifedipin or D-600 (not shown). Therefore, the activation of neonatal GABA$_A$ receptors results in a $[Ca^{2+}]_i$ increase due, at least partly, to Ca^{2+} entry through voltage-gated Ca^{2+} channels (**Figure 20.6d**).

Conversely, in the presence of antagonists of voltage-gated Ca^{2+} channels, synaptic GABA$_A$ receptor activation still induces a small increase of $[Ca^{2+}]_i$ that is abolished by APV, the selective antagonist of NMDA receptor channels (not shown). This suggests that the GABA$_A$-mediated depolarization can remove the voltage-dependent Mg^{2+} block of NMDA channels (see **Figure 20.7**). This was confirmed by recordings of single NMDA-channel activity and cell-attached recordings of the synaptic responses evoked in the presence of an AMPA receptor antagonist.

There is a synergy between GABA$_A$ and NMDA receptor channels in the immature hippocampus

Therefore, in immature neurons, there is a synergistic action between GABA$_A$ and NMDA receptors. This stands in contrast with the adult situation. In the neonatal hippocampus, GABAergic synapses act much like AMPA-receptor-mediated synapses at a later developmental stage: they provide the excitatory drive required to generate sodium and calcium action potentials as well as to activate NMDA receptors (see **Figure 20.7**).

There are, however, additional factors to take into account to fully comprehend the operative mode of the immature circuit. Even when GABA is depolarizing and excitatory, there is an inhibitory component due to a shunt mechanism. Also, at a given age of the rat, owing to the heterogeneity of hippocampal neurons, GABA will exert different actions in different neurons – excitatory in one and inhibitory in the other – presumably because they are in a different developmental stage. Thus, recording from different neurons in the same slice can reveal a cocktail of effects, including in some a net excitatory action, in others a dual effect (excitatory and inhibitory).

The developmental curve of the concentration of $[Cl^-]_i$ is exponential

Why does GABA depolarize and excite immature neurons? At least two hypotheses can be proposed: the GABA$_A$ channel is permeable to cations in immature neurons or the GABA$_A$ channel is permeable to Cl^- ions in immature neurons but the Cl^- driving force is reversed, due to an increased intracellular concentration of Cl^- ions.

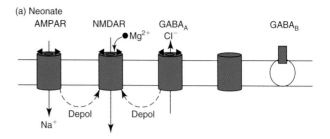

(a) Neonate

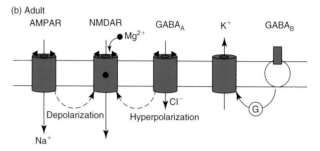

(b) Adult

FIGURE 20.7 Major developmental changes in the GABA–glutamate interactions.
See text for explanations. Adapted from Leinekugel X, Medina I, Khalilov I *et al.* (1997) Ca^{2+} oscillations mediated by the synergistic excitatory actions of GABA$_A$ and NMDA receptors in the neonatal hippocampus. *Neuron* **18**, 243–255, with permission.

To test the second hypothesis it is important to use a non-invasive technique that does not perturb $[Cl^-]_i$. Indeed, patch clamp or intracellular techniques modify $[Cl^-]_i$ and the polarity of the currents generated by GABA. In contrast, by recording in cell-attached configuration the openings of single GABA$_A$ channels, the concentration of $[Cl^-]_i$ can be determined at different ages. The value of the resting membrane potential is determined by means of recordings of the activity of single NMDA channels: since the NMDA current reverses at 0 mV, it is possible to calculate the genuine resting membrane potential. The difference between the two values enables to calculate the driving force for GABA (DF$_{GABA}$) (see Appendix 9.2). During maturation, neurons from the hippocampus show a progressive reduction of DF$_{GABA}$ with a shift of $[Cl^-]_i$ from 25–30 mM (embryonic) to 7–8 mM at the end of the first post-natal week (**Figure 20.8**). This corresponds to a shift of the reversal potential of GABA$_A$ current from 40 mV above V$_{rest}$ (embryonic) to a few mV close to V$_{rest}$ (adult). In the former case, GABA will generate spikes, in the latter it will inhibit their generation. This curve differs in various neuronal populations according to their developmental properties but even within the same population it will differ according to the degree of development and differentiation of the neuron. However, it is suggested that this curve is universal, i.e. observed in every animal species and brain structure.

This results from the early expression of the chloride importer NKCC1 in immature neurons whereas the main chloride exporter KCC2 has a delayed expression.

The developmental curve is interrupted around delivery

When the developmental decline of $[Cl^-]_i$ is investigated in detail, a brief interruption of the curve is observed around delivery time, shortly before and after. This abrupt fall of $[Cl^-]_i$ is associated with a powerful inhibitory action of GABA (**Figure 20.8**). This period is brief as the curve resumes its exponential decline subsequently with a higher $[Cl^-]_i$ and depolarizing and excitatory actions of GABA. Why is the curve interrupted transiently, $[Cl^-]_i$ dramatically reduced and GABA inhibitory during delivery? As delivery is initiated by an intense release of hormones, the possible role of oxytocine – an hormone released by the mother to trigger uterine contractions and labour – is tested (**Figure 20.8**). Blocking oxytocin receptors suppressed the transient reduction of $[Cl^-]_i$ and the excitatory to inhibitory shift of the actions of GABA (**Figure 20.8**). Therefore, the same maternal agent that triggers labour also induces a shift of the actions of GABA during that critical period. Although the exact mechanisms of this action have not been determined, they involve the co-transporters of chloride since blocking the main importer of chloride in immature neurons NKCC1 (see above) prevents the effects of the hormone.

20.2.2 GABA$_B$-receptor-mediated IPSCs have a delayed expression in immature neurons

Maturation is also associated with alterations in the development of inhibition mediated by GABA$_B$ (as well as adenosine or serotonin) receptors. In adults, activation of these receptors exerts a powerful control at both postsynaptic and presynaptic levels: at the postsynaptic level it generates a large hyperpolarization due to the activation of K^+ channels via a G-protein (it reverses at E_K). At the presynaptic level, activation of these receptors reduces the release of GABA and the amplitude of the GABAergic synaptic currents via a reduction of a presynaptic Ca^{2+} current and/or the activation of K^+ channels in axon termin-als. During maturation, the postsynaptic GABA$_B$ receptors are not functional at an early stage (at birth and until P5–P6 in pyramidal neurons) (**Figure 20.9**). Binding studies show that the receptors are present, but intracellular injection of GTP-γS to activate G-proteins fails to activate GABA$_B$-receptor-mediated currents. Therefore the absence of GABA$_B$-mediated currents is

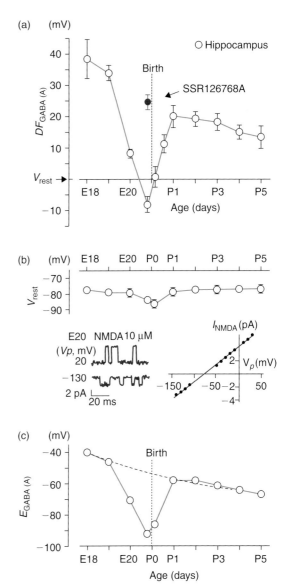

FIGURE 20.8   **Developmental profile of $E_{GABA(A)}$ in the rat.**
(a) Summary plot of the age-dependence of $DF_{GABA(A)}$ inferred from the recordings of single-channel GABA$_A$ currents [mena ± SEM.; 209 CA3 pyramidal cells (o); 6 to 24 patches for each point]. Red code – pretreatment with SSR126768A, an antagonist of oxytocin ($n = 25$ hippocampal patches). **(b)** Age-dependence of the resting membrane potential (V_{rest}) of CA3 pyramidal cells inferred from the reversal of single NMDA channels recorded in cell-attached mode ($n = 84$ cells; 4 to 12 patches for each point). **(c)** Age-dependence of the GABA$_A$ reversal potential ($E_{GABA(A)} = V_{rest} + DF_{GABA(A)}$). Note a transient hyperpolarizing shift of $E_{GABA(A)}$ near birth.
See Appendix 9.2 for the explanation of the method. Adapted from Tyzipo R, Cossart R, Khalilov I, Minlebaev M, Hubner CA, Represa A, Ben Ari Y, Khazipov R (2006) Maternal oxytocin triggers a transient inhibitory switch in GABA signaling in the fetal brain during delivery. *Science.* **314**, 1788–1792.

not due to a delayed expression of the receptors but more likely to a delayed coupling of the GABA$_B$ receptors to G-proteins and K^+ channels. In keeping with

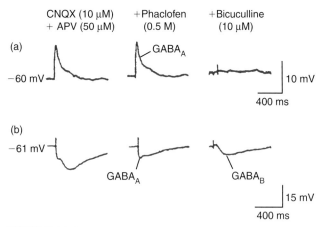

CNQX (10 µM)   +Phaclofen   +Bicuculline
+ APV (50 µM)   (0.5 M)   (10 µM)

FIGURE 20.9 **GABA$_B$-mediated response is absent in neonate and present in adult pyramidal cells.**
The activity of CA3 pyramidal cells recorded with an intracellular electrode filled with K methyl sulphate in the continuous presence of blockers of ionotropic glutamate receptors (CNQX + APV). **(a)** In the neonate: In response to electrical stimulation of stratum radiatum, a depolarizing response is recorded in the neonatal pyramidal cell. This response is mediated by GABA$_A$ receptors since it is unaffected by application of the GABA$_B$ antagonist phaclofen but is totally abolished by the GABA$_A$ antagonist bicuculline. **(b)** In the adult: In contrast, stimulation induces a biphasic hyperpolarization. Phaclofen reduces the late component (GABA$_B$) and leaves intact the early one (GABA$_A$) whereas bicuculline suppresses the early one and leaves intact the late one. Adapted from Gaiarsa JL, Tseeb V, Ben-Ari Y (1995) Postnatal development of pre- and postsynaptic GABA$_B$-mediated inhibitions in the CA3 hippocampal region of the rat. *J. Neurophysiol.* **73**, 246–255, with permission.

this, the other members of this family (metabotropic receptors) are also not operative at birth.

In contrast, the presynaptic mechanisms are operational already before birth in pyramidal neurons; the activation of GABA$_B$ (or adenosine and serotonin receptors) leads to a reduction of the PSCs in pyramidal neurons or interneurons (not shown). Therefore, the developing circuit operates with the two main postsynaptic receptor-mediated inhibitory mechanisms, GABA$_A$ and GABA$_B$, being poorly developed or acting in a reversed manner. In neonatal hippocampus, the principal mode of operation of transmitter-gated inhibition relies on a presynaptic control of transmitter release.

20.3 MATURATION OF COHERENT NETWORKS ACTIVITIES

20.3.1 Network-driven giant depolarizing potentials (GDPs) provide most of the synaptic activity in the neonatal hippocampus

Electrical activity in neonatal rats (postnatal days P0 to P8) is characterized by the presence of spontaneous

network-driven *giant depolarizing potentials* (GDPs) that provide most of the synaptic activity. GDPs are large and long-lasting (several hundreds of milliseconds) synaptic potentials giving rise to bursts of spikes that occur repetitively at a frequency varying between 0.05 and 0.2 Hz (**Figure 20.10a**). GDPs are recorded in the vast majority of neurons (both pyramidal neurons and interneurons) in hippocampal slices obtained from the brain of rats aged between birth and 2 weeks postnatal (**Figure 20.10c**). GDPs also prevail *in utero* in the monkey during the second half of gestation (E85 to E135). Therefore, GDPs and other similar patterns that provide the first coherent pattern of network activity, constitute a universal transient phase and are replaced subsequently by a more diversified panel of patterns that enable adults networks to generate a complex repertoire of diversified behaviours (see below).

20.3.2 Giant depolarizing potentials result from GABAergic and glutamatergic synaptic activity

GDPs are network-driven (in contrast to pacemaker patterns that are generated by the recorded neuron independently of its synaptic inputs), since (i) they are blocked by TTX (**Figure 20.10b**); (ii) their amplitude but not their frequency is modified by alterations of the resting membrane potential as expected from a synaptic current in contrast to an endogenous pacemaker oscillation; and (iii) they are often blocked by the GABA$_A$-receptor antagonist bicuculline (see **Figure 20.6a**) but also by ionotropic glutamate receptor antagonists (CNQX and APV), suggesting that both GABA and glutamate participate in their generation.

Determination of the currents underlying GDPs with patch clamp recordings (voltage clamp mode) reveals that GABA$_A$ currents are present either alone or in conjunction with AMPA and NMDA receptor-mediated currents. The relative participation of GABA and glutamate most likely depends on the maturational stage of the recorded neuron, reflecting the heterogeneity of the neuronal population.

The mechanism underlying the generation of GDPs therefore includes a network-driven barrage of depolarizing GABA and glutamate postsynaptic currents impinging on to pyramidal neurons and in turn leading to a recurrent GABAergic excitation from the GABAergic interneurons. This is also suggested by paired recordings from a pyramidal neuron and an interneuron that show synchronous GDPs in connected neurons and very few action potentials generated outside the GDPs (**Figure 20.11a**). Therefore, GDPs are triggered by the combined depolarizing effects of GABA- and glutamate-mediated currents and

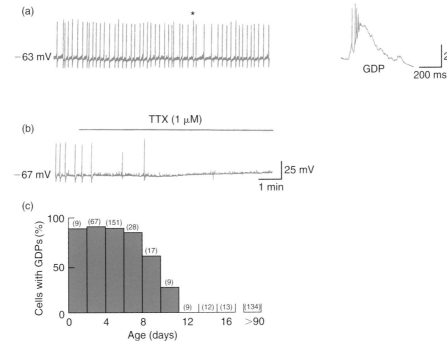

FIGURE 20.10   Spontaneous giant depolarizing potentials (GDPs) are generated in a polysynaptic circuit in neonatal hippocampus.

(a) Spontaneous GDPs are recorded from a CA3 pyramidal neuron with an intracellular electrode filled with K methyl sulphate. The inset shows a GDP on an extended timescale. **(b)** Spontaneous GDPs are blocked by TTX. **(c)** Histogram of the number of pyramidal cells with GDPs as a function of the age of the rat. The number of recorded cells is in parentheses. Part (a): JL Gaiarsa, personal communication. Parts (b) and (c) adapted from Ben-Ari Y, Cherubini E, Corradetti, Gaiarsa JL (1989) Giant synaptic potentials in immature rat CA3 hippocampal neurones. *J. Physiol. (Lond.)* **416**, 303–325, with permission.

by a basic circuit that includes feed-forward excitation of pyramidal neurons and interneurons followed by the recurrent depolarization produced by the recurrent collaterals of GABAergic interneurons (**Figure 20.11b**).

20.3.3 Giant depolarizing potentials are generated in the septal pole of the immature hippocampus and then propagate to the entire structure

Paired recordings in slices show that virtually all neurons have GDPs that are fully synchronous in neurons that are not too distant. To study the mechanism of the generation and propagation of GDPs, it is possible to record from an intact hippocampus superfused *in vitro*. In this preparation, the entire hippocampi are dissected and placed in a conventional *in vitro* chamber (**Figure 20.12a**). With multiple whole-cell (and extracellular field) recordings it is shown that *GDPs propagate with a septo-temporal gradient of automaticity*. This was shown by recording field potentials with a multiple electrodes array along the rostro-caudal axis. This type of experiment shows that there is a rostro-caudal latency (**Figure 20.12a**). In addition, transection of the

hippocampus in two along the longitudinal axis reveals that in both hemisected hippocampi there is a rostro-caudal latency (**Figure 20.12b,c**). Therefore, the rostral pole of the hippocampus, being the most active, paces the rhythm of the entire structure. Since GDPs are present in isolated portions of hippocampus, including mini-slices of CA3 or dentate gyrus subfields, the neuronal elements required for their generation are present in local neuronal circuits. However, the anterior parts of the hippocampus have a higher frequency of GDPs than the caudal ones.

This situation has some similarities with the generation of rhythmic activity in the cardiac muscle in which different parts, including sino-atrial node, atrioventricular node, His bundle and Purkinje fibres, have auto-rhythmic potentials that allow them to discharge periodically. However, the sino-atrial node is the normal pacemaker owing to a higher rhythm of activity. The mechanisms underlying the role of the anterior parts are not presently known but are likely to be due to a rostro-caudal gradient of maturation.

Using the two interconnected hippocampi *in vitro* it is also possible to determine the hemispheric propagation of GDPs. Paired whole-cell recordings from two

(a)

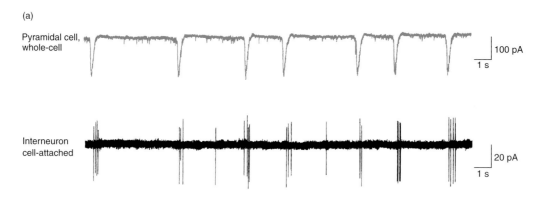

(b)

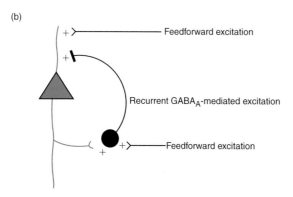

Feedforward excitation

Recurrent GABA$_A$-mediated excitation

Feedforward excitation

FIGURE 20.11   GDPs are syncronous in pyramidal cells and interneurons.
(a) Dual recordings of spontaneous currents (voltage clamp mode) in a CA3 pyramidal neuron (whole-cell configuration, upper trace) and a neighbouring interneuron (cell-attached configuration, lower trace). Note that bursts of currents of action potentials in the interneuron are synchronous with currents of GDPs in the pyramidal cell. **(b)** Schematic explaining the generation of GDPs. Part (a) adapted from Khazipov R, Leinekugel X, Khalilov I *et al.* (1997) Synchronization of GABAergic interneuronal network in CA3 subfield of neonatal rat hippocampal slices. *J. Physiol. (Lond.)* **498**, 763–772, with permission.

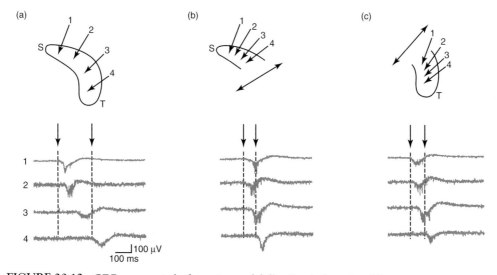

FIGURE 20.12   GDPs propagate in the rostro-caudal direction in immature hippocampus.
The spontaneous GDPs of four populations of hippocampal neurons are simultaneously recorded with four extracellular electrodes in **(a)** the intact hippocampus, or in **(b)** the isolated rostral or **(c)** caudal halves of the hippocampus of neonatal rats. In extracellular recordings, GDPs are recorded as inward field potentials. Adapted from Leinekugel X, Khalilov I, Ben-Ari Y, Khazipov R (1998) Giant depolarizing potentials: the septal pole of the hippocampus paces the activity of the developing intact septohippocampal complex *in vitro*. *J. Neurosci.* **18**, 6349–6357, with permission.

neurons in each hippocampus reveal that, even during the first few days after birth, GDPs can propagate from one hippocampus to the other and back, suggesting that the commissural connections are mature. This also suggests that the two hippocampi can synchronize each other at an early developmental stage.

In contrast, structures connected to the hippocampus, such as the septum or the entorhinal cortex, do not generate GDPs if they are disconnected from the hippocampus, but they do express GDPs originating in and propagating from the hippocampus. Therefore, the hippocampus must constitute a major source of network-driven synaptic activity that can modulate the electrical activity of brain structures with which it is connected.

20.3.4 Hypotheses on the role of the sequential expression of GABA- and glutamate-mediated currents and of giant depolarizing potentials

The earlier expression of GABAergic current, its depolarizing action and the resulting GDPs are key properties of developing networks. Similar effects of GABA have been described in several brain and peripheral structures, and GDP-like events also predominate in virtually all the structures studied so far, including spinal cord and neocortex. This raises the question of the role of this sequential expression of synapses, of the depolarizing actions of GABA, and of the GDPs. The following considerations should be taken into account:

- Owing to their long durations, $GABA_A$-mediated PSCs are highly suitable for summation (tens of milliseconds, in contrast to the milliseconds duration of AMPA-mediated PSCs). This is of importance in immature neurons that possess very few synapses initially, so that there are few spontaneous PSCs to summate if the excitatory drive is provided only by the brief AMPA-mediated PSCs. Furthermore, even if more depolarized than the resting potential, the reversal potential for GABAergic currents is closer to rest than that of glutamatergic PSCs. This and the shunting mechanism, which is inherent to the operation of GABAergic currents, prevent the occurrence of too strong depolarizations and of excitotoxic stimuli that occur when glutamatergic receptors are repetitively stimulated.
- The combined actions of GABA and glutamate facilitate the generation of a large increase of $[Ca^{2+}]_i$ as a result of the activation of voltage-gated Ca^{2+} channels and the removal of the Mg^{2+} blockade from NMDA channels (**Figure 20.7**). In keeping with this, studies in neonatal slices using Ca^{2+} imaging

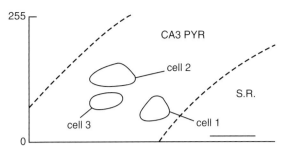

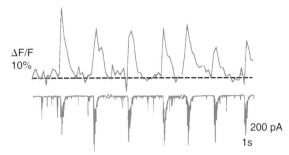

FIGURE 20.13 **Synchronous spontaneous Ca² oscillations in Ca3 pyramidal neurons.**
CA3 pyramidal neurons in slices are loaded with fluo-3 and their activity is simultaneously recorded in whole-cell configuration (voltage clamp model). Each spontaneous GDP current is concomitant to a transient increase of intracellular Ca^{2+} concentration. Adapted from Leinekugel X, Medina I, Khalilov I *et al.* (1997) Ca^{2+} oscillations mediated by the synergistic excitatory actions of $GABA_A$ and NMDA receptors in the neonatal hippocampus. *Neuron* 18, 243–255, with permission.

techniques indicate that GDPs are associated with important $[Ca^{2+}]_i$ oscillations (**Figure 20.13**). Other observations suggest that a rise of $[Ca^{2+}]_i$ is needed for dendritic growth and synapse formation.

- Network-driven oscillations like the GDPs may participate in the formation of functional neuronal units like the formation of visual columns. This may follow the Hebbian rule 'neurons that fire together wire together'. Network-driven oscillations provide a suitable way of organizing these units as both presynaptic and postsynaptic neurons will be excited in a synchronized manner. It is likely that these immature patterns of oscillations play an important role in electrically interconnecting neurons that will become part of an ensemble of neurons subsequently.

20.4 CONCLUSIONS

The first synapses to be established in the hippocampus are the GABAergic synapses between interneurons and pyramidal cells or between two interneurons. In these synapses, transmission is mediated by $GABA_A$ receptors.

Activation of $GABA_A$ receptors evokes a depolarization of the postsynaptic membrane, in contrast to what is observed in mature neurons where $GABA_A$ receptors mediate inhibition (i.e. membrane hyperpolarization or silent inhibition). This $GABA_A$-mediated depolarization is strong enough to activate Na^+, Ca^{2+} and NMDA voltage-gated channels and thus to evoke action potentials and Ca^{2+} entry.

$GABA_B$ receptors are not active in the postsynaptic membrane of immature GABAergic synapses owing to uncoupling to G-proteins and target K^+ channels. In contrast $GABA_B$ receptors are functional in the presynaptic membrane where they mediate presynaptic inhibition of GABA release.

As a consequence, $GABA_A$ receptors in the immature hippocampus play the role of AMPA receptors in adult networks, and the only transmitter-mediated synaptic inhibition present in immature hippocampal neurons is a $GABA_B$-mediated presynaptic one. The immature hippocampal network displays spontaneous discharges, owing to $GABA_A$-mediated giant depolarizations, which periodically allow Ca^{2+} entry and transient increases of intracellular Ca^{2+} concentration in hippocampal neurons.

FURTHER READING

Ben-Ari Y (2002) Excitatory actions of GABA during development: the nature of the nurture. *Nat. Rev. Neurosci.* **3**, 728–739.

Ben-Ari Y, Gaiarsa JL, Khazipov R (2007) A pioneer transmitter that excites immature neurons and generate primitive oscillations. *Physiological Rev.*, in press.

Hollrigel GS, Ross ST, Soltesz I (1998) Temporal patterns and depolarizing actions of spontaneous $GABA_A$ receptor activation in granule cells of the early postnatal dentate gyrus. *J. Neurophysiol.* **80**, 2340–2351.

LoTurco JJ, Owens DF, Heath MJ *et al.* (1995) GABA and glutamate depolarize cortical progenitor cells and inhibit DNA synthesis. *Neuron* **15**, 1287–1298.

Menendez de la Prida L, Bolea S, Sanchez-Andres JV (1998) Origin of the synchronized network activity in the rabbit developing hippocampus. *Eur. J. Neurosci.* **10**, 899–906.

Obrietan K and van den Pol AN (1998) $GABA_B$ receptor-mediated inhibition of $GABA_A$ receptor calcium elevations in developing hypothalamic neurons. *J. Neurophysiol.* **79**, 1360–1370.

Owens DF, Boyce LH, Davis MB, Kriegstein AR (1996) Excitatory GABA responses in embryonic and neonatal cortical slices demonstrated by gramicidin perforated-patch recordings and calcium imaging. *J. Neurosci.* **16**, 6414–6423.

Owens DF and Kriegstein AR (2002) Is there more to GABA than synaptic inhibition? *Nat. Rev. Neurosci.* **3**, 715–727.

Rivera C, Voipio J, Payne JA *et al.* (1999) The K^+/Cl^- co-transporter KCC2 renders GABA hyperpolarizing during neuronal maturation. *Nature* **397**, 251–255.

Rohrbough J and Spitzer NC (1996) Regulation of intracellular Cl^- levels by Na^+-dependent Cl^- cotransport distinguishes depolarizing from hyperpolarizing $GABA_A$ receptor-mediated responses in spinal neurons. *J. Neurosci.* **16**, 82–91.

Index

Acetylcholine, 120–1, 125, 127
  affinity constants, 174
  nicotinic receptors, 122, 160–85
  receptor binding, 166, 180–1
  spontaneous release, 156
Acetylcholinesterase, 9, 121–2
Acetylcholinesterase inhibitors, 180
Action potentials, 5, 45–82
  all or none, 47, 73
  Ca^{2+}-dependent action potentials, 83–110
    currents underlying, 90–4
    depolarization phase, 84–94
    in endocrine cells, 83–4
    initiation of, 97–9
    K^+ ions in, 94–7
    properties of, 83–4
    repolarization phase, 94–7
    role of, 99
  currents underlying, 70
  K^+ ions in, 46–7
  Na^+-dependent, 46, 47–62, 67–9
    backpropagating, 313–18
    characteristics of, 73
    currents underlying, 70
    depolarization phase, 38–9, 47–62
    initiation of, 69–71
    K^+ ions in, 46–7
    propagation of, 71, 72
    refractory period, 71, 73
    repolarization phase, 46, 62–7
    role in neurotransmitter release, 73
Activation rate, 55, 57
Active transport, 30–1
Active zones, 136, 143, 157
Adrenaline, 127–8
Adrenal medulla, 125–6
Agonists, 128
  competitive, 179
  of $GABA_A$ receptor, 199–200
  of ionotropic glutamatergic receptors, 209–10
  of nicotinic receptors, 178
Agrin, 123–4
All or none action potentials, 47, 73

Alternative splicing, 211
α-Amino-3-hydroxy-5-methyl-4-isoxazole propionate *see* AMPA
Ammon's horn, 368, 369–71
AMPA current, 215
AMPA receptors, 209, 210, 213–15
  contribution to EPSC, 229
  diversity of, 213
  electrophysiological properties, 214
  EPSPs, 380
  functional properties, 216
  ion permeability, 213–15
  long-term depression, 357–60
  long-term potentiation, 351–3
  unitary conductance, 213–15
Antagonists, 128
  competitive, 128
  of $GABA_A$ receptor, 199
  non-competitive, 165
Anterograde inhibition, 16
Anterolateral ascending sensory pathway *see* Spinothalamic tract
Antibodies, 129
Antigen-antibody complex, 131
Antigens, 129
Anti-idiotype antibodies, 130, 131
Antiport, 42
Apamin, 97
Aplysia neurons, 97
β-Arrestin, 264
Astrocytes, 21–4
  and blood-brain barrier, 22–3
  fibrillary, 22
  morphology, 21–2
  in neurotransmitter cycle, 23–4
  protoplasmic, 22
  regulation of extracellular fluid composition, 23
ATP, in axonal transport, 11
Avidin-biotin method, 130, 132
Axo-axonic cells *see* Chandelier cells
Axo-axonic synapses, 117
Axo-dendritic synapses, 117–19
Axonal transport, 9–16
  demonstration of, 9

  fast anterograde, 9–13
  of mitochondria, 16
  retrograde, 13–14
  slow anterograde, 15–16
Axons, 4–5
  arborization, 5
  lack of protein synthesis in, 8–9
Axon targeting signals, 116
Axon terminals, 5
Axo-somatic synapses, 117
Axo-spinous synapses, 112

Backpropagating Na^+ action potentials, 313–18
  activation of high-voltage-activated Ca^{2+} channels, 320–1
  dopaminergic neurons, 315
  Purkinje cells, 3, 315–17
  pyramidal cells, 314–315, 320–1
Baclofen, 232–3, 234, 243
Barbiturates, 195–9, 205
  $GABA_A$ current, 196, 198, 199
  specific receptor sites, 195
  structure, 195
Basal lamina, 121–2
Basket cells, 118–19, 370, 384
Benzocaine, 179, 180
Benzodiazepines, 195–9, 205
  $GABA_A$ current, 196, 198, 199
  specific receptor sites, 195
  structure, 195
Bernstein, Julius, 36
Bicuculline, 181, 194–5, 232–3, 381
Big K channels, 95–6, 100
Bistratified cells, 370
Blood-brain barrier, 22–3
Botulinum toxins, 151–2
Boutons en passant, 5, 117
Bromodeoxiurdine technique, 386
α-Bungarotoxin, 179, 185
Buzsaki, Gyorgy, 382

Ca^{2+}, 30, 31–2
  buffering, 144–5
  clearance, 143–5

synapses of, 119
tetrodotoxin-sensitive inward
currents, 312
Putative neurotransmitters, 114
Pyramidal cells, 2, 7, 369–70
backpropagating Na$^+$ action
potentials, 314–15, 320–1, 322–3
CA3, 381–2, 383
dendritic Na$^+$ channels, 310–13
backpropagating Na$^+$ action
potentials, 313–18
EPSP activation, 313
feedback inhibition, 380–1, 382
GABA$_A$-mediated IPSPs, 377–9
GABA$_B$-mediated IPSPs, 379–80
GABAergic synapses, 387–8
glutamatergic synapses, 387–8
high-voltage-activated Ca^{2+} channels,
318
high-voltage-activated Na$^+$ channels,
310
hyperpolarization-activated cationic
current, 303
low-threshold transient Ca^{2+} current,
298
Schaffer collaterals activation, 344–55
tetrodotoxin-sensitive inward
currents, 312

Quanta, 156, 157
Quaternary amine transmitters, 120–1,
125, 127
Quisqualate, 214, 215

Rabbit nerve action potential, 68, 70
Rapidly inactivating transient K$^+$
current (I$_A$), 287–8
gating properties and ionic nature,
287–8
pharmacology, 288
structure of main channel subunit, 287
Receptors, 128–33
AMPA, 209, 210, 213–15
anchoring to postsynaptic membrane,
116–17
complementarity with
neurotransmitters, 115–17
GABA see GABA$_A$ receptors; GABA$_B$
receptors
glutamate see Glutamate receptors
G-protein coupled, 114, 124, 234, 238
kainate, 209, 210, 215–18
localization of, 131
nicotinic, 122, 160–85
targeting to postsynaptic membrane,
114–15
Reciprocal synapses, 111, 112
Rectification, 191
Recurrent inhibition, 16
Refractory period, 71, 73
Relaxation experiments, 285
Resting membrane potential, 36–8
depolarization, 38–9
electrical circuit, 38
hyperpolarization, 39–40

Retrograde axonal transport, 13–14
direction of vesicle movement, 13
functions of, 13–14
minus-end motors, 13
Reversal potential of nicotinic response,
166
Reversible competitive antagonists, 128
Ribosomes, 7

Sarco-endoplasmic Ca-ATPse pumps,
144
Satellite oligodendrocytes, 24
Schaffer collaterals, 344, 372, 383
activation, 344–55
Schwann cells, 28
myelinating, 28
non-myelinating, 28
Secondary antibodies, 130
Second messengers, 360–2
Secretory cells, 1
Serotonin, 127–8
Sharp waves, 15–17
Short-term potentiation, 342–3
Shunting effect, 202
Silent inhibition, 278
Silent neurons, 36, 388
Single cell reverse transcriptase chain
reaction, 63
Single-channel current, 35
Single-channel tail current, 109
Single-spike EPSP, 136, 211
Single-spike IPSP, 186, 201–2
Single-stranded DNA probes, 132
Single-stranded RNA probes, 132–3
Slow anterograde axonal transport, 15–16
Slowing inactivating transient K$^+$
current (I$_D$), 288–9
gating properties, 288–9
pharmacology, 289
Small K channels, 95–6
Smooth muscle synapses, 124–5
synaptic cleft, 125
SNAP-25, 142, 148
SNAPs, 146, 149
SNARE motifs, 148
SNAREs, 142, 146, 147–8
Q-SNAREs, 150
R-SNAREs, 150
Sodium see Na$^+$
Sodium Green, 108
Soma, 1, 3, 5
Soma-dendritic synapses, 117
Somato-dendritic processing, 271–8
high-voltage-activated depolarizing
currents, 308–24
subliminal voltage-gated currents,
279–307
Somatosensory cortex, 18
Somatosensory pathways, 17–19
Spatial summation, 305–6
Spinothalamic tract, 17–19
Squid giant axon, 10, 31
action potential, 45, 47, 68
synaptic facilitation, 342–3
synaptic transmission, 135

Stellate cells, 118
Stratum lacunosum moleculare, 244
Stratum lucidum, 372
Suberyldicholine, 178
Subliminal voltage-gated currents,
279–93
depolarizing
hyperpolarization-activated
cationic current, 285–7
low-threshold transient Ca^{2+}
current, 282–5
persistent inward Na$^+$ current,
280–2
hyperpolarizing, 287–92
inward rectifier K$^+$ current, 291–2
K$^+$ currents activated by
intracellular Ca^{2+}, 289
K$^+$ current sensitive to muscarine,
289–91
rapidly inactivating transient K$^+$
current, 287–8
slowly inactivating transient K$^+$
current, 288–9
role of, 294–307
Subthalamic nucleus, 335
Supraliminal voltage-gated currents,
293
delayed rectifier K$^+$ current, 62–8
high-threshold Ca^{2+} current, 85, 86,
90, 96, 97, 99, 284, 293, 330–2
high-threshold Na$^+$ current, 308
voltage and Ca^{2+}-dependent K$^+$
current, 27, 63, 327
Symport, 42
Synapse-associated polyribosome
complexes, 7
Synapse-associated proteins, 116
Synapses, 1, 111–33
axo-axonic, 117
axo-dendritic, 117–19
axo-somatic, 117
axo-spinous, 112
chemical, 111–12
dendro-dendritic, 117
electrical, 112
GABAergic, 200
glutamatergic, 226–30
interneuronal, 117–20
mixed, 112
neuroglandular, 125–6
postganglionic, 125
reciprocal, 111, 112
smooth muscle, 124–5
soma-dendritic, 117
Synapsin, 398
Synaptic boutons, 5
Synaptic cleft, 111
neuromuscular junction, 121–2
smooth muscle synapses, 125
Synaptic complex, 112–17
generalized functional model, 114–15
neurotransmitter and receptors,
115–17
postsynaptic element, 19, 113–14
presynaptic element, 113–14